SOUND ANALYSIS
AND
NOISE CONTROL

D0930999

SOUND ANALYSIS AND NOISE CONTROL

JOHN E. K. FOREMAN, P.Eng.

Professor Emeritus

Sound and Vibration Laboratory
The University of Western Ontario
London, Canada

VNR VAN NOSTRAND REINHOLD
New York

WIDENER UNIVERSITY
WOLFGRAM LIBRARY
CHESTER, PA.

DISCARDED
WIDENER UNIVERSITY

Copyright © 1990 by Van Nostrand Reinhold

Library of Congress Catalog Card Number 90-11945

ISBN 0-442-31949-5

All rights reserved. No part of this work covered by the copyright hereon may be reproduced or used in any form or by any means—graphic, electronic, or mechanical, including photocopying, recording, taping, or information storage and retrieval systems—without written permission of the publisher.

Printed in the United States of America

Van Nostrand Reinhold
115 Fifth Avenue
New York, New York 10003

Van Nostrand Reinhold International Company Limited
11 New Fetter Lane
London EC4P 4EE, England

Van Nostrand Reinhold
102 Dodds Street
South Melbourne, Victoria 3205, Australia

Nelson Canada
1120 Birchmount Road
Scarborough, Ontario M1K 5G4, Canada

16 15 14 13 12 11 10 9 8 7 6 5 4 3 2 1

Library of Congress Cataloging-in-Publication Data

Foreman, John E. K., 1922–
    Sound analysis and noise control / by John E. K. Foreman.
      p. cm.
    At head of title: University of Western Ontario, Faculty of Engineering Science.
    ISBN 0-442-31949-5
    1. Noise control. 2. Sound. I. University of Western Ontario. Faculty of Engineering Science. II. Title.
TD892.F67 1990
620.2′3—dc20
                                      90-11945
                                      CIP

CONTENTS

PREFACE

This book has been written to provide an introduction to the fundamental concepts of sound and a comprehensive coverage whereby unwanted sound (noise) can be controlled. Although there are many notable textbooks which deal primarily with the physics (or theory) of sound, and others which treat noise control in a strictly practical (and sometimes even empirical) manner, there are few textbooks that provide a bridging between the necessary understanding of the fundamentals of sound (its generation, propagation, measurement) and the application of these fundamentals to its control. This book provides that link.

The text presents noise control primarily at the introductory level. As such, the book should be of interest to students in engineering programs at the university level and to students in technology and technician programs at community colleges who need to acquire a basic understanding of noise control principles and applications; it should be of value to the practicing engineer who may have no formal training in noise control but whose responsibilities now involve assessment of noise and implementation of methods to control it; and it should appeal to teachers of a first course in sound analysis and noise control who must rely upon an authoritative and comprehensive text book on the subject.

The development of the presentation of the material in the text has been designed so that the book can serve also as a means of self-study. To this end, worked examples are included at the appropriate sections throughout the text and in two chapters at the end of the text in order to demonstrate the principles which have been developed. The worked examples are typical of those problems which may be encountered in real situations.

Chapter 1 deals with the basics of sound, in order to inform the reader of some of the physics associated with sound phenomena.

Chapter 2 deals with the mechanism of hearing and the subjective rating of sound, including age-related and noise-induced hearing loss.

Assessment of any noise problem involves a knowledge of the instrumentation available for measurements, the limitations of this instrumentation, the appropriate procedures for making the measurements with the instrumentation, and the methods by which the measured data can be analyzed. Chapter 3 provides an up-to-date coverage of these requirements, including a section on one of the newest and most valuable tools in noise studies—sound intensity measurement. The capability of being able to measure sound intensity as compared with conventional sound pressure has given the noise engineer a means of dealing with noise problems which had heretofore been denied—particularly noise source identification.

Chapter 4 concerns the propagation and attenuation of sound in free fields (e.g., outdoors with no reflecting surfaces) and the attenuation of sound in an enclosure. Outdoor barriers are also discussed in this chapter.

Chapter 5, one of the most important in the book, deals with absorption of sound in enclosures, reverberation in enclosures (and its application) and sound transmission loss through barriers.

Noise is the result of surface radiation, i.e., the surface moves back and forth (a seemingly infinitesimally small amount at times), which in turn excites a pressure propagation (sound wave) due to molecular interaction and excitation of the air adjacent to the surface. This surface excitation is often the result of structure-borne vibration due to rotating or reciprocating machines. Where such conditions can exist, it is imperative that the vibrating machine be isolated (or decoupled) from its attachment so as to alleviate as much as possible the excitation of adjacent surfaces, as mentioned above. Chapter 6 deals with the theory of vibration and vibration control, including applications as to the appropriate vibrating isolators and/or damping material to be used to reduce structure-borne vibration.

The situation regarding noise criteria and regulations, as they have historically evolved in the U.S.A. and as they currently pertain in Canada, is discussed in Chapter 7. Criteria for indoor noise environment, hearing-damage risk and criteria at the workplace (industry), and outdoor noise criteria are extensively covered.

Notable aspects of this text are found in the last two chapters. Chapter 8 covers a review of general noise control measures and a program for noise control in industry; it also outlines a large number of practical examples involving noise situations and noise control techniques.

Chapter 9 discusses the use of sound power data in noise source diagnosis and presents a wide range of in-depth case studies of noise problems—all representative of situations encountered in practice.

Included in the appendices are a complete listing of acoustical standards organizations and acoustical standards, useful acoustical periodicals, and a list of suppliers of acoustical materials and measuring instruments.

The writing and completion of this book would not have been possible without the valuable assistance of the many experts in the field of sound analysis and noise control mentioned throughout the text and the references; the critical comments by uncountable numbers of students who have worked with the author in lectures, laboratories and tutorials on sound analysis and noise control over the years; the many typists who endured the largely indecipherable handwriting of the author, and who, with perseverance and admirable expertise, translated concepts into readable text; and the author's family who, besides being supportive, also endured many months of having the dining room table littered with all sorts of paper, books, periodicals, files, pencils, rulers, pens, scissors, glue bottles, and whatever was required in putting together a book of this nature.

ACKNOWLEDGMENTS

In the preface, I mentioned those people who played a part in the development of this text. I wish to single out, in particular, all of the ladies to whom I am indebted.

First of all, my wife, Jan, whose support during the arduous years of writing and finalizing the book meant so much to me.

Then come the many ladies who have translated my thoughts on paper into the text which now follows. These are Elizabeth, Cathie, Sandra, Jacquie, and Pennie.

And finally, I wish to acknowledge, in particular, Mary, who, when introduced to this project, quickly assessed the essence of what was required in making this book viable, and in fact became a welcome and valuable assistant to me.

SOUND ANALYSIS
AND
NOISE CONTROL

Chapter 1

BASICS OF SOUND

In its broadest sense, noise is simply unwanted sound. It can prevent people from performing at their maximum ability and efficiency, it can deprive them from enjoying their leisure to the fullest, it can interfere with their sleep, it can result in increased nervous tension with associated psychological effects, and, most particularly, it can result in damage to the audio-sensory mechanism and lead to premature loss of hearing. Noise, like air and water pollution (and it *is* a form of "air pollution"), is thus receiving more attention at the community, industrial, and governmental levels.

The following text presents an overall picture of the engineering fundamentals involved in the study of noise. It is intended to provide an introduction to the analysis of sound and to its control wherever noise may adversely affect the working or living condition, and the comfort and health, of man in modern society.

1.1. GENERATION OF SOUND

Sound requires a source, a medium for transmission, and a receiver. The source is simply an object which is caused to vibrate by some external energy source. The medium is the substance which carries the sound energy from one place to another. In this discussion the medium will generally be air, although it could be water or any solid substance such as metal or wood. The sound energy is transmitted through a medium back and forth in the same pattern as the vibration of the sound source. Fig. 1.1 illustrates the generation of sound waves in the air by the vibrating object. As the object moves to the right, the air molecules next to it are pushed together to form a compression. This slight increase in pressure is passed on to the molecules which are successively farther away from the object. This results in a slight high pressure area

moving out from the source.* When the object completes its motion to the right it begins to move back to the left. This results in a reduction of pressure next to the object allowing the air molecules to spread apart, producing a rarefaction. This slight decrease in pressure is passed on through the air away from the object. As the object moves back and forth it successively sends out a compression followed by a rarefaction to form the sound wave. Fig. 1.1(c) illustrates the variations of the pressure in the air along the sound waves.

The receiver is the human ear or possibly an electronic device used to measure the sound. The ear responds to the slight pressure changes in the air and through a series of energy transformations the brain interprets the energy fluctuations as sound. This will be discussed later under the mechanism of hearing.

A microphone measuring the pressure variations in the air at a fixed location in space relative to the travelling wave of Fig. 1.1, when connected to an oscilloscope display of the time-varying signal, would show that the sound pressure (the deviation from atmospheric or equilibrium pressure) varies sinusoidally with time, as shown in Fig. 1.2(a).

The pressure variation is periodic, one complete variation in pressure being called a *cycle*. The time T for one complete cycle is called the period of pressure oscillation. The frequency of pressure change f is defined as the number of cycles per unit time (cycles per second, or Hertz, abbreviated as Hz), i.e.,

$$f = \frac{1}{T}.$$

---

*Note that the molecules of the medium are not travelling with the waves; they only vibrate about their mean positions. It is the energy (pressure) which is moving away from the source.

1

(a) Generations of compressions and rarefactions by a vibrating source

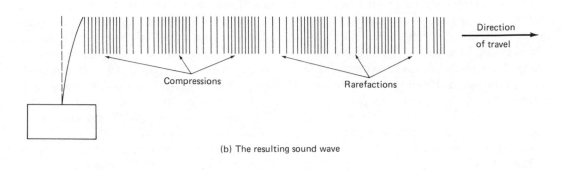

(b) The resulting sound wave

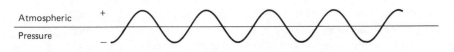

(c) The corresponding air pressure variations

Fig. 1.1. The generation of sound waves.

The pressure variation with time is given by the equation

$$p(t) = p_A \sin(\omega t + \phi) \qquad [1.0]$$

where

P_A = amplitude of pressure fluctuation,

ω = frequency of pressure fluctuation, in

$$\text{rad}/\text{sec} = \frac{2\pi}{T} = 2\pi f,$$

ϕ = phase of the sound signal being measured relative to some reference (say, the motion of the arm of the tuning fork).

The velocity of propagation c of the sound in Fig. 1.1 is given by

$$c = \frac{\Delta x}{\Delta t},$$

where Δx is the distance the sound would propagate during a time interval Δt [see Fig. 1.2(b)]. The wavelength of the periodic sound pressure wave is defined as the distance through which the sound propagates during time T. During the time T required for one oscillation of the sound pressure [Fig. 1.2(a)], the pressure wave in Fig. 1.1 would have moved one wavelength λ [Fig. 1.2(b)]. Hence, for a constant velocity c, the wave moves a distance $\Delta = cT$ during the time period T. Therefore,

$$\lambda = cT = \frac{c}{f}.$$

For example, the velocity of sound in air at sea level and 60°F is approximately 1100 ft/sec (335 m/sec). Hence, the wavelength of a 50 Hz sound in air is 22 ft (6.7 m). Similarly, the wavelength of a 5000 Hz sound in air is .22 ft (.067 m).

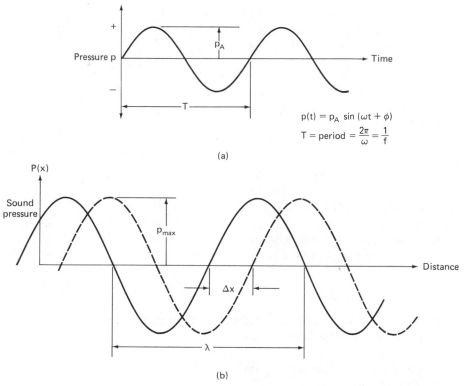

$$p(t) = p_A \sin(\omega t + \phi)$$
$$T = \text{period} = \frac{2\pi}{\omega} = \frac{1}{f}$$

(a)

(b)

Fig. 1.2. (a) Relation between pressure, period, and frequencies; (b) spatial characteristic of pure tone (space domain).

1.2. EQUATIONS OF MOTION OF PLANE WAVES

For a source of sound waves in the form of a vibrating sphere, the vibrations are directed radially and sound waves are propagated uniformly in all directions, with the result that the wavefronts are spherical. Where the source of sound waves is a plane surface (e.g., a piston generating a wave which travels along a duct which constrains it from spreading) vibrating in such a way that all points in the surface are in phase, the wavefronts are plane. In regions sufficiently far removed from a spherical source, the curvature of the wavefronts is so small that waves can effectively be regarded as being plane.

A sound wave at any given frequency is said to be freely propagating when the particle velocity is in the direction of wave propagation and when the pressure and particle velocity reach their maximum or minimum values simultaneously. This is said to occur in the far field. With regard to conditions in the near and far fields, it should be noted that most practical sources radiate nonplanar sound waves. Consequently, the acoustic pressure and particle velocities at points near most sources of sound radiation (even monopole, or single spherical surfaces) will not be in phase, and the above relationship does not apply. Also, values of sound levels close to a source may show appreciable fluctuations with position. Because of these fluctuations, it is not possible to use near-field data to predict sound-pressure levels far away from the source.

The extent of the *near field* is a function of the sound frequency of interest, the dimensions of the source, and the relative phasing of radiating surfaces of the source. As a rule of thumb, the near field usually extends outward from the "acoustic center" of the sound source for a distance equal to one and a half to two characteristic source dimensions (i.e., the length of a machine producing the noise). Further, the measurement location should be at least one wavelength away from the source in order for near-field effects to be unimportant.

As one moves away from the source in a free (nonreflecting) sound field, particle velocity and acoustic pressure become simply related, as in a plane wave. This is known as the *far field*.

In a freely propagating sound wave, where the pressure and velocity reach simultaneous maxima, it is necessary to evaluate the conditions that relate acoustic pressure and particle velocity. To do this, it is necessary to investigate the basic equations of motion of plane waves—and, as a first step, this will be done with one-dimensional sound waves from a single sound source.

Consider a plane source vibrating with simple harmonic motion described by the equation

$$y = y_0 \cos \omega t$$

where y_0 is the amplitude of the harmonic displacement,
   ω is the frequency of the harmonic displacement, and
   t is the time.

If plane waves are travelling from the source in the x direction with a constant speed c, they will arrive at some point A at a distance x from the source in a time $t' = x/c$. The phase of the vibrations at A will thus lag behind that of the source by an amount of $\omega t'$, and the particle displacement y in the medium at A will be given by the expression

$$y = y_0 \cos \omega (t - t') \qquad [1.1a]$$

or

$$y = y_0 \cos \omega \left(t - \frac{x}{c} \right) \qquad [1.1b]$$

In a single time period T, there will be a change in the phase angle (between the motion of the source and the particle displacement at A) by an amount 2π, during which the waves will have travelled the distance $\lambda = cT$ (where λ, as before, is the wavelength). Thus,

$$y = y_0 \cos \omega \left(t - \frac{x}{c} \right)$$

$$= y_0 \cos \frac{2\pi}{T} \left(t - \frac{xT}{\lambda} \right)$$

$$\left(\omega = \frac{2\pi}{T}, \quad c = \frac{\lambda}{T} \right)$$

$$= y_0 \cos 2\pi \left(\frac{t}{T} - \frac{x}{\lambda} \right)$$

or

$$y = y_0 \cos (\omega t - kx), \qquad [1.2]$$

where $k (= \omega/c = 2\pi/\lambda)$ is called the *wavenumber*.

The particle velocity u may be written as

$$u = \frac{\delta y}{\delta t} = -\omega y_0 \sin (\omega t - kx)$$

$$= \omega y_0 \sin (\omega t - kx + \pi). \qquad [1.3]$$

Putting $u_0 = \omega y_0$ and choosing a time scale so that the phase angle is suitably retarded,

$$u = u_0 \sin (\omega t - kx), \qquad [1.4]$$

where u_0 represents the velocity amplitude. It is seen that y and u differ in phase by 90°.

If Eq. [1.2] is differentiated twice with respect to x and t, respectively, then

$$\frac{\partial^2 y}{\partial x^2} = -k^2 y_0 \cos (\omega t - kx) \qquad [1.5a]$$

$$\frac{\partial^2 y}{\partial t^2} = -\omega^2 y_0 \cos (\omega t - kx) \qquad [1.5b]$$

thus giving

$$\frac{\partial^2 y}{\partial t^2} = c^2 \frac{\partial^2 y}{\partial x^2} \qquad [1.6a]$$

since $c = \omega/k$.

This is the general wave equation for plane waves. A similar expression may be obtained for the particle velocity u and another for the acoustic pressure p, i.e.,

$$\frac{\partial^2 u}{\partial t^2} = c^2 \frac{\partial^2 u}{\partial x^2} \qquad [1.6b]$$

and

$$\frac{\partial^2 p}{\partial t^2} = c^2 \frac{\partial^2 p}{\partial x^2}. \qquad [1.6c]$$

1.3. VELOCITY OF PLANE WAVES

Consider plane longitudinal waves propagated in the x direction in a homogeneous medium of uniform density ρ. Let AB (see Fig. 1.3) represent a thin, parallel-sided layer of the medium, with its surfaces A and B normal to the direction of propagation and at respective distances x and $x + \delta x$ from some source or origin. Now at some time t the wave motion causes A to be displaced to a position A' by an amount y, so that its distance from the origin along the x direction is increased to an amount $x + y$; B is then displaced to position B'. This latter displacement, which is assumed to be small, will be equal to $y + (\partial y/\partial x)\,\delta x$.

For a cross section of unit area, the mechanical strain suffered by the layer AB is seen to equal $\partial y/\partial x$. If p represents the acoustic pressure, i.e., the excess pressure caused by the passage of the sound waves, which gives rise to this strain, then in accordance with Hooke's law (7),

$$p = -q\,\frac{\partial y}{\partial x}, \qquad [1.7]$$

where q is the appropriate modulus of elasticity of the medium for the type of stress exerted. In this case, the layer is compressed and a positive stress will thus produce a negative strain. Adiabatic conditions are assumed to hold because no time is available for heat exchanges between adjoining layers of the medium (see any standard textbook on thermodynamics), and hence the adiabatic value of the elastic modulus must be used.

Consider another layer CD, similar to AB (but not shown in Fig. 1.3), in the medium and let the values of acoustic pressures at C and D be p and $p + (\partial p/\partial x)\,\delta x$, respectively. In accordance with Newton's third law of motion, equal and opposite pressures act on each of the boundaries C and D (7). This layer is thus compressed by an amount $+p$ at C and $-p + (\partial p/\partial x)\,\delta x$ at D, with the result that there is an unbalanced pressure equal to $-(\partial p/\partial x)\,\delta x$ acting on the layer. This imparts an acceleration, in accordance with Newton's second law, and the force causing this acceleration is

$$\rho\delta x \left(\frac{\partial^2 y}{\partial t^2} \right).$$

For unit area of the medium layer, this is also the pressure. Hence

$$-\left(\frac{\partial p}{\partial x} \right)\delta x = \rho\delta x\,\frac{\partial^2 y}{\partial t^2}$$

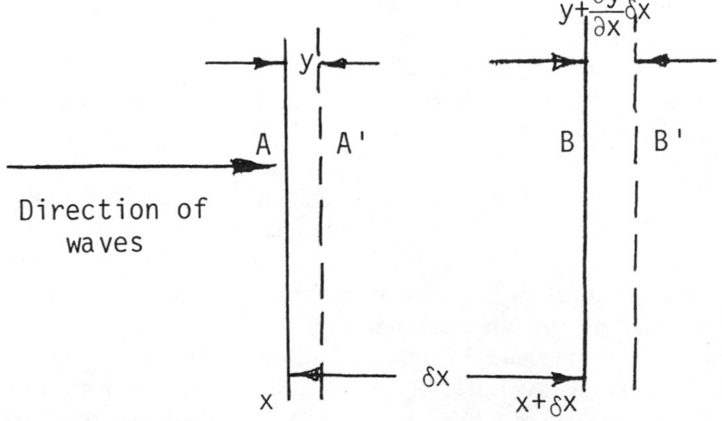

Fig. 1.3. Strain of a layer of homogeneous material due to the passage of longitudinal plane waves through it.

or

$$\frac{\partial p}{\partial x} = -\rho \frac{\partial^2 y}{\partial t^2}. \qquad [1.8]$$

Equation [1.7] can be written

$$\frac{\partial p}{\partial x} = -q \frac{\partial^2 y}{\partial x^2} \qquad [1.9]$$

by differentiating p with respect to x. Hence Eqs. [1.8] and [1.9] give

$$\frac{\partial^2 y}{\partial t^2} = \frac{q}{\rho} \frac{\partial^2 y}{\partial x^2}, \qquad [1.10]$$

and equating Eqs. [1.6a] and [1.10] results in

$$c^2 = \frac{q}{\rho}. \qquad [1.11]$$

1.4. SPECIFIC ACOUSTIC IMPEDANCE (OR CHARACTERISTIC IMPEDANCE)

The term *mechanical impedance* is defined by analogy with electrical systems: in a mechanical vibrating system, impedance is the ratio of the vibrating force to the resultant system velocity; similarly, in a simple electrical circuit, impedance is the ratio of potential difference (voltage) to the current. A similar analogy exists between acoustical and electrical systems, in which the acoustic pressure p, the particle displacement y, and the particle velocity u are equivalent to potential difference, electric charge, and electric current, respectively. The specific *acoustic impedance* can thus be defined as

$$Z_a = \frac{p}{u}. \qquad [1.12]$$

This is analogous to electrical impedance, and is equal to the mechanical impedance per unit area of cross section of the medium. Like both electrical and mechanical impedance, Z_a is complex and can be expressed as

$$Z_a = R_a + jX_a. \qquad [1.13]$$

For plane progressive waves, freely propagating,

$$p = -q \frac{\partial y}{\partial x} = -qky_0 \sin(\omega t - kx)$$

in accordance with Eqs. [1.7] and [1.2]. Using the value of u given by Eq. [1.3], then

$$Z_a = \frac{p}{u} = \frac{-qky_0 \sin(\omega t - kx)}{-\omega y_0 \sin(\omega t - kx)}$$

$$= \frac{qk}{\omega} = \frac{q}{c} = \rho c, \qquad [1.14]$$

since, from Eq. [1.11], $q = \rho c^2$.

Here, Z_a is real and, because there is no reactive component, it must be equal to R_a. The product ρc is called the *characteristic impedance* for the medium, and in units of $kg/m^2/sec$ (called mks rayls) is 415 mks rayls for air at standard temperature and pressure, where

$$\rho = 1.21 \text{ kg/m}^3$$

$$c = 343 \text{ m/sec}$$

Thus, the acoustic pressure p and the particle velocity u at a given distance x from the source are then related according to

$$\rho c = \frac{p(x, t)}{u(x, t)}$$

as indicated in Eq. [1.14].

1.5. SPHERICAL SOUND WAVES

Consider open space with no acoustic energy present. Individual air molecules move about in this space with random thermal motion. An equilibrium condition of the air results, characterized by an absolute temperature and pressure.

Now suppose that a small pulsating sphere is introduced into the space (surface vibrating at a frequency ω). As individual molecules collide with the vibrating surface of the sphere, they acquire additional momentum in the direction of motion of the surface. In subsequent collisions with neighboring molecules, this additional momentum is transferred, and a disturbance propagates into the space (i.e., a spherically spreading wave of single frequency results).

As this disturbance propagates outward, through momentum transfer, individual particles experience net displacements and velocities in the direction of propagation, as well as changes in absolute pressure and temperature. These small changes in equilibrium pressure, displacement and particle velocity of the elements of the medium can be positive or negative (depending on whether motion of the surface of the sphere is outward or inward). When these pressure variations occur at the eardrum or a microphone diaphragm, sound is perceived.

A spherical sound wave (of frequency ω) generated by the surface of a pulsating sphere (of radius R_0) will have a pressure amplitude and phase at radius r as follows (1):

$$p(r, t) = \frac{R_0 A}{r} \sin \left[k(r - R_0) - \omega t + \phi \right]$$
$$[1.15]$$

where A is the amplitude of the pressure wave at the surface of the sphere,

k is the wavenumber $2\pi/\lambda = 2\pi f/c = \omega/c$,

λ is the wavelength,

c is the propagating velocity in the medium, and

ϕ is the phase angle with respect to the motion of the pulsating sphere.

As the wave propagates from the sphere, its amplitude diminishes in proportion to its distance from the center of the sphere. $\rho c = p(r, t)/u(r, t)$ is also valid for the case of spherical waves (4).

1.6. SOUND INTENSITY

The intensity of a sound wave is defined as the average amount of acoustic power passing through a unit area of the medium that is perpendicular to the direction of sound propagation. Power is defined as:

$$\frac{\text{work}}{\text{unit time}} = \frac{\text{force} \times \text{displacement}}{\text{time}}$$

$$= \text{force} \times \text{velocity}$$

Therefore

$$\text{Sound intensity} = \frac{\text{sound power}}{\text{unit area}}$$

$$\approx \frac{\text{force}}{\text{area}} \times \text{velocity}$$

$$\approx \text{pressure} \times \text{velocity}$$

Hence for a freely propagating spherical sound wave of single frequency, sound intensity I is given by

$$I = \frac{1}{T} \int_0^T p(r, t) \cdot u(r, t) \, dt$$

$$= \frac{1}{T} \int_0^T p(r, t) \cdot \frac{p(r, t)}{\rho c} \, dt.$$

since $u = p(r, t)/\rho c$ from Eq. [1.14]. Therefore, sound intensity is

$$I = \frac{1}{T} \int_0^T \frac{p^2(r, t)}{\rho c} \, dt, \qquad [1.16]$$

where T is time interval of integration of the varying pressure $p(r, t)$.

The pressure fluctuation of a sound wave is periodic (or it may be random) and is usually expressed as a mean value (such as average pressure, peak or peak-to-peak pressure, root mean square pressure). See Fig. 1.4.

Defining the *effective pressure* as the root mean square (rms) acoustic pressure,

$$p = p_{\text{rms}} = \sqrt{\frac{1}{T} \int_0^T p^2(r, t) \, dt}.$$

Comparing this with Eq. [1.16] it can be seen that

$$I = \frac{p_{\text{rms}}^2}{\rho c} \qquad [1.17]$$

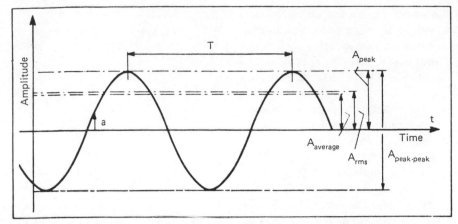

Fig. 1.4. Sinusoidal signal showing various measures of signal amplitude.

Note: Since p_{rms} is the effective pressure measured by a sound level meter, this gives a means for determining the intensity of sound from pressure measurements. (See also Section 3.95 on sound intensity measurements.)

Equation [1.17] is valid for both spherical and plane freely propagating waves. Putting Eq. [1.15] into [1.17] indicates that the inten-sity of the spherical wave diminishes with dis-tance, varying inversely with the square of the distance from the source (i.e., the inverse-square law).

An example of a point source radiating spherically is shown in Fig. 1.5. As shown, the sound energy spreads out equally in all direc-tions, so that as it travels further and further

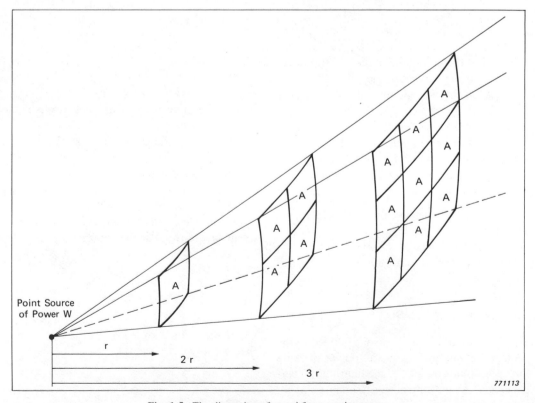

Fig. 1.5. The dispersion of sound from a point source.

from the source, its energy is received on an ever larger spherical area. The intensity is therefore the power of the source divided by the area of the spherical shell, i.e.,

$$I = \frac{W}{4\pi r^2}.$$

In a freely propagating sound wave confined to a long tube or duct of diameter D, if D is less than 1.2λ, the wave is "plane" (6). All properties of the sound wave (pressure, particle velocity, etc.) are uniform in a plane perpendicular to the direction of the sound propagation (the longitudinal axis of the tube). In this case (1),

$$p(x, t) = A \sin (kx - \omega t) \quad [1.18]$$

for a sound of frequency ω generated by a vibrating piston located in the duct at $x = 0$ (see Fig. 1.6).

Substitution of Eq. [1.18] into [1.17] shows that the intensity of a plane wave remains constant as it moves away from the source. This is why we can hear a child's voice when speaking into the other end of a garden hose or the heartbeat in a stethoscope.

1.7. LEVELS

The sound power of a very soft whisper is about 10^{-9} watt, while that of a large jet aircraft at take-off is 10^6 watts. This tremendous range is compressed when power is expressed in terms of the logarithm of the ratios. Thus, we define the *level* as the *logarithm of the ratio of two power-related quantities.*

The logarithmic ratio is designated as the Bel, and, for more practical purposes, the decibel. Therefore, the level of the sound power W in decibels is:

$$PWL = 10 \log_{10} \frac{W}{W_{\text{ref}}} \text{ dB re } W_{\text{ref}}$$

where W_{ref} is defined by international agreement (see below).

Inasmuch as power is proportional to the square of pressure ($I = p_{\text{rms}}^2 / \rho_0 c$), we can then define sound pressure level as:

$$SPL = 10 \log_{10} \left(\frac{p}{p_{\text{ref}}}\right)^2 \text{ dB re } p_{\text{ref}}$$

$$= 20 \log_{10} \left(\frac{p}{p_{\text{ref}}}\right) \text{ dB re } p_{\text{ref}}.$$

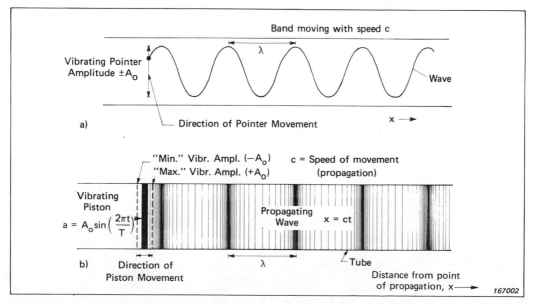

Fig. 1.6. The transformation of vibrations into waves: (a) by a vibrating pointer on a moving band; (b) by a vibrating piston in a fluid medium.

International agreement (i.e., ANSI Standard SI.8-1969) gives the following preferred reference quantities, all related to the threshold of hearing at 1000 Hz:

$$W_{ref} = 10^{-12} \text{ watts (W)}$$

$$I_{ref} = 10^{-12} \text{ W/m}^2$$

$$p_{ref} = 2 \times 10^{-5} \text{ N/m}^2 \ (0.0002 \ \mu\text{bar}).$$

Thus,

sound power level

$$PWL = 10 \log_{10} \left(\frac{W}{W_{ref}} \right) \text{ re } 10^{-12} \text{ W}$$

$$[1.19]$$

sound intensity level

$$IL = 10 \log_{10} \left(\frac{I}{I_{ref}} \right) \text{ re } 10^{-12} \text{ W/m}^2$$

$$[1.20]$$

sound pressure level

$$SPL = 20 \log_{10} \left(\frac{p}{p_{ref}} \right) \text{ re } 2 \times 10^{-5} \text{ N/m}^2.$$

$$[1.21]$$

1.8. WAVEFORM, FREQUENCY, PRESSURE CHANGE

The three factors which must be considered when examining the physical measurement of noise are frequency, intensity, and duration of the noise. When the psychological response to the noise is considered, the term *pitch* is used instead of *frequency*, and the term *loudness* is used instead of *intensity*, since the psychological response is not exactly the same as the physical measurement.

Most noise consists of tones of many frequencies at different intensities all mixed together without any consistent relationship between them. For this reason it is necessary to describe noise in terms of the frequencies and intensities of the pure tones of which it is comprised. The frequency composition of the noise gives an indication of the type of noise, that is, whether it is high pitched or low pitched. The frequency composition of a noise can be represented as shown in Fig. 1.7.

A conventional vibration trace is a time-domain view or time-history represented as amplitude versus time. Spectrum analysis "looks inside" this amplitude–time plot and shows it in the frequency domain as a plot of amplitude versus frequency.

Many types of noise, not easily recognized

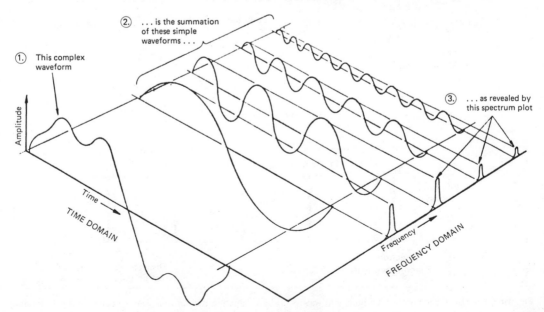

Fig. 1.7. Components of a complex waveform.

in the time domain, exhibit distinctive signatures in the frequency domain. Merely recognizing these signatures often suggests methods for stopping the noise. The loudness or intensity of a sound depends on the pressure changes exerted by the sound waves. Faint sounds are weak pressure changes and loud sounds are strong pressure changes. The power in sound is very small when compared with other everyday amounts of power around us. For instance, a noise with enough power to cause pain has a power of 0.01 W as compared with a standard light bulb which has a power of 60 W.

As has been discussed in Section 1.7 on Levels, the units used to measure pressure changes are Newtons per square meter or microbars (μbar), which is approximately one millionth of normal atmospheric pressure (standard atmospheric pressure = 1,013,250 μbar, and 1 bar = 10^6 dynes/cm^2).

If you elevate your head by one third of an inch the pressure change is equivalent to 1 microbar. However, a sound which changes the pressure by 1 microbar is relatively loud. At 1000 Hz the ear can detect a pressure of 0.0002 microbar (1). At 30 Hz a pressure of 0.2 microbar is necessary for the detection of sound by the ear.

The graph shown in Fig. 1.8 shows the relationship between the pressure change which can be detected by the ear and the frequency of sound. This will be discussed in Section 2.3 on subjective response of the human ear to sounds.

1.9. THE DECIBEL

As was pointed out in Section 1.7, the range of sound powers (and also sound pressures) is extremely large. Hence, the use of a logarithmic scale to compress this range. The decibel is a unit which measures the loudness of the sound on a relative scale: the faintest sound which the ear can hear is zero decibels, whereas the loudest man-made sound, the sound of a large rocket engine, is 175 decibels. All other sounds are measured within this range. The sound pressure of this weakest sound is 2×10^{-5} N/m^2 meter (or 0.0002 μbar) at 1000 Hz, and the loudest sound has a sound pressure of 10^5 N/m^2.

A healthy, unimpaired human ear has the ability to hear sounds over an extremely large range of sound pressure or acoustic energy. The sound pressure near a jet aircraft at takeoff can be 200 N/m^2, which is ten million times the sound pressure that a sensitive ear is capable of detecting. The decibel is able to express the range adequately and is at the same time an easy number to use.

The decibel scale can be thought of as representing points on a sharply rising curve. Thus, when measuring sound pressure level, 20 dB is 10 times the pressure at 0 dB, 40 dB is 100 times the pressure at 0 dB and 60 dB is 1000 times the pressure at 0 dB. Table 1.1 summarizes the relationship between sound pressure and sound pressure levels.

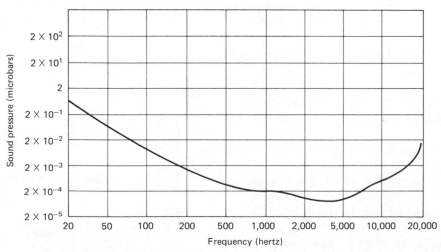

Fig. 1.8. The relationship between sound pressure detected by the ear and frequency.

Table 1.1. Average Sound Pressure Levels to Be Expected in Some Common Environments.

Sound Pressure (rms), N/m^2	Sound Pressure (rms), psi	Sound Pressure Level, dB	Typical Environment	Average Subjective Description
200	3×10^{-2}	140	30 m from military aircraft at takeoff	
63		130	Pneumatic chipping and riveting (operator's position)	Intolerable
20	3×10^{-3}	120	Ship's engine room (full speed)	
6.3		110	Automatic punch press (operator's position); sheet metal shop—hand grinding; textile weaving room	_____
2	3×10^{-4}	100	Automatic lathe shop; platform of underground station (maximum levels); printing press room	Very Noisy
6.3×10^{-1}		90	Heavy trucks at 6 m; construction site—pneumatic drilling.	_____
2×10^{-1}	3×10^{-5}	80	Curbside of busy street; office with tabulating machines	
6.3×10^{-2}		70	Loud radio (in average domestic room)	Noisy
2×10^{-2}	3×10^{-6}	60	Restaurant; department store; conversational speech at 1 m	_____
6.3×10^{-3}		50	General office	
2×10^{-3}	3×10^{-7}	40	Average suburban area at night, bedroom at night	Quiet
6.3×10^{-4}		30		_____
2×10^{-4}	3×10^{-8}	20	Background in TV and recording studios; leaves rustling	Very Quiet
6.3×10^{-5}		10		
2×10^{-5}	3×10^{-9}	0	Normal threshold of hearing	_____

An example of how sound pressure (rms) is converted to sound level (dB) follows. The sound pressure associated with a loud ratio is 6.3×10^{-2} N/m^2 (rms). Substitute this for p_{actual} in $SPL = 20 \log_{10} p_{act}/p_{ref}$, where $p_{ref} = 2 \times 10^{-5}$ N/m^2, and the SPL solves for 70 dB.

The decibel scale also provides a convenient method for comparing the sound pressure level of one noise to another. For example, a noise which has a level of 80 dB has a sound pressure which is one thousand (10^3) times as much pressure as a noise of 20 dB. A noise which has a level of 100 dB has a sound pressure which is ten times as much as a noise which is 80 dB.

Sound pressures on the decibel scale do not add linearly, but must be converted to their energy equivalents and added. For example, if two sounds of 60 dB are added together the resultant sound pressure level will be 63 dB and not 120 dB.

Consider the effect of adding another machine in an area where other equipment is op-

erating. Assume that the ambient sound level due to other equipment is $SPL_1 = 90$ dB and the level from the machine to be added is $SPL_2 = 88$ dB. Estimate the combined level.

Solution:

$$SPL = 10 \log \left(\frac{p}{p_{ref}}\right)^2 \text{ or } \left(\frac{p}{p_{ref}}\right)^2$$

$$= \text{antilog} \left(\frac{SPL}{10}\right)$$

Thus,

$$\left(\frac{p_1}{p_{ref}}\right)^2 = \text{antilog} \frac{90}{10} = 10 \times 10^8$$

$$\left(\frac{p_2}{p_{ref}}\right)^2 = \text{antilog} \frac{88}{10} = 6.31 \times 10^8$$

$$\left(\frac{p_{tot}}{p_{ref}}\right)^2 = \left(\frac{p_1}{p_{ref}}\right)^2 + \left(\frac{p_2}{p_{ref}}\right)^2$$

$$= 10 \times 10^8 + 6.31 \times 10^8$$

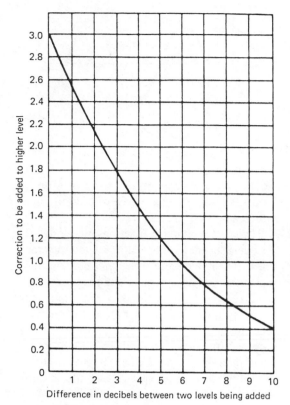

Fig. 1.9. Adding sound-pressure levels.

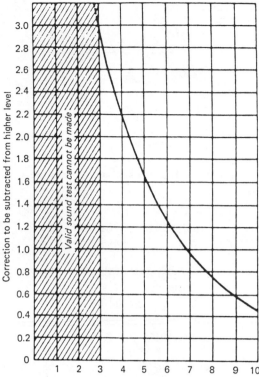

Fig. 1.10. Subtracting sound-pressure levels.

$$SPL_{tot} = 10 \log (16.31 \times 10^8)$$

$$= 92.12 \ (92) \ \text{dB.*}$$

Rather than going through this cumbersome procedure each time it is required to combine decibels, a graphic procedure using Figs. 1.9 and 1.10 is recommended.

Considering the previous example, note that the difference in the two *SPL*s is 2 dB. From Fig. 1.9, the correction to be added to the highest level is approximately 2.12, and hence the combined sound pressure level is $90 + 2.12 = 92.12$ (92) dB.

As another example, assume the sound-pressure level at the operator's position, near a commercial cutoff saw, is measured at 98 dB. It is suspected that most of the sound is being generated by air turbulence around the teeth of the circular saw blade. When the saw is operated with a toothless blade, a new level of 91

---

*Note that it is seldom necessary to express dB levels to a greater accuracy than 1 dB.

dB is measured. What is the sound-pressure level of the sound being generated by air turbulence?

In this case the level both before and after one source was removed are known, and it is required to determine the level of the source which was removed. The difference in the two levels is 7 dB. Using this difference, Fig. 1.10 indicates that about 1 dB should be subtracted from the higher level. Hence, the contribution due to air turbulence is approximately 97 dB.

The results can be checked by asking what the sound-pressure level would be if the original sawblade were returned. In this case, we combine two levels: 97 dB due to air turbulence and 91 dB due to the machine without the turbulence. The difference between these levels is 6 dB. Figure 1.9 indicates that about 1 dB should be added to the higher of the levels in order to determine the combined level of the saw plus blade. Adding this quantity to the higher level gives approximately 98 dB.

For an example of determining the overall level when more than two individual levels are

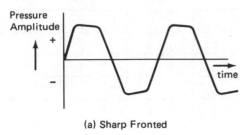

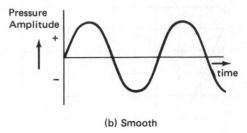

(a) Sharp Fronted   (b) Smooth

Fig. 1.11. Typical waveforms.

involved, see the problem example in Section 4.10.7.

1.10. FREQUENCY SPECTRA OF SOUND

Returning to the model of Fig. 1.1, the motion of the vibrating object was periodic, but the manner in which it moved between the maximum and minimum positions was not stipulated. The pressure wave generated would assume a shape dependent upon the fork's movement and would maintain that shape as it propagated. Fig. 1.11 shows typical waveforms. Although the waveforms of Fig. 1.11 may have the same period, they would sound quite different to the ear because of different *harmonic content*.

If the motion of the object in Fig. 1.1 was simple harmonic, the pressure wave generated would be simple harmonic, i.e., $p = P \sin \omega t$ where ω = circular frequency in rad/sec. Such a wave is shown in Fig. 1.12(a), and the representation of energy of this wave on the frequency spectrum is a single vertical line of amplitude p^2 at a frequency f_0, all the energy content being concentrated at this frequency. When two sine waves are added together, one being 3 times the frequency of the other, the distorted waveform shown in Fig. 1.12(b) would be obtained; this is represented in the frequency domain by two vertical lines. Analysis of a far more complex but still periodic function such as the square wave in Fig. 1.12(c) yields an infinite number of lines in the frequency domain at the odd harmonics of the repetition frequency of the signal. A mathematical analysis called Fourier Analysis (1, 4, 5)

would give a description of the wave form in terms of a sum of sine waves of differing amplitude and frequencies.

The spectra of all the functions mentioned so far have been harmonically related discrete lines, and are therefore termed *periodic*, as they repeat themselves at regular and predictable intervals. More commonly, however, practical everyday noise is nonperiodic, and contains a large number of frequency components which are not harmonically related, at infinitesimally small intervals, forming a continuous spectrum as in Fig. 1.12(d). A special case of the nonperiodic signal is *white noise*, which has equal energy at all frequencies (or a continuous flat spectrum); it is of particular importance in studies on building acoustics (Section 5.3.1) and psychoacoustic testing (Section 2.4).

REFERENCES

(1) Beranek, L. L. 1971. *Noise and Vibration Control.* New York: McGraw-Hill.
(2) Lord, Gatley, and Evensen. 1980. *Noise Control for Engineers.* New York: McGraw-Hill.
(3) Harris, C. M., ed. 1957. *Handbook of Noise Control.* New York: McGraw-Hill.
(4) Reynolds, D. D. 1981. *Engineering Principles of Acoustics.* Toronto: Allyn and Bacon.
(5) Blitz, J. 1964. *Elements of Acoustics.* London: Butterworth.
(6) Porges, G. 1977. *Applied Acoustics.* London: Edward Arnold.
(7) Newman, F. H., and V. H. L. Searle. 1957. *The General Properties of Matter*, 5th ed., London: Edward Arnold.
(8) Kinsler, L. E., and A. R. Frey. 1950. *Fundamentals of Acoustics.* New York: John Wiley & Sons.
(9) Hassall, J. R., and K. Zaveri. 1979. *Acoustic Noise Measurement.* Naerum, Denmark: Bruel and Kjaer.

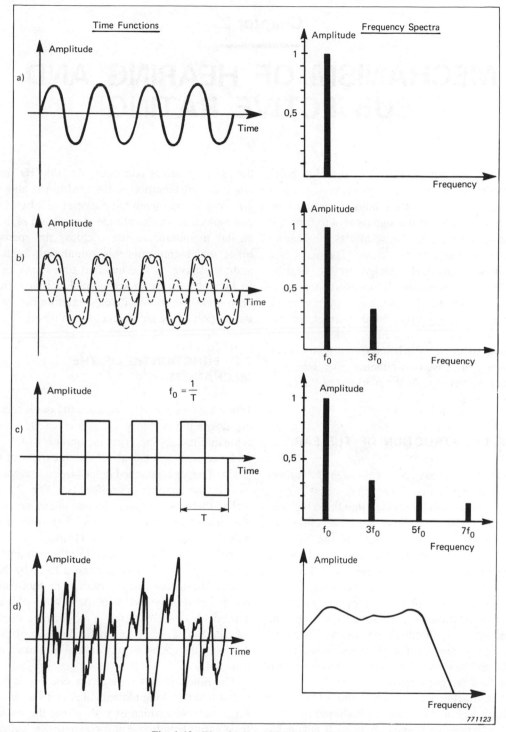

Fig. 1.12. Waveforms and frequency spectra.

Chapter 2

MECHANISM OF HEARING AND SUBJECTIVE RATING

If one is concerned with noise control or noise reduction it is very useful to know something about the working of the human ear and how the brain interprets the signals sent to it by the hearing mechanism. Noise affects the hearing function of the ear in two ways: [1] it may cause permanent physical damage to the hearing mechanisms, rendering it insensitive to important components of sound, and [2] it may mask or drown out desirable sound. Further, annoyance is a highly subjective, unfavorable response to noise which interferes with human activity, or sense of wellbeing. For these reasons a knowledge of the working of the ear is most useful.

2.1. CONSTRUCTION OF THE EAR

A simplified cross section through the human ear is shown in Fig. 2.1.

The ear is usually divided into three regions: outer, middle, and inner ear. The *outer ear* consists of the fleshy *pinna* and the *auditory canal*, which is about an inch long and ends at the *tympanic membrane* or ear drum. The *middle ear* consists of the air space containing three small bones or *auditory ossicles*. The *inner ear* consists of fluid-filled passageways inside the bone of the skull called the *cochlea*. Some of these passageways, known as *semicircular canals*, are associated with the balance mechanism and are not related to hearing.

The hearing mechanism is one of the most delicate organs in the human body and there are some facts often quoted about it. It is capable of sensing minute sound pressures and also responding to very large sound pressures without damage. There is a range of about 10^7 from the smallest to the largest sound-pressure amplitude which can be detected, and for this reason

the ear response is nonlinear. At 1000 Hz the amplitude of vibration of the eardrum is about 10^{-9} cm, or one-tenth the diameter of a hydrogen molecule. The amplitude of vibration of the basilar membrane in the cochlea, the membrane which transmits this stimulation to the auditory nerve, is one hundred times smaller. In fact, the ear is so sensitive that if it were only slightly more sensitive the random Brownian motion of the air molecules could be heard.

2.2. FUNCTIONING OF THE MECHANISM

When a sound wave approaches the ear it travels down the auditory canal and excites the tympanic membrane. This membrane which is under tension and has about the thickness of a piece of paper is attached to the largest auditory ossicle, the hammer or *malleus* (see Fig. 2.2).

The three ossicles are situated at the end of the *eustachian tube*. In swallowing, pressure may be equalized across the tympanic membrane by the eustachian tube. The three ossicles act as a lever system to transmit a force to the base of the stirrup, the *footplate*. The footplate is connected to the oval window in the cochlea. The cochlea is constructed in the form of a spiral (see Fig. 2.3). A schematic diagram of the ear, with the cochlea "unwound," is shown in Fig. 2.4.

The liquid-filled cavity of the cochlea is divided into two longitudinal canals by the *basilar membrane*, which extends along the cochlea's entire length except for a small gap called the *helicotrema* at the cochlea's far end or apex. When the stirrup, responding to an acoustic stimulus, moves the oval window, the resultant fluid disturbance passes along the upper canal (*scala vestibuli*), around the helicotrema, into

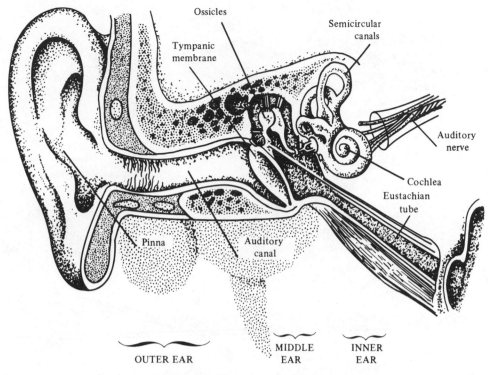

Fig. 2.1. The main parts of the ear.

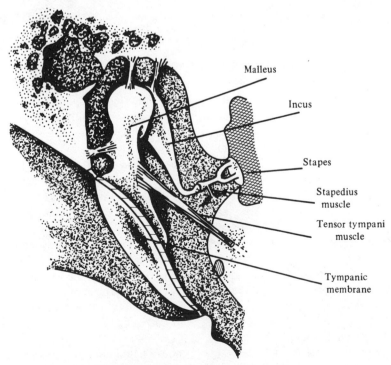

Fig. 2.2. Eardrum and three auditory ossicles.

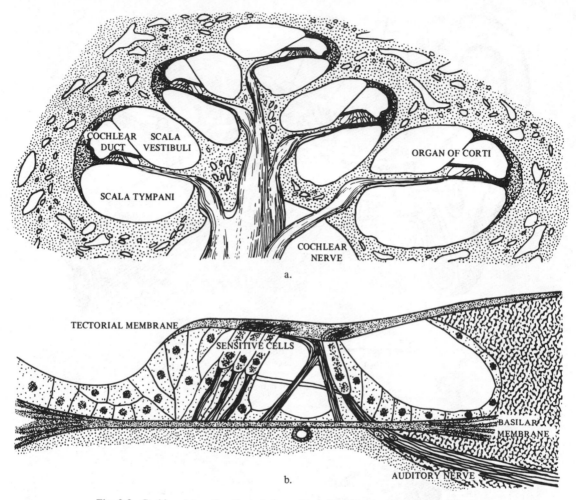

Fig. 2.3. Cochlea: (a) section through the cochlea; (b) enlargement of Organ of Corti.

the lower canal (*scala tympani*), and ultimately the round window deflects to accommodate (or absorb) the fluid disturbance. During its passage through the lower canal, the disturbance distorts the basilar membrane, on whose upper surface (Fig. 2.3) there are about 25,000–30,000 *hair cells*. These hair cells are extremely sensitive and are known to have piezoelectric properties. The distortion of these cells is transformed into an electrical impulse which travels along the auditory nerves to the brain cortex. The brain interprets these neural impulses to give the sensation of hearing. The way in which this is done is not too well understood.

The distortion of the hair cells appears to be frequency dependent, i.e., certain cells respond to certain frequencies of disturbance, while other cells respond to other frequencies. The frequency sensitivity varies with distance along the basilar membrane, the maximum response at high frequencies occurring near the oval window and that at low frequencies near the helicotrema (Fig. 2.4). By this system of canals, levers, membranes, and hair cells, the ear is able to detect sounds over enormous ranges of frequency and intensity. The highest-frequency sound the healthy human ear can hear is 1,000 times the frequency of the lowest-frequency sound, and the loudest sound can have a sound pressure one million times that of the quietest sound that can be heard (an intensity ratio of $10^{12} : 1$).

The outer and middle ears are best understood. The middle ear increases the pressure on the fluid in the cochlea relative to that in the

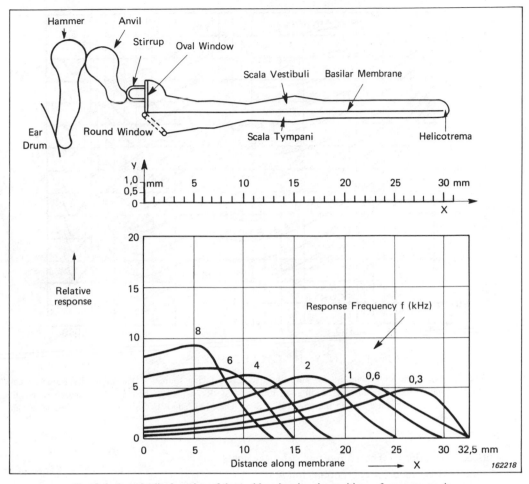

Fig. 2.4. Longitudinal section of the cochlea showing the positions of response maxima.

tympanic membrane about 22 times. This is achieved partly by the mechanical advantage of the lever system and partly by the reduction in area from the tympanic membrane to the oval window. It is thought that there is an approximate matching of the air impedance and the inner ear fluid impedance by this change from low-pressure, large-amplitude motion at the tympanic membrane to high-pressure, small-amplitude motion in the cochlea. Of course, impedance matching would ensure the maximum of energy transmitted into the ear from a sound wave. It is interesting to note that there is a safety device built into the three ossicles. If a high-level sound increases in level (but not too rapidly) this is sensed and muscles rotate the ossicles so that the motion passed onto the oval window does not increase correspondingly. This can prevent damage. For a more

complete description of the ear mechanism the reader is referred to Ref. 1 at the end of this chapter.

2.3. SUBJECTIVE RESPONSE AND UNITS FOR RATING OF NOISE

The sensitivity of the hearing mechanism is highly dependent upon the frequency content of the received sound. Fig. 2.5 summarizes a series of tests conducted on thousands of young people with good hearing. Subjects were placed in a free sound field (an anechoic chamber, with no reflections) and were asked to listen to a reference tone at 1000 Hz. They were then asked to adjust the loudness of a second tone to equal the loudness of the reference tone. After statistical analysis, a family of curves were produced (Fig. 2.5), where each contour curve

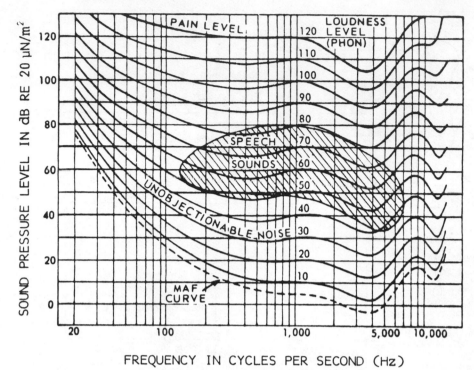

Fig. 2.5. Free-field equal-loudness contours for pure tones (observer facing source). (Determined by Robinson and Dadson at the National Physical Laboratory, Teddington, England [ISO/R226-1961]. The majority of speech sounds lie in the 300–5000 Hz frequency range and in the 45–75 dB range. The MAF curve is the minimum audible to normal ears).

represents equal loudness. The sound pressure level of the 1000 Hz tone for each equal-loudness contour is defined as the *loudness level* in phons.

2.4. LOUDNESS; PHONS AND SONES

It is seen that people rate low-frequency pure tones not as loud as mid-frequency tones produced by the same sound pressure level. For instance, the average young person would rate a pure tone of 40 Hz at 70 dB as loud as a pure tone of 1000 Hz at 40 dB. For this reason, both of these sounds are said to have a loudness level of 40 phons.

It is found that when a sound increases in loudness level by about 10 phons it appears to double in loudness. Hence a linear loudness scale (in sones) has been created to correspond to the logarithmic phon scale (5). A doubling in sones represents a doubling in loudness.

$$S = 2^{(P-40)/10}$$

where

S = loudness (sones),
P = loudness level (phons).

A graph showing the relationship between loudness in sones and loudness in phons is shown in Fig. 2.6.

It is interesting to note that the ear appears quite insensitive to low-frequency noise. If the ear were any more sensitive at low frequency we would be able to hear low-frequency noise conducted through the bones of the skull due to muscle contractions and bone movements (1).

In practice, most noise is not pure tone in character but has a continuous spectrum. Similar subjective experiments have been conducted to find the loudness of *bands* of noise based upon still another measure of loudness, called the *loudness index* (3). Fig. 2.7 shows contours joining bands of noise with equal loudness index.

Probably the most widely used method for

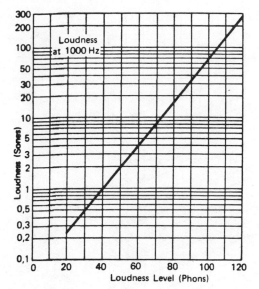

Fig. 2.6. The relationship between loudness in sones and loudness level in phons.

establishing the loudness of a complex noise is that developed by Stevens. The version called Mark II has been adapted by the American National Standards Institute as ANSI S3.4-1968 (8). The method is based upon the measurement of octave and $\frac{1}{3}$-octave band sound-pressure levels. The measured band pressures are used in conjunction with the equal-loudness-index contours shown in Fig. 2.7 to determine the loudness level by means of a simple calculation:

1. The loudness index for each octave or $\frac{1}{3}$-octave band pressure level (from the center frequencies) is determined from Fig. 2.7.
2. Loudness (in sones) $= I_m(1 - k) + k(\Sigma_i I_i)$

   where I_m is the maximum value of all the bands,

   I_i is the same value of a band i (including I_m),

   k is a constant (0.3 for octave bands, 0.15 for $\frac{1}{3}$-octave bands).

The loudness so obtained may be converted to phons by the use of Fig. 2.6. The prominence of the maximum band value I_m is be-

cause that band produces a partial masking of the other bands. (Equal-loudness contours are not sufficient to allow an assessment of the loudness of composite sounds unless account is taken of the so-called "critical bandwidths" in the masking process. For a good account of masking, see Refs. 7 and 9). The use of sone values also allows the addition of each band arithmetically, which would not have been possible in phons (which add logarithmically).

A limitation of the Stevens Mark II method is that it is only designed to be used for diffuse sound fields and when the spectrum is relatively flat and does not contain pure tones. Zwicker (10) devised a loudness rating scheme that can be used for both diffuse and free field conditions and for broadband noise even when pronounced tones are present. However, the method is somewhat time-consuming and involves measuring the area under a curve.

Figure 2.8 shows many well known sources of sound approximately placed with regard to the sound pressure level at which they are normally heard and at their major frequencies. The equal-level contours (and loudness levels in phons) are shown for comparison.

2.5. LOUDNESS OF SHORT DURATION SOUNDS

In the previous sections, the loudness of sound as perceived by the ear was shown to be dependent on both its amplitude and its frequency. If, in addition, the sound stimulus is of relatively short duration (less than about 200 msec), its loudness is reduced compared with the same sound heard continuously. The shorter the duration becomes, the less loud the sound appears to be. Many experimenters have investigated this phenomenon, and research on the subject has recently been given a boost by the desire to define more closely the relationship between sounds of short duration and both hearing damage and annoyance.

The results of Fig. 2.9 were obtained from psychoacoustic experiments in which tone-burst impulses of different durations were compared with steady tones. The ordinate of the figure is the sound pressure level difference $L_i - L_D$ between the level of the impulse L_i and the steady

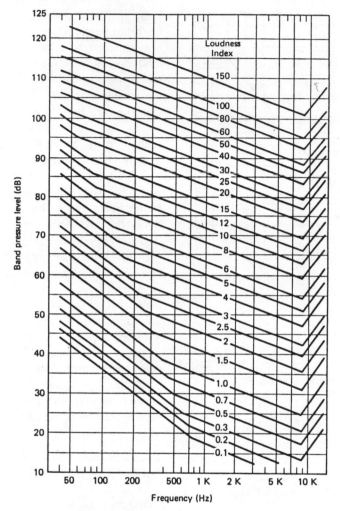

Fig. 2.7. Equal loudness index contours.

sound level L_D which was judged equally loud. The abscissa is simply the pulse duration. As long as its duration is greater than a certain length, the loudness of a pulse is judged to be equal to that of a steady sound of the same sound pressure level, i.e., $L_i - L_D = 0$. For shorter pulses, however, the loudness decreases with pulse length. The breakpoint corresponds to the effective averaging time of the ear, and the slope of most of the curves show that an increase of 3 dB in sound pressure level (a doubling of pulse intensity) is necessary to maintaining the same perceived loudness when the pulse duration is halved. As the product of intensity and time is energy, the ear appears to act as an energy-sensitive device, as far as the perception of loudness is concerned. The im-

pulse characteristic of sound measuring systems has been standardized on this basis (see Fig. 2.9), the time constant being designed to give a measure of the loudness of single impulses. For the correct measurement of actual peak levels for purposes other than loudness evaluation, an instrument is required which is capable of detecting and holding peaks with rise times of less than 50 microsecs.

2.6. AGE-RELATED HEARING LOSS — PRESBYCUSIS

The normal aging process leads to hearing losses, which tend to be small up to the age of 30 years or so and increases rapidly as old age is reached. Low frequencies below, say, 1 kHz

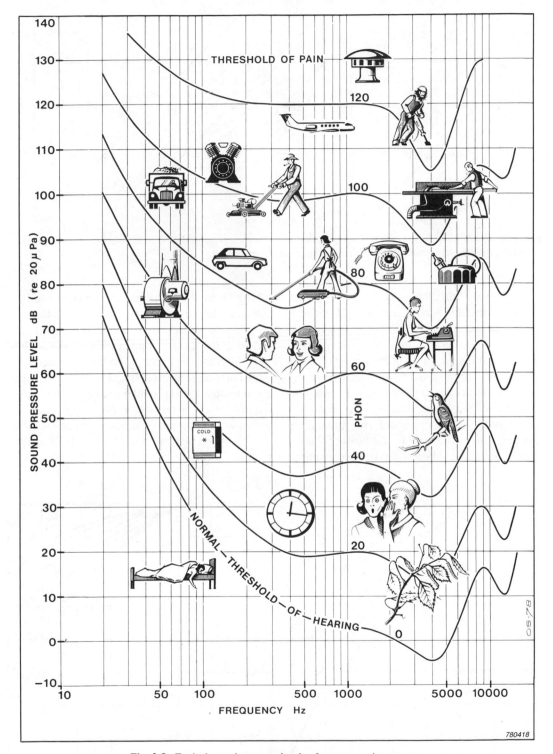

Fig. 2.8. Typical sound pressure levels of common noise sources.

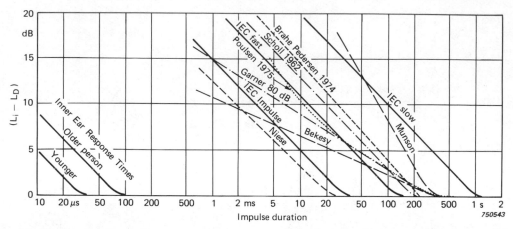

Fig. 2.9. Results of studies by different researchers of the subjective perception of short impulses compared with the standardized sound-level meter characteristics and inner-ear response times.

are relatively little affected, but the loss increases steadily with frequency as demonstrated by the results of surveys shown in Fig. 2.10. Not all this loss may be directly attributable to aging; some of it may be due to the noise which bombards modern man from all sides in his normal day to day activities. But such nonoccupational hearing loss contributions (dubbed *sociocusis*) are very difficult to quantify. Measurements of the hearing thresholds of Nomadic tribesmen living in exceptionally quiet rural conditions indicate their age-related hearing loss to be far smaller than that of comparable groups in industrialized countries, although this may be due at least in part to the vast difference in other, nonacoustic, factors.

A correction at each frequency must be applied to audiograms to take account of this age-dependent characteristic, in order to be able to arrive at the amount of hearing damage which can fairly be attributed to the effects of noise exposure.

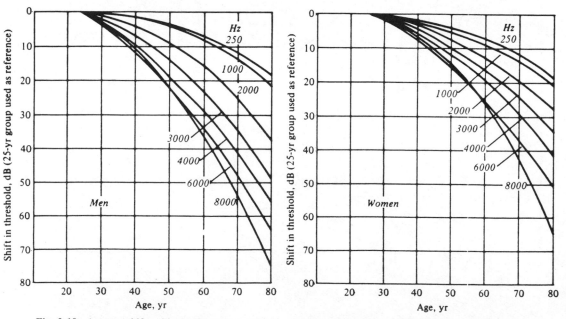

Fig. 2.10. Average shifts with age (in persons with "normal" hearing) of threshold of hearing for pure tones.

2.7. NOISE-INDUCED HEARING LOSS

2.7.1. Steady Noise

If the ear is exposed to a high level of noise for a short period, a test of the sensitivity of that ear taken immediately afterward reveals a small hearing loss known as a *temporary threshold shift*. The hearing threshold is the lowest sound pressure level which can be detected by the subject, and this may rise by up to 20 dB at certain frequencies after even a relatively short exposure. That is, a sound would have to be much louder than it was before the exposure to be heard afterwards. Fortunately, the phenomenon is temporary in nature, the ear recovering its original sensitivity after a relatively short time, any permanent threshold shift being too small to measure.

However, as the level and/or the exposure time increase, so too does the temporary threshold shift and the length of time the ear takes to recover from it. As long as the exposure times are relatively short and the intervals between them are long, then the permanent effects are not significant. Unfortunately, a large number of people work in factories and workshops where the noise levels are consistently high; and when the exposure takes place regularly for up to 8 hours every day, year after year, the effects cease to be temporary. A permanent hearing loss develops which in time may become severe enough to make normal conversation difficult to follow, and ultimately lead to chronic disability. The damage is by this time permanent and irreversible, for no amount of ''rest'' will lead to any significant recovery.

The form which a loss of hearing takes is remarkably independent of the mechanism which brings it about. There is almost always an initial dip in the audiogram at approximately 4 kHz., whatever the frequency content of the noise exposure which caused the damage, and the greatest shift is invariably at a frequency above that of the noise. This is characteristic of noise-induced hearing loss, and occurs for both temporary and permanent threshold shifts.

Permanent damage, like temporary damage, begins with a drop in sensitivity around 4 kHz,

and as the exposure time increases the shift becomes greater and extends gradually down to include the lower frequencies. The course which this takes is very clearly shown in Fig. 2.11, which is taken from a classic survey of jute weavers in Dundee, Scotland. The workforce was particularly stable, many of them having worked at the same factory, and in some cases at the same loom, for up to 50 years. During this time virtually the same machinery was in continuous use and the noise had therefore remained remarkably constant. This provided an excellent opportunity to obtain all the necessary information at one time and in one place. The sample even included people who had left the industry or retired after a significant period of exposure and therefore had had a long time for any recovery process to have ended. Of particular interest is that the disability shown in Fig. 2.11 rarely exceeds 20 dB below 1 kHz, and is less severe at frequencies of 6 kHz and above than it is between 1 kHz and 5 kHz, which is the frequency range of most importance to the understanding of speech.

2.7.2. Impulsive Noise

Impulsive noise is particularly important both as the cause of annoyance and as a significant hazard to hearing. It occurs widely in industry and construction, where it arises from impact-producing operations such as rivetting, bottling, and materials handling, and also from the explosive release of energy, e.g., firearms, explosive forming, blasting, and cartridge guns. Otherwise apparently harmless objects such as toy pistols, clicking toys, and fireworks are also capable of emitting dangerously high levels of impulsive noise.

Based on audiometric studies and close examination of the physical characteristics of impulsive sources, some factors important in causing damage have been identified. These include peak level, duration, rise time, and in the case of repeated events or reverberant fields, the repetition rate or reflection intensity respectively. The problem is further complicated because the loudness of a pulse reduces with its length below a certain critical duration (see Fig. 2.9). Acoustic impulses of extremely short du-

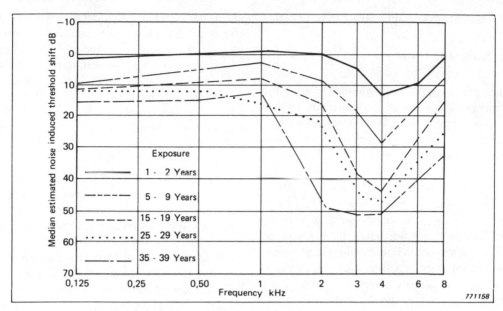

Fig. 2.11. The development of noise-induced hearing loss.

ration, which may be intense enough to contribute to hearing damage, may sound relatively quiet to the observer, and therefore the possibility of a hearing hazard may be overlooked. Normal sound-level meters are designed for steady-state conditions and cannot respond to the very rapid rise times of impulsive sounds, and so will in general underestimate the extent of a potential hearing hazard (see Chapter 3). To measure the absolute sound pressure level of an impulsive noise correctly, it is necessary to use a precision meter with a very rapid rise time (50 μsec or less) and a peak hold facility, so that the true maximum peak level can be identified and the value retained.

The noise specialist should always approach a noise measurement aware of the need to check for characteristics in the noise which might necessitate the use of special instruments or techniques of measurement or analysis. Impulsiveness is perhaps the most important of these, and will be mentioned in specific instances in later Chapters.

REFERENCES

(1) Von Bekesy, C. 1957. "The Ear." *Scientific American*, August.

(2) Robinson, D. W., and R. S. Dadson. 1956. "A Redetermination of the Equal Loudness Relations for Pure Tones." *British Journal of Applied Physics*, **7** (May), 161–181.

(3) Beranek, L. L. 1971. *Noise and Vibration Control.* New York: McGraw-Hill.

(4) Kryter, K. D. 1970. *The Effects of Noise on Man.* New York: Academic Press.

(5) Harris, C. M., ed., 1957. *Handbook of Noise Control.* New York: McGraw-Hill.

(6) Hassall, J. R., and K. Zaveri. 1979. *Acoustic Noise Measurement.* Naerum, Denmark: Bruel and Kjaer.

(7) Crocker, M. J., and A. J. Price. 1975. *Noise and Noise Control*, Volume 1. Cleveland, Ohio: CRC Press.

(8) ANSI. 1968. "Procedure for the Computation of Loudness of Noise." American National Standard USAS 53.4-1968. New York: American National Standards Institute.

(9) May, D. N., ed., 1978. *Handbook of Noise Assessment.* Toronto: Van Nostrand Reinhold.

(10) Zwicker, E. 1960. "Ein Verfahren zur Berechnung der Lautstarke (A Means for Calculating Loudness)," *Acustica,* **10,** 304.

Chapter 3

INSTRUMENTATION FOR NOISE MEASUREMENT

3.1. GENERAL

The instrumentation used for measuring noise may range from a simple sound-level meter to a sophisticated signal analysis processing system.

The sound that one hears is generally transmitted through the air to the ear. However, to understand and solve a noise problem it may be necessary to measure not only the sound in the vicinity of the source of interest but also the vibration of various parts of the source of interest. Thus, sound and vibration measurements go hand in hand to assist a noise control engineer in understanding and solving his noise control problems. Microphones and vibration measuring devices are referred to as *transducers*, because they change either the sound or vibrational energy into an electrical signal that can be processed. With most instrumentation the microphone or pickup is usually connected directly to the input of a sound-level meter. This constitutes the simplest form of noise-measuring instrumentation. However, the signal output from the sound-level meter may be fed into a magnetic tape recorder and stored for detailed analysis at a later time or the signal may be fed directly from the sound level meter into one of a variety of signal analyzers and data readout devices, including digitizing of data for computer processing.

Microphones and vibration pickups (transducers) are the most crucial part of any measuring system and should receive the most care and attention. These next sections are a discussion of some of the attributes of transducers commonly used for the measurement of noise and vibration.

3.2. MICROPHONES

The ear is sensitive to sound pressure and fortunately there are available several types of sound pressure measuring microphones suitable for acoustical measurements. It is sometimes desirable to also determine the sound intensity (see Section 3.9.5).

There are three basic types of pressure sensitive microphones used for sound measurements: the ceramic microphone, the standard condenser microphone, and more recently the electret condenser microphone. (Another microphone type, called electrodynamic or moving coil, is used extensively in the performing arts—stage, TV, etc.—and is primarily noted for its rugged construction and endurance).

The ceramic microphone contains a piezoelectric crystal element that produces an electrical signal in the presence of a pressure field or when a force is exerted on the crystal element (see Figs. 3.1 and 3.2).

The crystal element itself is small and is not very sensitive to sound pressure. For this reason most crystal or ceramic microphones (as they are called) contain a diaphragm approximately $\frac{1}{2}$ inch to 1 inch in diameter with a connecting pin in the center that is attached to the crystal or ceramic element to mechanically amplify the force exerted upon the crystal due to the sound field to which the microphone is exposed. This type of microphone works much in the same manner as the eardrum and mechanical linkage in the middle ear. The crystal or ceramic microphone is one of the oldest and most reliable type of microphone that has been in common use for sound measurements. Its major disadvantages are its relatively uneven

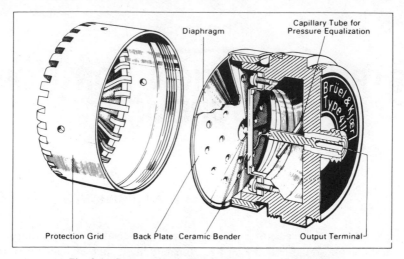

Fig. 3.1. Cross-sectional view of piezoelectric microphone.

and limited frequency response at high frequency and the temperature limitation of the crystal or ceramic element or adhesives used in construction. Contemporary ceramic microphones are useful from very low frequencies to between 10,000 and 20,000 Hz and temperatures covering ±60°C. The chief advantages of the ceramic microphones are their ruggedness, low susceptibility or immunity to high

humidity, and relatively high electrical capacitance. Typical frequency responses for ceramic microphones are shown in Fig. 3.3.*

During the past fifteen to twenty years con-

*Deviations from linear response at high frequency are due to a combination of mechanical resonance (diaphragm, etc.) and acoustical interaction between small wavelengths and geometric size of microphone (1).

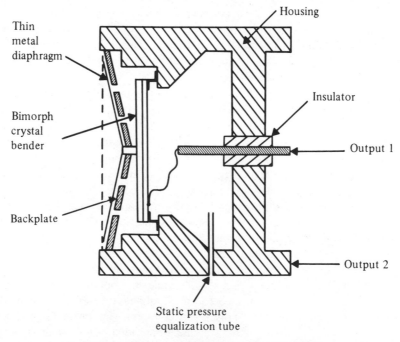

Fig. 3.2. Simplified cross section of piezoelectric microphone.

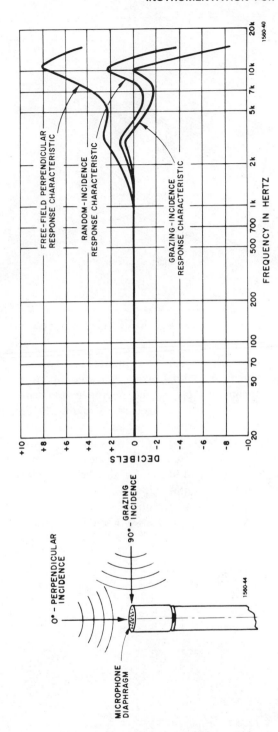

Fig. 3.3. Typical response curves for a ceramic microphone.

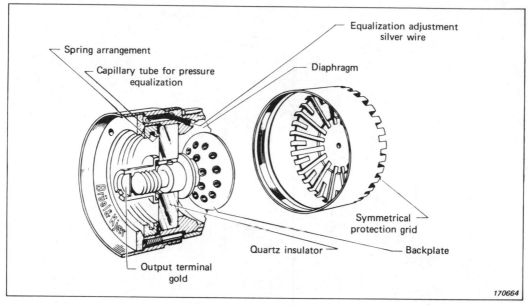

Fig. 3.4. Sectional view of a condenser microphone cartridge.

denser microphones have gained widespread acceptance for acoustical measurements. Construction of a condenser microphone is shown in Fig. 3.4.

The condenser microphone has a broad frequency range and a much smoother response than the ceramic microphone at high frequency. The smaller the diaphragm element in the condenser microphone the higher is the upper frequency limit with an attendant penalty in the microphone sensitivity. Thus, with any of the microphones discussed here, there is always a tradeoff between frequency response and microphone sensitivity. The microphone sensitivity indirectly determines the inherent lower noise-level limit that can be measured. The frequency response of typical condenser microphones is shown in Fig. 3.5.

Condenser microphones generally require a high-voltage DC polarizing voltage for operation. The condenser microphone has proved to have good long-term stability and relatively

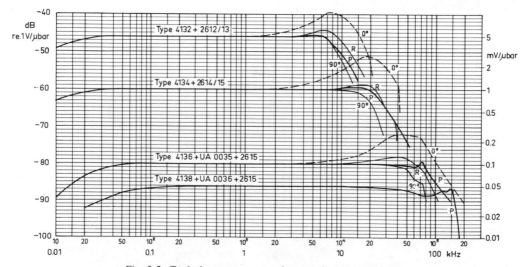

Fig. 3.5. Typical response curves for a condenser microphone.

high sensitivity to sound fields and low sensitivity to vibration excitation. The major disadvantage of the condenser microphone is its susceptibility to high humidity and its relatively low electrical capacitance which puts more stringent requirements on the input of the sound level meter, or preamplifier associated with the condenser microphone. Because of the stretched metal diaphragm used in a condenser microphone it also makes it more susceptible to damage from shock such as dropping the microphone on a hard surface. The condenser microphone has proved to be the best choice for measuring extreme temperatures ($-50°C$ to $150°C$) and extreme pressures.

The most recent microphone introduced for precise sound measurements has been the electret condenser microphone. The electret condenser microphone is constructed somewhat similar to the standard condenser microphone that uses the stretched metal diaphragm (see Fig. 3.6). However, the electret condenser microphone utilizes a plastic diaphragm with a metallized coating on the exterior surface. The advantage of an electret condenser microphone is its simplicity of construction, ruggedness, and relative immunity to high humidity. The electret condenser microphone diaphragm actually touches the back electrode intentionally at many locations, whereas the conventional condenser microphone with a metal diaphragm has a required but fixed small air space (approximately 0.001 inch) between the diaphragm and the back electrode. A further advantage of the electret condenser microphone is that it requires no external polarizing voltage

as the plastic diaphragm is self-polarized during manufacture. The electret condenser microphone appears superior to the ceramic microphone in all respects. However, the electret condenser microphone does not have as smooth a frequency response at high frequencies as the conventional condenser microphones and it has a maximum temperature limitation of approximately 50°C.

The cross section of a typical moving-coil microphone is shown in Fig. 3.7. The moving coil microphone suffers from several resonances and by proper design, these resonances may be damped so that the pressure frequency response between about 50 and 8,000 Hz is relatively flat.

The moving-coil microphone works on the reverse principle of the common loudspeaker (Fig. 3.7). A light diaphragm whose center portion is stiffened by giving it a slight spherical deformation is attached to a coil. The coil is situated in the annular gap of a permanent magnet. The terminal connections are made to each end of the coil. When sound pressure deflects the diaphragm and consequently displaces the coil, a voltage E is generated across the coil.

The types of microphones that have been discussed thus far are round cylinders available in diameters of 1 inch, $\frac{1}{2}$ inch and in some cases $\frac{1}{4}$ inch, and $\frac{1}{8}$ inch. All of these microphones are considered omnidirectional (nondirectional). They do indeed have directional characteristics because of their physical size irrespective of the type of microphone or manufacturer. This directionality phenomenon is shown in Figs.

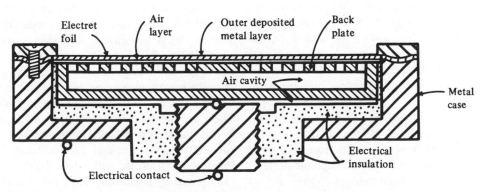

Fig. 3.6. Cross section of electret microphone.

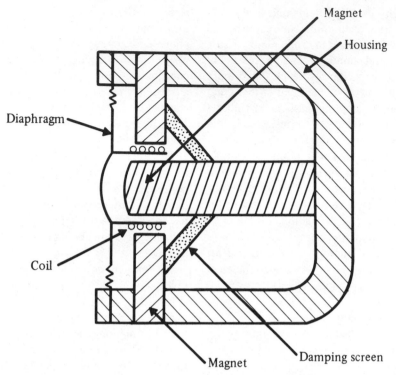

Fig. 3.7. Cross section of a typical moving-coil microphone.

3.8(a) and (b). For a 1-inch microphone a good rule of thumb to follow is to assume that a microphone is more sensitive for sounds arriving perpendicular to the diaphragm versus parallel to the diaphragm by 1 decibel per 1000 Hertz.

For example, at 1000 Hertz the difference between normal incidence (0°) and grazing incidence (90°) is approximately $\frac{1}{2}$ decibel; at 4000 Hertz it is approximately 3 decibels; at 8000 Hertz it is approximately 8 decibels. For a $\frac{1}{2}$-inch microphone this (pressure doubling) problem moves up one octave. For a $\frac{1}{4}$-inch microphone the pressure doubling effect moves up two octaves relative to the pressure doubling or diffraction problems associated with a 1-inch microphone. Therefore, the microphone orientation with respect to the sound source becomes of very real significance at high frequencies. It is best to follow the manufacturer's instructions for obtaining the frequency response desired. This diffraction effect for a 1-inch microphone can be reduced to one-half by replacing the condenser microphone protection grid with a special random incidence corrector.

Sound measurements out of doors generally should not be attempted when the wind velocity is in excess of 20 km/hr. Wind velocity in excess of this level may cause wind noise to dominate the sound field of interest or even overload the sound level meter or other instrumentation at subaudible frequencies. An example of wind-induced noise for a Bruel and Kjaer $\frac{1}{2}$-inch microphone fitted with a rain cover and a nylon mesh wind screen is shown in Fig. 3.9.

It is suggested that wind screens recommended by the microphone manufacturers be utilized for all out-of-doors measurements unless the wind velocity is less than approximately 8 km/hr. Even then their use is suggested to minimize damage to the microphone from the microphone being dropped and to protect it from dust and moisture contamination. There are two types of suitable wind screen: a 6-inch-diameter open wire frame covered with silk cloth, and open-cell polyurethane foam spheres. Windscreens will reduce the important low-frequency noise (due to turbulence at the microphone) between 10 and 20 dB. This mag-

Fig. 3.8(b). Directional characteristics of typical half-inch condenser microphones.

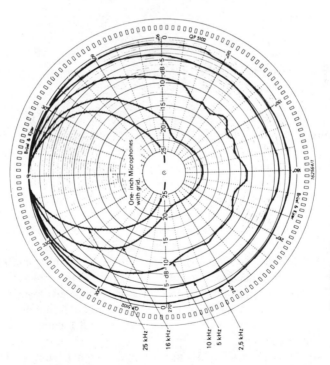

Fig. 3.8(a). Directional characteristics of typical one-inch condenser microphones.

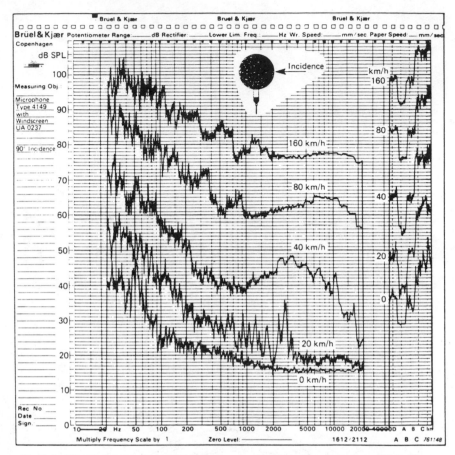

Fig. 3.9. Wind-induced noise of half-inch microphones fitted with windscreen—90° incidence.

nitude of wind noise reduction may not be sufficient for measurement of low sound levels.

An important aspect of the disturbance of the sound field at the measuring microphone is often overlooked. The presence of a person observing the reading on a sound-level meter can produce errors ranging from 5 to 10 decibels. For this reason the sound-measuring microphone should be located on a tripod at least 10–15 feet away from the sound-level meter, observers, or other large reflecting surfaces if one is to measure the free-field sound environment with an accuracy better than 5–10 decibels.

Whenever the microphone is remotely located from the sound-level meter with an extension cable there will be a significant drop in overall sensitivity of the microphone/sound-level meter system unless a preamplifier is remotely located with the microphone. This latter situation is the preferred technique and the

newer sound-level meters incorporate a feature for remote operation of the microphone preamplifier. Without a preamplifier between the microphone and the sound-level meter, not only will there be a loss in system sensitivity but there will be an increase in the noise floor of the measuring system. The extension cable itself may be significantly microphonic (i.e., sensitive to vibration or shaking due to local wind). Again these problems are virtually eliminated when the microphone preamplifier is remotely located from the sound-level meter (with the microphone). With remote microphone and preamplifier operation there is one precaution that must be kept in mind: the maximum signal handling capacity of the microphone and preamplifier may be reduced due to the shunt capacitance of the extension cable. Manufacturers' literature should be consulted in this regard.

The most significant step in any sound measurement is acoustic calibration of the system immediately prior to and following each series of sound level measurements. Typical acoustical calibrators for performing this task are an electromechanical driver that utilizes two small reciprocating pistons to develop a known sound pressure (124 dB SPL at 250 Hz) when the unit is placed over the microphone, and calibrators that generate an acoustic signal of known frequency and sound-pressure level with the aid of a built-in electronic oscillator and electromechanical transducer. These calibrators are accurate to ±0.3 dB when well maintained. If ever the calibrator and meter reading disagree by more than approximately 0.5 dB one should suspect a problem and investigate the cause before continuing with the sound measurements. Fig. 3.10 illustrates a typical application of the calibrator for checking the accuracy of the microphone and sound level meter. (See next section for discussion of sound-level meters). This particular calibrator produces 110 dB sound

Fig. 3.10. Calibration of sound-level meter.

pressure level at frequencies of 125, 250, 500, 1000, and 2000 Hz.

3.3. SOUND-LEVEL METERS

All transducers, whether they be microphones or vibration pickups, require some form of signal amplification. For noise control this is generally done with a portable sound-level meter. All sound-level meters for noise analysis should conform with ANSI S1.4-1971 "Specification for Sound Level Meters" (either the Type Two General Purpose or the Type One Precision Meter). A sound-level meter meeting the requirements for a precision instrument obviously will not qualify as a precision instrument when it is hand-held with its microphone only 2–3 feet from the observer as discussed earlier. Thus, a hand-held sound-level meter may only be used as a rapid method for determining approximately the sound level at a specific location. Again it should be emphasized that the microphone must be remote from the meter and the observer at least 10 feet away (in most cases) to minimize disturbance of the sound field at the microphone. A sound-level meter that is specifically designed to meet the ANSI S1.4 Type One Precision Meter is shown in Fig. 3.11.

The Occupational Health and Safety Act has stimulated the development of several miniature noise exposure instruments that can be worn by workers in noisy areas. Such an instrument, shown in Fig. 3.12(b), is worn by the operator in Fig. 3.12(a). The method by which readings of the noise-level meter are assessed is discussed later in Chapter 7 on Criteria and Regulations.

3.4. WEIGHTING CURVES

The apparent loudness that we attribute to a sound varies not only with the sound pressure but also with the frequency (or pitch) of the sound. In addition, the way it varies with frequency depends on the sound pressure. This effect can be taken into account to some extent for pure tones by including certain "weighting" networks in an instrument designed to measure sound-pressure level. These networks

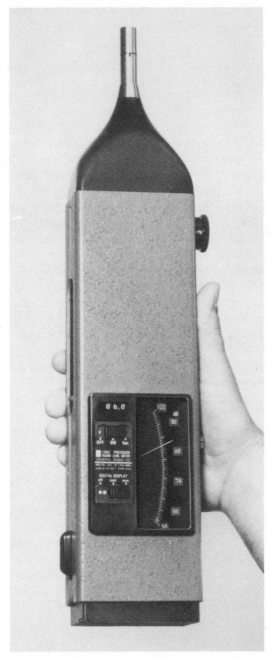

Fig. 3.11. Precision sound-level meter with octave and $\frac{1}{3}$-octave analyzers.

Fig. 3.12(a). Personal noise dosimeter in use.

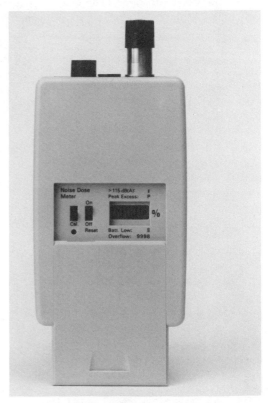

Fig. 3.12(b). Personal noise dosimeter.

are the inverse of some of the equal-loudness contours in Fig. 2.5. In order to assist in obtaining reasonable uniformity among different instruments of this type, the International Standards Organization and the American National Standards Institute, in collaboration with scientific and engineering societies, have established standard weighting networks. When incorporated in a sound-level meter, the response of the meter simulates the subjective response of the ear to loudness values. Fig. 3.13(a)

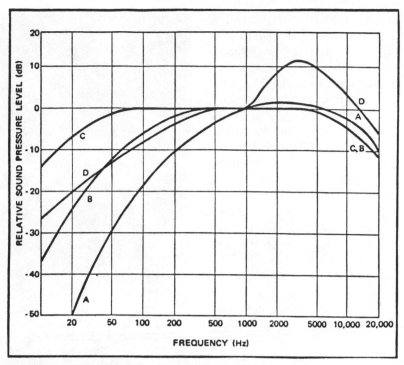

Fig. 3.13(a). International Standard A-, B-, and C-weighting curves for sound-level meters. Also shown is the D-weighting curve for monitoring jet aircraft noise.

shows the characteristics of these filters; the A-filter approximately corresponds to the 40-phon contour, the B-filter to the 70-phon contour and the C-filter to the 100-phon contour. The D-filter is used for measuring jet aircraft noise, where frequency content varies from 1500 Hz to 8 kHz.

3.5. BASIC SOUND-LEVEL METER SYSTEM

Figure 3.13(b) shows a block diagram of the circuitry of a typical sound-level meter. The microphone transforms the sound pressure into an electrical signal which is proportional to the amplitude of the fluctuating sound wave. The capacitive attenuator and the main attenuator are interconnected and can be operated by a single range switch to give 10 dB steps in the attenuation; this adapts the instrument for measuring noise over a wide range of sound levels. The amplifiers increase the relatively weak microphone signal to a value sufficient to drive the indicating meter or to display the signal on a recorder. The weighting networks are filters modifying the strength of the signal at specific

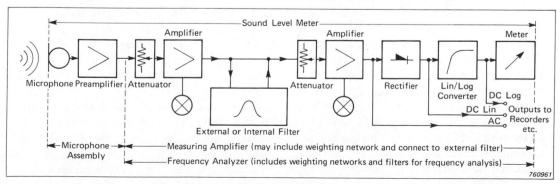

Fig. 3.13(b). Block diagram of typical sound-level meter.

frequency bands, as mentioned previously; these give greater importance to sounds in some frequency ranges than other frequency ranges in determining the overall reading of the meter, in order to approximate the subjective response to noise. The meter circuit incorporates a rectifier to convert the oscillating signal into a DC signal for display on the meter; the internationally standardized "fast" (200 ms response time) and "slow" (1 sec response time) meter dampings are provided in this circuit. The response time is that required for an instrument to respond to a full-scale step input; the "slow" characteristic is intended for use when the fluctuations of the meter needle in the "fast" mode are too great (more than some 4 dB) to give a well defined value for the sound level. The meter circuit also incorporates a linear to logarithmic converter, so that the readings taken from the meter will be in decibels. A full-scale meter reading is available for each 10 dB step of the attenuator(s).

If the sound to be measured consists of impulsive or impact noise, the normal "fast" and "slow" time constants of the ordinary meter are not sufficiently short to give a meter indication which is representative of the subjective human response. To measure such signals properly, a more sophisticated instrument, called the *impulse precision sound-level meter*, is available. This contains the special circuitry necessary to respond to such impulsive sounds. These characteristics are standardized in ANSI specifications for precision sound-level meters (12). The instrument also has facilities for holding the impulsive value or peak value on the meter, which can be reset by a press switch once the readings are taken. A D-frequency weighting is included (specifically for aircraft noise measurement), as well as the other standard weightings, and the DC level or an AC signal can be output to external equipment.

Most sound-level meters now have digital LED (light emitting diode) or liquid crystal readouts in place of the meter dial. Further, several manufacturers have incorporated plug-in modules to the basic precision sound level meter which, in addition to carrying out $\frac{1}{1}$-octave and $\frac{1}{3}$-octave analysis of the sound being measured (see Section 3.7), provides for analysis of many statistical parameters such as

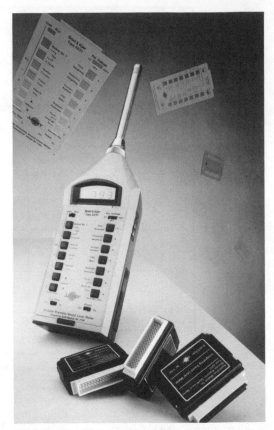

Fig. 3.14. Precision sound-level meter with optional plug-in modules.

probability and cumulative distributions, exceedance levels, energy equivalent level, etc. (see Sections 3.9.1 and 3.9.2). Plug-in modules are also available for the determination of reverberation time (a required function in the analysis of building acoustics—see Section 5.2). An example of such a meter is shown in Fig. 3.14.

3.6. VIBRATION TRANSDUCERS

Thus far we have discussed sound-pressure measuring devices. Most machine-oriented noise problems originate with the vibration of one or more structures within the machine which in turn radiate the sound that one hears. To determine the relative contribution of the various vibrating surfaces to the overall noise field it is usually desirable to measure the vibration at specific locations on a machine. Today most vibration measurements are done with

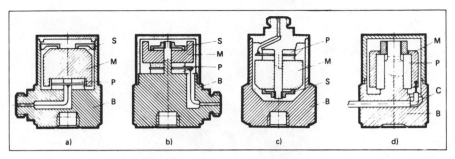

Fig. 3.15. Schematic cross sections of four common piezoelectric accelerometer designs: (a) peripherally mounted compression design; (b) center-mounted compression design; (c) inverted center-mounted compression design; (d) shear design. *Legend:* S is spring, M is mass, P is piezoelectric crystal, B is base, C is conductor.

accelerometers. Accelerometers are used because of their convenience and compatibility with the same instrumentation (sound-level meter) that receives a microphone signal. The basic elements making up an accelerometer are shown in Fig. 3.15. The accelerometer contains its own inertial mass, thus eliminating a reference point as is necessary for measuring displacement (13). The piezoelectric disks held down by the inertial mass in the accelerometer produce an output voltage when the entire unit is vibrated. The electrical connections may be on top or on the side of the unit. The lower opening is to receive the mounting stud. Normally these accelerometers may be operated to temperatures up to 260°C without special cooling. These accelerometers may be operated on surfaces with temperatures in excess of 1000°C with cooling water flowing through the base of the accelerometer.

Accelerometers have a uniform frequency response from less than 0.1 Hz to approximately half their resonant frequency. The sensitivity of the accelerometer is increased from 15 dB to 40 dB at its resonant frequency. Generally, the accelerometers that are of interest for noise and vibration work weigh between 1 gram and 100 grams and have a resonant frequency between 5000 and 100,000 Hz. The larger and heavier units are more sensitive (up to 1 V/g) and have the lowest resonant frequency. The lighter-weight accelerometers may have a sensitivity of only 1 mV/g. Calibrating the accelerometer connected to the measuring system is as essential as calibrating the microphone with an acoustic calibrator. Although a frequency-response curve and calibrated sen-

sitivity are supplied by the manufacturer with each accelerometer, responses and sensitivities may change with age, or as a result of shock damage due to dropping on hard surfaces, etc. Portable vibration calibrators, suitable for recalibration, are available from the manufacturers.

Figure 3.16 shows a precision sound-level meter which has been converted to a vibration meter by means of an electrical integrator with a connected accelerometer. Interchangeable scales are provided in the meter dial so that acceleration, velocity, and displacement (by successive integration of the acceleration signal) can be read directly off the meter scale. An octave or one-third octave filter set can be attached to the meter in order to analyze the vibration signals into frequency bands. (See the next section, Frequency Analyzers and Filtering). Thus, this single instrument can be used not only as a sound-level meter (with frequency discrimination), but as a vibration meter (acceleration, velocity, or displacement) with frequency discrimination.

The instrument shown in Fig. 3.16, although a good example of earlier combination sound/vibration measuring systems, is somewhat dated. There are now sophisticated vibration measuring and monitoring instruments which have capabilities of integration and narrowband spectrum analysis of the vibrating signal (see Section 3.9.4 and Appendix VII for list of instrument suppliers).

Displacement pickups are used on certain applications, especially where the load of an accelerometer influences the motion of the surface of interest or where it is not practical to

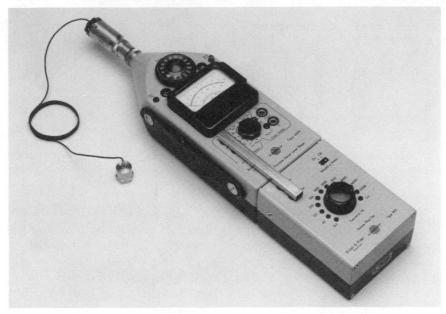

Fig. 3.16. Precision sound-level meter with octave-band filters, integrator, and accelerometer connected.

mount an accelerometer. Capacitance-type displacement pickups, magnetic or inductive displacement pickups, or optical sensors are available for this purpose (8,9).

3.7. FREQUENCY ANALYZERS AND FILTERING

As previously noted, the sound-level meter (reading on the linear scale) responds equally to all frequencies. A breakdown into components at various frequencies is made by feeding the signal which has been amplified by the sound-level meter into a frequency-analyzing instrument. The frequency analyzer contains electrical or crystal filters which reject signals at all frequencies except selected frequencies or frequency ranges, and thus the instrument reading indicates the magnitude of the sound pressure only at those frequencies which have been allowed to pass in the filter.*

---

*This applies only to analog instruments using electrical circuit or crystal filters which operate in finite time (i.e., signal must be steady and continuous to perform frequency analysis); real-time analyzers (which perform frequency analyses almost instantaneously or in "real time") are discussed in Section 3.9.4.

If an analyzer allows frequencies from 300 to 600 Hz to pass, the 300–600 Hz region is called the *passband* and the difference (300 Hz), is called the *bandwidth*. The limiting frequencies, in this case 300–600 Hz, are called *cutoff frequencies*. Analyzers with bandwidths as low as 1 or 2 Hz are available. It might also be noted that a practical filter does not completely reject signals outside the passband, because of the slope of the passband curve of conventional meters. Fig. 3.17 shows the relationship between a typical filter and an ideal filter. According to ANSI standards, the deviation in the ripple at the filter top should be within ± 0.5 dB, and the bandwidth is defined as the half-power point on the filter (i.e., the 3 dB down point or -3 dB from the filter top; note that the ordinate in Fig. 3.17 is pressure amplitude squared, i.e., proportional to power).

Constant-percentage bandwidth analysis is the most widely used. In this type of analysis, the bandwidth is a constant percentage of the center frequency to which the filter is tuned. The center frequency is defined as the geometric mean of the bandwidth, i.e.,

$$f_c = \sqrt{f_l f_u},\qquad [3.1]$$

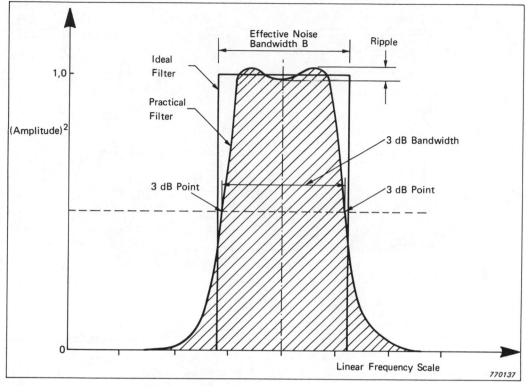

Fig. 3.17. Practical versus ideal filter characteristics.

where

f_c = center frequency,

f_l = lower cutoff frequency (-3 dB, down),

f_u = upper cutoff frequency (-3 dB, down),

$f_u - f_l$ = bandwidth.

Hence, as the center frequency increases, the filter bandwidth increases in proportion; a family of constant-percentage filter response characteristic curves are identical when plotted on a log-frequency scale (Fig. 3.18).

For an octave filter,

$$f_u = 2f_l \qquad [3.2]$$

For a fractional octave filter,

$$f_u = 2^n f_l \qquad [3.3]$$

where n is the fraction of the octave. Substituting $f_l = f_c^2/f_u$ (Eq. [3.1]) into Eq. [3.3], then

$$f_u = 2^{n/2} f_c. \qquad [3.4]$$

Similarly,

$$f_l = \frac{f_c}{2^{n/2}}. \qquad [3.5]$$

The bandwidth Δf is

$$\Delta f = f_u - f_l = \left(2^{n/2} - \frac{1}{2^{n/2}} \right) f_c$$

$$= \left(2^{n/2 - 1} \right) f_c, \qquad [3.6]$$

which confirms that the bandwidth is a constant percentage of the center frequency.

From the above, it can be determined that a full-octave bandwidth is 70.7%, one-third octave is 23.6%, and one-tenth octave is 7% of the center frequency.

In the majority of studies involving environmental noise, we are concerned with wideband noise, i.e., where appreciable sound power exists over a large portion of the audible frequency range. For this reason, octave and one-third octave bandwidth analyses are usually sufficiently discriminatory as to give reasonable

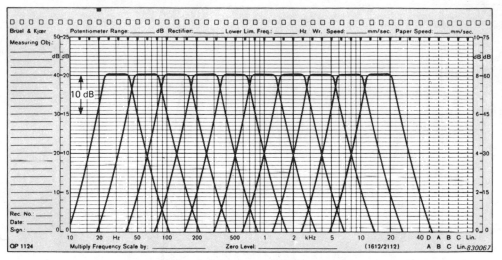

Fig. 3.18. Octave-band analyzer filter characteristics.

and meaningful data for these types of noise studies.

Examples of standard octave center frequencies and cutoff frequencies for octave and one-third octave filters are shown in Table 3.1, and an example of typical filter characteristics of a contiguous octave-band analyzer is shown in Fig. 3.18. Contiguous analyzes utilize several overlapping filters where the signal is analyzed in "steps," as compared with a continuous filter in which the signal is analyzed by a continuous "sweeping" filter.

An example of attenuation of filters in a constant percentage bandwidth analyzer is shown in Fig. 3.19.

Other bandwidths, such as $\frac{1}{10}$-octave and narrowband analyzers, have been used in special cases when a finer frequency analysis is required. The narrowband analyzers have either a constant bandwidth throughout the frequency range of operation (used mainly for diagnostic analysis of harmonics in machinery) or have bandwidths which are a fixed percentage (say 10, 3, or 1%) of the center frequency to which they are tuned. However, with the exception of measurement in $\frac{1}{1}$ and $\frac{1}{3}$-octave bandwidths (which correlates best with the human response to environmental noise—see Chapter 7 on criteria and standards), most analyses are now carried out using FFT (fast Fourier transform) analyzers which provide 200-, 400-, and 800-"line" (fixed bandwidth) spectra. Such analyz-

ers are discussed in Section 3.9.4., under real time analyzers. The complexity (and cost) of instrumentation and analysis increases substantially as the bandwidth is reduced. It is therefore practical that the frequency breakdown should not be finer than is necessary to make a reasonable engineering analysis of the noise.

An example of a typical continuous-sweep constant-percentage bandwidth frequency analyzer, and a typical measurement arrangement, are shown in Figs. 3.20(a) and 3.20(b), respectively. A schematic diagram showing the method by which the instrumentation in Fig. 3.20(b) carries out a frequency analysis of a typical sound fluctuation is shown in Fig. 3.21. The effect which each of the two types of frequency analysis, constant-percentage bandwidth and the constant bandwidth, has on the width of the passband at different frequencies is shown in Fig. 3.22. The comparison is made on a linear frequency scale (top diagram) and a logarithmic scale (bottom diagram). Note that, on the logarithmic scale, the constant-percentage filter appears to have a constant width at all frequencies (see also Fig. 3.18).

Examples of the sound-pressure level spectrum as obtained with a narrowband analyzer (6% of center frequency) on various machine-induced noises are shown in Figs. 3.23(a) and 3.23(b). An example of frequency analyses of a particular noise, by means of octave, one-third octave, and constant bandwidth (4 cps)

Table 3.1 Center and Approximate Cutoff Frequencies for Standard Set of Contiguous Octave and One-Third Octave Bands Covering the Audio Frequency Range.

| | Frequency Hz | | | | | |
| | Octave | | | One-third Octave | | |
Band	Lower Band Limit	Center	Upper Band Limit	Lower Band Limit	Center	Upper Band Limit
12	11	16	22	14.1	16	17.8
13				17.8	20	22.4
14				22.4	25	28.2
15	22	31.5	44	28.2	31.5	35.5
16				35.5	40	44.7
17				44.7	50	56.2
18	44	63	88	56.2	63	70.8
19				70.8	80	89.2
20				89.1	100	112
21	88	125	177	112	125	141
22				141	160	178
23				178	200	224
24	177	250	355	224	250	282
25				282	315	355
26				355	400	447
27	355	500	710	447	500	562
28				562	630	708
29				708	800	891
30	710	1,000	1,420	891	1,000	1,122
31				1,122	1,250	1,413
32				1,413	1,600	1,778
33	1,420	2,000	2,840	1,778	2,000	2,239
34				2,239	2,500	2,818
35				2,818	3,150	3,548
36	2,840	4,000	5,680	3,548	4,000	4,467
37				4,467	5,000	5,623
38				5,623	6,300	7,079
39	5,680	8,000	11,360	7,079	8,000	8,913
40				8,913	10,000	11,220
41				11,220	12,500	14,130
42	11,360	16,000	22,720	14,130	16,000	17,780
43				17,780	20,000	22,390

filters is shown in Fig. 3.24. Note that although the approximate shape of the sound spectrum is apparent with the one-third octave analysis, the pure tone components are not identifiable; a narrowband analysis is necessary in this case.

3.8. TAPE RECORDERS

Tape recording has been found to be a very convenient technique for gathering extensive field data for more detailed analysis at a later time. It is also useful for recording transient events that would otherwise be difficult, if not impossible, to analyze.

High-quality audio recorders for broadcast use and high-fi systems for the home can be adapted for recording acoustical data. However, this should only be done by personnel very experienced with the inherent limitations of tape recorders. Acoustical data generally cover a very wide dynamic range. Often there are transients and peaks that one would like to store on magnetic tape. Most professional broadcast recorders are not suitable for this application. They usually lack in adequate low-frequency response, lack adequate dynamic range at low frequencies, have excessive pre-emphasis at high frequencies (causing high fre-

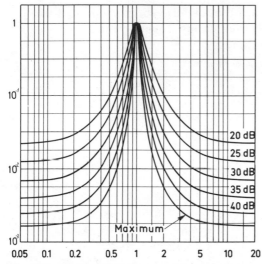

Fig. 3.19. Attenuation of filters used in a constant-percentage frequency analyzer for different bandwidth settings. Bandwidths vary from the narrowest (maximum attenuation) of 6% to the widest of 29%. Note that at the $2 \times$ and $0.5 \times$ center frequency points, the narrowest bandwidth attenuates the signal by 45 dB while the maximum bandwidth attenuates the signal by only 20 dB.

quency overload of acoustic signals), and, most important, lack an indicator to show that transient or peak events have been faithfully recorded.

Fig. 3.25 illustrates a tape recorder that has

been designed specifically for recording noise and vibration data. The reel-to-reel instrument shown in Fig. 3.25 is a sophisticated four-channel data recorder that has a flat frequency response from 0 to 20,000 Hertz. It operates at two different speeds to permit frequency scaling of the recorded data for more convenient analysis. The frequency modulation system utilized to obtain the 0–20,000 Hertz bandwidth also permits faithful reproduction of the relative phase information recorded on the FM data channels.

Cassette tape recorders, although at one time inferior to reel-to-reel tape recorders in dynamic range and frequency response, have now improved to the extent that the convenience (and protection from dust, etc.) of cassettes make them attractive instruments. Furthermore, companies like SONY and TEAC (see instrumentation summary and guide to manufacturers in Appendix VII) have developed instrumentation recorders that record on tape cassettes in digital (binary) form (called digital audio tapes)—to be later transformed on playback to analog data and processed in the usual fashion. By appropriate interface, the data can also be fed directly to a computer for further processing. These recorders provide enhanced

Fig. 3.20(a). Typical continuous-sweep frequency analyzer.

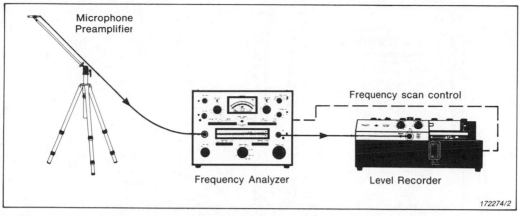

Fig. 3.20(b). Arrangement for continuous constant-percentage bandwidth analysis.

dynamic range (i.e., the noise floor is suppressed to a minimum), tape speed variation is eliminated (as compared with analog tape recorders), and wide frequency response range and long-time recording per tape are also featured. They can simultaneously record many channels of data (up to 21 in the TEAC unit).

3.9. SIGNAL PROCESSING

When a noise level varies unpredictably with time, as it does in communities and in factories, it is usually impractical to describe the noise by a complete graph of dBA versus time. Scientists have developed several single-number criteria for this kind of environment.

The assessment of community noise levels frequently relies upon two basic measurements. These are the *energy equivalent continuous sound level*, and the *statistical distribu-* *tion of the noise levels* through their probability distribution and cumulative distribution plots (see Sections 3.9.1 and 3.9.2).

3.9.1. L_{eq} Concept

The "equivalent sound level," denoted L_{eq}, is the value of the constant sound which would result in exposure to the same total A-weighted energy as would the specified time-varying sound, if the constant level persisted over an equal time interval. It is typically measured in dBA.

The way in which people respond to sounds in terms of annoyance has been found to better correlate with the square of the rms pressure associated with the sound (i.e., energy) rather than pressure directly. Hence L_{eq} provides a good measure of the intensity of a fluctuating noise in that it puts more emphasis on the oc-

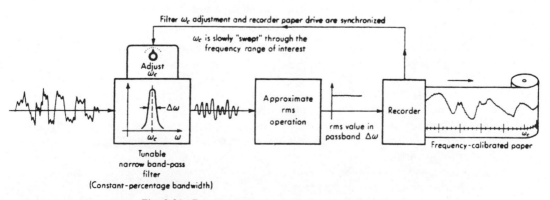

Fig. 3.21. Frequency-spectrum analyzer schematic diagram.

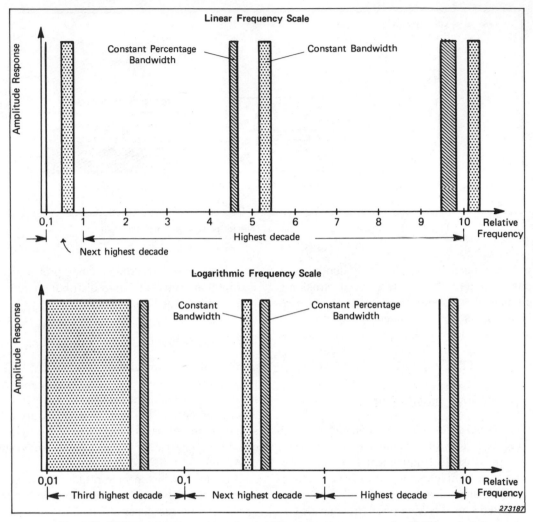

Fig. 3.22. Difference between constant-bandwidth and constant-percentage bandwidth analysis.

casional high (or disturbing) noise levels which can accompany more general lower noise levels.

The mathematical description of L_{eq} is as follows:

$$L_{eq} = 10 \log_{10} \frac{1}{T} \int_0^T \frac{p^2(t)}{(p_r)^2} \, dt, \quad [3.7]$$

where $p(t)$ is the time-varying A-weighted sound pressure, T is the time period over which L_{eq} is determined, and p_r is the reference pressure of $2 \times 10^{-5} \, \text{N/m}^2$.

The equivalent sound level can be measured directly using an integrating sound-level meter which performs a continuous-time integration of the sound being sampled over a specified period. Alternatively, one can estimate L_{eq} from a recording of A-weighted sound level versus time by using the definition:

$$L_{eq} = 10 \log_{10} \left[\frac{1}{T} \int_0^T 10^{0.1 L_A(t)} \, dt \right], \quad [3.8]$$

where $L_A(t)$ is the time-varying noise level in dBA, and T is the specified period during which the sound is sampled.

By breaking the sound-level record into N equal increments of time Δt_i, Eq. [3.8] can be approximated by

$$L_{eq} \cong 10 \log_{10} \left[\frac{1}{T} \sum_{i=1}^{i=N} 10^{0.1 L_{Ai}} \Delta t_i \right], \quad [3.9]$$

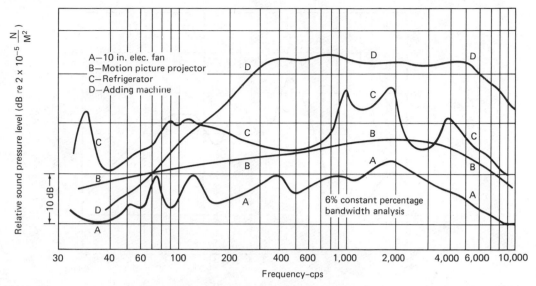

Fig. 3.23a. Sound spectra of four mechanisms: fan, projector, refrigerator, and adding machine. Energy source of broad spectrum of projector is particularly difficult to identify because there are no peaks around any given frequency.

when N is the total number of samples taken, and L_{Ai} is the noise level in dBA of the ith sample.

This is the form of computation that takes place in integrating sound-level meters and statistical analyzers. See Chapter 7 for discussion of 3-dB change in equivalent level (which represents a doubling or halving of energy) and 5-dB tradeoff for noise exposure (according to U.S. Occupational Safety and Health Act).

To illustrate how much emphasis is put upon the occasional high noise levels, consider the

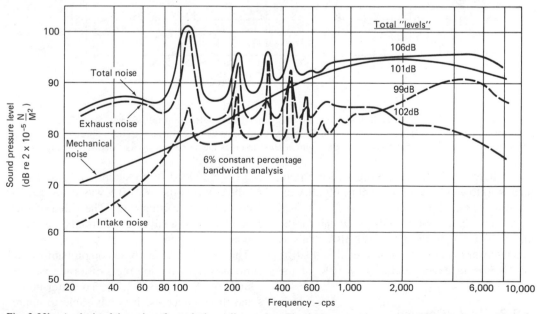

Fig. 3.23b. Analysis of the noise of a typical gasoline engine. The three sources have much different spectra, although each has about the same total level.

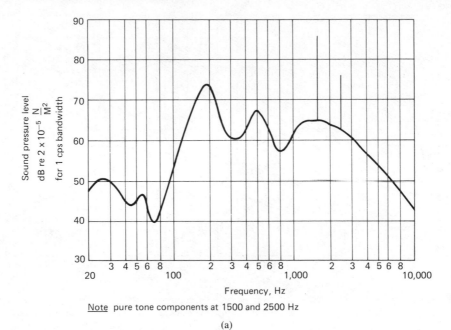

Note pure tone components at 1500 and 2500 Hz

(a)

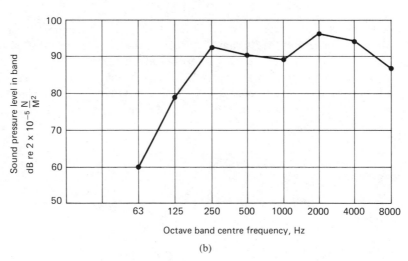

(b)

Fig. 3.24. (a) Actual spectrum, comparable to using filter bandwidth of 1 Hz at all frequencies. *Note:* Pure tone components at 1500 and 2500 Hz. (b) Spectrum of noise in (a) as measured by octave band filters. (c) Spectrum of noise in (a) as measured by $\frac{1}{3}$-octave band filters. (d) Spectrum of noise in (a) as measured by constant bandwidth filter of 4 Hz.

case where a noise level of 90 dBA for 5 minutes is followed by a noise level of 60 dBA for 50 minutes (see Fig. 3.26).

A simple calculation would show that for the entire 55 minutes the L_{eq} is 80 dBA. This is much closer to the higher noise level of 90 dBA which was in existence for only 10% of the time.

3.9.2. The L_N Concept

The parameter L_N is a statistical measure which indicates how frequently a particular sound level is exceeded. For example, a value $L_{40} = 72$ dBA means that 72 dBA was exceeded for 40% of the measuring time. The value of such a parameter in analyzing environmental noise can be easily seen.

The values of L_N are based on probability and cumulative distribution plots of a time-varying sound (see Section 3.9.2.2.). To better understand these concepts, it is advisable to appreciate some of the principles of probability analysis and the operation of statistical noise analyzers.

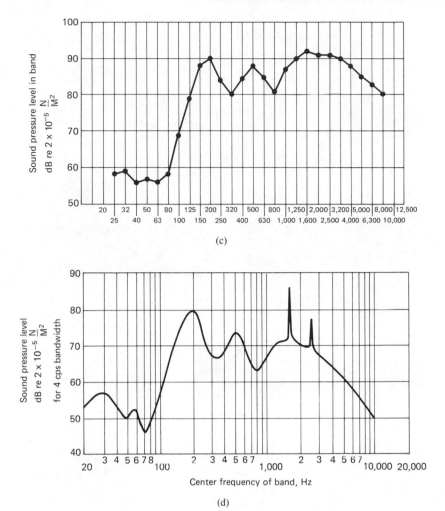

(c)

(d)

Fig. 3.24. (*Continued*)

3.9.2.1. Stationary Random Signals.

As demonstrated in Chapter 1, a stationary sinusoidal signal can be completely described by three quantities: the amplitude, the frequency, and the phase. It can be further shown that a complex, but still periodic signal, composed of many such sinusoids, could be exactly represented by a Fourier series (1). The instantaneous amplitude could therefore always be predicted from the signal's describing equation. A process which can be thus described is known as *deterministic*, i.e., its instantaneous amplitude can always be precisely determined from its equation. Deterministic signals are made up entirely of sinusoidal components at discrete frequencies, which in periodic signals, are always multiples of some fundamental frequency, but in quasi-periodic signals, e.g.,

from several independently rotating shafts, are not harmonically related.

A great number of practical noise sources give rise to signals which can be considered stationary, but are not periodic. Examples are jet noise, aerodynamic noise, traffic noise, crowd noise, most music, and many types of industrial noise. The amplitude of these signals does not vary in a periodic, and therefore predictable, manner. In order to describe them adequately it is necessary to introduce the concept of amplitude density instead of amplitude. Instead of being able to predict an instantaneous amplitude, a probability that the random signal will fall within a certain amplitude range is now defined.

The amplitude scale must be divided up into small intervals Δx; then the proportion of the

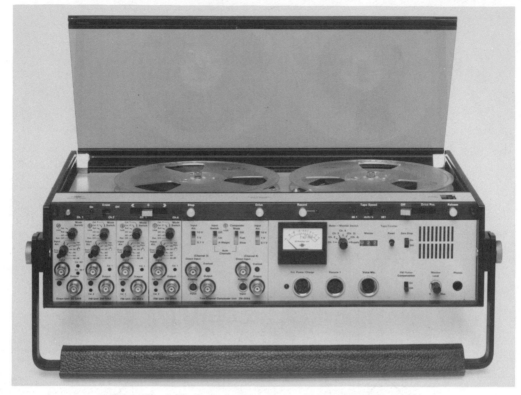

Fig. 3.25. Typical multichannel analog tape recorder.

total time which the signal spends between the values of x and $x + \Delta x$ can be defined in relation to the total time over which the signal is being studied (see Fig. 3.27). If $P(x, x + \Delta x)$ represents the probability that the signal amplitude lies in the interval $(x, x + \Delta x)$ then

$$P(x, x + \Delta x) = \lim_{T \to \infty} \frac{\sum \Delta t_n}{T}$$

where each Δt_n represents one of the time intervals in the sample T where the signal lies between the amplitude values x and $x + \Delta x$.

The probability density $p(x)$ at the amplitude level x is this probability divided by the interval width Δx, thus giving a density. Then

$$p(x) = \lim_{\Delta x \to 0} \frac{P(x, x + \Delta x)}{\Delta x}.$$

By varying the value of x from $-\infty$ to $+\infty$, $p(x)$ can be plotted as a function of x. The probability density curve describing the instantaneous amplitude of an acoustic pressure can vary considerably from problem to problem. However, the best known amplitude density curve is that from a normal (Gaussian) process, which is used as the model for many of the random processes encountered in practice. The curve is shown in Fig. 3.28 and the amplitude density distribution is given by

$$p(x) = \frac{1}{\sigma\sqrt{2\pi}} \exp\left(\frac{(x - \mu)^2}{2\sigma^2}\right),$$

where

σ = standard deviation,
μ = mean value.

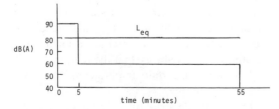

Fig. 3.26. Example of varying sound level and L_{eq}.

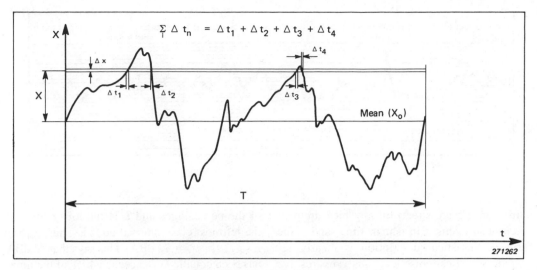

$$\sum_i \Delta t_n = \Delta t_1 + \Delta t_2 + \Delta t_3 + \Delta t_4$$

Fig. 3.27. Time record of a random process.

This curve may be considered as an e^{-x^2} curve centered on the mean value μ and scaled as follows:

1. Along the x-axis it is scaled in terms of σ, the standard deviation from the mean μ. For a zero mean, σ is the root-mean-square value of the signal and σ^2 the variance, i.e.,

$$\sigma = \frac{1}{T} \int_0^T x^2 \, dt = A_{\text{rms}}$$

2. Along the $p(x)$-axis it is scaled so that the integral under the curve for all x, i.e., the probability that x must lie between $\pm\infty$, is 1.

3.9.2.2. Statistical Level Analysis.

In the preceding section it was shown how the RMS value (or level) of a nondeterministic fluctuating signal could be determined by applying statistical techniques to the instantaneous amplitude values of the signal. The important overall characteristics of a random signal can therefore be described adequately in terms of its average properties. Statistical methods can be applied, not only to the instantaneous value of the signal itself, but also to its RMS value, i.e., its level which, because of the fluctuating nature of the signal and the finite observation duration, also varies statistically.

It is not sufficient to describe the sound-level history shown in Fig. 3.29 (which was re-

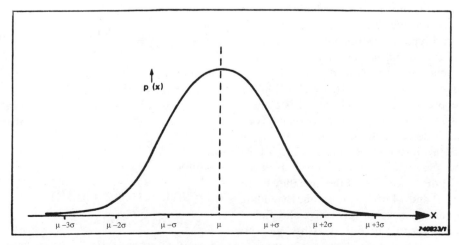

Fig. 3.28. The normalized Gaussian amplitude density curve.

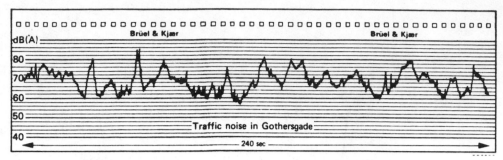

Fig. 3.29. Typical statistically varying sound level.

corded in a busy street) by any measurement taken at any single instant in time, and some form of statistical description is clearly required. Statistical noise-level analyzers resolve the dynamic measurement range into equal class intervals of width Δx. The instrument samples the incoming sound level at regular intervals, records these samples in a digital memory, and calculates the proportion of time the sound level spends in each of the amplitude intervals. This is the *level distribution histogram* and can be read out directly from the digital display on the instrument's front panel, or automatically to a level recorder or printer. The shapes of these histograms vary widely from one type of noise to another, as shown in Fig. 3.30. Fig. 3.30(a) is taken from a motorway noise measurement and shows a typical Gaussian shape. Fig. 3.30(b) shows the skew distribution typical of congested city-center traffic. Fig. 3.30(c) shows the highly skew distribution spread over a wide amplitude range measured in an office.

Allied to the probability distribution curve is the *cumulative distribution curve*. For example, certain types of instruments provide probability distribution data which corresponds with 32 2-dB-class intervals spread across their dynamic range. See Fig. 3.31(a). (This appears as a stepped function which would become a continuous, smooth curve as the class intervals are reduced to zero.) In the example shown in Fig. 3.31(a), it can be seen that for 22% of the time the measured noise level dwelt between 70 and 72 dBA, for 17% of the time they dwelt between 72 and 74 dBA, and so on. By adding the percentages given in successive class intervals from right to left [Fig. 3.31(b)], a corresponding L_N can be derived where N is the sum

of the percentages and L is the lower limit of the leftmost class interval added. Thus, $L_{40} = L_{(2+7+14+17)} = 72$ dBA. This says that 72 dBA has been equalled or exceeded 40% of the time.

A typical cumulative probability plot is reproduced in Fig. 3.32, and should be compared with the amplitude distribution plot of Fig. 3.30(b) for the same data. From this graph, important values such as L_{90}, L_{50}, L_{10}, etc. can be read directly.

An example of statistical analyzers used to measure L_N, L_{eq}, probability, and cumulative distribution is shown in Fig. 3.33.

3.9.2.3. Example of Determination of L_{eq}.
At the fence of an industrial plant the data shown in Table 3.2 were obtained using statistical analysis instrumentation. Ten equal intervals from 75 dBA to 100 dBA were employed. It is required to determine the equivalent sound level L_{eq}.

Reference 36 gives the following alternate method of computing energy equivalent level:

$$L_{eq} = 10 \log_{10} \left[\sum_{i=1}^{i=N} (p_i)(10^{L_i/10}) \right] \text{ dBA}$$

where
$N =$ number of intervals,
$p_i =$ fraction of time spent in interval i,
$L_i =$ A-weighted level at center of interval i.

Therefore

$$L_{eq} = 10 \log \left[(0.002)(10^{7.625}) \right.$$
$$+ (0.003)(10^{7.875})$$
$$+ (0.05)(10^{8.125}) + (0.124)(10^{8.375})$$

$+ (0.196)(10^{8.625}) + (0.213)(10^{8.875})$

$+ (0.167)(10^{9.125}) + (0.138)(10^{9.375})$

$+ (0.102)(10^{9.625})$

$+ (0.005)(10^{9.875})]$

$= 10 \log (1.2963 \times 10^9)$

$= 91.127$ dBA.

3.9.3. Level Recorders

The instrumentation discussed thus far requires the user to read the data from a meter and record the information manually on a graph or on a preprinted form. If considerable data are to be analyzed some means of analyzing and recording the data is most helpful. Figure 3.34 illustrates a graphic level recorder with differ-

ent plug-in analyzers. A constant frequency bandwidth analyzer of adjustable bandwidth is available with the graphic level recorder. One-third octave analysis, etc. can be obtained with interfacing with selective analyzers. The level (rms, average, peak) in selected frequency bands is plotted against frequency calibrated paper to give an amplitude-frequency plot of the signal being analyzed. Typical frequency spectra of noise samples are shown in Figures 3.23 and 3.24.

3.9.4. Real-Time Analyzers (Digital and FFT)

For analysis of sound, one-third octave and octave constant-percentage bandwidth filters are usually preferred, as a percentage proportional

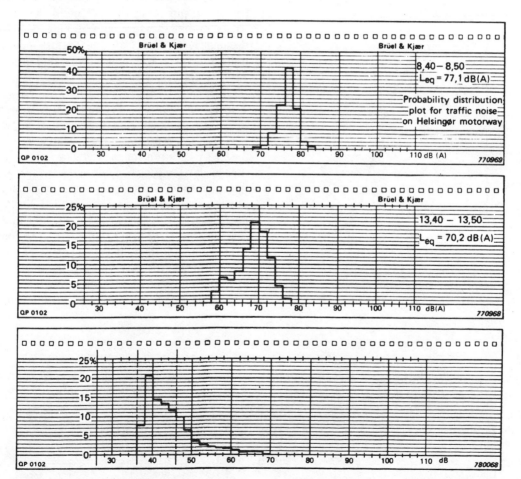

Fig. 3.30. Typical amplitude density plots: (a) Gaussian—motorway noise; (b) skew—congested traffic noise; (c) highly skew—office noise.

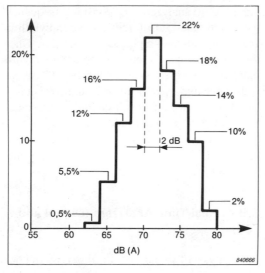

Fig. 3.31(a). Probability distribution plot. (How often between L and $L + 2$ dBA?)

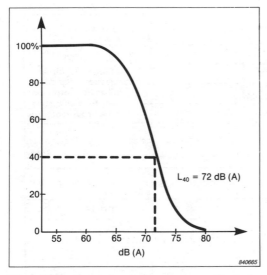

Fig. 3.31(b). Cumulative distribution curve. (How often above a certain level?)

bandwidth agrees well with the nature of sound and the human perception of it. The use of constant bandwidth analysis is primarily of importance when noise sources, such as faulty or worn machine bearings, gear teeth meshing, structural resonance and harmonics, etc. have to be traced. At high frequencies it provides better resolution than constant-percentage bandwidth analysis, enabling very closely spaced frequency components to be separated. This is especially useful when detailed analysis of high-order harmonic components of one or several pure tones have to be made. However, it should be remembered that when the bandwidth of an analog filter is decreased to improve the resolution, the time required to complete the analysis is increased. For this reason real-time frequency analyzers have become extremely popular. These use digital filtering techniques providing analyses in all frequency bands, over the entire analysis range, almost instantaneously (15), or utilize fast Fourier transform (FFT) techniques to obtain constant narrowband analyses (35, 40, 41).

3.9.4.1. Digital Frequency Analyzers. An example of a dual-channel digital real-time frequency analyzer is shown in Fig. 3.35. This instrument provides accurate real-time analysis in octave, $\frac{1}{3}$-octave, $\frac{1}{12}$-octave, and $\frac{1}{24}$-octave bandwidths. A large internal memory buffer can store, for example, up to 1000 $\frac{1}{3}$-octave spectra, with additional storage on a built-in floppy disc. The disc is formatted to be compatible with transfer of data to a computer. Frequency range, for example, covers 18 octave channels with center frequencies from 0.125 Hz to 16 kHz, or 54 $\frac{1}{3}$-octave channels with center

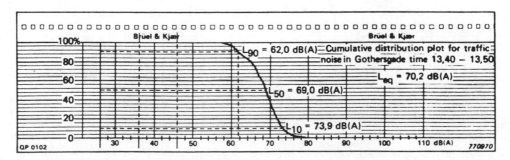

Fig. 3.32. Typical traffic noise cumulative probability plot.

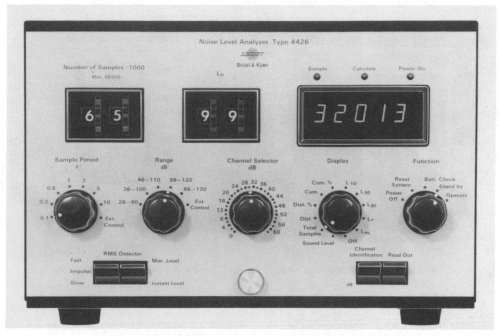

Fig. 3.33. Typical noise level analyzer and statistical processor.

frequencies from 0.1 Hz to 20 kHz. The large TV-like display has 80-dB amplitude range plus alphanumeric display of a wide range of relevant analysis procedure and data. Standard IEC interface allows the analyzer to be matched to a desktop computer for expanded computational processing. Special-purpose application packages can be run within the analyzer (e.g., building acoustic testing: sound transmission, absorption, reverberation time—see Chapter 5). With an appropriate sound-intensity analyzing system, the instrument can be used for noise-source identification, sound-intensity contour plotting, sound-power measurement, etc. (see Section 3.9.5). The instrument has, consequently, extensive applications in acoustical research, building acoustics, machine diagnostics, psychoacoustics, phonetics and speech therapy, product design and quality control, instrument calibration, sound intensity analysis, etc.

Various output options are available with real-time analyzers, as shown in Fig. 3.36. By means of special software programs for use with desktop computers, various calculations and procedures can be carried out. Fig. 3.37

Table 3.2. Noise Statistics at a Plant Boundary.

Interval No.	A-Weighted Level of the Interval, L_i	Percentage of Time Spent in Interval $P_i \times 100$	Percentage of Time A-Weighted Level Exceeded Interval Level
1	76.3	0.2	100.0
2	78.8	0.3	99.8
3	81.3	5.0	99.5
4	83.8	12.4	94.5
5	86.3	19.6	82.1
6	88.8	21.3	62.5
7	91.3	16.7	41.2
8	93.8	13.8	24.5
9	96.3	10.2	10.7
10	98.8	0.5	0.5

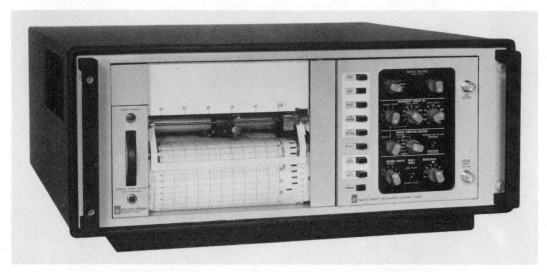

Fig. 3.34. Graphic level recorder with optional plug-ins for frequency analysis and time recording.

shows three-dimensional plots (from a hard-copy printer linked with a computer) of reverberation time and voice analysis (see discussion of reverberation time in Chapter 5).

Originally, frequency analyzers used analog techniques whereby the analysis was carried out using a bank of parallel analog filters. In the mid-1970s, digital filtering began to take over from analog filtering for acoustical measurements. A digital filter is a digital network which simulates the response of an analog filter (15, 37). Digital filtering should not be confused

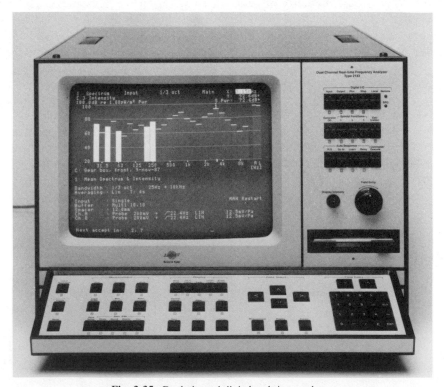

Fig. 3.35. Dual-channel digital real-time analyzer.

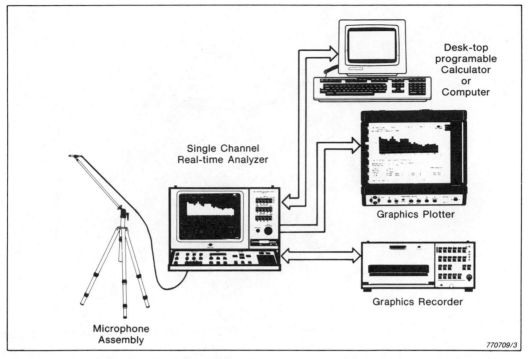

Fig. 3.36. Output options with real-time analyzer.

with fast Fourier transform filtering techniques (FFT), which use a completely different form of processing (see the next section).

The reasons for the switch to digital instead of analog filters are numerous. In addition to being able to carry out multifrequency analyses in so-called real time (i.e., almost instantaneously) with accompanying real-time display, there are three other major reasons: stability of the filter bank in that digital filters do not drift, greatly simplifying calibration; the possibility of using real-time digital detectors, allowing averaging times from milliseconds to days; and the fact that by reprogramming the filter bank the filter characteristics could be changed, such that, for example, the described analyzers can operate in $\frac{1}{1}$-, $\frac{1}{3}$-, $\frac{1}{12}$-, and $\frac{1}{24}$-octave bands.

Digital filtering techniques of the type employed in real-time frequency analysis automatically produce data on an octave or fractional octave basis, i.e., constant percentage bandwidth basis. This is in contrast to FFT analysis, which automatically produces constant bandwidth data on a linear frequency scale (see the next section and references). For this reason, where octave or fractional octave data

are required (as is the case where human response is correlated with environmental noise), then digital processing is preferred (37).

3.9.4.2. Fast Fourier Transform (FFT) Analyzers.

Narrow constant bandwidth frequency analyzers, utilizing fast Fourier transform (FFT) techniques, are available from a large number of manufacturers (Ref. 39 and Appendix VII). These are used for analyzing both continuous and transient signals, in particular where extremely fine resolution in frequency discrimination is required (35), such as in the closely spaced, high-frequency harmonics associated with the noise from rolling contact bearings and the meshing of gear teeth. These instruments are available with a wide range of capabilities. For instance, narrowband spectra of 100–800 lines can be captured on one to four (or more) channels; time and frequency domains of any of the channel inputs can be displayed simultaneously, and, by means of a cursor, any of the frequency lines can be selected and its level displayed by means of alphanumerics on the instrument screen. Frequency ranges typically vary from 0–10 Hz to

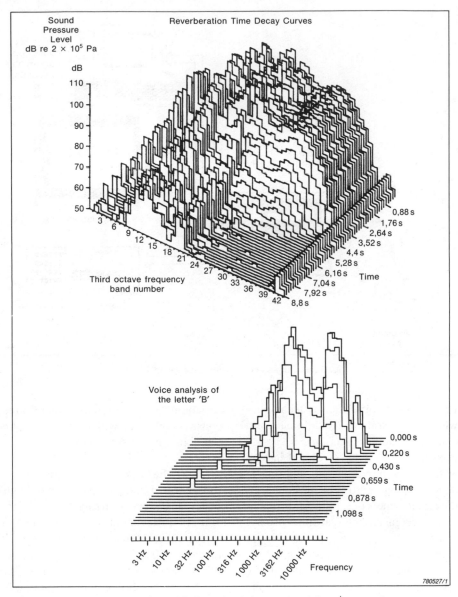

Fig. 3.37. Examples of 3-dimensional plots produced from $\frac{1}{3}$-octave data.

0–80 kHz. Most instruments have buffer memories for nonvolatile recording of signals for later analysis, and can usually be interfaced to a companion computer for extended signal processing. Many instruments have been designed to enable most calculations routinely required in acoustics to be done in the analyzer, thereby freeing the user from the complications of data transfer to a computer. Three-dimensional display are common in these instruments, e.g., amplitude (dB) versus frequency (Hz) versus time (sec or machine rpm), pre- and post-trig-

gering of transients, arithmetic functions, and special packages (for statistical analysis, $\frac{1}{1}$ and $\frac{1}{3}$-octave analysis, zooming, differentiation/integration, and sound intensity measurement). For a coverage of the principle of FFT analysis and certain applications, the reader is referred to Refs. 35, 38, 40, 41, 42, 43, and 44.

An example of a typical FFT analyzer is shown in Fig. 3.38(a). Typical plots (and printouts from a companion printer) are shown in Fig. 3.38(b). Although this analyzer is movable it is not conveniently portable. Examples

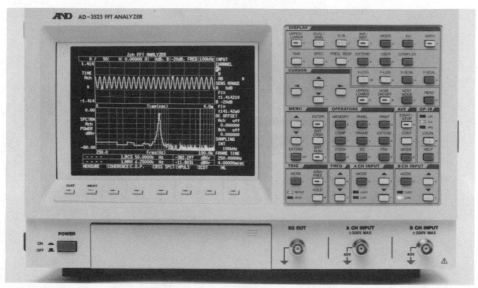

Fig. 3.38(a). Typical narrowband spectrum analyzer.

of light-weight portable analyzers are shown in Figs. 3.39(a) and 3.39(b). The analyzer in Figure 3.39(a) (Scantek Inc.) has two-channel capability and measures in $\frac{1}{3}$-octave bands, provides various L-levels, can measure sound power and intensity, has floppy disc data storage, and an interface with external computers for systems enhancement. With optional software, many sophisticated room acoustic measurements are possible.

The FFT analyzer shown in Fig. 3.39(b) is also portable. It operates on AC, DC, or optional battery, has 60-dB dynamic range, time function, frequency spectrum, histograms and arithmetic functions, phase measurement, averaging, transient capture, storage memory; with appropriate software (and a PC) it provides predictive maintenance vibration spectrum monitoring, acoustical measurements, and structural analysis. The unit can also be provided with an optional printer, as shown in the figure.

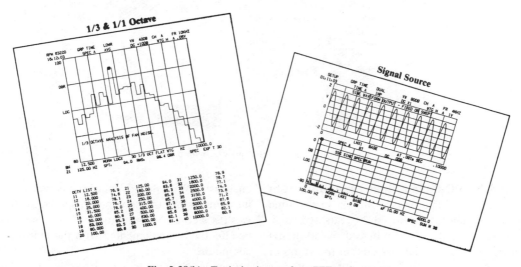

Fig. 3.38(b). Typical printouts from FFT analyzer.

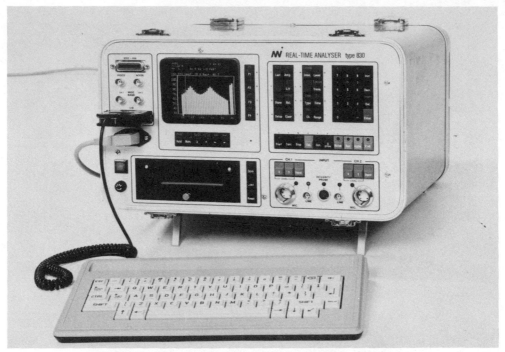

Fig. 3.39(a). Example of light-weight, portable real-time analyzer.

Fig. 3.39(b). Portable FFT analyzer with printer.

3.9.4.3. PC-Based Real-Time Spectrum Analyzers.

The first revolution in real-time spectrum analysis occurred 20 years ago when analog analyzers were replaced by all-digital instruments using the fast Fourier transform or FFT. Now, it appears, the second revolution is underway. With the advent of the powerful yet inexpensive personal computer, spectrum analysis (as well as other instrument functions) can be effectively implemented with the aid of the computer.

Compared with dedicated hardware instruments, PC-based spectrum analyzers, whether simple programs or complete instruments, have certain universal advantages. First, and most obvious, is their lower price. Mass production of the PC and its standard peripherals produce a cost-performance combination that cannot be matched by the lower-volume dedicated instruments.

A second advantage is flexibility. Since the user can add programs of his/her own, signal processing and formal variations are limited only by the programmer's creativity (and programming skill). The unit can be customized for a particular application (such as go/no-go testing of noise from a washing machine on a production line), and then be reconfigured (with appropriate software) for an entirely different function. This flexibility also allows for growth as future needs arise. The manufacturer need only provide revised software to improve, change or expand the unit's operation.

A third advantage is the availability of large amounts of inexpensive digital storage. Random-access, high-speed memories are used for data buffering, averaging, and display. Discs serve not only for storing setups and programs, but for archiving thousands of sets of data, any of which can be quickly recalled. With such a large database, signal changes over time or from one test to another can be recognized by comparing the past with the present, as in machine maintenance trend analysis.

An example of a PC-based FFT analyzer is shown in Fig. 3.40. This instrument was manufactured by Signology (recently acquired by Tektronix) and incorporates all of the above-mentioned capabilities. An internal output function generator, swept sine analysis, one-third-octave analysis, statistical function determination, arithmetical measurements, machinery diagnostics, three-dimensional plots (dB versus frequency versus time) are all available through software packages.

3.9.4.4. Standards for Octave and Fractional-Octave Filters.

Octave and fractional-octave filters have become standardized to ensure consistent measurement of data (see ANSI SI.11-1966, rev. 1971, Appendix III-9). When the original standards were written, only analog filters were in use, and the standards consequently concentrated on analog filter response. When FFT analyzers were introduced, the filter shapes required by the standards were synthesized by adding together narrow bands to make a $\frac{1}{1}$ or $\frac{1}{3}$ octave from the narrowband data. However, the problem with the process was, and still is (at least at this writing), that if an appreciable frequency range is to be covered, the process cannot be carried out in real time

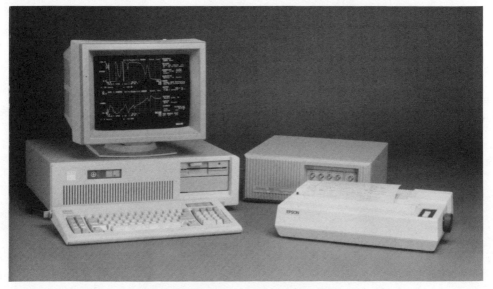

Fig. 3.40. Example of PC-based FFT analyzer.

with today's generation of FFT analyzers (37, 46). Therefore, one could not be sure that the correct results were being obtained.

The above problem is now being addressed by the new generation of octave and fractional octave filter standards. ANSI SI.11-1986 already exists and contains a test for the time-domain response of the filter bank. IEC 225 is in the process of being updated, and presumably will incorporate a similar type of test. These tests effectively prevent the use of synthesis of octave and fractional-octave data from FFT for all but the narrowest and lowest of frequency ranges, at least until a newer and much faster generation of FFT analyzers appears. In the meantime, analog and digital parallel filter analyzers can continue to be used for octave and fractional-octave measurements.

3.9.5. Sound-Intensity Measurements

Traditional sound-pressure measurements indicate the total sound pressure level at the receiver. Intensity measurements, on the other hand, can reveal the contributions of individual sources.

Sound intensity is a vector which describes the magnitude and direction of the flow of sound energy at any given point. This is in contrast to sound pressure, which is a scalar quantity, that is, it has magnitude but no direction. The measurement of intensity, however, has only recently started to supplant the measurement of pressure in certain applications. This has been mainly due to improvements in technology which have made the measurement of sound intensity more practical.

3.9.5.1. Theory.* The theory associated with sound intensity measurements is as follows. Sound intensity is power per unit area, i.e.,

$$\text{Intensity} = \frac{\text{Power}}{\text{Area}} = \frac{\text{Work}}{\text{Area} \times \text{Time}}$$

$$= \frac{\text{Force} \times \text{Distance}}{\text{Area} \times \text{Time}}$$

$$= \text{Pressure} \times \text{Velocity}$$

---

*All theory and diagrams are courtesy of Bruel and Kjaer and Ref. 7.

Therefore, sound intensity is calculated by time averaging the product of the instantaneous sound pressure p and the instantaneous particle velocity u, i.e.,

$$I = \overline{p \times u}$$

where the bar indicates time averaging.

Thus, if the sound pressure and the particle velocity can be measured, the sound intensity can be calculated. The sound pressure can be measured easily; the particle velocity can be approximated by making two closely spaced sound-pressure measurements.

From Newton's (second) law:

$$\rho \frac{du}{dt} = -\Delta p = -\frac{dp}{dr} \text{ (in one direction } r\text{)},$$

where ρ is the density of the air medium. The pressure gradient dp/dr can be approximated by measuring the pressures p_A and p_B at two closely spaced points and dividing the pressure difference by the separation between the two points Δr. This is valid as long as the separation is small compared with the wavelength of the sound being measured (i.e., $\Delta r \ll \lambda$). Therefore,

$$\rho \frac{du}{dt} \approx (p_B - p_A)/\Delta r,$$

and solving for the particle velocity,

$$u \approx \frac{1}{\rho \Delta r} \int (p_B - p_A) \, dt.$$

The sound intensity can now be obtained by multiplying the expression for the particle velocity by the sound pressure.

The sound pressure is the mean pressure between the two microphones:

$$\frac{p_A + p_B}{2}.$$

Hence the sound intensity is

$$I = \frac{1}{2\rho \Delta r} (p_A + p_B) \int (p_B - p_A) \, dt,$$

where, once again, the bar indicates time averaging.

Thus, sound intensity calculations can be made by measuring the sound pressure at two closely spaced points by conventional microphones.

3.9.5.2. Example of Response.

Fig. 3.41 shows the response of the two microphones when measured in an anechoic chamber with a solid spacer between the microphones. The two microphones see the same sound pressure to within 0.5 dB, and the responses are almost flat.

The directional response of the probe is shown in Fig. 3.42. The maximum in the response is along a line through the center of the microphones (where $p_A + p_B$ and $p_B - p_A$ are some positive value), and it is this orientation of the probe which is used in making absolute measurements of sound intensity. The maximum in the response is quite broad, and it is possible to be as much as $\pm20\%$ off the direction of the intensity vector (or more at lower frequencies), before any significant errors are introduced. The minimum in the response is located in the place midway between the two microphone diaphragms and parallel to them (where $p_B - p_A \rightarrow 0$) and is very sharp. Because of its sharpness, it is the minimum response of the probe and the associated change

in the sign of the intensity which is used in locating the direction of the sound intensity vector; the probe can be used as an acoustic "pointer" to indicate the location of a noise source.

3.9.5.3. Sound-Intensity Analysis Systems.

From the discussion in the previous section, it is obvious that in any sound-intensity analysis system a very high degree of phase matching is required between the two channels. This has effectively prevented the design and manufacture of sound-intensity analysis systems based on analog techniques, although as a result of the use of the most modern technology, analog-based sound-intensity meters are now beginning to appear. The frequency analysis of sound intensity, however, rests very firmly in the digital domain.

A breakthrough in the analysis of sound intensity occurred with the availability of relatively low-cost two-channel FFT analyzers, since it can be shown that sound intensity is related to the imaginary part of the cross spectrum between two microphones (14). Today, this forms a commonly used method of measuring sound intensity, a computer being used with the FFT analyzer to carry out the final calculations. However, although this method can offer a narrowband analysis of the sound inten-

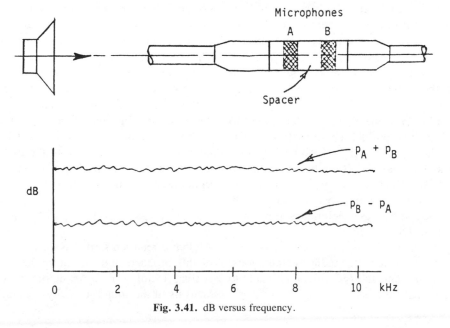

Fig. 3.41. dB versus frequency.

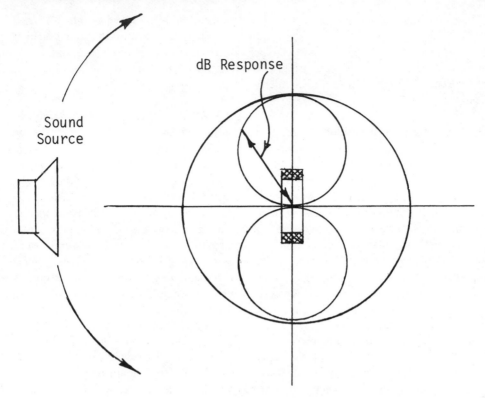

Fig. 3.42. Directional characteristics.

sity, it does have certain drawbacks. One of these is that sound measurements are normally specified in octaves and $\frac{1}{3}$ octaves. The calculation of these from narrowband spectra is a long and drawn-out procedure, usually requiring multipass analyses, which cannot be carried out in real time (hence restricting the procedure to use with stationary nonimpulsive signals). Another drawback is that since two devices, the analyzer and the computer, are involved, the system can become rather complex to operate.

The solution to the above problem is not to base the analysis on FFT techniques, but to use digital filtering instead. A high-performance octave and $\frac{1}{3}$-octave analyzer based on digital filtering (the Bruel and Kjaer 2131) has been available for some years (15). It requires only an extension of the basic techniques to produce a system which will measure sound intensity directly in octaves, $\frac{1}{3}$ octaves, and $\frac{1}{12}$ octaves without the need for intermediate calculations. Equally important, the system can be made to operate in real time.

The block diagram of a sound-intensity analyzing system (the Bruel and Kjaer Type 2134) is shown in Fig. 3.43. The signals from the two microphones are input via two phase-matched input channels to the analog-to-digital converters, and then into the digital filters. Because digital filters are used, their characteristics are identical—including their phase responses (15). The output of the digital filters are the analyses of the pressure signals from microphone A and microphone B. The difference between the two is then entered into an integrator to carry out the necessary time integration, and the output of the integrator now becomes one input to a multiplier. The other input to the multiplier is the sum of the digital filter outputs. The multiplier hence forms the product

$$(P_A + P_B) \int (P_B - P_A)\, dt.$$

All that is required further is the time averaging of this product, and the multiplication by the scaling factor $-1/2\rho\Delta r$ before a frequency analysis of the sound intensity is obtained.

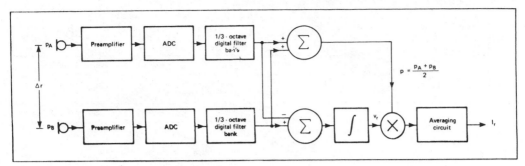

Fig. 3.43. Block diagram of the Type 2134 sound-intensity analyzer.

3.9.5.4. Frequency Range of Sound-Intensity Measurements.

Measurement of intensity using two closely spaced microphones introduces some errors not normally encountered in sound-pressure measurements. At high frequency, they are a function of the spacing between the microphones, while at low frequency, they are a function of the phase mismatch between the two measurement channels, and, indirectly, the microphone spacing (7). This means that at the frequency extremes, the true intensity will always be underestimated by an amount L_ϵ which is a function of the microphone spacing.

Figure 3.44 plots L_ϵ against frequency for the various microphones and spacers of the Type 3519 probe. The probe comes with a matched pair of $\frac{1}{2}$-in. microphones and a matched pair of $\frac{1}{4}$-in. microphones (phase matched to $0.2°$). Four spacers are provided, namely 6 mm and 12 mm for the $\frac{1}{4}$-in. microphones, and 12 mm and 50 mm for the $\frac{1}{2}$-in. microphones. Fig. 3.44 shows that if the underestimate L_ϵ is to be kept to less than -1 dB, then the upper frequency limits for the 50 mm, 12 mm, and 6 mm spacers are 1.24 kHz, 5 kHz, and 10 kHz, respectively. (These are independent of microphone type).

From the above, it is evident that the measurement of sound intensity at high frequencies requires small spacers. At low frequencies, the reverse is true, since measurements at lower and lower frequencies require larger and larger spacers. Fig. 3.45 plots the error for the var-

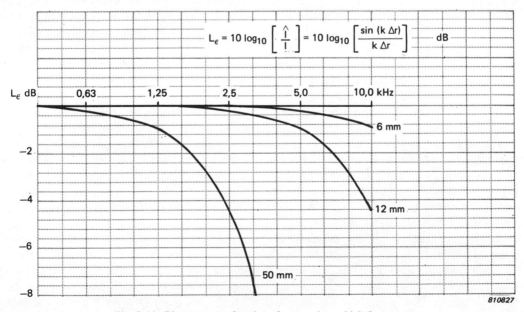

Fig. 3.44. Bias error as a function of spacer size at high frequency.

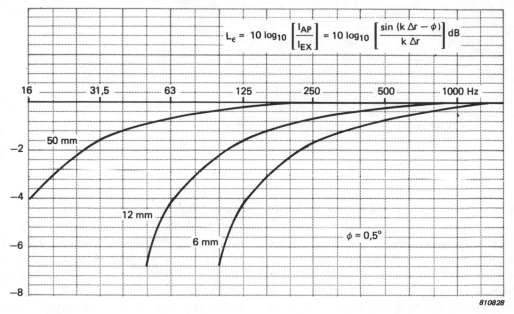

$$L_\epsilon = 10 \log_{10} \left[\frac{I_{AP}}{I_{EX}} \right] = 10 \log_{10} \left[\frac{\sin (k\, \Delta r - \phi)}{k\, \Delta r} \right] dB$$

Fig. 3.45. Bias error as a function of spacer size at low frequency, assuming a phase mismatch of less than 0.5°.

ious microphones and spacers when measuring at low frequencies. Note that this error can be positive or negative, and if it is to be kept less than ± 1 dB, then the lower frequency limits for the 50 mm, 12 mm, and 6 mm spacers are 50 Hz, 200 Hz, and 400 Hz, respectively.

3.9.5.5. Example of a Sound-Intensity Analyzer.
The Bruel and Kjaer Type 3360 sound-intensity analyzing system is shown in Fig. 3.46. This consists of the probe (Type 3519), the Type 2134 sound-intensity analyzer, and the Type 4175 display unit. Additionally, there is a remote indicating unit (Type ZH-0250). The remote indicating unit can be used to, say, display intensity in octave or $\frac{1}{3}$-octave bands selected by the cursor of the 3360. The averaging processes and the readout of results from the 3360 can be automatically controlled from the ZH-0250 at up to 50 m from the 2134.

Figure 3.47 shows an example of a $\frac{1}{3}$-octave intensity spectrum displayed on the 3360. Note how three different screen brightnesses are used. The brightest bar indicates the cursor position, the frequency and level being given in the alphanumeric display at the top of the screen. The next brightest bars indicate "negative" intensity, that is, in these $\frac{1}{3}$ octaves the sound flow is coming from behind the probe.

The darkest bars are used to indicate "positive" intensity, that is the sound flow is coming from in front of the probe.

The averager of the 2134 allows linear averages of sound pressure or intensity to be added together automatically. This is an important feature when intensity measurements are used to obtain the sound power of a source, where multiple measurements must be made and averaged to allow for the directionality of the source. These measurements may be added in the 2134 to give a result which is directly related to the sound power of the source (see later section).

3.9.5.6. Noise-Source Identification.
Sound-intensity measurements can be used to locate noise sources in a quick, simple, and direct manner. This is in contrast to conventional techniques, such as isolating sources by enclosing them in barrier materials, switching noise sources on and off (in a multiple-noise-source system) for individual analysis, etc.

The continuous sweep method of noise source location uses the minimum in the response of the 3519 probe and the real-time feedback of the 3360 display. As was mentioned previously, the 3519 probe exhibits a sharp minimum in its response in a plane mid-

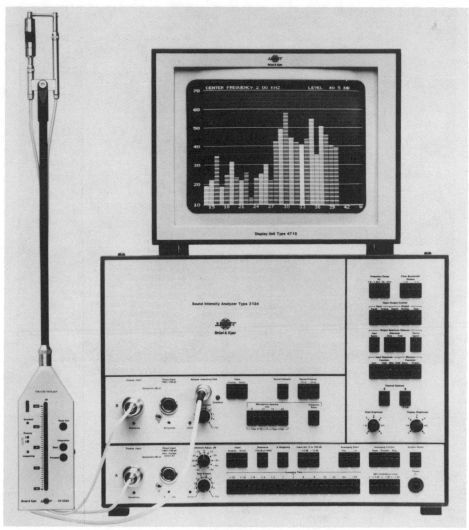

Fig. 3.46. Bruel and Kjaer Type 3360 sound-intensity analyzing system.

way between the two diaphragms and parallel to them.

A passage through the minimum in any octave or $\frac{1}{3}$ octave is indicated by a rapid change in the brightness of the corresponding bar on the 3360 display screen, showing the change from "positive" to "negative" intensity, and vice versa.

The method is indicated in Fig. 3.48. For instance, when searching for a noise source on a machine, the probe is held parallel to the machine under investigation and is steadily moved in a line while the display unit is observed. When the display indicates a sudden change of the direction of the intensity, this fixes the noise source on a plane passing perpendicularly through the center of the probe's spacers. The sweep is then repeated along a line at right angles to the first, and a further plane is identified. The noise source will be situated along the line of intersection of the two planes.

Note that, because sound intensity measurements respond only to flow of sound energy, any effects due to a reactive part of the sound field (such as part of the sound field being diffused or the presence of standing waves) will be rejected from the measurement. For instance, the sound field inside the cabin of a car or truck will be largely reactive (or reverberant). Any attempt to identify noise sources

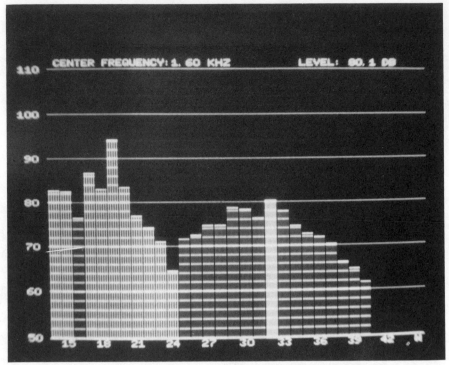

Fig. 3.47. Display of an intensity spectrum on the 3360.

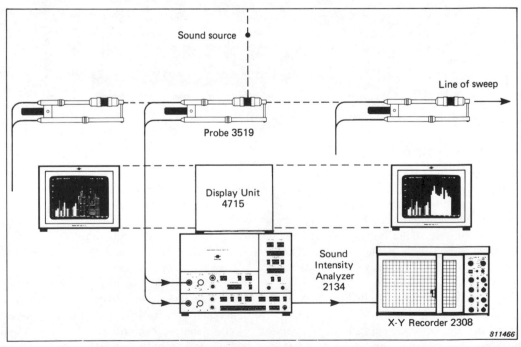

Fig. 3.48. Continuous sweep method for locating noise sources using the 3360. As the probe passes the source, the intensity spectrum on the display screen changes brightness to indicate a change in the direction of the intensity, i.e., from the rear of the probe assembly and not from the front. Note that the change occurs in the middle to upper frequency range, indicating that the sound associated with the lower frequencies is overall background noise.

within the cabin using pressure measurements alone will usually be frustrated by this reactive part. Intensity measurements, on the other hand, will reject the reactive part of the sound field, allowing sources of noise to be identified. Such measurements can, of course, be extended to many other applications.

Example. Noise from an Automatic Lathe. As a further example of noise source identification, consider the measurement results for a small automatic lathe in a machine

shop. As shown in Fig. 3.49, a continuous sweep with the probe in the direction of the axis of the lathe revealed a major noise source, i.e., the pulley and gear housing, where the middle frequencies of the displayed intensity spectrum changed in brightness.

Example. Intensity Measurements in Noise Isolation. Noise isolation by an acoustic barrier is shown in Figs. 3.50 and 3.51. Fig. 3.50 shows a lathe (on the right) with a quality con-

Fig. 3.49. Continuous sweep method for locating sound sources on a small lathe. As the median plane of the probe is swept past the source, i.e., the pulley and gear housing, the middle frequencies of the displayed intensity spectrum change in brightness.

Fig. 3.50. Large lathe (right), showing the proximity of the quality control bench and the positions of the feed-tube, the hood, and the screen.

trol bench and worker on the left. The sound-intensity spectrogram measured at the control bench without any acoustic barrier is shown in Fig. 3.51.

The effect of placing an acoustic barrier between the lathe and the operator's quality control bench is shown in Fig. 3.51. The barrier was 2.2 m high by 1.5 m wide, and consisted of a 3-cm-thick blanket of mineral wool, backed by a thin sheet of lead. The whole barrier was enclosed in heavy, perforated plastic and supported in a tubular metal frame. The measurements were averaged by sweeping a probe back and forth over an area of 30 cm by 30 cm, using a 4-second linear average on the sound-intensity analyzer.

Note that in Fig. 3.51, the unhatched and hatched parts of the curves indicate that the measured sound intensity is incident from the front and rear of the probe, respectively. It is interesting to note that without the barrier, the main noise source is the lathe (noise incident from the front of the probe) while with the barrier in place, the noise source is from the background (i.e., noise incident from the rear part of the probe).

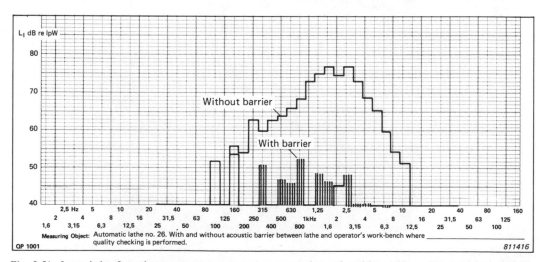

Fig. 3.51. Large lathe. Intensity spectrogram measured at operator's usual position at his quality control bench, with and without an intervening acoustic barrier.

Example. Intensity Measurements on a Personal Computer. To illustrate the use of intensity measurements on business machines, the technique was used to measure the sound power of a personal computer at a workplace. The station consisted of a visual display unit, keyboard, and computer unit. Intensity mapping was also made over the personal computer.

For the measurement of sound power on the personal computer, the measurement surface chosen was a rectangular box, just enclosing the computer and visual display units and extending down to a reflecting plane (i.e., the floor—see Fig. 3.52).

The measurement was made by sweeping the sound-intensity probe over each surface of the box, while carrying out a linear integration (Fig. 3.53). A 16-sec averaging (integration) time was used for each surface, resulting in a total measurement time of two to three minutes. Fig. 3.54 shows the resulting sound-intensity spectrum. The total A-weighted sound power is calculated as 52.6 dB. The maximum A-weighted sound intensity occurs at the 400 Hz $\frac{1}{3}$-octave band; the 100-Hz component can be neglected, since it will be attenuated by the A-weighting.

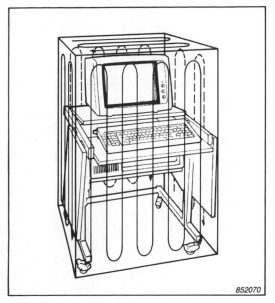

Fig. 3.53. Measurement of sound power on a personal computer, showing the intensity probe's path over the measurement surface.

Sound intensity can, of course, be used to locate a noise source (see previous section). If the sources on the machine under test are few and well defined, then it might be possible to locate them by using a so-called real-time method, i.e., use is made of the figure-eight directional characteristic of the sound-intensity probe (see Sect. 3.9.5.2).

On machines where there are several sources of noise, it is usually more informative to use an off-line method of source location. Here, a series of intensity measurements is made over a measurement grid and then intensity plots are generated to reveal the location of the major noise source(s).

These principles can be applied to the measurements on the personal computer. Returning to the original measurements of sound power, the various measurement surface elements on the computer (i.e., the front, sides, back and top) can be ranked in terms of their contribution to the overall sound power. Fig. 3.55 shows that the major contribution comes from the rear of the computer.

In this case, the noise source on the rear of the computer is fairly obvious, since there is a ventilation fan. (This could be readily verified

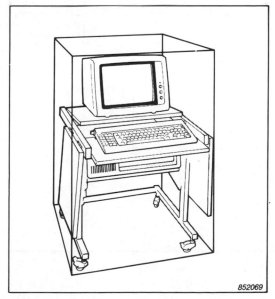

Fig. 3.52. Measurement surface about a personal computer.

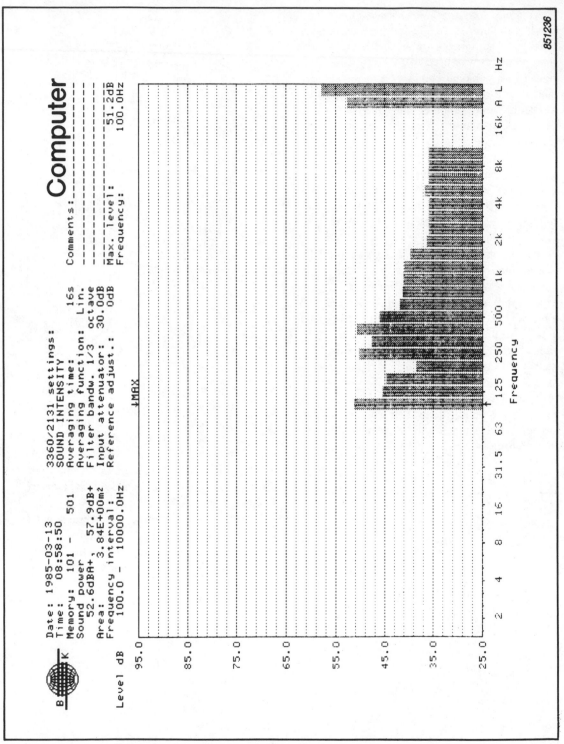

Fig. 3.54. Sound power of a personal computer.

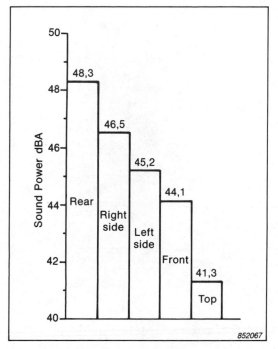

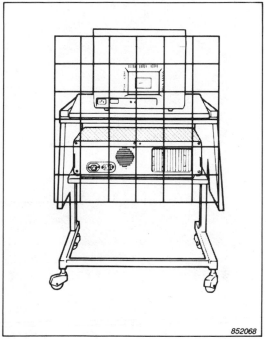

Fig. 3.55. Ranking of the sound power passing through the various measurement surface elements on a personal computer.

Fig. 3.56. Setting up the measurement grid over the rear of the personal computer.

using real-time location.) However, the procedure will be illustrated by means of an intensity map. A measurement grid is set up across the rear of the computer, and the sound intensity is measured at each point on the grid. All measurements are made with the probe perpendicular to the plane of the grid (Fig. 3.56).

Figure 3.57 shows a number plot with contours for the measurement in the frequency band where maximum sound power was found, i.e., 400 Hz. These are sound intensities measured at each point on the grid in that $\frac{1}{3}$ octave. Note that some of the numbers are negative, thus indicating that at that point on the grid and in that $\frac{1}{3}$ octave the direction of flow of sound energy is into the grid rather than out of it. This contour plot, in the 400 Hz $\frac{1}{3}$ octave, clearly shows that the major noise source is the ventilation fan, as expected.

A further means of representing the intensity data is by a 3-dimensional plot, which can be viewed from different observation corners, and with different angles of view (Fig. 3.58). This plot was obtained by means of the instrumentation shown in Fig. 3.59, in which a special applications package has been combined with a graphics recorder.

The above instrumentation enables analyses to be performed in $\frac{1}{1}$-, $\frac{1}{3}$-, and $\frac{1}{12}$-octave bands. If one is interested in narrower bands, for instance locating a pure tone emitted from a product, then a sound intensity probe, together with a narrowband FFT analyzer, a desktop computer, and a special software package (23) provide a narrowband sound-intensity analyzing system. A comparison of sound-intensity measurements using these two systems, and the compromises involved, are discussed in Ref. 29.

An example of a rather sophisticated digital analyzer, FFT analyzer, and sound-intensity analyzer is shown in Fig. 3.60. This instrument is made by Larson Davis (Appendix VII) and incorporates: $\frac{1}{1}$-, $\frac{1}{3}$-, and $\frac{1}{12}$-octave digital filters; 100–200–400–800 line FFT analysis; a dynamic range in excess of 80 dB; measurement bandwidths as high as 100 kHz; on-board intensity analysis for the measurement of sound

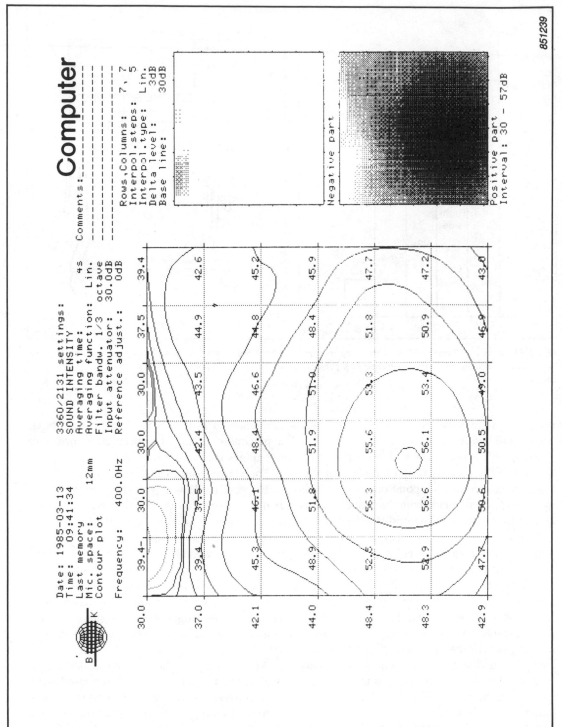

Fig. 3.57. Contour plot over rear of personal computer in the 400-Hz one-third octave band.

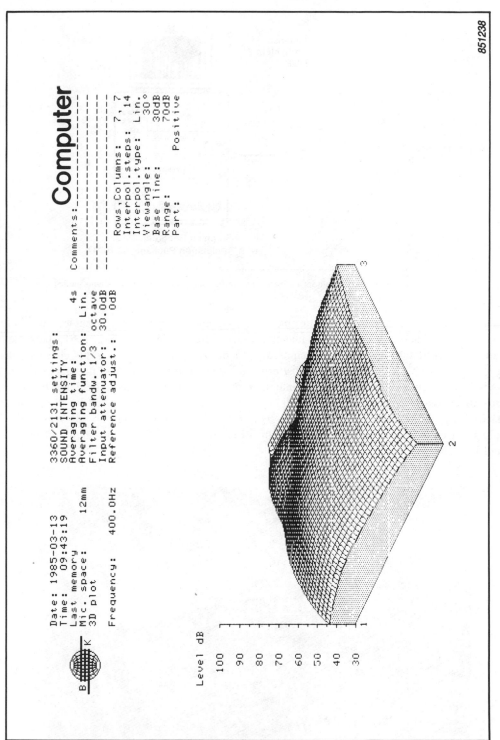

Fig. 3.58. Three-dimensional plot of personal computer in the 400-Hz one-third octave band.

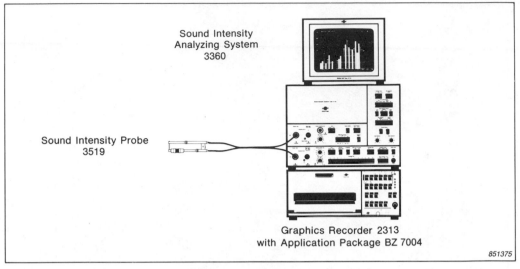

Fig. 3.59. A system for semi-automatic measurement of sound power based on sound intensity.

intensity, sound power, sound pressure, particle velocity, and numerous related parameters. No outboard computer or software is required. The unit is portable (25 pounds) and battery powered. Available options include printers, floppy disc drives, and standard computer interfaces. Fig. 3.60 shows the instrument as would be used in noise source location, with sound intensity probe.

Another example of a portable sound-intensity meter is shown in Fig. 3.61. This meter, made by Caldon Inc., is a low-cost and easy-to-use instrument for obtaining practical sound-intensity measurements in field and factory applications. It was developed by the Acoustics and Noise Control Group at Westinghouse Electric Corporation's Research and Development Center.

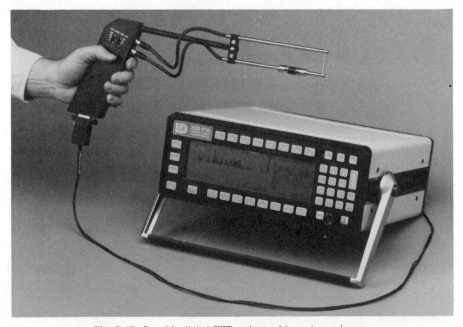

Fig. 3.60. Portable digital FFT and sound intensity analyzer.

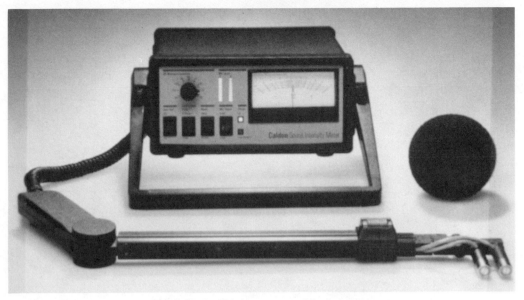

Fig. 3.61. Portable low-cost sound-intensity meter.

Sound intensity is measured over the range of 50 Hz to 4 kHz and displayed continuously in real time on a single analog meter, showing both magnitude and direction. The instrument consists of a light-weight, battery-powered unit containing computing circuits and display, and a probe containing two matched microphones and preamplifiers.

In addition to measuring sound power in the presence of background noise, the manufacturer states the following practical applications:

- Verifying on the shop floor that manufactured equipment meets noise specs and standards;
- Locating noise sources and checking noise levels both in the shop and in the field;
- Checking noise levels of equipment operating in the field for compliance with regulations;
- Periodic monitoring of equipment noise profiles and levels over time for detecting changes which could lead to maintenance decisions;
- Evaluating results of design and construction changes on equipment noise levels.

Further examples of the use of sound-intensity measurements in noise-source identification,

contour plotting, and sound-power determination are given in the references.

3.9.5.7. Measurement of Sound Power.

The sound intensity coming from a noise source is directly related to its sound power. Hence, a convenient method of measuring the sound power of a source could be to set up some arbitrary closed test surface surrounding the source and then average the sound intensity passing through the surface. The sound power of the source can then be calculated from the average intensity using the following equation:

$$PWL = I + 10 \log_{10} A$$

where PWL is the sound-power level in dB re 1 pW, I is the intensity from the source in dB re 1 pW/m^2, and A is the area of the test surface in m^2 (see Chapter 4).

The chief advantage of this method is that it does not require the use of an anechoic or reverberation chamber, and can be carried out with high background noise. The reason for the above can be seen in Fig. 3.62, which shows that if a noise source is going to have any influence on the average intensity, it has to be inside the test surface. As long as there is no absorption within the test surface, the contribution to the average intensity from an external

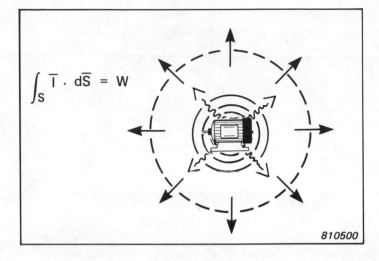

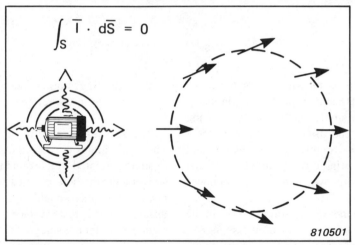

Fig. 3.62. Illustration that sound intensity from a source outside a test surface will integrate to zero.

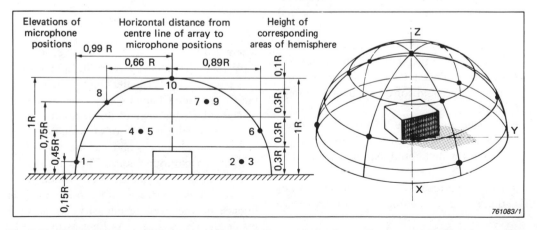

Fig. 3.63. The measurement of sound power with a hemisphere enclosing the sound source and the microphone positions according to ISO 3745.

noise source will always be zero, since any intensity coming from the external noise source which enters the surface at some region must also exit from the surface at some other region, hence cancelling out. It does not make any difference if the measurements are made in the near field or the far field. All that is required is that the background noise remain the same throughout the entire averaging process.

In a practical measurement, the procedure is first to set up a suitable test surface around the object (Figure 3.63).

Although the shape of the surface is completely arbitrary, since its surface must be known, it is usual to make a simple geometrical shape. Also, in order to take account of the directional characteristics of the source, the test surface should be divided into a number of equal areas, and a measurement made in each area. The intensity passing through each area is then measured at the geometric center of each area or by sweeping the probe across the area, noting an average reading. Note that in either case, the probe should always be kept perpendicular to the test surface. The results are then averaged together to give a final figure for the average intensity, from which the sound power can be calculated.

REFERENCES

(1) Beranek, L. L. 1971. *Noise and Vibration Control*. New York: McGraw-Hill.

(2) Harris, C. M., ed. 1957. *Handbook of Noise Control*. New York: McGraw-Hill.

(3) Reynolds, D. R. 1981. *Engineering Principles of Acoustics*. Toronto: Allyn and Bacon.

(4) Crocker, M. J., and A. J. Price. 1975. *Noise and Noise Control*, Vol. 1. Cleveland, OH: CRC Press.

(5) Peterson, A. P. G., and E. E. Gross, *Handbook of Noise Measurement*. 1974. Concord MA: General Radio Co.

(6) Crocker, M. J., ed. October 1972. *Internoise '72 Tutorial Papers*. Washington, DC.

(7) Upton, R. 1982. "Sound Intensity—A Powerful New Measurement Tool." *Sound and Vibration*, **16**, (10), 10–18.

(8) Foreman, J. E. K. 1985. "Experimental Measurement and Analysis." Tutorial Notes, Faculty of Engineering Science, The University of Western Ontario, London, Ontario, Canada.

(9) Holman, J. P. 1978. *Experimental Methods for Engineers*. Toronto: McGraw-Hill.

(10) Lord, H., W. S. Gatley, and H. A. Evensen. 1980. *Noise Control for Engineers*. Toronto: McGraw-Hill.

(11) Hassall, J. R., and K. Zaveri. 1979. *Acoustic Noise Measurement*. Naerum, Denmark: Bruel and Kjaer. 4th edition.

(12) ANSI. 1971. "Specifications for Sound Level Meters," S1.4-1971. New York: American National Standards Institute.

(13) Irwin, J. D., and E. R. Graff. 1979. *Industrial Noise and Vibration Control*. Englewood Cliffs, NJ: Prentice-Hall.

(14) Fahy, J. J. 1977. "Measurement of Acoustic Intensity using the Cross-Spectral Density of Two Microphone Signals," *JASA*, **62**, 1057–1059.

(15) Bruel and Kjaer. 1977. "Digital Filters in Acoustic Analysis Systems." Bruel and Kjaer Technical Review No. 1, Naerum, Denmark.

(16) Bruel and Kjaer Product Data Sheets. "Sound Intensity Analyzing System Type 3360." Naerum, Denmark.

(17) Bruel and Kjaer Product Data Sheets. "Sound Intensity Analyzer, Type 4433." Naerum, Denmark.

(18) Bruel and Kjaer Application Notes. "Sound Power Determination Using Sound Intensity Measurements," Parts 1 and 2. Naerum, Denmark.

(19) Bruel and Kjaer Application Notes. "Intensity Measurements in Building Acoustics." Naerum, Denmark.

(20) Bruel and Kjaer Application Notes. "Sound Power Determination of Household Appliances on the Production Line," Naerum, Denmark.

(21) Bruel and Kjaer Application Notes. "Production Testing of Noise from Business Machines." Naerum, Denmark.

(22) Bruel and Kjaer Application Notes. "On Road Location of Sound Services Using a Portable Intensity Analyzer." Naerum, Denmark.

(23) Bruel and Kjaer Application Notes. "Business Machine Measurements Using Sound Intensity." Naerum, Denmark.

(24) Bruel and Kjaer Application Notes. "Sound Intensity Measurements Inside Aircraft." Naerum, Denmark.

(25) Bruel and Kjaer Application Notes. "Gated Sound-Intensity Measurements on a Diesel Engine." Naerum, Denmark.

(26) Bruel and Kjaer Application Notes. "Sound Intensity Measurements Inside a Motor Vehicle." Naerum, Denmark.

(27) "Sound Intensity." Three Papers Presented at *Senlis Congress on Sound Intensity*, September 30–October 2, 1981. Naerum, Denmark: Bruel and Kjaer.

(28) Rasmussen, Gunnar. 1986. *Intensity Measurements*. Naerum, Denmark: Bruel and Kjaer.

(29) Ginn, K. B., and Upton, R. 1985. "Comparison of Sound Intensity Measurements Made by a Real-Time Analyzer Based on Digital Filters and by a 2-Channel FFT Analyzer." *Proceedings Internoise*, 1079–1082.

(30) Bruel and Kjaer Product Data Sheets. "Personal Noise Dose Meter, Type 4428 (ISO) and Type 4434 (OSHA)." Naerum, Denmark.

(31) International Electrotechnical Commission. 1979. "Standard 651, Sound Level Meters."

(32) International Electrochemical Commission. 1985. "Standard For Integrating-Averaging Sound Level Meters."

(33) Ontario Ministry of the Environment. 1983. "Procedures for the Measurement of Sound." Publication NPC-203.

(34) Ontario Ministry of the Environment. 1983. "Instrumentation for the Measurement of Sound." Publication NPC-202. (General purpose sound-level meter, impulse sound-level meter, peak pressure-level detector, integrating sound-level meter, vibration velocity detector, acoustic calibrator.)

(35) Bruel and Kjaer Product Data Sheets. "Single and Dual Channel FFT Analyzers, Types 2123 and 2133." Naerum, Denmark.

(36) Magrab, E. B. 1975. *Environmental Noise Control*. New York; John Wiley and Sons.

(37) Upton, R., and D. L. Strebe. 1988. "The Digital Filter Analyzer Comes of Age." *Sound and Vibration Magazine*, March.

(38) Randall, R. B. 1987. *Frequency Analysis*. Naerum, Denmark: Bruel and Kjaer.

(39) Sound and Vibration. 1988. "Instrumentation Buyer's Guide." *Sound and Vibration Magazine*. March.

(40) Oppenheim, A. V., and R. W. Schafer. 1975. *Digital Signal Processing*. Englewood Cliffs, NJ: Prentice-Hall.

(41) Hewlett-Packard. 1985. "The Fundamentals of Signal Analysis." Application Note 243. Palo Alto, CA.

(42) Hewlett-Packard. 1983. "Dynamic Signal Analyzer Applications." Application Note 243-1. Palo Alto, Ca.

(43) Bendat, J. S., and A. G. Piersol. 1971. *Random Data: Analysis and Measurement Procedures*. New York: Wiley-Interscience.

(44) Bendat, J. S. and A. G. Piersol. 1980. *Engineering Applications of Correlation and Spectral Analysis*. New York: Wiley-Interscience.

(45) GenRad. 1986. "Microphones and Accessories." Concord, MA: General Radio Co.

(46) Rothschild, R. S. 1987. "The Second Revolution in Real-Time Spectrum Analyzers." *Sound and Vibration*, March.

Chapter 4

SOUND FIELDS

4.1. REFLECTED SOUND

In making sound pressure measurement, the effects of reflected sound must be considered. If the measurements are made very close to the source, however, the incident sound will predominate. Reflections will also occur off objects within the room as well as the walls. If a sound is produced at one point in the room and then stopped, the waves will reflect off various surfaces in a manner similar to light. With no surface absorption reflections would, theoretically, continue indefinitely. However, the movement of air particles converts some of the energy into heat by viscous action and heat conduction and also by the effects of molecular absorption and dispersion involving an exchange of translational and vibrational energy between colliding molecules (1); in addition, energy is absorbed by the reflecting material surface. Attenuation and decay of the reflections thus results.

The attenuation of a sound wave traveling through air due to molecular absorption is dependent upon frequency and relative humidity; at a relative humidity of 50% and a sound frequency of 100 cps the attenuation is approximately 0.001 dB per foot (1). This change, it can be seen, is extremely small compared with the reduction obtained by the expanding wave front where the pressure amplitude decreases in inverse proportion to the distance from the source (see the next section on the ideal sound source). Much greater attenuation of the sound occurs from absorption by the reflecting surfaces in the path of the wavefront.

At any instant when the sound source is constant, the incident sound or direct sound from the source and the reflected sound will reach an equilibrium sound energy density at any point in the room with the amount of sound that is absorbed. The time required for the sound energy density in an enclosure to decay by 60 dB (i.e., to 10^{-6} of its original value) is referred to as the *reverberation time*. The reverberation time depends not only upon the frequency of the sound source and the humidity of the air in the enclosure, but (mostly) on the absorbing or reflecting quality of the enclosure walls. In large enclosures, the mean free path of the sound is much greater between reflections; hence frequency and humidity will have a proportionately greater effect on the reverberation time. All of the characteristics of reflected sound will be discussed in detail in later sections.

4.2. IDEAL SOUND SOURCE

Consider an ideal source of acoustical energy. Such an ideal source can be pictured as a small sphere pulsing in and out at a certain frequency.

Such an ideal source would radiate acoustic energy equally in all directions in a free field (i.e., in the absence of any reflecting surfaces). It has been shown that the acoustic energy flowing through a unit area is proportional to the sound pressure squared. Thus, if we halve the distance to our ideal sound source we may expect an increase in the sound-pressure level (SPL) of 6 dB (see Fig. 4.1). The reason for this is that all the sound energy must pass through a sphere drawn around the source. Halving the distance to the source means that a sphere drawn around the source at the new radius would have a surface area of only one-fourth the original. Since the same total sound energy must pass through each sphere, the pressure squared must increase by four times. The definition of

$$\text{SPL} = 20 \log_{10} \left(\frac{p}{p_{\text{ref}}} \right) \text{ dB}$$

$$= 10 \log_{10} \left(\frac{p^2}{p_{\text{ref}}^2} \right) \text{ dB}$$

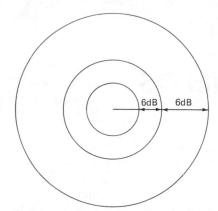

Fig. 4.1. Ideal sound source.

where p is the RMS sound pressure and p_{ref} is the reference, will show that the *SPL* increases by 6 dB.

4.3. PRACTICAL SOUND SOURCE

In practice, sound sources do not behave exactly in the way just described. It was assumed in the discussion of an ideal source that all parts of the surface of the sphere moved in and out, in phase. Such a phenomenon does not happen with a complicated device such as a machine (Fig. 4.2). Here different parts of the machine are moving out of phase with each other. Also, the sound wave from the machine is often not freely propagating, because the particle velocity is not in the direction of wave propagation, and pressure and particle velocity do not reach their maximum and minimum values simultaneously (see Section 1.2). Very close to such a practical source of sound we find that the air particles move from the in-phase to the out-of-phase part of the source. Here, the acoustic in-

Fig. 4.2. Practical noise source.

tensity is not simply related to pressure squared. The extent of these deviations depends upon frequency, characteristic source dimensions and the phase of the various radiating surfaces.

Briefly, the net result of these effects is that near to the source the SPL does not increase at 6 dB for each halving of distance but could increase at a greater rate (see Fig. 4.3). Others (10, 11) have found the SPL tends to level off.

This region close to the machine is called the *near field*. The region outside the near field is called the *far field*.

Another phenomenon is experienced in practice: it can never be assumed that there will be no reflections. Eventually, moving far enough away from the machine which is located in a building, the reflections can be greater in magnitude than the sound received directly from the machine. This is the transition between the *free field* and the *reverberant field*. In the reverberant field the SPL remains almost constant as we move away from the machine.

Looking at Figs. 4.1 and 4.2 again, it is to be noted that a practical sound source does not radiate sound energy equally in all directions. The difference in the SPL measured at a point compared with the SPL created by an ideal sound source which would be measured is known as the *directivity index* (see Section 4.6). However, with a practical sound source, provided we are in the far field, and not in a reverberant environment, the SPL decreases by 6 dB for each doubling of distance (Fig. 4.3). It is wise to check this phenomenon, and it can usually be done easily (e.g., with a hand-held sound-level meter). However, it should be noted that it is sometimes impossible to measure in the far field because of the physical limitations in the size of the room where the machine is mounted.

The two extremes of a free (or nonreflective) field and the reverberant (or reflective) field are represented in the anechoic and reverberant chambers, respectively. Both of these types of rooms play an important role in acoustic studies, and are described in the next two sections.

4.4. ANECHOIC ROOMS

As the name suggests, *anechoic* comes from a Greek word meaning *without echoes*. An ane-

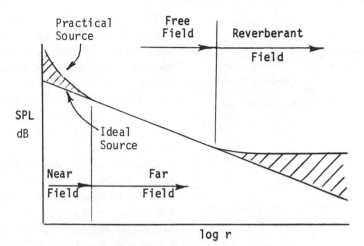

Fig. 4.3. Ideal and practical sound sources.

choic room is simply a means of providing a work space which is essentially a simulation of free-field conditions. A section through a typical anechoic room is shown in Fig. 4.4.

The first problem is to ensure that there is low background noise in the room. This is normally achieved by building a room with fairly massive walls. If the room is built in an environment where there is a high noise level (e.g., in a noisy factory area or near to road traffic noise) it may be necessary to use a double-wall construction.

It may also be necessary to vibration-isolate the room if there is a high level of vibration caused by road traffic or rail traffic or if machines such as presses or compressors are operating nearby. This is normally achieved by mounting the room on metal springs or rubber pads or by suspending the room from many wires. The springs or pads are normally designed to have a resonance frequency of between 5 and 10 Hz (see Chapter 6 for a discussion on vibration isolation).

The second problem is to provide sufficient sound-absorbing material on the walls of the room to keep reflections from the noise source under investigation to a minimum in the frequency range of interest. This is normally achieved by mounting absorbing wedges on the walls of the room. Typical wedges are made of

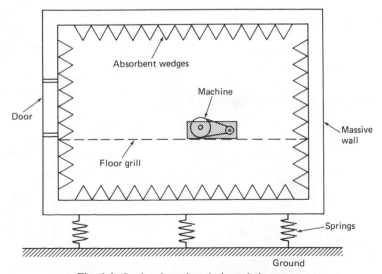

Fig. 4.4. Section through typical anechoic room.

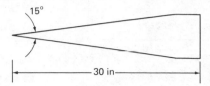

Fig. 4.5. Typical absorbent wedge used in an anechoic room.

fiberglass or foam rubber and normally have apex angles of between 13° and 17° and lengths of about 30 inches (see Fig. 4.5). The length of the wedge is determined by the lowest working frequency which is desired for the room (1, 5).

The effectiveness of the anechoic room is normally determined by measuring the deviation from the inverse square law (6 dB decrease for doubling of distance) when a noise source is placed in the room (see Fig. 4.6). Deviations of 1 or 2 dB from the inverse square law are tolerable for most purposes.

Measurements should not be made too close to the noise source (because of the near-field effect discussed earlier) nor too close to a wall. Also, the noise source should not be placed too close to a wall (say not closer than half a wavelength, $\lambda/2$, since its sound power output will be affected by the presence of the wall which will constructively or destructively interfere with the original source) (1). For these reasons the room should have certain minimum dimensions. These dimensions are dependent on the lowest frequency of interest. The acoustic wavelength λ is given by

$$\lambda = 1120/f \text{ ft.}$$

where f is the frequency in Hz (see Chapter 1).

Thus for a lower frequency of 100 Hz, λ is approximately 10 feet and $\lambda/2$ is 5 feet. If the machine has dimensions of about 2 feet then measurements should not be made closer to it than about $2 \times 2 = 4$ feet. Perhaps the room should have dimensions no smaller than about $4 + 5 = 9$ feet. Typical rooms are constructed to have dimensions of about 10 ft $\times$ 10 ft $\times$ 10 ft.

Unless company personnel have great experience in acoustics, if they are considering building an anechoic room, they often find it helpful to call in a company which specializes in anechoic rooms and absorbent wedges. Such a company will either build a complete anechoic room to order, or provide advice in the construction of the room. Several companies specialize in such a service.*

Anechoic rooms are ideal for measurement of machine noise under ideal conditions. Using the same measurement position (or several measurement positions on a hemisphere) the effect of changes of machine operation on the noise produced can be determined. It should be noted that noise could be reduced in one direction by a change but increased in another if the modification caused a change in directivity of the noise source. The anechoic room may also be used to compare the noise from different production models of the same machine. Since reflections are small in the room, the original directivity pattern is preserved. If this is carefully mapped out by making a large number of measurements, in a particular frequency band, it is often possible to determine where the predominant noise source is on a machine.

---

*See the annual issue of *Sound and Vibration* magazine where acoustical materials and systems, together with suppliers, are listed.

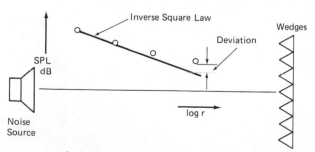

Fig. 4.6. Measuring the effectiveness of an anechoic room.

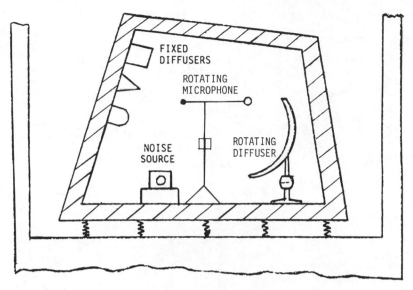

Fig. 4.7. Reverberant room schematic.

4.5. REVERBERANT ROOMS

A reverberant room is opposite to an anechoic room. Absorption is kept to a minimum so that not far from the noise source, the sound field is diffuse and almost the same reading is obtained with a microphone anywhere in the room (Fig. 4.7).

The room should be built with massive walls and isolated from the rest of the building (in a similar way to that described for an anechoic room) to keep the background noise level low. The walls of the room are made highly reflecting by covering them with plaster and painting them with a gloss paint.

Reverberant rooms are ideal for measuring the sound power of a machine, provided that it does not produce much sound energy at discrete frequencies (usually called *tones*). Tones will cause what is known as *standing waves* to be set up in the room. Standing waves are patterns of sound waves in a room which do not appear to travel with time, but remain stationary (1). It is obvious that standing waves are unwanted because the sound energy in this case is not diffuse in the room. If strong tones are present in the machine noise to be measured, special precautions must be taken such as the installation of a rotating vane device (to break up the standing waves and diffuse the sound) and a traveling microphone boom which will give an average reading for the varying sound pressure level throughout the room.

It should be noted that the measurement of machine sound power is much more rapidly achieved with the use of a reverberant room than in a free field or with an anechoic room since fewer measurements are needed. However, obviously with such a reverberant room measurement, information on the directivity of the noise source is lost.

For a more complete discussion on measurement of noise from machines the reader is referred to the references at the end of this chapter and to individual standards. Important standards are listed in Appendix II. Note that these standards are periodically revised.

4.6. DIRECTIVITY INDEX AND DIRECTIVITY FACTOR

Many noise sources exhibit a feature called directivity, i.e., there is a departure from spherical wave spreading in a free field, which varies with frequency. Many sources are nondirectional at low frequencies, and become directional at higher frequencies. An example of the polar plot of sound-pressure level (SPL) directivity of a noise source is shown in Fig. 4.8.

The *directivity index* DI_θ of a sound source in free space at an angle θ and for a given fre-

Directivity index in direction θ
$$DI_\theta = SPL_\theta - SPL_s$$
Directivity factor in direction θ
$$Q_\theta = \text{antilog}_{10}\left(\frac{DI_\theta}{10}\right)$$

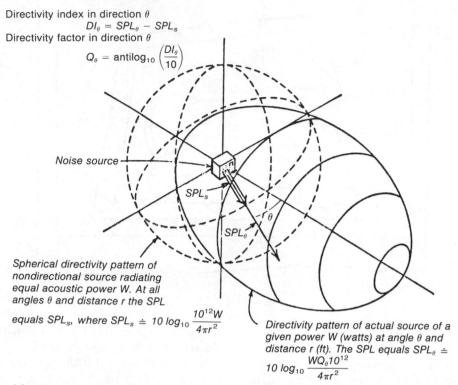

Noise source

SPL_s

θ

SPL_θ

Spherical directivity pattern of
nondirectional source radiating
equal acoustic power W. At all
angles θ and distance r the SPL
equals SPL_s, where $SPL_s \doteq 10\ log_{10}\ \dfrac{10^{12}W}{4\pi r^2}$

Directivity pattern of actual source of a
given power W (watts) at angle θ and
distance r (ft). The SPL equals $SPL_\theta \doteq$
$10\ log_{10}\ \dfrac{WQ_\theta 10^{12}}{4\pi r^2}$

Fig. 4.8. Directivity pattern, directivity index, and directivity factor of a noise source radiating in free space.

quency is defined as

$$DI_\theta = SPL_\theta - SPL_s \qquad [4.1]$$

where

SPL$_\theta$ = sound-pressure level, measured at a distance r and angle θ from a source radiating power W into a free field (Fig. 4.8),

SPL$_s$ = sound-pressure level, measured at a distance r from a nondirectional source of power W radiating into a free field (Fig. 4.8).

In the case of a nondirectional radiation, DI$_\theta$ = 0 at all angles of θ.

The directivity factor Q_θ is defined as the ratio of the mean-square sound pressure (N/m^2) measured at angle θ and distance r from an actual source radiating W watts to the mean-square sound pressure measured at the same distance from a nondirectional source radiating the same acoustic power W. Alternatively, Q_θ is defined as the ratio of the intensity (watts/m^2) measured at angle θ and distance r

from an actual source to the intensity measured at the same distance from a nondirectional source, both sources radiating the same power W. Thus,

$$Q_\theta = \frac{p_\theta^2}{p_s^2} = \frac{I_\theta}{I_s} = \frac{\text{antilog }(SPL_\theta/10)}{\text{antilog }(SPL_s/10)}.$$

Therefore

$$Q_\theta = \text{antilog}\left(\frac{SPL_\theta - SPL_s}{10}\right)$$

$$= \text{antilog}\left(\frac{DI_\theta}{10}\right)\ (\text{dimensionless}).$$

$$[4.2]$$

Note that in the case of a nondirectional radiation, as before DI$_\theta$ = 0, and hence Q_θ = 1.

The directivity index of a sound source on a rigid plane radiating into a hemispherical space is computed from

$$DI_\theta = SPL_\theta - SPL_H + 3\ dB$$

The 3 dB is added to SPL_H because measurements of sound-pressure levels were made over a hemisphere instead of a full sphere; the intensity at radius r is twice as large if the source radiates into a hemisphere, as compared to a sphere. In other words, for uniform radiation (nondirectivity) into a hemisphere space,

$$DI_\theta = 3 \text{ dB}$$

and

$$Q_\theta = \text{antilog } \frac{3}{10} = 2.$$

Similarly, if a source were in a dihedral corner (intersection of a wall and the floor), the directivity index DI_θ would be 6 dB and the directivity factor Q_θ would be 4. The comparable value for a source in a three-plane corner would be 9 dB and 6, respectively.

An example of noise source placement for the four directivity factors, 1 for spherical spreading, 2 for hemispherical, 4 for a dihedral (2 surface spreading) and 6 for a three-plane corner, is shown in Fig. 4.9.

Fig. 4.10 shows the case of a factory room in which a number of noisy machine tools are placed in four lines, two of them against the walls, with three access lanes between them. This increases the noise from the two lines of machines placed next to the walls. In order to overcome this problem, the machines along the walls are moved beside the other two lines so that there are only two lines of machines. The space along the walls is used as access lanes, of which there are still three. With this arrangement the overall noise in the room is reduced.

4.7. ATTENUATION OF SOUND PRESSURE WITH DISTANCE IN A FREE FIELD

In the case of a nondirectional radiation (spherical spreading), the sound-pressure level is

$$SPL_s = 20 \log_{10} \frac{p(\text{at } r)}{p_{ref}}$$

$$(p_{ref} = 2 \times 10^{-5} \text{ N/m}^2)$$

$$= 10 \log_{10} \frac{p^2(\text{at } r)}{p_{ref}^2}$$

$$= 10 \log_{10} \frac{p^2(\text{at } r)}{4 \times 10^{-10}} \text{ dB.} \qquad [4.3]$$

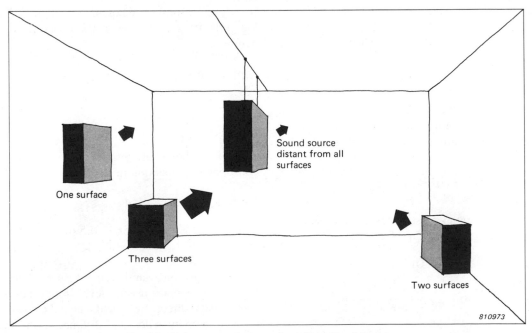

810973

Fig. 4.9. Noise source placement for various directivity factors.

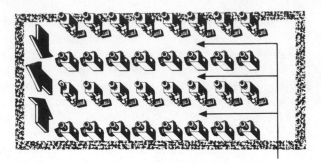

Fig. 4.10. Room noise reduction by appropriate machine placement.

From Section 1.2,

$$I = \frac{W}{4\pi r^2} = \frac{p_{rms}^2(\text{at } r)}{\rho c},$$

where I is in spherical spreading at radius r. Substituting $W\rho c/4\pi r^2$ for p^2 (at r) in Eq. [4.3], and assuming $\rho c \approx 400$ (for air at normal temperature and pressure), then

$$\text{SPL}_s = 10 \log_{10} \frac{W \times 10^{12}}{4\pi r^2} \text{ dB.} \quad [4.4]$$

Now, the sound-pressure level for nonspherical spreading is, from Eq. [4.1],

$$\text{SPL}_\theta = \text{SPL}_s + \text{DI}_\theta$$

$$= 10 \log_{10} \frac{W \times 10^{12}}{4\pi r^2} + 10 \log_{10} Q_\theta.$$

$$[4.5]$$

In level form, recognizing that $10 \log_{10}(W/10^{-12}) = $ power level (PWL) by definition,

$$\text{SPL}_\theta = \text{PWL} + \text{DI}_\theta - 20 \log_{10} r - 11 \text{ dB}$$

$$[4.6a]$$

(11 dB is $-10 \log_{10} 4\pi$). Note that r is the radius from the radiating source in meters. The corresponding equation in English units is

$$\text{SPL}_\theta = \text{PWL} + \text{DI}_\theta$$

$$- 20 \log_{10} r - 0.5 \text{ dB}, \quad [4.6b]$$

where r is in feet.

These equations can therefore be used to determine the sound pressure level from a noise source in the direct (free) field where the power level, directivity index and distance from the source are known.

Example: What power output level is required from a spherically radiating noise source to produce a sound pressure level of 60 dB at a distance of 100 m if both source and hearer are at ground levels? Assume that the ground is flat, unobstructed, and nonabsorbing.

From Eq. [4.6a] (where DI_θ for hemispherical spreading is 3),

$$PWL = SPL - DI_\theta + 20 \log_{10} r + 11$$

$$= 60 - 3 + 20 \log_{10} 100 + 11$$

$$= 60 - 3 + 40 + 11$$

$$= 108 \text{ dB}.$$

4.8. ATTENUATION OF SOUND PRESSURE WITH DISTANCE IN AN ENCLOSURE

In order to discuss the variations in sound pressure with distance, it is necessary to deal with sound power (and consequent energy absorption) from a radiating sound source. As has been mentioned previously, sound power is not normally directly measurable, and we therefore use sound pressures (which are related to sound power, i.e., sound power is proportional to the square of pressure).*

From Chapter 1, intensity I was defined as

$$I = \frac{p_{\text{rms}}^2}{\rho c} \text{ watts}/\text{m}^2,$$

where ρc is the characteristic resistance of the medium (air), or the impedance. If we divide intensity I by c, the acoustic velocity (in units of m/sec), then

$$I \left(\frac{\text{watts}}{\text{m}^2} \right) \times \frac{1}{c} \left(\frac{\text{sec}}{\text{m}} \right)$$

$$= \text{energy density} \left(\frac{\text{watts sec}}{\text{m}^3} \right),$$

i.e., the energy (watt sec) stored in a volume (m^3). This energy density is defined as

---

*See discussion on sound-intensity (and hence sound-power) measuring instruments in Chapter 3.

$$E = \frac{p^2}{\rho c} \left(\text{intensity} \right) \times \frac{1}{c}$$

$$= \frac{p^2}{\rho c^2} \frac{\text{watt sec}}{\text{m}^3}. \qquad [4.7]$$

Consider a small area ΔS chosen arbitrarily in a reverberant space (i.e., a reflective room), as shown in Fig. 4.11. E of the reverberant field is due to waves traveling in all directions with equal probability. Hence, the power flow due to waves traveling from left to right is only one-half of the total power flow through ΔS. Further, waves normal to ΔS contribute a power flow equal to $I\Delta S$, whereas those incident at an angle θ provide a reduced amount of power flow, equal to $I\Delta S \cos \theta$. It can be seen that the average amount of power flowing from left to right is one-half the power at normal incidence.

Thus, gathering up these two factors of one-half, the power flow per unit time in one direction through ΔS in terms of energy density E is $(Ec/4) \Delta S$. (Note that, from before, $I = Ec$.)

Now assume that ΔS lies in one of the walls of the enclosure. The total power absorbed from the reverberant field can be obtained by summing over the whole area S. Hence, the power W which falls on area S is given by

$$W = \frac{Ec}{4} S$$

It will be shown in a later chapter on absorption that, in an enclosure, the ratio of sound energy absorbed by the walls of the enclosure to the sound energy incident upon the walls is defined as the absorption coefficient $\bar{\alpha}$. Hence, in an enclosure, the total energy per second (power)

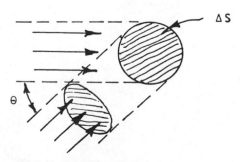

Fig. 4.11. Power flow through elemental area.

absorbed by the walls is

$$W = \frac{Ec}{4} S\overline{\alpha}.$$

The total energy per second *removed* from enclosure can also be derived as follows. If E is sound energy density stored in a small (unit) volume of the room, c is the acoustic velocity, mpf (mean free path) is the average distance between reflections, and $\overline{\alpha}$ is the absorption coefficient of room surfaces, then total energy removed per second from the enclosure is

$$EV \frac{c}{\text{mfp}} \overline{\alpha},$$

where

$$\frac{c}{\text{mfp}} = \frac{\text{distance traveled/sec}}{\text{average distance between reflections}}$$

$$= \text{number of reflections/sec}.$$

Note that both of the expressions for energy per second just derived are equal. Hence,

$$\frac{Ec}{4} S\overline{\alpha} = EV \frac{c}{\text{mfp}} \overline{\alpha}. \qquad [4.8]$$

From this, the mean free path (mfp) can be calculated as $4V/S$.

By definition, the reverberant-field energy (steady state) is the energy in the enclosure after the first reflection. (The energy that has not been reflected is the direct field energy). Assume E_R = sound energy density in reverberant field. Then

$$E_R V \frac{c}{\text{mfp}} \overline{\alpha} = E_R V \frac{cS}{4V} \overline{\alpha} = W(1 - \overline{\alpha}),$$

where W is total power supplied by source, and $W(1 - \overline{\alpha})$ is the power in the first reflection.

Rearranging:

$$E_R = \frac{4W}{c} \left(\frac{1 - \overline{\alpha}}{S\overline{\alpha}} \right) = \frac{4W}{cR} \qquad [4.9]$$

where R is defined as room *constant* = $S\overline{\alpha}/(1 - \overline{\alpha})$,

S is room surface area, and

$\overline{\alpha}$ is average room absorption coefficient (see Section 5.1.1 for discussion on absorption coefficient and typical values).

Room constant R must be modified to take into consideration the effects of air absorption in large rooms; this is discussed under a subsequent section on absorption.

In an enclosure where reflection is present, at a point located a certain distance from the sound source, the total energy density is the sum of that from the direct field plus that from the reverberant field, i.e.,

$$E_T = E_\theta + E_R. \qquad [4.10]$$

Inasmuch as energy density is proportional to mean-square pressure (Eq. [4.7]), we can then deal with the latter. From Section 4.7, for a case of spherical spreading in the direct field

$$p_s^2 \text{ (at distance } r) = \frac{W\rho c}{4\pi r^2}.$$

Where the sound is directional, then, from Section 4.6,

$$p_\theta^2 = p_s^2 Q_\theta$$

Therefore,

$$p_\theta^2 \text{ (at distance } r) = \frac{W\rho c Q_\theta}{4\pi r^2} \qquad [4.11]$$

For the reverberant field, the energy density is

$$E_R = \frac{4W}{cR}.$$

Also E_R by definition = $p_R^2/\rho c^2$. Therefore,

$$p_R^2 = \frac{4\rho c W}{R}. \qquad [4.12]$$

Thus, the total mean-square sound pressure in an enclosure (where both the direct and reverberant field pertain) is the sum of Eqs. [4.11] and [4.12], i.e.,

$$p^2 = p_\theta^2 + p_R^2$$

$$= \frac{W\rho c Q_\theta}{4\pi r^2} + \frac{4\rho c W}{R}$$

$$= W\rho c \left(\frac{Q_\theta}{4\pi r^2} + \frac{4}{R}\right) \qquad [4.13]$$

To convert to a log scale, divide mean-square pressure by p_{ref}^2 ($= (2 \times 10^{-5} \text{ N/m}^2)^2 = 4 \times 10^{-10} \text{ N/m}^2$) and multiply by $10 \log_{10}$. Note also that $W_{ref} = 10^{-12}$ watts. Hence $p_{ref}^2 = 400 W_{ref}$. Therefore

$$10 \log_{10} \frac{p^2}{p_{ref}^2} = 10 \log_{10} \left[\frac{W}{W_{ref}}\right]$$

$$+ 10 \log_{10} \frac{\rho c}{400}$$

$$+ 10 \log_{10} \left(\frac{Q_\theta}{4\pi r^2} + \frac{4}{R}\right) \text{ dB}.$$

The quantity $10 \log_{10} (\rho c/400)$ is very small, and can be neglected. Hence we are left with

$$20 \log_{10} \frac{p}{p_{ref}} = 10 \log_{10} \frac{W}{W_{ref}}$$

$$+ 10 \log_{10} \left(\frac{Q_\theta}{4\pi r^2} + \frac{4}{R}\right) \text{ dB}$$

$$[4.14]$$

Therefore

$$\text{SPL} = \text{PWL} + 10 \log_{10} \left(\frac{Q_\theta}{4\pi r^2} + \frac{4}{R}\right)$$

$$\text{for metric units.} \qquad [4.15a]$$

For English units:

$$\text{SPL} = \text{PWL} + 10 \log_{10} \left(\frac{Q_\theta}{4\pi r^2} + \frac{4}{R}\right)$$

$$+ 10.5 \text{ dB.} \qquad [4.15b]$$

Note that English and MKS units differ by approximately 10 dB because r^2 and R are each larger in English units by a factor of 10.5.

The above Eqs. [4.15] can be interpreted as follows. If, for instance, a machine was located in a reverberant enclosure, if the room constant R was known, if the acoustic power level of the machine (in any frequency) was known, if the directivity factor Q_θ was known, and if the distance r from the acoustic center of the machine to some receiving point in the room was known, then the sound-pressure level at the point could be determined (for any particular frequency in which the power level was known). Conversely, if the sound-pressure level could be measured, then the power level of the machine could be calculated for a given distance, directivity factor and room constant.

It can be seen that if $Q_\theta/4\pi r^2$ is large with respect to $4/R$, then SPL is largely due to direct radiation, and changes in R will have little effect on SPL. On the other hand, if $4/R$ is large with respect to $Q_\theta/4\pi r^2$, the room absorption is the predominant factor in determining SPL (because the point under consideration lies in the reverberant field). The term SPL − PWL is referred to as the relative sound-pressure level. A generalized graph of SPL − PWL versus $r/\sqrt{Q_\theta}$ is shown in Fig. 4.12. Note that the above derivations and Fig. 4.12 are based on the assumption of a diffuse field. There are examples (such as offices) where all absorption is on one surface, and consequently the sound field is not necessarily diffuse. Deviations of measured values from those calculated might be expected in these cases.

Example: Attenuation of Sound Pressure in a Room. A line printer for a computer is located in a room 40 ft long, 25 ft wide and 12 ft high. The printer is installed on the floor at one end of the 40-ft length of the room, at the interface between the floor and the end wall; it is located at the midpoint of the end wall. The computer manufacturer has submitted PWL (power levels) for the printer; the PWL in the 1000 Hz octave band is 85 dB. The overall absorption coefficient for the room is 0.30. What is the SPL (sound-pressure level) in the

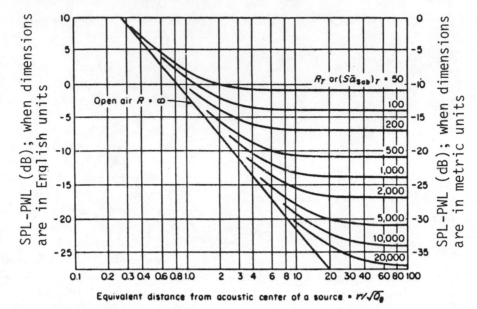

Fig. 4.12. Chart for determining sound-pressure level in an enclosure at a distance r from the center of a source of directivity factor Q_θ for various room constant values R.

1000 Hz band at a point 25 ft from the printer at the midpoint between the side walls?

$$\text{Room constant } R = \frac{S\bar{\alpha}}{1 - \bar{\alpha}} \text{ (Eq. [4.9])}$$

$$S = 2(40' \times 25') + 2(40' \times 12')$$
$$+ 2(25' \times 16')$$
$$= 2000 + 960 + 600$$
$$= 3560 \text{ ft}^2.$$

Therefore

$$R = \frac{3560 \times 0.30}{1 - 0.3} = \frac{1068}{0.7} = 1526.$$

From Section 4.6, $Q_\theta = 4$, hence $r/\sqrt{Q_\theta}$ (in Fig. 4.12) is $25/\sqrt{4} = 12.5$. Hence, interpolating for R and $r/\sqrt{Q_\theta}$ in Fig. 4.12, relative sound-pressure level (or SPL $-$ PWL) is -15 dB. Now PWL (as given) is 85 dB. Therefore

$$\text{SPL} = \text{PWL} - 15 = 85 - 15 = 70 \text{ dB}.$$

This result can be verified by substituting the appropriate values in Eq. [4.15b].

A term which is often used in room acoustics is *critical distance*. This is the distance from the noise source in a reverberation room where the energy density in the reverberant field equals the energy density in the direct field, i.e., when

$$E_R = E_\theta.$$

Substituting from the foregoing for these quantities,

$$\frac{4W}{cR} = \frac{WQ_\theta}{4c\pi r^2}.$$

Therefore,

$$r^2_{\text{critical}} = \frac{RQ_\theta}{16\pi},$$

where, as previously noted, $R = S\bar{\alpha}/(1 - \bar{\alpha})$ and $Q_\theta = $ directivity factor. Hence

$$r = \left(\frac{RQ_\theta}{16\pi}\right)^{1/2}$$
$$= 0.14(Q_\theta S\bar{\alpha})^{1/2}$$

for small values of $\bar{\alpha}$. The critical distance can be said to be that distance from a noise source where the reverberant field becomes more dominant than the direct field.

4.9. PROPAGATION OF SOUND IN AIR*

Any source of noise has a characteristic sound power, a basic measure of its acoustic output, but the sound-pressure level that it gives rise to depends on many external factors, such as the distance and orientation of the receiver, the temperature, humidity and velocity gradients in the medium, and the environment (i.e., absorption characteristics of sound path and the point of reception). Sound power, on the other hand, is a fundamental property of the source alone, and is therefore an important absolute parameter which is widely used for rating and comparing sound sources (see Sections 1.6, 1.7, and 3.9.5).

The above various factors which offset the propagation of sound in the air will be considered next.

4.9.1. The Plane Source

Consider (see Fig. 4.13) an elemental tube of the medium, with unit cross-sectional area and length equal to the distance traveled by the sound wave in one second, i.e., numerically equal to the speed of sound (c). If a piston source of power W is constrained by hard walls to radiate all its power into the elemental tube to produce a plane wave, the tube will contain

*The diagrams in Sections 4.9.1 to 4.9.12 have been taken from Ref. 9 with permission.

a quantity of energy numerically equal to the power output of the source. Assuming no other losses, the intensity, i.e., the acoustic energy flowing through unit area anywhere along the tube in unit time, is independent of the distance from the source and numerically equal to its sound power. Apart from duct systems, plane waves and plane sources are rarely encountered in normal noise-measurement situations.

4.9.2. The Point Source

Sound sources can be considered as point sources if their dimensions are small in relation to their distance from the receiver, and many common noise sources, including industrial plant, aircraft, and individual road vehicles can normally be treated in this way. As shown in Fig. 4.1 of this chapter, the ideal point source can be considered to produce a series of spherical wavefronts resulting from successive disturbances at the point source. For a pure sinusoidal disturbance, the distance between wavefronts representing the successive peak pressures will of course be the wavelength, a fact which is important when considering the effects of reflections within the sound field. As shown in Fig. 1.5 of Chapter 1, the sound energy spreads out equally in all directions so that as it travels further and further from the source its energy is received on an ever larger spherical area. If the medium is assumed to be non-dissipative then, as was pointed out in Section 1.5, the entire power output of the source passes through a spherical shell of radius r. The intensity is therefore the power of the source divided by the area of this shell. Thus we have

$$I = \frac{W}{4\pi r^2}.$$

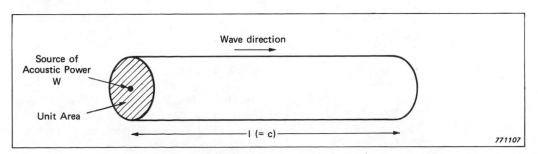

Fig. 4.13. Acoustic radiation into an elemental tube.

Hence, as shown in Section 4.2, the intensity, and hence the sound-pressure level, attenuates 6 dB per doubling of distance from the source.

4.9.3. The Line Source

A line source may be continuous radiation, such as from a pipe carrying a turbulent fluid, or may be composed of a large number of point sources so closely spaced that their emissions may be considered as emanating continuously from a notational line connecting them. In this category are included such factory sources as closely spaced machines and conveyors, and two extremely important sources of environmental noise, namely roads and railways. A road which has a traffic flow high enough to be a noise nuisance can usually be considered as a line source rather than a succession of single events. Railways are often treated as line sources at the distances from the track which are usually the most important from the point of view of community annoyance. Very close or very far from the track, the field is rather more complex. Consider the diagram in Fig. 4.14 of part of an infinite line source which has a constant power per unit length. The wavefront spreads out from the line in only one dimension perpendicular to its direction of travel, so that any two points at the same distance from the line are on the same wave front and have the same properties. The wavefronts therefore form concentric cylindrical surfaces about the surfaces about the line source as axis. The energy released from a unit length of the source in unit time passes through the same length of cylindrical surface at all radii. The intensity at

a given radius is therefore the power emitted by this element, divided by the area of the cylindrical elemental surface. Thus

$$I = \frac{W}{2\pi r \times 1}.$$

The intensity is therefore inversely proportional to the distance from the source i.e. it attenuates 3 dB per doubling of distance. Typical examples of point source and line source propagation are shown in Fig. 4.15.

4.9.4. General Factors Affecting Sound and Propagation

In addition to the reduction in intensity by distance, discussed in the previous sections, there are many other factors which can significantly affect the propagation of sound in a real medium like the atmosphere. Velocity and temperature gradients alter the direction of the wave; turbulence distorts it, and viscosity causes absorption. This latter effect is far greater for high than for low frequencies (see Sections 5.1 and 5.3.3. on absorption), so the atmosphere tends to act as a lowpass filter, attenuating high frequencies and thus distorting the frequency spectrum of a noise, as well as reducing its strength and changing its propagation path. In addition, most measurements are made near ground level where people live and work and where noise is invariably received and, with the notable exception of aircraft noise, produced. For this reason the reflection and absorption of the ground under the path between source and receiver is very important, and must be taken into account as a matter of course whenever studying the transmission of outdoor noise. (For a comprehensive discussion on ground absorption, see Refs. 1 and 5).

4.9.5. The Effects of Wind

The atmosphere is in a state of continuous motion over the earth's surface and is also a real fluid with all the normal physical properties including viscosity. Because the air is viscous the velocity of molecules at the ground must be

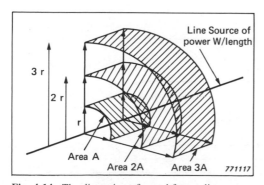

Fig. 4.14. The dispersion of sound from a line source.

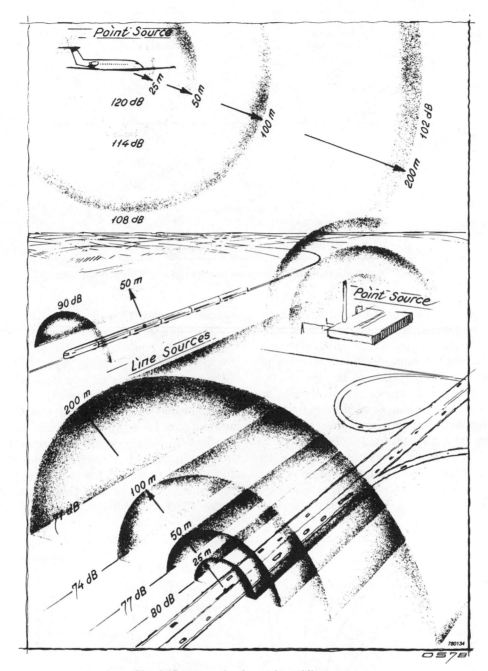

Fig. 4.15. Attenuation from point and line sources.

zero and a boundary layer is formed near the surface, in which the wind speed gradually increases with height until the speed of the main air mass is attained. This region may be as much as several hundred meters thick so it can, and does, affect measurements made of most noise sources. When a sound wave impinges on a layer of air which has a different speed, the wave's direction of travel changes, as represented by the sound "rays" and vector constructions in Fig. 4.16. This happens because the speed of sound depends only upon the medium in which it is propagated, so any movement of that medium must necessarily impose

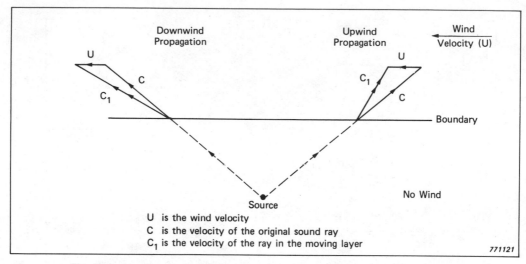

U is the wind velocity
C is the velocity of the original sound ray
C_1 is the velocity of the ray in the moving layer

771121

Fig. 4.16. Sound propagation across a boundary between layers with different velocities.

a similar movement on the sound wave as seen from the ground. If it has a component in the same direction as the wind, the ray representing the direction of propagation is therefore refracted towards the interface between the two different velocity regions when entering that with the higher speed, or away from the interface when entering a region of lower speed. The effects are simply reversed if the direction of sound propagation is opposite to that of the wind. Although the lower atmosphere is not a series of discrete layers with fixed velocities, but a transition region in which the velocity changes continuously with altitude, it can be seen from this simple approach how the sound rays will be continuously refracted as they progress through the boundary layer.

The overall effect as far as a stationary observer on the ground is concerned, is to bend the downwind sound rays back towards the earth and bend the upwind rays away from it as shown in Fig. 4.17. A region of noise reinforcement is therefore formed downwind of the source and a sound shadow, a region of reduced intensity, occurs on its upwind side. Refraction effects can only occur because there is

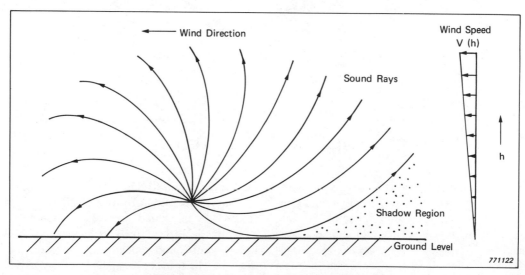

771122

Fig. 4.17. Sound refraction in a boundary layer.

a wind gradient, i.e., because the wind speed varies with altitude, and are not the result of sound being convected along by the wind. The magnitudes of changes in sound intensity which can be attributed to this phenomenon are dependent only on the rates of change of wind speed with altitude. Attenuations in the shadow region may be as high as 30 dB but the increases which are due to reinforcement downwind of the source are usually rather less.

4.9.6. The Effects of Temperature

The velocity of sound in air increases with temperature, and in a normal atmosphere the temperature itself decreases with height as in Fig. 4.18(a). A rising sound ray, on entering a layer with a lower temperature, undergoes a reduction in propagation velocity and is refracted away from the interface between the two layers. The result is that, in the absence of wind, the rays are continuously bent away from the ground surface as shown in Fig. 4.19(a) and a shadow region is found beginning at a distance from the source which depends on the strength of the temperature gradient. As with wind gradients, the effects of temperature gradients are made less distinct by the inhomogeneity of the atmosphere in its normal state, with turbulence and local heat exchange scattering sound into the shadow regions.

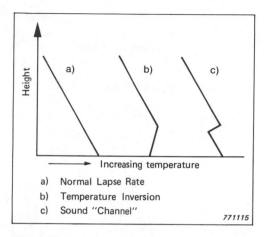

a) Normal Lapse Rate
b) Temperature Inversion
c) Sound "Channel"

Fig. 4.18. Typical atmospheric temperature gradients.

Sometimes, however, the temperature gradient near the ground is positive, i.e., the temperature increases with height up to a point where it reverts to the normal lapse rate, as in Fig. 4.18(b). This situation is called a *temperature inversion* and leads to effects opposite to those described above for a normal lapse rate. A sound ray becomes refracted downward towards the ground as it progresses through the warmer layer of air, reinforces the sound field at surface level around the surface, and, as Fig. 4.19(b) shows, no shadow region will be formed.

A double temperature gradient, such as that in Fig. 4.18(c), is rarely encountered, but can

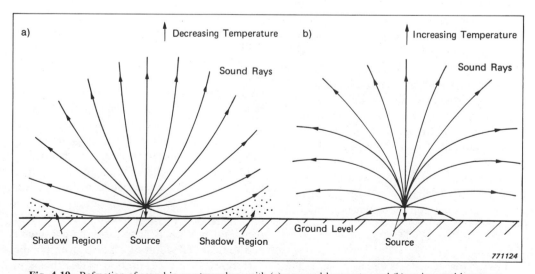

Fig. 4.19. Refraction of sound in an atmosphere with (a) a normal lapse rate, and (b) an inverted lapse rate.

trap slightly inclined sound waves in the inversion layer and channel them over considerable distances with only low attenuation.

4.9.7. Humidity and Precipitation

The absorption of sound in air varies with frequency, humidity, and temperature in an extremely complicated fashion, the only general trend being that it is higher at high frequencies, and shows a tendency to increase with temperature but decreases at higher relative humidities. (For the effects of humidity on absorption, see Section 5.3.3, Fig. 5.38 in Chapter 5). If it is imperative to include these effects when carrying out research into distant noise sources, the current literature on the subject, and the available tables which contain the relationships between all the relevant parameters, should be consulted. The oft-mentioned ability of sound to "carry" in fog or light precipitation of any kind is not due to any changed physical property of the medium which is conducive to better propagation. Reduced human activity and still air conditions often combine to produce a lower than normal background noise level during these periods.

4.9.8. Absorption by Natural Features

If the ground surface below a sound wave is perfectly flat and reflecting, the wave would propagate without any excess attenuation over that attributable to the spreading of the acoustic energy throughout an ever increasing volume. Even a man-made hard surface such as concrete is not perfectly reflecting, however, and most natural ground covers have significant absorption. This causes a significant reduction in intensity, which is most marked when source or receiver, or both, are near the ground and relatively distant. As would be expected, the attenuation is greater for high frequencies than for lower ones* and depends to a large extent on the effective "roughness" of the surface, i.e., the ratio between the wavelength and the dimensions of the irregularities of the ground.

---

*See discussion on absorption in Chapter 5.

The values are, therefore, low for ordinary grassland, but may rise to as much as 20 dB per 100 meters for long grass, corn, or low shrubs and trees.

4.9.9. Reflection

When sound waves come in contact with a surface, part of the energy is reflected from it, part is transmitted through it and part is absorbed by it. The instantaneous sound pressure at any point in the field is therefore due to the direct radiation from the source, and the sound arriving indirectly after one or more reflections from the surfaces, where a part of its energy, however small, is absorbed. If absorption and transmission are low, and therefore most of the sound energy incident on the surface is reflected, it is said to be *acoustically hard*, and can be considered to reflect sound in much the same way as a mirror reflects light. The ray reflected from a flat rigid surface then takes up the position as shown in Fig. 4.20 and the sound ray and wave fronts can be considered as coming from the image. The reflected wave fronts and those arriving directly from the source reinforce or cancel each other where they cross, giving rise to problems when making noise measurements in the presence of reflecting surfaces such as hard ground, roads, and building facades. (This will be discussed in greater detail in Chapter 5).

The effect of curved surfaces, parallel flat surfaces, and corners, on the sound field, is shown diagrammatically in Fig. 4.21. If the reflecting surface is curved then the rays will be focused when the surface is concave, and dispersed when convex. A ray entering a right-angled corner will be reflected from it, after two reflections, back along a path different from, but parallel to, its incoming one. Parallel surfaces cause two important effects. First, the formation of *standing waves*, which occur at frequencies such that an integral number of half-wavelengths occur between the two surfaces, leading to a very large variation in sound pressure from node to antinode. The second effect, *flutter echo*, is caused by the continuous and regular reflection of a pulse from parallel surfaces with low absorption. These phenom-

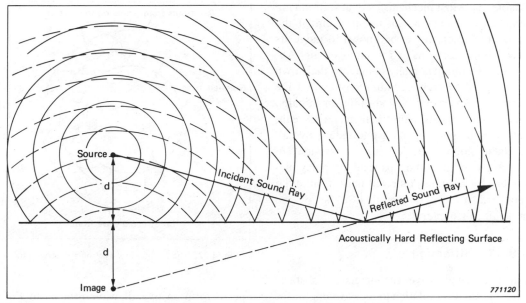

Fig. 4.20. Reflections from a plane surface.

ena are undesirable in architectural acoustics (concert halls, lecture rooms etc.) and in acoustic testing chambers, where uniformity of the sound field is generally required.

4.9.10. Absorption

Whenever a sound wave meets a surface, some small amount of its energy is lost. The absorp-

tion of a surface is a function of many parameters, including its effective roughness, its porosity, its flexibility, and in some cases its resonant properties. The efficiency of an absorbing surface is expressed as a number between 0 and 1, called the *absorption coefficient*: 0 represents no absorption, i.e., perfect reflection, which is never encountered in practice, and 1 represents perfect absorption (see

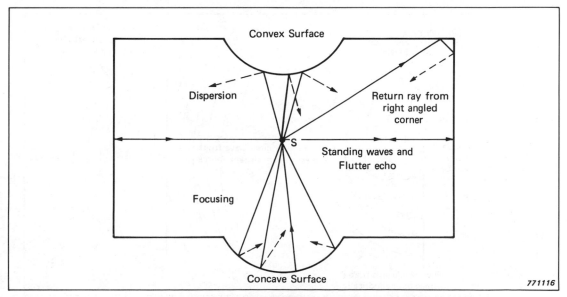

Fig. 4.21. Reflections from surfaces of various shape.

Chapter 5). Most mechanisms of absorption are frequency dependent, so the spectrum of the noise concerned has to be known to judge its effect, both in rooms and in the open air. Absorption techniques really come into their own with regard to architectural acoustics and the control of reverberation and diffusion in halls, rooms, and offices. Absorbent materials can, however, be put to good use to reduce overall levels in noisy factories, and to provide acceptable conditions near particularly noisy machinery by reducing reflected noise from adjacent hard surfaces. (Again, this will be covered in detail in Chapter 5.)

4.9.11. Diffraction

When a sound wave encounters an obstacle which is small in relation to its wavelength, the wave passes round it almost as if it did not exist, forming very little shadow. But, if the frequency of the sound is sufficiently high and the wavelength is therefore sufficiently short, a noticeable shadow is formed. These phenomena can be explained by first introducing a method of wavefront construction, which states that a source may be considered as an infinite number of point sources covering its surface, and radiating in all directions. At an instant in time, each point emits a sound wave and these com-

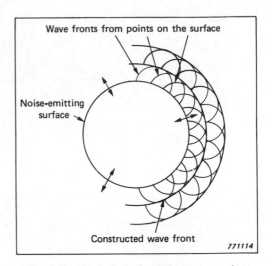

Fig. 4.22. The method of wavefront construction.

bine to form an overall wavefront as in Fig. 4.22. Similarly, each point on a wave front can be considered as a new sound source, so the next position of the wavefront can be constructed from the last. Extending this concept to the two cases of Fig. 4.23(a) and 4.23(b), we can see what happens when the wavefronts from a distant source impinge on the edge of, or an opening in, an otherwise infinite wall.

In the case of the situation in Fig. 4.23(a) the wall edge can be considered as a source of secondary wavelets radiating away from it in

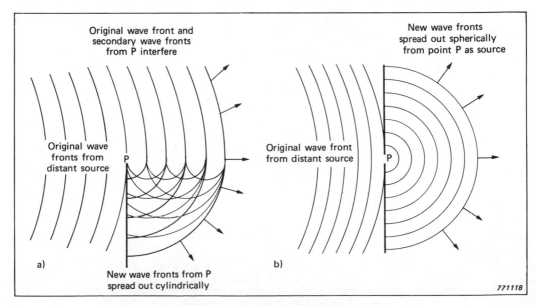

Fig. 4.23. Effects of diffraction at low frequencies.

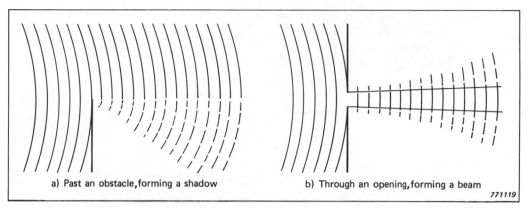

a) Past an obstacle, forming a shadow

b) Through an opening, forming a beam

771119

Fig. 4.24. Effects of diffraction at high frequencies.

all directions according to the previously mentioned principle. These secondary wavelets combine to form wavefronts which spread out cylindrically in the quadrant behind the wall, in the so-called shadow region. For the second case the opening becomes, in effect, a new point source, radiating hemispherically into the space beyond the wall, but with a lower intensity (depending on the size of the opening) than the incident sound (1, 5, 9).

A large ratio of wavelength to obstacle size causes the diffraction pattern as shown in Fig. 4.23(a) and 4.23(b). A small value leads to the formation of a more distinct shadow behind a barrier, Fig. 4.24(a), or a beam of sound through an opening as shown in Fig. 4.24(b). The greatest attenuation behind a barrier occurs if the angle between the ray from the source to the barrier top, and the line from there to the receiver, is as small as possible. Practically, this means that the barrier should be as near as

possible to either the source or the receiver for greatest effect. When taking measurements of noise sources in the field, unobstructed situations are always to be preferred unless the barrier effect is of direct interest. (See also Section 4.10).

4.10. OUTDOOR BARRIERS

4.10.1. Introduction

Noise barriers are solid structures (walls, buildings, fences, panels) that provide attenuation of sound by shielding the receiver from the source of the sound. Fig. 4.25 shows the progression of sound waves passing through or crossing over a barrier. The sound waves near the top of the figure pass clearly over the barrier and are not affected by it. Sound waves which pass over the top edge of the barrier are diffracted and hence bent down as shown in the

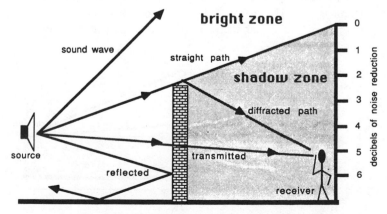

Fig. 4.25. The effect of a barrier on sound waves.

figure. A receiver standing in the way of the diffracted sound waves (i.e., in the shadow zone) will hear sound at a reduced level than a receiver standing in the area of the unaffected sound (i.e., in the bright zone). This process of sound-wave diffraction is similar to the process of diffraction in the study of the behavior of light.

4.10.2. Path-Length Difference

From optical diffraction theory, the sound attenuation of an acoustical barrier is found by calculating the *path length difference*, the extra distance the sound is forced to travel because of the presence of the barrier.

4.10.3. Thin Barriers

Figure 4.26 shows a barrier placed between a source and a receiver. In the absence of the barrier, the sound travels directly between the source and the receiver, shown as the distance d. However, with the barrier in place, the sound must travel from the source to the top of the barrier (distance a) and from the top of the barrier of the receiver (distance b), for a total distance of $a + b$. The path-length difference is obtained by subtracting d from $a + b$.

The geometrical aspects of sound propagation over an outdoor barrier (assuming there are no adjacent surfaces present to reflect sound into the protected "shadow" zone) is shown in Fig. 4.26. The source-to-barrier distance is D_{SB}, the source-to-receiver distance is D_{RB}, the source height is H_S, the receiver height is H_R, and the barrier height is H_B. The path-length difference, usually referred to as δ, can then be calculated from:

$$\delta = a + b - d.$$

where
$$a = \sqrt{(H_B - H_S)^2 + D_{SB}^2},$$
$$b = \sqrt{(H_B - H_R)^2 + (D_{BR} - T)^2}, \text{ and}$$
$$d = \sqrt{(H_S - H_R)^2 + (D_{SB} + D_{BR})^2}.$$

4.10.4. Thick Barriers

It is often possible to use a row of townhouses or a high-rise apartment block as a noise barrier (with care being taken during design of the building that the noise-sensitive areas within the building, such as bedrooms, living rooms, balconies, etc., are on the opposite side of the building from the noise source). In this case, some extra sound reduction is achieved as the sound is diffracted around the *two* upper edges of the building (see Fig. 4.27). Again, the path length difference can be used as a simple indicator of the sound reduction.

If the building thickness is defined as T, then the calculation of *PLD*, δ is as follows:

$$\delta = a + b + T - d,$$

where
$$a = \sqrt{(H_B - H_S)^2 + D_{SB}^2},$$
$$b = \sqrt{(H_B - H_R)^2 + (D_{BR} - T)^2}, \text{ and}$$
$$d = \sqrt{(H_S - H_R)^2 + (D_{SB} + D_{BR})^2}.$$

Care should be taken in calculating δ that the receiver is in the shadow zone, that is, the source cannot be seen by the receiver. In this situation, δ is given a positive sign and a significant sound attenuation will result. If, how-

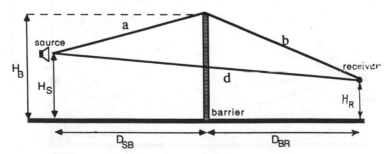

Fig. 4.26. Calculation of path-length-difference for a thin barrier.

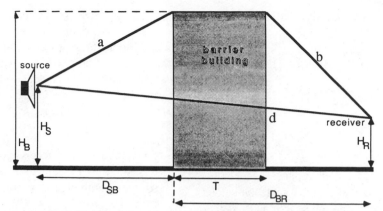

Fig. 4.27. Calculation of path-length-difference for a thick barrier.

ever, the source *can* be seen by the receiver despite the presence of the barrier, then the receiver is in the bright zone and δ should be given a negative sign. In the latter case, the barrier will give very little, if any, sound reduction at the receiver position.

4.10.5. Fresnel Number

Once the path length difference δ has been calculated, the sound attenuation of the barrier can only be calculated with a knowledge of the wavelength or frequency of the sound wave under consideration. This is necessary, as the low-frequency sound (or long wavelengths) can be diffracted or "bent" more easily than high-frequency (or short-wavelength) sound. Following the optical diffraction theory (12, 13), a Fresnel Number N must be calculated before the sound attenuation can be found:

$$N = \frac{2\delta}{\lambda}$$

where λ is the wavelength of the sound, and
δ is the path-length difference.

As we usually work with frequency rather than wavelength, it is useful to convert the formula for N into a form involving frequency f. Since $f = c/\lambda$ (Section 1.1), where c = speed of sound in air, then $\lambda = c/f$ and

$$N = \left(\frac{2f}{c}\right) \delta. \qquad [4.16]$$

Care should be taken when using this equation to use a consistent set of dimensions. If the path-length difference is calculated in meters, then the speed of sound must be in meters/second. If it is calculated in feet, then the speed of sound must be in feet/second. Thus,

$$N = \left(\frac{2f}{344}\right) \delta \quad (\delta \text{ in meters})$$

or

$$N = \left(\frac{2f}{1130}\right) \delta \quad (\delta \text{ in feet}). \qquad [4.17]$$

It is clear that in order to calculate N, either the frequency of the sound must be known, or if the sound covers a wide frequency range, some representative frequency must be assumed. [The Ontario Ministry of the Environment uses a frequency of 500 Hz to represent both road and rail transportation noises in the prediction of environmental noise impact on proposed adjacent subdivisions (14).]

4.10.6. Attenuation for Point and Line Sources

A noise barrier provides different values of sound attenuation for stationary point sources and line sources. Moving point sources, for which the attenuation of a passby is required, should be considered as line sources. In general, the attenuation for a line source is less than that for a point source.

The barrier attenuation for a point source Δ,

assuming an infinite barrier, can be expressed as (12, 13):

$$\Delta = 20 \log \left(\sqrt{2\pi N} / \tanh \sqrt{2\pi N} \right) + 5 \text{ dB},$$

$$-0.19 < N \le 5.03$$

$$= 20 \text{ dB}, \qquad\qquad N > 5.03$$

[4.18]

where $N = N_0 \cos \phi$, with N_0 the Fresnel Number, $N_0 = 2\delta/\lambda$, and ϕ is the angle subtended by the line connecting the source and receiver and a line perpendicular to the barrier. (Note that a line connecting the source and receiver is not necessarily perpendicular to the barrier.)

For a line source, the attenuation provided by an infinite barrier is obtained by integrating Eq. [4.18] from $-\pi/2$ to $\pi/2$ (12, 13), and is given as

$$\Delta_1 = 20 \log \left(\frac{1}{\pi} \int_{-\pi/2}^{\pi/2} 10^{-\Delta/10} \, d\phi \right),$$

[4.19]

where Δ is the point-source attenuation defined in Eq. [4.18].

Figure 4.28 gives the attenuation provided by an infinite barrier for both a point and a line source (14). The upper curve on Fig. 4.28 applies to a point source such as a stationary ma-

chine, car or truck, or an idling locomotive. The lower curve applies to a line source such as a continuous flow of traffic along a roadway or the wheel–rail noise from a long freight train. It can be seen that the attenuation provided by a barrier is less for line sources than for point sources. The graphs in Fig. 4.28 assume that:

- The barrier is very long (at least eight times the source to receiver distance) or is turned through a right angle away from the source at both ends to protect the receiver from the sides (Fig. 4.29);
- The barrier does not allow sound to penetrate through it and thus raise sound levels in the shadow zone [to avoid this effect, the barrier can be made of any material, such as solid wood, masonry, or earth in the form of a berm so long as it has a surface mass density of 20 kg/m^2 (or 4 pounds/feet^2) and has no holes or gaps (14) ;
- The barrier is at right angles to a line drawn from the receiver to the source (for a point source) or a line drawn from the receiver perpendicular to the source (for a line source).

In many practical cases, the conditions just described are not met, and the barrier cannot be considered infinite. Therefore, the attenuation provided by a finite barrier must be cal-

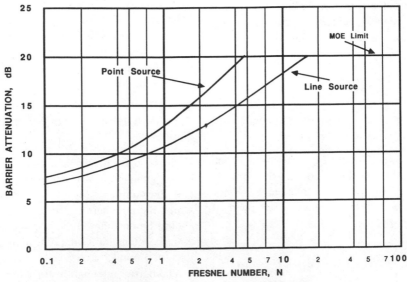

Fig. 4.28. Barrier attenuation versus Fresnel Number N.

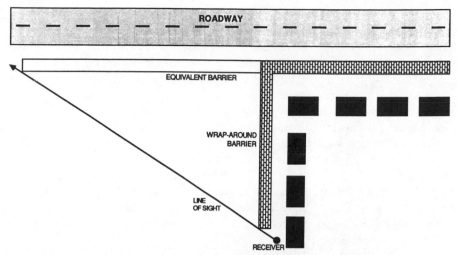

Fig. 4.29. Equivalent configuration for wraparound barriers.

culated. Finite barrier attenuation is obtained from Eq. [4.19], replacing the angles $-\pi/2$ and $\pi/2$ (i.e., $-90°$ and $90°$) by appropriate values.

It was indicated earlier that in some instances, a barrier is extended, or wrapped around a corner lot or a row of houses, to simulate an infinite barrier. A simple example is shown in Fig. 4.29. The barrier attenuation calculation depends on the angle subtended at the receiver by the end of the barrier. For the receptor location shown in Fig. 4.29, the wraparound section of the barrier can be represented by an equivalent barrier section which considerably extends the overall length of the barrier parallel to the roadway.

4.10.7. Ground Attenuation

As a sound wave passes over a region of absorptive surface, the source sound level decreases by more than the expected 3 dBA per doubling of distance for a line source, or the 6 dBA per doubling of distance for a point source. This extra sound-level decrease is due to ground-surface attenuation. If, however, the sound wave has to travel over a barrier in order to reach the receiver, it must necessarily travel further above the ground and hence this results in a smaller ground attenuation.

The effect of interposing a barrier between a source and a receiver is thus twofold. First, the sound level is increased because at least a part of the ground attenuation is removed. (If, however, the distance from the barrier to the receiver is considerable, ground attenuation may still predominate due to "bending" of the sound waves toward the ground between the barrier and the receiver.) Second, the sound level is decreased by the presence of the barrier and ensuing barrier attenuation. Thus, this removal of part of the ground attenuation makes the barrier less effective, a fact which must be taken into consideration when calculating the barrier attenuation. (See Refs. 14 and 16, which have been based on a model developed by the Federal Highways Administration in the United States, presented in Ref. 15).

Example: Attenuation Due to Distance, Air Absorption, and a Barrier. A cooling tower associated with a commercial air-conditioning unit for a manufacturing building is located approximately 630 ft from a residential building. The manufacturer of the cooling tower has given the following power levels (PWLs) and sound-pressure levels as measured at 50 ft from the side of the tower facing the residential building (SPLs):

	\multicolumn{8}{c}{Octave band center frequency, Hz}
	63	125	250	500	1k	2k	4k	8k
PWL (dB)	97	96	92	93	87	83	79	73
$SPL_{50'}$ (dB)	67	66	61	63	56	53	48	42

To reduce the possible impact of the sound pressure levels from the cooling tower at the residential building, an isolation barrier 20 ft high, 50 ft wide is proposed, 30 ft from the cooling tower (or 600 ft from the residential building). (Note that the barrier must be designed so that the sound transmission loss through the barrier is at least 10 dB greater than the barrier noise reduction values in each frequency band; see Section 5.4 for discussion of transmission loss.)

Assuming a relative humidity of the air at 50% and a temperature of 20°C, predict the sound pressure levels in each of the above octave-band center frequencies at the residential building. What is the overall A-weighted sound level at the building?

The total sound pressure attenuation is the sum of the distance attenuation, the air absorption attenuation, and the barrier attenuation. (This assumes that there are no other modifying factors such as wind gradients, temperature gradients, fog or snow, etc.)

With regard to distance attenuation, from Eq. [4.6b],

$$\text{SPL}_{\theta 50'} \simeq \text{PWL} + \text{DI}_\theta$$
$$- 20 \log_{10} r_{50} - 0.5 \text{ dB}$$

and

$$\text{SPL}_{\theta 630'} \simeq \text{PWL} + \text{DI}_\theta$$
$$- 20 \log_{10} r_{630} - 0.5 \text{ dB}.$$

$20 \log_{10} (50/630)$ is -22 dB, a distance attenuation which is the same for all frequency bands.

Now for air absorption attenuation, Beranek (1) recommends that air attenuation at a temperature of 20°C is

$$\text{Atten}_{\text{air}} = 7.4 \frac{\omega^2 r}{\phi} \times 10^{-8} \text{ dB}$$

where

ω = octave-band center frequency, Hz;
ϕ = relative humidity, %;
r = distance between source and receiver, m.

In air attenuation, energy is extracted from a sound wave by vibrational and rotational activities of the oxygen molecules in the air (1). The vapor content of the air determines the time constant of the vibration, which is more important than the rotational activity. In addition, this "molecular absorption" depends in a major way on temperature.

For the 63-Hz band (where $r = 630' \times 0.305 = 192$ m)

$$\text{Atten}_{\text{air}} = 7.4 \frac{(63)^2(192)}{50} \times 10^{-8}$$
$$= 0.001 \text{ dB}.$$

Other air attenuations at center frequencies are shown below:

	Octave-band center frequency, Hz							
	63	125	250	500	1k	2k	4k	8k
Atten$_{\text{air}}$	0.001	0.004	0.017	0.07	0.248	1.14	4.55	18.19

Subtracting $\text{SPL}_{\theta 630'}$ from $\text{SPL}_{\theta 50'}$, and recognizing directionality of the source,

$$\text{SPL}_{50'} - \text{SPL}_{630'}$$
$$= -20 \log_{10} 50 + 20 \log_{10} 630.$$

Therefore

$$\text{SPL}_{630'} = \text{SPL}_{50'} + 20 \log_{10} 50$$
$$- 20 \log_{10} 630$$
$$= \text{SPL}_{50'} + 20 \log_{10} \frac{50}{630}.$$

Air attenuation, as it refers to reverberation time, is discussed in Section 5.2.3 of Chapter 5.

The barrier attenuation can be determined from Section 4.10.6.* Assuming that the noise source acoustic center and the receiver sensitive area are both 10 ft above the ground, then from Fig. 4.26, the path length difference δ is

---

*The barrier is assumed to be sufficiently wide as to eliminate any "end flanking" of sound between the cooling tower and the residential building.

$$\delta = \sqrt{(H_B - H_S)^2 + D_{SB}^2}$$

$$+ \sqrt{(H_B - H_R)^2 + D_{BR}^2}$$

$$- \sqrt{(H_S - H_R)^2 + (D_{SB} + D_{BR})^2}$$

$$= \sqrt{(20 - 10)^2 + (30)^2}$$

$$+ \sqrt{(20 - 10)^2 + (600)^2}$$

$$- \sqrt{(10 - 10)^2 + (30 + 600)^2}$$

$$= \sqrt{100 + 900} + \sqrt{100 + 360,000}$$

$$- \sqrt{396,900}$$

$$= 31.6 + 600.1 - 630$$

$$= 1.7 \text{ ft.}$$

From Eq. [4.18] and Fig. 4.28, barrier insertion loss is as follows:

Octave-band center frequency, Hz							
63	125	250	500	1k	2k	4k	8k
Insert loss (dB) 8.0	10.0	12.5	15.0	17.5	20.0	20.0	20.0

Hence, the resultant SPL at the residential building is as follows:

In order to get the total (or overall) A-weighted sound level, it will be necessary to sum the dBA values in each frequency band logarithmically. This necessitates converting their levels to energy equivalents for summation.

Recalling that $SPL_i = 10 \log_{10} (p/p_{ref})_i^2$ dB, then

$$\left(\frac{p}{p_{ref}}\right)_i^2 = \text{antilog } (SPL_{i/10})$$

The total sound pressure level SPL_T is

$$SPL_T = 10 \log_{10} \left[\sum_{i=1}^{n} \left(\frac{p}{p_{ref}}\right)_i^2 \right] \text{dB}$$

or in terms of the sound pressure levels,

$$SPL_T = 10 \log_{10} \left[\sum_{i=1}^{n} \text{antilog} \left(\frac{SPL_i}{10}\right) \right] \text{dB}$$

Upon further simplification,

$$SPL_T = 10 \log_{10} \left(\sum_{i=1}^{n} 10^{SPL_i/10} \right) \text{dB}$$

	Octave band center frequency, Hz							
	63	125	250	500	1k	2k	4k	8k
$SPL_{50'}$ (dB)	67	66	61	63	56	53	48	42
Distance attenuation (dB)	−22	−22	−22	−22	−22	−22	−22	−22
Air absorption attenuation (dB)	−0.001	−0.004	−0.017	−0.07	−0.284	−1.14	−4.55	−18.19
Barrier attenuation* (dB)	−8.0	−10.0	−12.5	−15.0	−17.5	−20.0	−20.0	−20.0
$SPL_{600'}$ (dB)	37.0	34.0	26.5	26.0	16.2	15.0	15.0	15.0**

*Although the presence of the barrier reduces the effect of ground attenuation near the barrier (except where the ground is reflective—Section 4.9.4), the distance involved in this case is such that ground attenuation is the predominant factor.
**It is recommended that the minimum SPL at any frequency be 15 dB, commensurate with a very quiet background.

Now A-weighting relative response from flat linear is as follows (from Fig. 3.13):

	Octave-band center frequency, Hz							
	63	125	250	500	1k	2k	4k	8k
Rel dB	−26.2	−16.1	−8.6	−3.2	0	+1.2	+1.0	−1.1

Numerically adding to $SPL_{600'}$ gives

	Octave-band center frequency, Hz							
	63	125	250	500	1k	2k	4k	8k
$SPL_{600'}$ A-wt.	10.8	17.9	17.9	22.8	16.2	16.2	16.0	13.9

Therefore, SPL $= 10 \log_{10} (10^{10.8/10}$

$$+ 10^{17.9/10} + 10^{17.9/10}$$

$$+ 10^{22.8/10}$$

$$+ 10^{16.2/10} + 10^{16.2/10}$$

$$+ 10^{16.0/10}$$

$$+ 10^{13.9/10}) \text{ dBA}$$

$$= 10 \log_{10} (473.62)$$

$$= 26.8 \text{ dBA}$$

The overall level can also be obtained by adding the individual octave band SPLs two at a time, starting with the lowest levels and continuing until the final level is reached.

Considering the above, using Figure 1.9:

This dBA level due to the cooling tower, impacting on the residential building, is extremely low—undoubtedly below the usual background noise experienced in a residential area. In fact, even without the barrier attenuation, the dBA level would appear to be within acceptable limits.

It should be noted that the power levels from the cooling tower (PWLs), although stated by the manufacturer, were not used in any of the previous calculations. However, PWLs would be required in computing the directivity factor of the cooling tower in each frequency band. From Eq. [4.6b],

$$\text{SPL} = \text{PWL} + 10 \log_{10} Q_\theta$$

$$- 10 \log_{10} (r)^2 - 0.5$$

$$10 \log_{10} Q_\theta = \text{SPL} - \text{PWL} + 10 \log_{10} (r)^2$$

$$+ 0.5$$

At 63 Hz,

$$10 \log_{10} Q_\theta = 67 - 97 + 10 \log_{10} (50)^2$$

$$+ 0.5$$

$$10 \log_{10} Q_\theta = 67 - 97 + 34 + 0.5$$

$$= 4.5.$$

Hence $Q_\theta \simeq 2.8$

Performing similar calculation for other frequencies:

	Octave-band center frequency, Hz							
	63	125	250	500	1k	2k	4k	8k
Q_θ	2.8	2.8	2.2	2.8	2.2	2.8	2.2	2.2

For a noise source radiating hemispherically into the air, the usual value used for $Q_\theta = 2$. Nonhemispherical radiation from the cooling tower has contributed to different values, as indicated above.

4.11. ATTENUATION PROVIDED BY TREES

Heavy dense growths of woods provide a small amount of sound attenuation. To be effective both winter and summer, there should be a reasonable mixture of both deciduous and evergreen trees. Also, the ground cover should be sufficiently dense that sound cannot pass under the absorbent upper portion of the trees. For

Table 4.1. Approximate Noise Reduction (in dB) Provided by Dense Woods. (Mixed Deciduous and Evergreen trees; 20–40 ft height, visibility penetration of 70 to 100 ft).

Octave Frequency Band (Hz)	Excess Attenuation (dB per 100 ft of woods)
63	$\frac{1}{2}$
125	1
250	$1\frac{1}{2}$
500	2
1000	3
2000	4
4000	$4\frac{1}{2}$
8000	5

1. For average 10–20 ft height, use one-half the rate given in the table.
2. For sparse woods of 200–300 ft visibility penetration, use one-half the rate given in the table.

dense woods of several hundred feet depth, the sound may pass over the tops of the trees, in which case the attenuation through the trees should never be considered greater than the excess attenuation over the trees, as determined from the application of Table 4.1.

Table 4.1 gives the approximate excess attenuation of sound through dense woods, where dense woods are taken as having an average "visibility penetration" of about 70–100 ft. Occasional trees and hedges give no significant attenuation. "Visibility penetration" is the average maximum distance in the woods at which some small portions of a large (3-ft square) white cloth can still be seen.

REFERENCES

(1) Beranek, L. L. 1971. *Noise and Vibration Control*. New York: McGraw-Hill.
(2) Lord, H., W. S. Gatley, and H. A. Evensen. 1980. *Noise Control for Engineers*. New York: McGraw-Hill.
(3) Beranek, L. L. 1969. *Noise Reduction*. New York: McGraw-Hill.
(4) Irwin, J. D. and E. R. Graff. 1979. *Industrial Noise and Vibration Control*. Heightstown, NJ: Prentice-Hall.
(5) Reynolds, D. D. 1981. *Engineering Principles of Acoustics—Noise and Vibration Control*. Toronto: Allyn and Bacon.
(6) Crocker, M. J., ed. 1972. *Internoise 72*, October 4–6. Tutorial Papers on Noise Control, Washington, D.C.
(7) Bruel and Kjaer. 1986. *Noise Control—Principles and Practice* 1st edition. Naerum, Denmark.
(8) Miller, L. N. 1981. *Noise Control for Buildings and Manufacturing Plants*. Cambridge, MA: Bolt, Beranek and Newman.
(9) Hassall, J. R. and K. Zaveri. 1979. *Acoustic Noise Measurements*. Naerum, Denmark: Bruel and Kjaer.
(10) Jones, R. S. 1984. *Noise and Vibration Control in Buildings*. New York: McGraw-Hill.
(11) Personal correspondence, Bolt Beranek and Newman, Cambridge, MA, 1988.
(12) Maekawa, Z. 1965. "Noise Reduction by Screens." *Memoir of Faculty of Engineering*, Vol. II, pp. 29–53, Kobe University, Japan.
(13) Kurze and Anderson. 1971. "Sound Attenuation by Barriers." *Applied Acoustics*, **4**, 56–74.
(14) "Environmental Noise Assessment in Land Use Planning," Ontario Ministry of the Environment, Toronto, February 1987.
(15) Barry, T. M. and J. A. Reagan. 1978. "FHWA Highway Traffic Prediction Model." U.S. Federal Highway Administration, FHWA-RD-77-108.
(16) Hajek, J. J. and F. W. Jung. 1982. "Simplified FHWA Noise Prediction Method." Ontario Ministry of Transportation and Communications, AE-82-05.

Chapter 5

ABSORPTION, SILENCERS, ROOM ACOUSTICS, AND TRANSMISSION LOSS

In studying acoustic phenomena in enclosures, it is necessary to distinguish between sound absorbing materials (i.e., materials which absorb some of the energy in an incident sound wave and reduce reflections) and materials which reduce the transmission of sound through the materials. Allied with the absorption of a room is its reverberation characteristics, i.e., the rate at which sound decays in the room.

5.1. ABSORPTION

5.1.1. Introduction

Acoustic absorption materials are used widely in architectural acoustic (e.g., acoustic ceiling tiles) and engineering noise control (e.g., in lining for enclosures or in absorbent mufflers).

Most materials absorb some of the acoustic energy of an incident sound wave falling upon the materials. With some materials, the fraction of the incident acoustic energy absorbed is very small, while with other materials the fraction of incident acoustic energy absorbed approaches one. From the above discussion, it can be seen that the absorption coefficient α is:

$$\alpha = \frac{\text{absorbed acoustic energy}}{\text{incident acoustic energy}}$$

The absorption coefficient α can have values between 0 and 1.

The mechanisms by which energy is dissipated by absorbing materials are:

(a) *Viscous flow*. An effective absorber consists of a series of interconnected pores and voids through which sound waves propagate. During propagation, the par-

ticle velocity associated with the sound wave causes relative motion between the air molecules and the surrounding material. As a result, friction generates heat, which is dissipated into the atmosphere as lost energy from the sound wave.

(b) *Internal friction*. Some absorptive materials have resilient fibrous or porous structures that are compressed or flayed by sound wave propagation. In these structures, dissipation of energy occurs not only from the viscous flow losses, but also from the internal friction of the material itself.

(c) *Panel vibration*. Some increase in low-frequency absorption can often be obtained by mounting the absorption material at a suitable distance from the walls of a room. This is because the energy in the low-frequency incident sound causes the material to vibrate like a panel, and in so doing some energy is removed from the incident sound wave. (For example, drapes should be hung away from a wall, not touching it, if it is desired to increase the low frequency absorption of the drape).

The absorptive characteristics of an acoustical material are determined to a large extent by the pore or void size, interconnections between pores and voids, and material thickness. The acoustical impedance at the material surface and at various frequencies and angles of incidence best describes this relationship. (Recall that the acoustical impedance defines the magnitude and phase relationships between acoustic pressure and particle velocity). If a particular combination of frequency and material structure results in a high impedance, the reflected wave amplitude will be relatively large and the absorption coefficient will be correspondingly small. On the other hand, material with an impedance roughly equal to the impedance of the medium (air) will produce no reflected waves. In other words, when the impedance of the medium and the material of the structure are matched, all of the incident energy is dissipated within the material structure.

Dissipation of acoustic energy in a material requires that the relative velocity between the movement of the molecules in the medium and the material structure be as large as possible.

For a material placed directly over a solid backing, as in Fig. 5.1(a), it can be seen that the particle velocity within the material (the vector sum of the incident and reflected wave particle velocities) is zero at the solid backing. As the t/λ ratio increases, the particle velocity approaches a maximum within the material and is greatest when the material thickness is equal

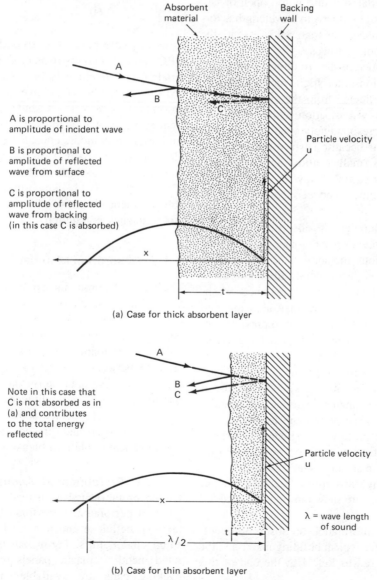

A is proportional to amplitude of incident wave

B is proportional to amplitude of reflected wave from surface

C is proportional to amplitude of reflected wave from backing (in this case C is absorbed)

(a) Case for thick absorbent layer

Note in this case that C is not absorbed as in (a) and contributes to the total energy reflected

(b) Case for thin absorbent layer

Fig. 5.1. Sound waves, material thickness, and particle velocity distribution.

to a quarter wavelength ($\lambda/4$) of the incident sound, at which point maximum relative motion between air molecules and material structure occurs, giving maximum energy dissipation. It can also be seen that the normal incidence absorption of a thin material [Fig. 5.1(b)] increases dramatically if it is separated by an air space of $\lambda/4$ dimension.

If the resistance of the structure to acoustic wave propagation is too low, energy dissipation will be small and most of the energy will be reflected from the backing material into the surroundings. Low resistance can occur because the material structure is too open or because the ratio of thickness to wavelength is too small, in which case the material is essentially transparent to sound waves.

If the resistance of the structure to acoustic-wave propagation is too high, most of the energy will be reflected from the surface of the structure. Energy dissipation within the structure will be correspondingly small. High resistance can occur because surface openings are too few or too small to allow adequate wave propagation (or particle "flow"), or because of insufficient openness between internal voids or pores.

Surface treatment of absorptive material (e.g., thin coating or paint, coarse-weave cloth, a thin impervious membrane, or a perforated cover fabricated from metal, fiber board, or plastic) is used to improve appearance, washability, and resistance to weather, abrasion, and erosion. Treatment is also used to prevent contact between the material (especially glass fiber) and people or animals.

An impervious membrane (e.g., Mylar, usually $\frac{1}{2}$–$1\frac{1}{2}$ mil in thickness) bonded to or stretched over the material surface will enhance low-frequency absorption (due to energy dissipation as a result of vibrating panel effect) and degrade high frequency absorption (due to increase of flow resistance—see Fig. 5.3).

Representative absorptive characteristics of typical materials are shown in Figs. 5.2 and 5.3 and Table 5.1.

For further information on absorption coefficients for other typical building materials, the reader is referred to Ref. 1 at the end of this chapter.

Many manufacturers of acoustical material cite an absorption value which is called the *noise reduction coefficient* (NRC). NRC is a single-number rating which is the arithmetic average of the absorption coefficients at the four intermediate octave-band frequencies of 250, 500, 1000, and 2000 Hz. For instance, considering the $\frac{5}{8}$-in. perforated aluminum-faced mineral-fiber tile in Table 5.1, the noise reduction coefficient for this material is:

$$\text{NRC} = \frac{0.32 + 0.51 + 0.81 + 0.77}{4}$$

$$= 0.60$$

However, care must be exercised when using NRC values for comparison or selection of materials (they should only be used as a guide), because, as can be seen above, their individual absorption characteristics can vary widely with frequency, even though NRC values may be similar. Furthermore, consider 1 lb/ft^3 fiberglass insulation board, $1\frac{1}{2}$ in. thick (Table 5.1). Here the NRC = 0.80, but the absorption at 125 Hz is only 0.11—which may be considerably deficient if control of noise in a room at this frequency is required.

5.1.2. Absorption Material

The types of porous materials in common use are fibrous materials (such as rock wool, mineral wool, fiberglass), perforated loose-texture board, foam rubber (in particular, reticulated polyurethane foam—where there is a common air path between cells in the foam), fabrics, carpets, drapes, upholstery, etc. Commercial porous materials can be divided into three categories: (1) prefabricated acoustical units, (2) acoustical plasters and sprayed-on material, and (3) acoustical (isolation) blankets.

5.1.2.1. Prefabricated Acoustical Units.
Of the prefabricated acoustical units, various types of perforated, unperforated, fissured, or textured cellulose and mineral fiber tiles, fissured cinder units, lay-in panels, and perforated metal or plastic panels with absorbent backing pads are available. They' can be

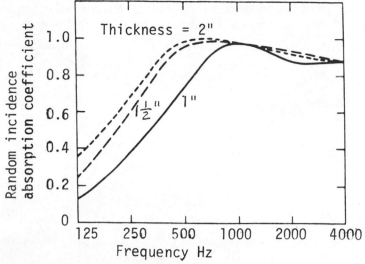

Fig. 5.2(a). Absorption coefficients of porous rigid glass fiber formboard mounted directly on hard backing.

mounted in several ways, according to manufacturers' instructions—for example, cemented to a solid backing, nailed or screwed to a wood framing, or laid in a ceiling suspension system. Special prefabricated units are used on walls or ceiling surfaces in decorative spaced arrangements or in patches, installed either with cement or with simple mechanical attachments.

A form of the prefabricated unit is the *space absorber*. In ceiling areas where it is difficult to carry out usual acoustical treatment because of complication of ceiling configuration, lighting, beams, ducting, etc., these absorbers can be suspended as individual units. Space absorbers can be made from perforated sheet (steel, aluminum, hardboard) or from nylon netting or other textures in the shape of panels, prisms, cubes, spheres, cylinders or single or

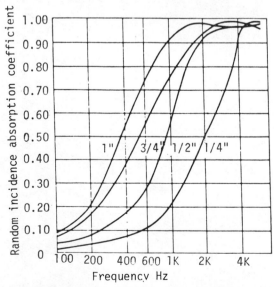

Fig. 5.2(b). Random-incidence sound absorption as a function of frequency for polyurethane foams of various thicknesses.

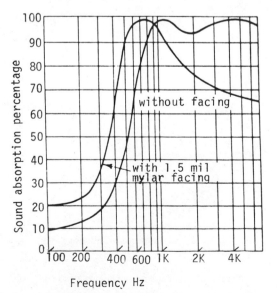

Fig. 5.3. Effect of a 1.5-mil Mylar facing on the random sound absorption of 1-inch polyurethane foam.

Table 5.1.   Absorption Coefficients for a Variety of Common Building Materials and Finishes. (See list of material suppliers in Appendix VI for more details.)

Material	Frequency, Hz					
	125	250	500	1,000	2,000	4,000
Acoustic tiles						
$\frac{5}{8}$-in. perforated aluminum-faced mineral fiber tiles #7 mounting (16-in. air space)	0.32	0.32	0.51	0.81	0.77	0.55
Mat-faced textured fiberglass, $\frac{5}{8}$-in. #7 mounting	0.63	0.90	0.68	0.90	0.96	0.91
Pin perforated fiberglass, $\frac{5}{8}$-in. #7 mounting	0.85	0.86	0.64	0.84	0.90	0.89
Sonocor fiberglass 1-in. #7 mounting	0.60	0.61	0.92	0.83	0.71	0.46
Sonoglas—nubby $\frac{3}{4}$-in. #7 mounting	0.89	0.78	0.83	0.97	0.99	0.91
Perforated $\frac{1}{2}$-in. wood fiber tiles, #7 mounting	0.14	0.30	0.51	0.67	0.68	0.56
Audience and seats						
100% occupied audience—upholstered seats	0.52	0.68	0.85	0.97	0.93	0.85
Unoccupied—average well-upholstered seating areas	0.44	0.60	0.77	0.89	0.82	0.70
Unoccupied leather-covered upholstered seating areas	0.40	0.50	0.58	0.61	0.58	0.50
Wooden pews—100% occupied	0.57	0.61	0.75	0.86	0.91	0.86
Wooden chairs—100% occupied	0.60	0.74	0.88	0.96	0.93	0.85
Wooden chairs—75% occupied	0.46	0.56	0.65	0.75	0.72	0.65
Brick wall—painted	0.01	0.01	0.02	0.02	0.02	0.02
Brick wall—unpainted	0.02	0.02	0.03	0.04	0.05	0.07
Carpets						
Ozonite	0.05	0.13	0.20	0.42	0.48	0.48
Heavy, on concrete	0.02	0.06	0.14	0.37	0.60	0.65
Heavy on 40 oz. hairfelt or foam rubber	0.08	0.24	0.57	0.69	0.71	0.73
Heavy, with impermeable latex backing on 40 oz. hairfelt or foam rubber	0.08	0.27	0.39	0.34	0.48	0.63
Concrete						
Poured/unpainted	0.36	0.44	0.31	0.29	0.39	0.25
Poured/painted	0.10	0.05	0.06	0.07	0.09	0.08
Blocks/unpainted	0.36	0.44	0.31	0.29	0.39	0.25
Blocks/painted	0.10	0.05	0.06	0.07	0.09	0.08
Cork						
Floor tiles ($\frac{3}{4}$-in. thick) glued down	0.08	0.02	0.08	0.19	0.21	0.22
Wall panels (1-in. thick)	0.25	0.55	0.70	0.75	0.75	0.75
Cotton						
Fabric, 14 oz/yd^2 draped to half area	0.07	0.31	0.49	0.81	0.66	0.54
Fabric, draped to $\frac{7}{8}$ area	0.03	0.12	0.15	0.27	0.37	0.42
Curtains						
Light velour, 10 oz/yd^2, hung straight in contact with wall	0.03	0.04	0.11	0.17	0.24	0.35
Medium velour, 14 oz/yd^2 draped to half area	0.07	0.31	0.49	0.75	0.70	0.60
Heavy velour, 18 oz/yd^2 draped to half area	0.14	0.35	0.55	0.72	0.70	0.65
13 oz/yd^2 fiberglass full	0.29	0.38	0.57	0.53	0.72	0.89
Fiberglass insulation blankets and boards on #6 mounting						
0.5 lb/ft^3 PF 334 $\frac{1}{2}$ in.	0.09	0.40	0.32	0.43	0.64	0.70
$\frac{3}{4}$ in.	0.10	0.46	0.44	0.61	0.73	0.75
1 in.	0.10	0.51	0.54	0.73	0.79	0.79
$1\frac{1}{2}$ in.	0.11	0.58	0.69	0.90	0.87	0.83

Table 5.1. (*Continued*)

Material	Frequency, Hz					
	125	250	500	1,000	2,000	4,000
Fiberglass insulation blankets and boards on #6 mounting (*Continued*)						
1 lb/ft^3 PF 336 $\frac{1}{2}$ in.	0.09	0.40	0.36	0.55	0.71	0.75
1 in.	0.10	0.51	0.58	0.86	0.85	0.83
1$\frac{1}{2}$ in.	0.11	0.58	0.73	0.96	0.92	0.87
1.5 lb/ft^2 PF 338 $\frac{1}{2}$ in.	0.09	0.40	0.38	0.63	0.76	0.78
1 in.	0.10	0.51	0.61	0.90	0.89	0.85
AF board AF 110 1 in.	0.16	0.47	0.61	0.83	0.81	0.76
AF 160 1 in.	0.19	0.51	0.63	0.88	0.83	0.78
AF 530 1 in.	0.15	0.56	0.65	0.88	0.88	0.78
AF 545 1 in.	0.17	0.62	0.75	0.91	0.91	0.84
AF 530 1 in. behind burlap	0.15	0.30	0.65	0.85	0.75	0.40
Floors						
Concrete or terrazzo	0.01	0.01	0.01	0.02	0.02	0.02
Linoleum, asphalt, rubber, or cork tile on concrete	0.02	0.03	0.03	0.03	0.03	0.02
Parquet flooring with subfloor on sleepers	0.05	0.03	0.06	0.09	0.10	0.20
Parquet flooring in asphalt on concrete	0.04	0.04	0.07	0.06	0.06	0.07
Varnished wood joist floor	0.15	0.11	0.10	0.07	0.06	0.07
Wood platform with large space beneath	0.40	0.30	0.20	0.17	0.15	0.10
Glass						
Large panes of heavy plate	0.18	0.06	0.04	0.03	0.02	0.02
Windows glazed with up to 32-oz glass	0.35	0.25	0.18	0.12	0.07	0.04
Gypsum wall board						
$\frac{1}{2}$ in. gypsum wall board on 2 × 4 studs at 16 in. on center	0.29	0.10	0.05	0.04	0.07	0.09
$\frac{5}{8}$ in. gypsum wall board on 2 × 4 studs at 16 in. on center	0.20	0.08	0.05	0.05	0.05	0.05
Glazed tile/marble	0.01	0.01	0.01	0.01	0.02	0.02
Panels						
$\frac{3}{8}$-in. plywood paneling	0.28	0.22	0.17	0.09	0.10	0.08
$\frac{3}{4}$-in. pine sheathing	0.10	0.11	0.10	0.08	0.08	0.05
$\frac{1}{4}$-in. plywood, 3-in. air space with absorptive blanket	0.60	0.30	0.10	0.09	0.09	0.09
$\frac{1}{4}$-in. plywood, no absorption in air space	0.13	0.15	0.10	0.09	0.09	0.09
Thin wood (0.2- to 0.4-in.) paneling with air space	0.42	0.21	0.06	0.05	0.04	0.04
Wood paneling $\frac{3}{8}$ in. to $\frac{1}{2}$ in. thick mounted over 2-in. to 4-in. air space	0.30	0.25	0.20	0.17	0.15	0.10
Cedar wall	0.20	0.15	0.15	0.10	0.10	0.10
Plaster						
Plaster, gypsum, or lime, rough finish on lath	0.14	0.10	0.06	0.05	0.04	0.03
Plaster, gypsum, or lime, smooth finish	0.14	0.10	0.06	0.04	0.04	0.03
Acoustic plaster, $\frac{1}{2}$-in. trowel application	0.31	0.32	0.52	0.81	0.88	0.84
Acoustic plaster, 1 in. thick	0.25	0.45	0.78	0.92	0.89	0.87
1-in. damped plaster walls or ceiling	0.14	0.10	0.06	0.05	0.04	0.04
Spray-on materials						
Spray-acoustic, on metal lath, 3$\frac{5}{8}$-in. air space	0.47	0.88	0.87	0.95	0.95	0.92
Spray-acoustic, on gypsum wall board	0.15	0.47	0.88	0.92	0.87	0.88
Limpet asbestos, $\frac{3}{4}$ in.	0.08	0.19	0.70	0.89	0.95	0.85

Table 5.1. (*Continued*)

Material	Frequency, Hz					
	125	250	500	1,000	2,000	4,000
Spray-on materials (*Continued*)						
Limpet asbestos, 1 in.	0.30	0.42	0.74	0.96	0.95	0.96
Limpet asbestos, $\frac{3}{4}$ in. on metal lath	0.41	0.88	0.90	0.88	0.91	0.81
K-13 1 in. on metal lath	0.47	0.90	1.10	1.03	1.05	1.03
1 in., solid backing	0.08	0.29	0.75	0.98	0.93	0.76
$\frac{5}{8}$ in., solid backing	0.05	0.16	0.44	0.79	0.90	0.91
Stage Openings	0.30	0.40	0.50	0.60	0.60	0.50
Water (surface of pool)	0.01	0.01	0.01	0.015	0.02	0.03

double conical shells; they are invariably filled or lined with sound-absorbing materials such as rock wool or glass wool etc.; see Fig. 5.4(a). They are rugged, reasonably impervious to oil mist, smoke, dust, etc. when made from perforated hard material, and are easily installed or removed without interfering with existing fixtures or equipment. These features make space absorbers a particularly suitable treatment for noisy industrial areas, gymnasiums, certain auditoriums and lecture halls, cafeterias, pubs, etc.

An example of the use of hanging sound absorbing baffles is shown in Fig. 5.4(b). A workshop with a high noise level, especially at low frequencies, has to be treated to reduce the noise levels over the entire frequency range. Hanging panel absorbers can be used in a large part of the workshop where the ceiling is free of obstructions. These are very efficient, having two absorbing sides to each panel. A traversing crane makes it impossible to use these in the other part of the workshop. Instead, horizontal absorbent panels are mounted well below the ceiling to obtain improved low-frequency absorption. Except in regions close to a noise source it is possible to reduce the overall noise level by up to 10 dB (see Section 5.2.2).

5.1.2.2. Acoustical Plasters and Sprayed-On Material.

With regard to spray-on absorbers, these include a range of materials formed from mineral or synthetic organic fibers mixed with a binding agent to hold the fibrous material together. During spraying the fibrous material is mixed with the binding material and water to produce a soft, lightweight material of coarse surface texture with high sound-absorbing characteristics. This material may be ap-

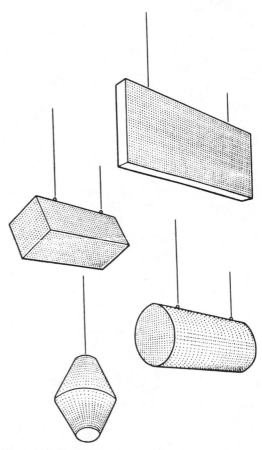

Fig. 5.4(a). Space absorbers can be suspended as individual units from the ceiling. They are used when the area of the room surfaces is not adequate for conventional acoustical treatment.

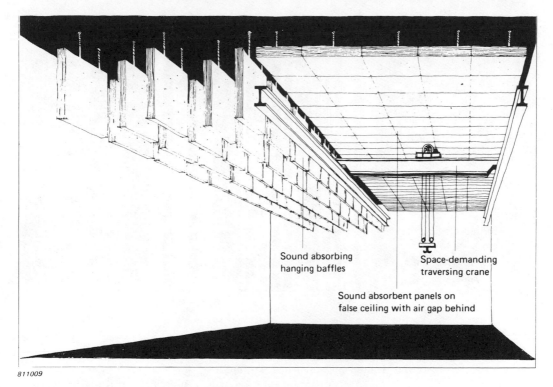

811009

Fig. 5.4(b). Use of sound absorbing baffles.

plied directly to a wide variety of surfaces including wood, concrete, plaster, metal lath, steel, and galvanized metal.

When sprayed onto a solid backing, the materials exhibit good mid- and high-frequency absorption, and when applied to a metal lath with an air space behind it the material also exhibits good low-frequency absorption.

Spray-on materials of this type can be used successfully for broadband absorption in a variety of architectural spaces including schools, gymnasiums, auditoriums, shopping centers, pools, and sports stadiums, and in a variety of industrial applications such as machine shops and power plants. As far as keeping the surface of the material clean, some manufacturers claim that their product can be spray painted without loss of acoustical performance.

5.1.2.3. Isolation Blankets.
Acoustical (isolation) blankets are manufactured from rock wool, glass fibers, wood fibers, hair felt, etc. Generally installed on a wood or metal framing

system, these blankets are used for acoustical purposes in thicknesses varying between 1 in. and 5 in. (25 and 125 mm). Their absorption increases with frequency, particularly at low frequencies. Low-frequency absorption (which is usually difficult to achieve with most absorbent materials) can be enhanced by using 3–5 in. (75–125 mm) thick isolation blanket. Since acoustical blankets do not constitute an aestetically satisfactory finish, they are normally covered with a suitable type of perforated board, wood, slats, textile screening, spaced brick, etc.

In order to improve the appearance of absorption material in a building it is possible to choose from a wide variety of materials to cover large areas of the absorbent. The only requirement is that the uncovered area be sufficiently large so as to allow the absorbent material to do its job satisfactorily. Usually, a perforation ratio of 15% is sufficient for thin panels; the ratio must be greater the thicker the panel becomes (11). Textile, wood strips, expanded metal, and various types of partially open panel

Porous absorbent

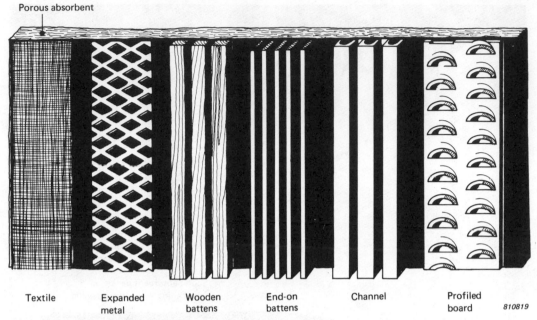

| Textile | Expanded metal | Wooden battens | End-on battens | Channel | Profiled board | 810819 |

Fig. 5.5. Typical materials used as aesthetic facing for sound-absorbent backing.

can be used to create the desired appearance (see Fig. 5.5).

It should be understood that isolation blankets are used primarily to improve the acoustic qualities in the space of concern, and are not to stop transmission of sound from one space to another (see Section 5.4.1 for the latter).

5.1.3. Panel Absorption

As indicated at the beginning of this chapter, panels mounted on a framework absorb sound at low frequencies in fairly narrow ranges, whose frequency depends on the size and thickness of the panel and its distance from the wall. The effective absorption bandwidth depends on the internal damping of the panel: high internal damping (such as by laminates or fiber particleboard) provides a wider frequency range of absorption. For a porous absorbent to be as good at low frequency, it would have to be extremely thick.

Examples of absorption versus frequency for different panel constructions and mountings are shown in Fig. 5.6.

The mechanism of panel absorption is well established (3, 12, 13). The incident sound wave on the plate causes the plate to vibrate at its fundamental frequency (and also higher-or-

der modes); the energy in the sound wave which causes this vibration is dissipated in internal viscous damping in the plate, and the reflected sound wave thus has less energy. In all practical cases, this effect takes place over the frequency range of 40–300 Hz.

If a panel is hung in front of a hard wall at a small distance from it, then the air space acts as a compliant element (spring) giving rise to a resonant system comprised of the panel's lumped mass and the air compliance. Fig. 5.7 gives the resonant frequencies of such air-backed panels as a function of their surface weight and depth of air space (3). The values from Fig. 5.7 are applicable only to simply supported edges. If the plate is clamped at the edges, the fundamental mode of vibration occurs at approximately twice the frequency for the simply supported plate (12, 13).

As in the case of a simply supported panel, a spaced panel absorbs energy through its internal viscous damping. Since its vibration amplitude is largest at resonance, its sound absorption is maximum at this frequency. This absorption can be both further increased in magnitude and extended in its effective frequency range (giving a broader resonant peak) by including a porous sound-absorbing material, such as fiberglass blanket, in the air space

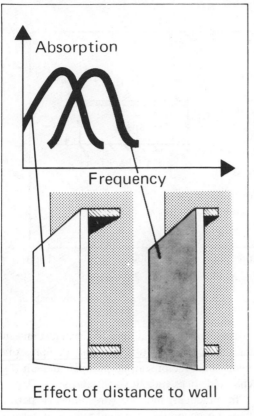

Absorption

Frequency

Effect of distance to wall

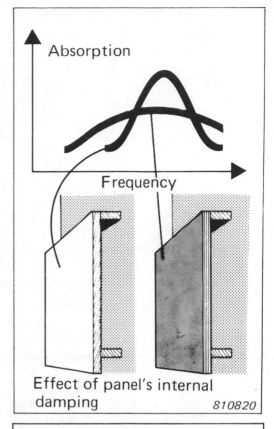

Absorption

Frequency

Effect of panel's internal
damping

810820

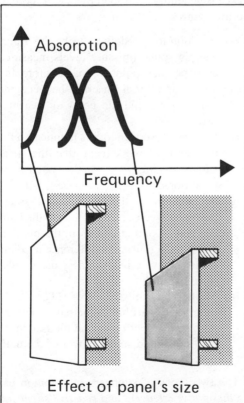

Absorption

Frequency

Effect of panel's size

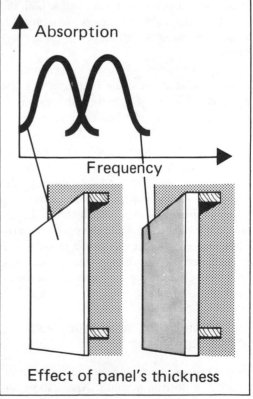

Absorption

Frequency

Effect of panel's thickness

Fig. 5.6. Examples of absorption of different panel configurations and mounting.

119

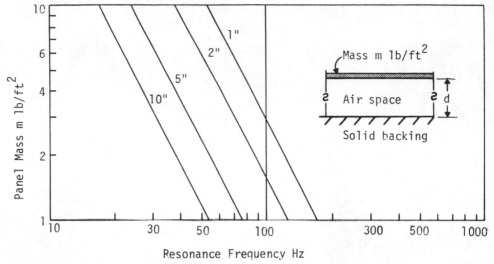

Fig. 5.7. Resonance frequency of sound-absorbing flexible panels as a function of mass and air cavity depth, *d*.

contained by the panel. This effectively introduces more damping into the resonant system. Fig. 5.8 shows the effect of introducing a 1-in.-thick fibrous blanket into the $1\frac{3}{4}$-in. air space contained by a $\frac{1}{8}$-in. plywood sheet. The change is quite significant at low frequencies in the region of the resonant peak where both the magnitude and width of the peak are increased. In contrast, there is little effect at higher frequencies.

A further example of a panel application to absorb a sound at a resonant frequency of a test engine is outlined in Fig. 5.9. In this example, a resonance is excited when the test engine is run at or near its normal speed. At speeds away from the normal running speed the resonance disappears completely. In order to correct this resonance, the walls may be covered in panels fixed to a wooden frame, the dimensions being chosen so that frequency range includes the unwanted resonance. To make the panel effective at frequencies at either side of the resonance, a panel material with high internal damping should be chosen, e.g., a laminate or fiberboard.

5.2. DISSIPATIVE, REACTIVE, AND ACTIVE SILENCERS

Acoustic silencers attenuate sound waves propagated in a flowing medium. Applications include ventilating systems (where fan noise is usually dominant); automotive exhaust and intake systems; gas turbines and compressors; rotary and reciprocating pumps; and air-discharge devices (nozzles and jets).

In practice, muffler (or silencer) performance is generally described by one or more of the following specifications:

(a) Insertion loss is defined to be the difference in sound-pressure levels measured at a specified point in space before and after a muffler is inserted between the noise source and the point of measurement.

(b) Dynamic insertion loss is identical to the definition above except that the measurements are made under rated flow conditions.

(c) Attenuation is the decrease in sound power between a point at inlet to the muffler and a point at exit from the muffler.

(d) Noise reduction is the difference in sound pressure levels measured at the muffler inlet and outlet.

(e) Muffler transmission loss (TL) is 10 times the logarithm to the base 10 of the ratio of incident power on the muffler to the transmitted sound power of the muffler.

The two types of silencers in common use are *dissipative silencers* and *reactive silencers*, both of which will be discussed next.

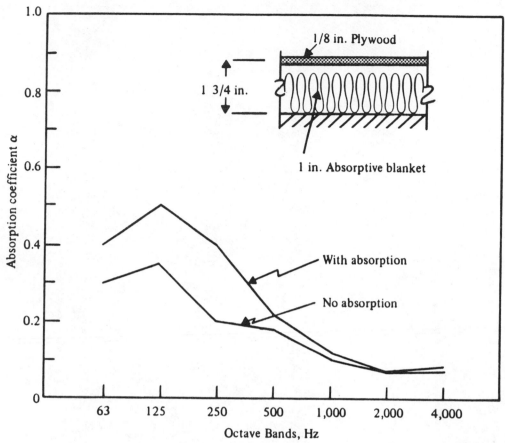

Fig. 5.8. The effect on the absorption coefficient of placing a 1-in.-thick sound-absorptive blanket in the air space ($1\frac{3}{4}$ in. deep) behind a flexible $\frac{1}{8}$-in. plywood panel.

5.2.1. Dissipative Silencers

Dissipative silencers are devices incorporating a flow channel (or several flow channels) lined with absorptive material (see Fig. 5.10).

In general, the noise reduction of this type of muffler occurs through the use of absorption material which lines all or part of each flow channel—where sound waves, infringing upon the absorbing material, create friction with the material fibers and hence, through generation of heat, degrade the energy in the soundwave. The absorption material may require a special protective facing if placed in a duct with a high-velocity or high-temperature gas stream (e.g., mineral wool with perforated stainless steel facing may be required). If the gas stream contains an oil (or other chemical) mist, then thin $1\frac{1}{2}$-mil Mylar can be employed to protect the sound-absorbing material.

Dissipative silencers, lined ducts and bends, and (to a lesser extent) plenums are commonly used devices to control fan and ducting component noise. For instance, Fig. 5.11 shows a typical situation in ventilation where noise reduction is required to reduce fan noise to some desired criterion level. The upper curve shows the amount of noise in a room if the fan were allowed to produce its total noise without the benefit of any noise reduction treatment. The lower curves show the sound levels required to meet, for example, NC-35 or NC-40 criteria (see Chapter 7 on Noise Criteria). The difference between the upper noise curve and the lower criteria curves represents the amount of noise reduction required in the ductwork to meet this condition.

The self-noise (as a result of air flow and turbulence) of dissipative silencers, air valves, dampers, diffusers, and grilles must be taken

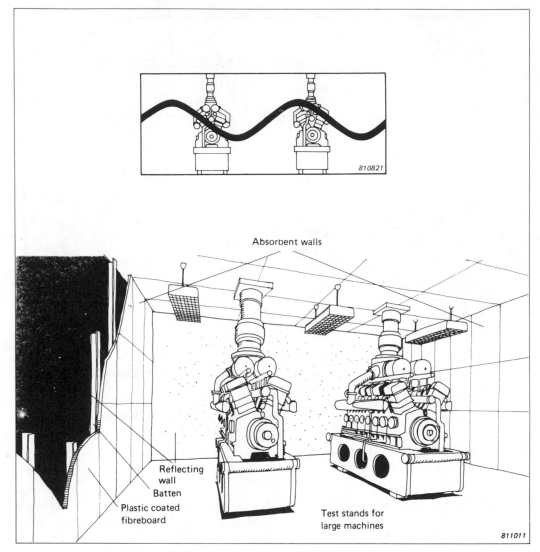

Fig. 5.9. Application of panel resonance to absorb discrete frequency sounds.

into consideration when estimating the noise level(s) in a room due to a ducted ventilation system providing air to that room. The power levels of self-generated noise can usually be obtained from the manufacturers of the various components, or can be estimated from published data in the literature (4, 15, 28).

Duct lining is a commercial material, typically glass-fiber insulation (1.5 lb/ft^3 density) with a treated surface to minimize erosion due to air flow. Recommended thicknesses are 1–2 inches for ducting and 2–4 inches for plenums. Typical random incidence absorption coefficients and representative attenuation data are listed in Table 5.2 (2). By contrast note that duct attenuation in dB per foot of duct length for bare, unlined, straight metal ducts is considerably less than that for the lined duct. (For complete data on unlined ducts, see Refs. 28, 34, and 36).

Right-angle square or rectangular bends are normally fabricated with turning vanes to reduce turbulent noise generated by flow around the bend. A lined bend is not acoustically effective unless the lining extends at least three (preferably four) transverse duct dimensions downstream and upstream from the bend. Representative attenuations for lined round and

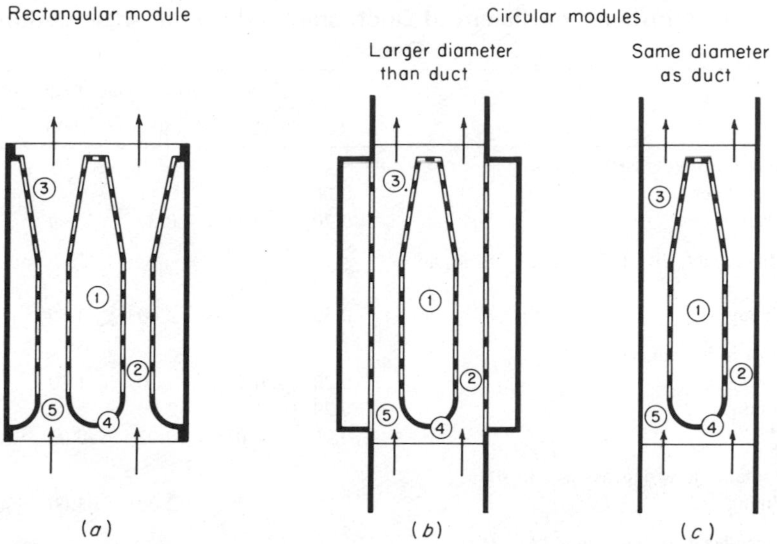

Rectangular module   Circular modules

Larger diameter   Same diameter
than duct   as duct

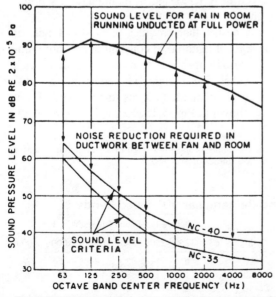

(a)   (b)   (c)

Fig. 5.10. Schematic of some types of dissipative silencers for use in air-handling systems: (a) Rectangular module which may be stacked in parallel to increase the cross-sectional area: (b) circular module in which the diameter of the muffler exceeds the diameter of the inlet and outlet ducts; and (c) circular module for use when the muffler diameter may not exceed the diameter of the duct. (1) Absorbing material; (2) air passage; (3) expansion passage at outlet, the taper is typically linear or exponential; (4) nose, usually solid; (5) bell mouth.

square/rectangular bends are listed in Table 5.3 (2). If the bend and adjacent duct work are unlined, the attenuation is negligible below 500 Hz.

Attenuation data for lined 180° bends are presented in Fig. 5.12 (4, 33). Such devices are often used as ventilation ducts on enclosures around noisy machinery (where air flow is required with minimum noise egress); the insertion loss can be significant, but space and pressure-drop requirements may preclude the use of 180° bends in some applications.

A plenum is a relatively large, lined chamber that is normally used to attenuate acoustic energy generated by a fan or transmitted by ductwork (Fig. 5.13). An approximate expression for the insertion loss of a plenum is given by (4)

$$A = \log_{10} \left| \frac{1}{S_e \left(\dfrac{\cos \theta}{2\pi d^2} + \dfrac{1 - \alpha}{\alpha S_w} \right)} \right| \text{ dB} \quad [5.1]$$

where
α = random-incidence absorption coefficient of plenum lining,
S_e = plenum exit area, ft^2,
S_w = plenum wall area, ft^2,
d = distance between entrance and exit, ft,
θ = the angle of incidence at the exit, i.e., the angle that the direction d makes with the normal to the exit opening, degrees.

Fig. 5.11. Representation of the noise reduction requirement for fan noise to meet typical noise criterion for a room.

Table 5.2. Absorption Coefficient of Duct Liner and Sound Attenuation of Lined Ducts.

Description	Octave-based center frequencies. Hz					
	125	250	500	1000	2000	4000
Random-incidence absorption coefficient:						
1-in-thick duct liner	0.06	0.24	0.47	0.71	0.85	0.97
2-in-thick duct liner	0.20	0.51	0.88	0.99	0.99	0.99
Attenuation of lined duct, dB/ft. Round, 5 to 10 in. flow diameter						
1-in-thick lining	0.30	1.20	2.00	3.00	3.00	2.40
Round, 18 to 24 in flow diameter						
1-in-thick lining	0.22	0.78	1.30	1.80	1.80	1.24
2-in-thick lining	0.29	1.05	2.00	2.80	2.70	2.30
Unlined	0.06	0.03	0.02	0.02	0.02	0.02
Rectangular, smaller flow dimensions 5 to 10 in						
1-in-thick lining	0.30	1.20	2.00	3.00	3.00	2.40
Rectangular, smaller flow dimensions 18 to 24 in						
1-in-thick lining	0.25	0.50	1.20	1.80	1.80	1.40
2-in-thick lining	0.33	0.87	1.80	2.40	2.30	1.90
Unlined	0.12	0.06	0.05	0.04	0.04	0.04

Many companies market duct silencers in various modular sizes. The manufacturers can provide flow rates and silencer insertion loss and self-noise for their products. (Silencers having a large proportion of acoustic absorption material have high values of insertion loss, but also relatively high pressure drop—hence the need for classification according to pressure drop).

Table 5.4 give some typical insertion loss data for duct silencers. For specific applications, manufacturers' data should be consulted. Conventional silencers are approximately 35–75% full of sound-absorption material (placed in parallel baffles inside the silencer), leaving about 65–25% of the volume for air passage through the silencer. In ventilation duct silencers, the baffles are usually about 4 in. thick.

For special applications where an insertion loss greater than that provided by commercial silencers must be obtained (e.g., the baffle must be made thicker to produce large values of the insertion loss in, say, the lower frequency ranges), silencers can be custom designed.* Representative insertion losses for various baffle arrangements (e.g., baffle number and thickness, baffle spacing, and baffle length) are given in Ref. 15 and in manufacturer's literature.

A design that provides parallel channels of absorbent material of constant width, but without visual openness through the unit (i.e., a "blocked line-of-sight") is shown in Fig. 5.14. Typical attenuation curves for this type of unit are shown in Fig. 5.15. Example of typical ab-

*See example of custom design of a commercial silencer in case studies of Sect. 9.2.3.

Table 5.3. Attenuation of Round and Rectangular 90° Lined Bends.*

Description	Dimension D, in	Octave-band center frequency, Hz						
		63	125	250	500	1000	2000	4000
Round duct of diameter D, or rectangular duct with dimension D in plane of bend	5–10	0	0	1	2	3	4	6
	11–20	0	1	2	3	4	6	8
	21–40	1	2	3	4	5	6	8

*Lining thickness equals 0.1 D.
  Lining extends for minimum distance 3 D on both sides of bend. Rectangular bends have turning vanes.

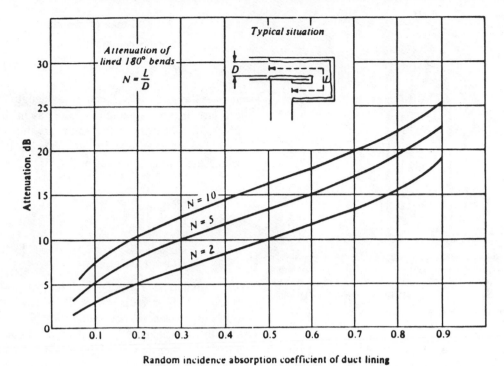

Fig. 5.12. Design curves for lined 180° bends.

sorbing characteristics of different thicknesses and spacing of absorbent material is shown in Fig. 5.16. A further example of the use of baffle silencers is shown in Fig. 5.17. Gas-turbine-powered standby generator sets are used extensively both as emergency power supplies and to complement normal generating plant in peak periods. It can be necessary to quieten the set by anything up to 70 dB, over a wide frequency range. Extensive use is made of absorbent materials in the form of splitters, baffles, and linings on the walls of plenum chambers on both the intake and exhaust sides. An example of the use of silencers and absorption material in ventilation ducting is given in Chapter 9, Section 9.4.

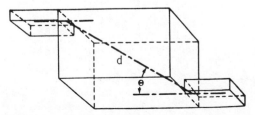

Fig. 5.13. Plenum for reduction of airborne duct noise.

5.2.2. Reactive Silencers

Reactive silencers usually consist of one or more nondissipative elements arranged in parallel or series. These silencers reflect sound waves back toward the sound source, energy being dissipated in the source sound wave through cancellation with the reflected wave. Typical elements include expansion chambers,

Table 5.4. Approximate Insertion Loss (in dB) of Various Lengths of Commercial Duct Silencers.

Octave Frequency Band (Hz)	Low Pressure Drop			High Pressure Drop		
	Muffler Length			Muffler Length		
	3 ft	5 ft	7 ft	3 ft	5 ft	7 ft
63	4	8	10	8	11	13
125	7	12	15	10	14	18
250	9	14	19	15	23	30
500	12	16	20	23	32	40
1000	15	19	22	30	38	44
2000	16	20	24	35	42	48
4000	14	18	22	28	36	42
8000	9	14	18	23	30	36

Refer to manufacturers' literature for more specific data.

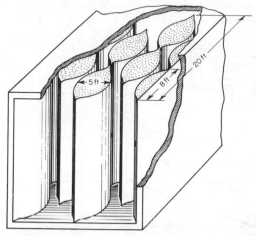

Fig. 5.14. Isometric drawing of a muffler designed especially for jet engine test cells and industrial gas turbine installations (see Fig. 5.17). (Most industrial silencers nowadays are of the parallel baffle type. See Fig. 9.6 in Section 9.2.3.)

side-branch resonators, bends, and perforated tubes (see Fig. 5.18(a) and (b)). They are useful when the noise contains discrete tones because in general they operate on the basis of narrowband attenuation.

5.2.2.1 Expansion Chambers.

An expansion chamber as shown in Fig. 5.19, which is the simplest type of reactive muffler, has a length and an abrupt change in cross-sectional area at each end. Its behavior can be described in terms of two parameters m and kl, where

$$m = \frac{\text{cross-sectional area of chamber}}{\text{cross-sectional area of duct}} = \frac{S_2}{S_1}$$

$$[5.2]$$

and $kl = 2\pi l/\lambda$, where λ is the wavelength of the sound at the temperature of the gas in the muffler. Its transmission loss in the absence of a steady air flow is given by the equation describing the family of curves in Fig. 5.20 (18):

$$L_{TL} = 10 \log_{10} \left[1 + \frac{1}{4} \left(m - \frac{1}{m} \right)^2 \right.$$

$$\left. \cdot \sin^2 kl \right] \text{ dB} \qquad [5.3]$$

We see that L_{TL} is a periodic function in kl, repeating every π radians (180°). This result is valid at frequencies below that for which 0.8 times the wavelength is equal to the largest transverse dimension.

The performance shown by Fig. 5.20 may be interpreted in terms of the wave system existing inside the muffler. At very low frequencies ($kl \to 0$) or whenever the length of the muffler equals $\lambda/2$, λ, $3\lambda/2$, etc., a standing-wave system is produced with enhanced sound pressures at the end walls of the cavity. This has the effect of increasing the characteristic impedance of the duct of cross section S_2 from $\rho c/S_2$ to $m\rho c/S_2$, which from Eq. [5.2] is exactly the value for the inlet and outlet pipes,

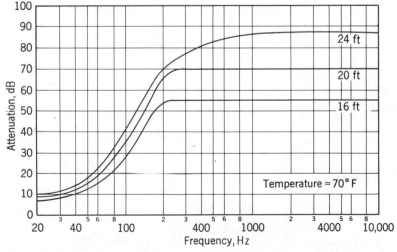

Fig. 5.15. Performance curves for three lengths of absorbing units constructed as shown in Fig. 5.14.

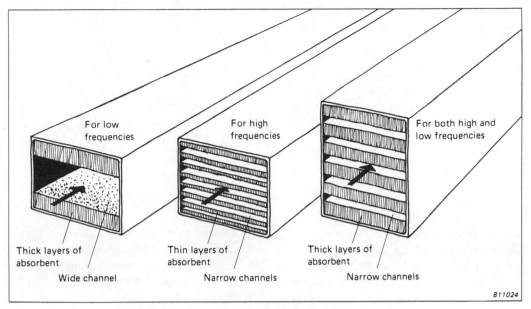

Fig. 5.16. Types of absorbent material used as duct liners and baffles.

i.e., $\rho c / S_1$. Thus, at these resonance frequencies the muffler is a perfect impedance match for the pipe and its L_{TL} is 0 dB. At intermediate frequencies (and wavelengths) reflected waves inside the muffler interfere destructively with incident waves at the inlet pipe—leading to the reflection of sound energy back along the inlet pipe toward the source of sound.

At higher frequencies, where the wavelength is equal to or less than the transverse dimension of the chamber, L_{TL} is dependent on other parameters. For example, when the diameter d_2

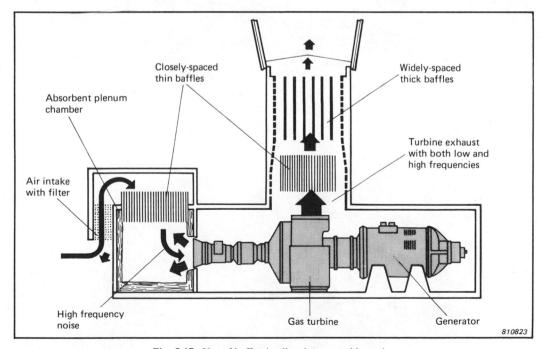

Fig. 5.17. Use of baffles in silencing gas turbine noise.

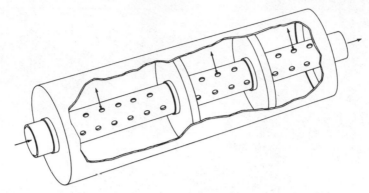

Fig. 5.18(a). Artist's sketch of a low-priced, low-performance, straight-through-type of muffler.

(corresponding to the area S_2) equals λ, a significant reduction of L_{TL} is found for a very large value of m (18). However, for $m = 4$ and $\lambda = d_2$ in a muffler consisting of a series of coupled expansion chambers, the transverse wave motion may be utilized to enhance L_{TL}. Theoretically, there should be a maximum of L_{TL} when the difference in the diameters of expansion chamber and inlet pipe is an odd number of half-wavelengths (19), i.e., when $d_2 - d_1 = \lambda/2, 3\lambda/2, 5\lambda/2$, etc.

The L_{TL} of a simple expansion chamber is

not affected by the presence of a superimposed steady flow (20), at least up to a velocity of 35 m/sec. At very high velocities the flow noise may become large enough to render the muffler ineffective.

Example: Simple Expansion Chamber. Suppose that the noise produced by an engine occurs primarily at 125 Hz. The particular frequency of the noise is governed by the operating speed of the engine. Determine the design parameters of an expansion chamber which,

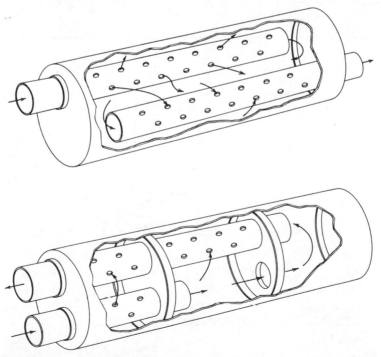

Fig. 5.18(b). Artist's sketch of two practical, high-performance mufflers for trucks or automobiles. (Designs courtesy Nelson Muffler Corporation.)

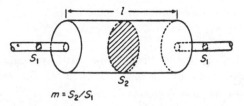

$$m = S_2/S_1$$

Fig. 5.19. Diagram of a single-expansion-chamber muffler.

when attached to the exhaust pipe of the engine, will give a 25-dB transmission loss.

From Fig. 5.20, for 25 dB TL, m is 40. Maximum transmission loss occurs when $kl = n\pi/2$, where $n = 1, 3, 5$, etc. For $kl = \pi/2$, then,

$$l = \frac{\pi}{2k} = \frac{\pi}{2\left(\dfrac{2\pi}{\lambda}\right)} = \frac{\lambda}{4} = \frac{c}{4f} \text{ (ft or m)}.$$

The acoustic velocity is a function of temperature and is given by $c = 49.03\sqrt{R^\circ}$ ft/sec

where R° is the absolute temperature of the gas in degrees Rankine ($= 459.7 + F^\circ$) (4). If the temperature at the point where the expansion chamber is to be inserted is 200°F, then

$$c = 49.03\sqrt{459.7 + 200}$$
$$= 1259 \text{ ft/sec.}$$

Therefore chamber length is

$$l = \frac{1259}{4(125)} = 2.52 \text{ ft.}$$

From Fig. 5.20, note that for $m = 40$, the required transmission loss spans the range

$$1.09 \le kl \le 2.0$$

Therefore, for $l = 2.52$ ft,

$$1.09 \le \frac{2\pi f}{c}(2.52) \le 2.0$$

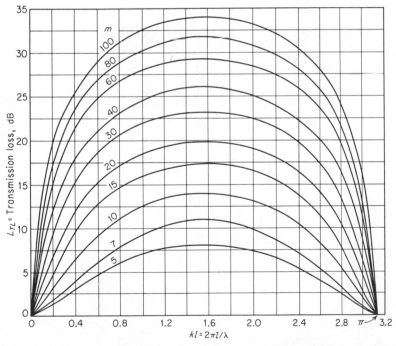

Fig. 5.20. Transmission loss L_{TL} of an expansion chamber of length l and $S_2/S_1 = m$ (see Fig. 5.19). The cross section of the muffler need not be round, but its greatest transverse dimension should be less than 0.8λ (approximately) for the graph to be valid. For values of kl between π and 2π, subtract π and use the scale given along the abscissa. Similarly, for values between 2π and 3π, subtract 2π; etc. Note that when $kl = \pi$, then $l = \lambda/2$; when $kl = 2\pi$, then $l = \lambda$; etc.

or

$$87 \text{ Hz} \leq f \leq 159 \text{ Hz}.$$

Note, therefore that the chamber not only provides a transmission loss of 25 dB at 125 Hz, but that it is effective in the narrow frequency range from 87 Hz to 159 Hz.

The foregoing discussion may be extended to include two or more expansion chambers in series, either with or without external connecting tubes, as shown in Fig. 5.21. It is seen (21) that L_{TL} increases as the number of chambers is increased, although the addition of a third chamber represents only a small improvement over two chambers. From Fig. 5.21(d) we see

that L_{TL} increases as the length of the tube separating the two cavities increases.

5.2.2.2. Projection Tube of Expansion Chambers in Series.

Another design for joining a single expansion chamber to an inlet and outlet tube, or for the connecting tube between two chambers, is to permit the tubes to project into the chamber from one or both ends, as shown in Fig. 5.22. With this design higher L_{TL} over the whole frequency range is obtained than with the same sizes of expansion chambers without projecting pipes.

The pipe resonator cannot provide an exact impedance match $\rho c / S_1$ simultaneously to both the inlet and outlet tubes, except when the orifices of the inlet and outlet tube are at similar

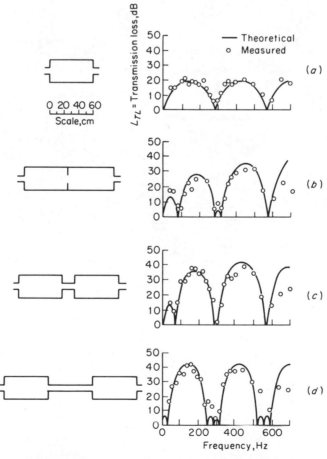

Fig. 5.21. A comparison of theoretical and experimental results for $m = 16$, without flow, for two expansion chambers in series compared to one expansion chamber. The L_{TL} is greater for two chambers than one, and also increases as the length of the connecting tube increases. Note that the passbands, where $L_{TL} \approx 0$, are increased in width when the connecting tube has a length of $\lambda/2$, λ, etc.

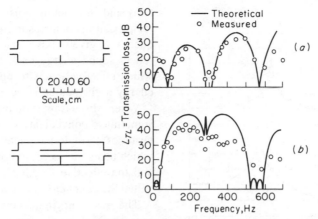

Fig. 5.22. A comparison of theoretical and experimental values of L_{TL} for (a) two chambers in series and (b) two pipe resonators in series, where the pipes terminate in the center of each chamber. The transmission loss of the system remains large at the frequency of the $\lambda/2$ resonance of each chamber, since the connecting-pipe ends are then at $\lambda/4$ points, i.e., points of low impedance compared with that of the inlet duct.

locations in relation to the standing-wave pattern in the resonator. Thus in the absence of the tube connecting the two cavities, L_{TL} would be 0 dB at about 280 Hz, i.e., when $l = \lambda/2$, as shown in Fig. 5.22. With the connecting tube terminating at the center of each cavity as shown in (b), i.e., at the $\lambda/4$ point at this frequency, there is a maximum mismatch of impedance, and L_{TL} remains large (21).

5.2.2.3. Volume-Resonator Muffler. A
volume-resonator muffler (or *cavity resonator*)

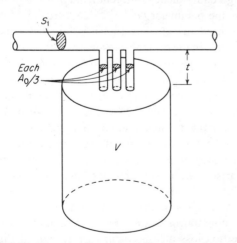

Fig. 5.23. Diagram of a volume (side-branch) resonator. The volume V is joined to a duct of cross-sectional area S_1, by three tubes in parallel, each of length t and area $A_0/3$. The parameters V, A_0, and t are all important in determining the properties of the resonator.

differs from those discussed previously in that no gas flows through the chamber and the muffler connects to the main duct through one or more small openings or tubes (see Fig. 5.23). In the literature, these mufflers are also called *side-branch resonators* or *Helmholtz resonators*.

The resonator consists of an air cavity which acts like a spring and a neck which acts like a mass. Excitation of the cavity volume (by pressure propagation due to the source sound) can cause motion of the air mass in the neck. Such an acoustic mass-spring system has a particular frequency at which it becomes resonant (see Chapter 6 on vibration). At this frequency, energy losses in the system due to frictional and viscous forces acting on the air molecules in and close to the neck reach their maximum, and so absorption characteristics also peak at that frequency. These resonators can be useful as acoustic absorbers at low frequencies where other methods often fail.

Although this device has a sharply defined absorption peak at low frequencies (see Fig. 5.24), this peak can be flattened and extend into the high-frequency range by filling the neck (and to a lesser extent, the cavity) with some form of porous material such as fiberglass. Also, the attenuation of a wide range of frequencies can be accomplished by employing a number of these devices, each tuned to a different frequency within the range.

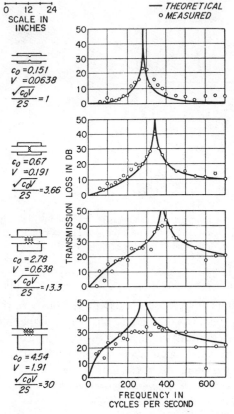

Fig. 5.24. Transmission loss for single-chamber resonators.

The resonant frequency f_0 of the resonator is given by (3, 4)

$$f_0 = \frac{c}{2\pi} \sqrt{\frac{A_0}{t^1 V}} \qquad [5.4]$$

where c is the acoustic velocity (343 m/sec),

A_0 is the cross-sectional area of the neck (m^2),

t^1 is the effective length of the neck = $t + 0.82d$ (m),

d is the diameter of the neck (m),

V is the cavity volume (m^3).

Examples of the attenuation (absorption) due to a volume-resonator muffler are shown in Fig. 5.24.

Side-branch resonators in reactive mufflers have many applications, particularly in reducing noise from automotive engine exhaust sys-

tems and in quieting noisy machines such as reciprocating compressors. At the resonant frequency f_0 given above, the resonator tends to reflect sound energy back toward the source, which effectively acts to oppose and dissipate energy in the source wave.

When discussing volume resonators it is sometimes convenient to write formulas in terms of a single parameter for the openings, which takes into account both the total area A_0 and the effective length t^1. This parameter is called the *conductivity* c_0 and is equal to A_0/t^1.

The resonant frequency can thus be expressed as

$$f_0 = \frac{c}{2\pi} \sqrt{\frac{c_0}{V}} \text{ Hz}$$

The transmission loss of a resonator such as in Fig. 5.23 is given as (18)

$$\text{TL} = 10 \log_{10} \left[1 + \left(\frac{\dfrac{c_0 V}{2S_1}}{\dfrac{f}{f_0} - \dfrac{f_0}{f}} \right)^2 \right].$$

$$[5.5]$$

The resonant frequency is controlled by the ratio $\sqrt{c_0/V}$. The amount of transmission loss at a given frequency different from f_0 is controlled by the parameter $(c_0 V)^{1/2}/2S_1$. Eq. [5.5] can be used to calculate transmission loss curves for use in the design of resonator filters. Curves of this type are presented in Fig. 5.24 (21).

As these curves show, the resonators act as low-pass filters, and by use of sufficiently large values of $(c_0 V)^{1/2}/2S_1$ a broad stop-band can be obtained. Thus filters of this type have great practical usefulness in noise control work. The accuracy of the theory is indicated by the comparison with experimental measurements in Fig. 5.24 (21).

Examples of instrumentation used in determining transmission-loss measurements and insertion-loss measurements are shown in Figs. 5.25 and 5.26 (21).

Example: Design of Resonator Muffler. The noise generated at one end of a com-

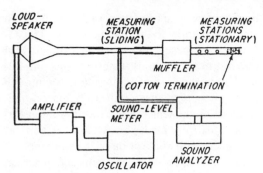

Fig. 5.25. Schematic diagram of experimental apparatus for transmission-loss measurements.

pressed-air duct has a predominant component at 200 Hz. Design a spherical cavity resonator to eliminate the noise at this frequency.

From Eq. [5.4]

$$f_0 = \frac{c}{2\pi} \sqrt{\frac{A_0}{t^1 V}} \text{ Hz}$$

where
f_0 = 200 Hz,
c = 1128 ft/sec.

Hence

$$200 = \frac{1128}{2\pi} \sqrt{\frac{A_0}{t^1 V}}$$

This equation has three unknowns. Hence, assume two of them, solve for the third and see if the result is reasonable. Rearranging and

simplifying yields

$$1.241 = \frac{A_0}{t^1 V}.$$

Assume that the neck (cylindrical) has a radius of 0.5 in. (0.042 ft) and an effective length of 1.0 in. (0.083 ft).
Therefore

$$1.241 = \frac{\pi(0.042)^2}{(0.083)V}$$

or

$$V = 0.054 \text{ ft}^3.$$

The radius of the spherical chamber which is required is

$$r = \left(\frac{3V}{4}\right)^{1/2}.$$

Substituting for V,

$$r = 0.11 \text{ ft or } 1.36 \text{ in.}$$

This spherical cavity radius seems reasonable when compared with the assumed dimensions of the neck.

5.2.2.4. Slotted Concrete Block Resonator.
Individual cavity resonators made of empty clay vessels of different sizes were used in medieval churches in Scandinavia, Russia,

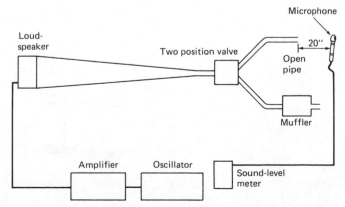

Fig. 5.26. Experimental arrangement for measuring insertion loss due to a muffler.

France, and Yugoslavia. These vases were located in the walls and ceilings of the churches forming resonator absorbers, the frequency depending on the size of the vases and their openings. Absorption could be obtained over a wide range of 80–400 Hz.

Standard concrete hollow blocks, using a regular concrete mixture but with slots cast or cut in the sides of the blocks and communicating with the interior cavity, constitute a contemporary version of the medieval clay vessel resonator. The blocks are available in various sizes, and are closed at the top; when stacked on one another, the slots allow the closed cavities to act as a cavity resonator (see Fig. 5.27). Typical absorption coefficients are also shown in Fig. 5.27.

As can be seen, the block resonators have a pronounced absorption at relatively low frequencies, but absorption can be improved at higher frequencies (and flattened at the pre-

dominant resonant frequency) by adding absorption material to the internal cavity. The advantage of these blocks lies in durability in contrast to soft, porous materials; this durability permits their use in gymnasiums, swimming pools, bowling lanes, industrial plants, mechanical equipment rooms, transportation terminals, barriers for highways, etc. A particularly successful application of the slotted block resonator is in transformer rooms and electrical power stations where the blocks are tuned to absorb the strong 120-Hz noise produced therein.

5.2.2.5. Perforated-Panel Resonators.

Perforated panels, spaced away from a solid backing, provide a widely used practical application of the cavity-resonator principle. They contain a large number of necks, which are the perforations of the panel, thus functioning as an array of cavity resonators. The perforations

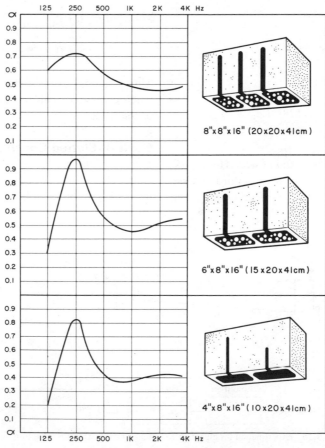

Fig. 5.27. Typical Soundblox units used as individual cavity resonators.

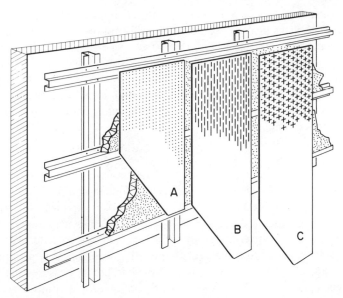

Fig. 5.28. Typical installation of a perforated-panel resonator using various types of perforated facings and with an isolation blanket in the air space: (A) perforated board; (B) slotted hardboard; (C) perforated metal or plastic.

are either circular, slotted, or arranged to form a decorative panel facing. The air space behind the perforation forms the undivided body of the resonator, separated into bays by horizontal and vertical elements of the framing system.

Perforated panel resonators do not provide absorption as selective as do single-cavity resonators, particularly if an absorption blanket is installed in the air space behind the exposed perforation board (see Fig. 5.28).

If the perforated panel is selected with adequate open area (called *transparency*), the absorption blanket increases the overall absorption efficiency by broadening the frequency region in which absorption can be expected (see Fig. 5.29).

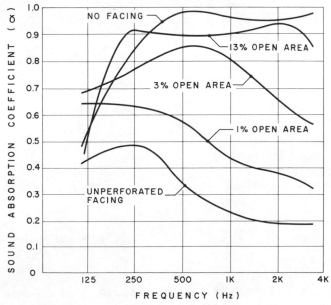

Fig. 5.29. Sound absorption of perforated-panel resonators with isolation blanket in the air space. The open area (sound transparency) of the perforated facing has a considerable effect on the absorption.

5.2.2.6. Slit Resonators. In designing auditoriums, the desired acoustical effect can often be accomplished by using relatively inexpensive absorption blankets along the room surfaces. However, because of their porosity, absorption blankets need protection against abrasion. Thus, the designer can create a decorative protective surface treatment or screen, with openings to allow the sound waves to penetrate into the porous backing. Such protective screens can be wood, metal, or rigid plastic slates, cavity blocks, or bricks with a series of openings such as gaps or exposed slots. The protective screen with openings and absorbent blanket behind are in effect a slit-resonator absorber. The total area between the elements should be at least 35% of the total area of the acoustical treatment (11). Fig. 5.30(a) shows the use of slit-resonator absorbers in the treatment of an auditorium. The details of typical applications of spaced-brick and wooden-slat treatments are shown in Fig. 5.30(b).

It was mentioned earlier in this chapter that the use of absorption material in conjunction with cavity resonators can effectively broaden the absorption values over a wider frequency range. An example of this is shown in Fig. 5.31 where a combination of slotted blocks, fiberglass batts, and perforated metal facing produce high absorption coefficients over the frequency range from 125 Hz to 4000 Hz.

5.2.3. Active Silencers

A rather innovative procedure for controlling low-frequency noise (both tonal and random) in ducts associated with HVAC fans, compressors, and pumps uses electronic sound cancelling technology to reduce and reshape sound traveling through ducts or pipes. An input microphone probe measures the unwanted noise, and an electronic controller calculates the required cancelling waveform almost instantaneously. The calculated signal is directed to matched speakers which are mounted on the outside of the duct. Because the speakers are not mounted in the line of the duct flow, there is no pressure drop or regenerated noise (as

Fig. 5.30(a). View of acoustical treatment with open brick and wood slats in auditorium.

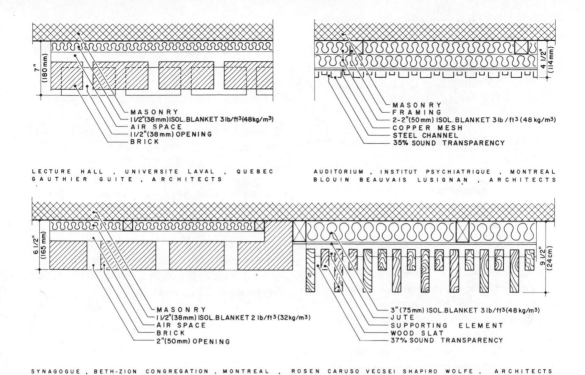

MASONRY
1 1/2"(38mm) ISOL.BLANKET 3 lb/ft³(48kg/m³)
AIR SPACE
1 1/2"(38mm) OPENING
BRICK

LECTURE HALL , UNIVERSITE LAVAL , QUEBEC
GAUTHIER GUITE , ARCHITECTS

MASONRY
FRAMING
2-2"(50mm) ISOL.BLANKET 3 lb/ft³ (48 kg/m³)
COPPER MESH
STEEL CHANNEL
35% SOUND TRANSPARENCY

AUDITORIUM , INSTITUT PSYCHIATRIQUE , MONTREAL
BLOUIN BEAUVAIS LUSIGNAN , ARCHITECTS

MASONRY
1 1/2"(38mm) ISOL.BLANKET 2 lb/ft³ (32kg/m³)
AIR SPACE
BRICK
2"(50mm) OPENING

3"(75mm) ISOL.BLANKET 3 lb/ft³(48kg/m³)
JUTE
SUPPORTING ELEMENT
WOOD SLAT
37% SOUND TRANSPARENCY

SYNAGOGUE , BETH-ZION CONGREGATION , MONTREAL , ROSEN CARUSO VECSEI SHAPIRO WOLFE , ARCHITECTS

Fig. 5.30(b). Details of typical applications of separated-brick, steel-channel, and wooden-slat facings with absorbent backing.

Sound Absorption Coefficients

125 Hz	250 Hz	500 Hz	1,000 Hz	2,000 Hz	4,000 Hz	NCR Range
0.80	0.97	1.02	0.90	0.77	0.71	85–95

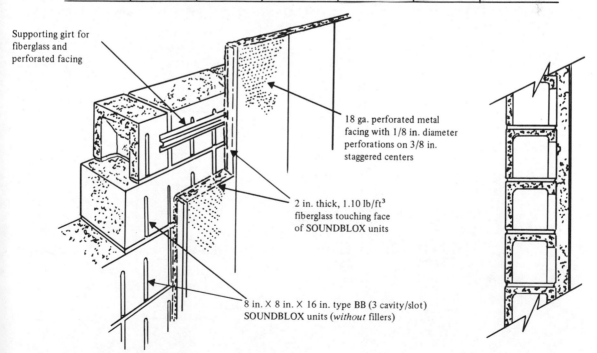

Supporting girt for fiberglass and perforated facing

18 ga. perforated metal facing with 1/8 in. diameter perforations on 3/8 in. staggered centers

2 in. thick, 1.10 lb/ft³ fiberglass touching face of SOUNDBLOX units

8 in. × 8 in. × 16 in. type BB (3 cavity/slot) SOUNDBLOX units (*without* fillers)

Fig. 5.31. Slotted concrete blocks faced with fiberglass and covered with a perforated metal provide good low- and high-frequency absorption characteristics.

could be the case with a silencer inserted in the line). Also, reduced pressure drop helps minimize fan operating costs. For further information on this subject, refer to Ref. 35.

5.3. THE BEHAVIOR OF SOUND IN ROOMS

5.3.1. Reverberation and Decay of Sound in a Room

If a sound source of power is turned on in a quiet room, sound will travel outward from the source and eventually impinge on a surface boundary of the room. Some of the sound energy will be absorbed (or transmitted) by the boundary, and the remainder will be reflected back into the room. After several reflections from scattering objects, the flow of energy will approach diffuseness and the energy density in the room will build up until an equilibrium, or steady-state, condition exists.

If the sound source is now suddenly shut off, the sound in the room will continue to reverberate, with some of the acoustical energy absorbed with each reflection, until the level of intensity decays to a value equal to the ambient intensity level of the room. The time required

for the sound intensity to decay to one millionth of its original value (60 dB) is called the *reverberation time* T_{60} of the room.

Calculation of reverberation time is based on the initial slope of the sound-level decay, which is usually obtained for each $\frac{1}{3}$-octave band of interest. A typical setup for measuring decay is shown in Figure 5.32(a). Random noise* is broadcast into the space, then shut off; the decay of the sound is filtered before it is displayed on a graphic level recorder. Alternatively, the decay of the sound field can be tape recorded for later playback and display on the level recorder; again, the sound-field decay must be filtered in order to obtain a decay curve for each frequency band of interest.

A typical decay curve is shown in Fig. 5.32(b). The decay time t for a decay of 40 dB can be extrapolated to give the decay time T_{60} for 60 dB.

Certain manufacturers (see Ref. 39 of Chapter 3) offer sophisticated precision sound-level meters with a wide range of plug-in modules for sound-level integration, statistical measurements, frequency analysis, and reverberation

---

*An impulsive, wideband sound source, such as from a handgun or a large balloon bursting, can also be used.

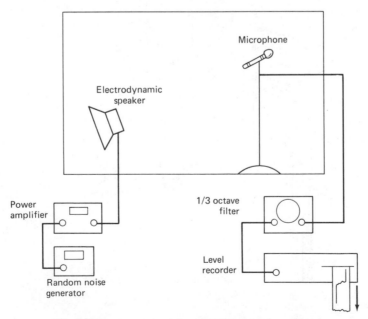

Fig. 5.32(a). Arrangement for measuring decay in a room.

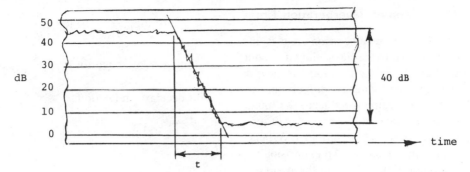

Fig. 5.32(b). Typical room decay curve.

processing. A typical reverberation module allows measurement of T_{60} based on decay intervals 0 to -10 dB, -5 to -25 dB, and -5 to -35 dB. Results can usually be printed out on portable printers.

The reverberation time is the single most important parameter used in assessing the acoustical qualities of a room. If T_{60} is large (i.e., it takes many reflections to absorb the acoustical energy), then there will be a relatively large buildup of sound from a source in that room. A certain amount of buildup is desirable in a lecture room so that the speaker's voice can be more easily heard. However, in a large room, the reflected syllables may be sufficiently retarded that, when combined with the direct speech, they result in making the speaker's words unintelligible. Reverberation is desirable in a music room for aesthetic values because the reflected sounds cause a "blending" from the different musical instruments which is pleasing to listeners.

Recommendation for reverberation times for rooms and halls and for different functions (ranging from lecture rooms to churches) are given in Fig. 5.33 (from Refs. 4, 8, and 9).

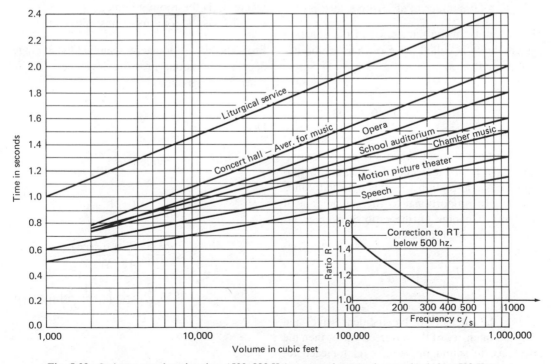

Fig. 5.33. Optimum reverberation time (500–800 Hz) versus volume (and correction below 500 Hz).

5.3.2. Calculation of Reverberation Time

It can be seen that reverberation time is inversely proportional to the absorption of a room (large absorption results in shorter time for sound to decay) and directly proportional to room volume (if volume is reduced, the sound does not have to travel as far between absorptive reflections, resulting in shorter reverberation time for a given decay). An expression for the reverberation time, knowing the room absorption and the room volume, will be developed next.

The room-averaged sound absorption coefficient $\overline{\alpha}$ is defined as:

$$\overline{\alpha} = \frac{\sum\limits_{i=1}^{n} \alpha_i S_i}{S}$$

$$= \frac{\alpha_1 S_1 + \alpha_2 S_2 + \cdots + \alpha_n S_n}{S_1 + S_2 + \cdots + S_n} \quad [5.6]$$

where

S_i = area of ith surface,
α_i = absorption coefficient of ith surface,
S = total surface area of room = ΣS_i,
n = total number of absorptive surfaces.

The quantity in the numerator of the equation given above is called the room absorption denoted by A:

$$A = \sum\limits_{i=1}^{n} \alpha_i S_i = \overline{\alpha} S \; (\text{units } \text{m}^2 \text{ or } \text{ft}^2).$$

Each time a sound wave strikes the boundaries of a room, a fraction $\overline{\alpha}$ of energy is absorbed, and a fraction $1 - \overline{\alpha}$ is reflected. Since pressure is proportional to the square root of intensity (Chapter 1), the ratio of average reflected to incident pressures is

$$\frac{p_r}{p_i} = (1 - \overline{\alpha})^{1/2}.$$

If I_i is the intensity of the incident wave, and I_r is the intensity of the reflected wave, then the decrease in sound pressure level for each reflection is given by:

$$10 \log_{10} \frac{I_i}{I_r} = 10 \log_{10} \frac{p_i^2}{p_r^2}$$

Substituting from above for p_r/p_i

dB decrease/reflection

$$= 10 \log_{10} \frac{p_i^2}{p_i^2 (1 - \overline{\alpha})^{(1/2) \times 2}}$$

$$= 10 \log_{10} \left(\frac{1}{1 - \overline{\alpha}}\right) \quad [5.7]$$

Number of reflections/sec

$$= \frac{\text{distance traveled/sec}}{\text{average distance between reflections}},$$

where the average distance between reflections is defined as the mean free path. The mean free path (mfp) has been determined in Section 4.6, Eq. [4.8]:

$$\text{mfp} = \frac{4V}{S}$$

where V is the room volume,
S is the total room absorptive area.

From the above, it can be seen that the average decrease of sound-pressure level/sec (decay rate) can be given as:

$$\text{Decay rate} = \frac{\text{dB decrease}}{\text{reflection}}$$

$$\times \frac{\text{reflections}}{\text{sec}} \, \text{dB/sec}$$

$$= 10 \log_{10} \left(\frac{1}{1 - \overline{\alpha}}\right)$$

$$\times \frac{c(= 1130 \text{ ft/sec})}{4V/S}$$

$$= 2825 \frac{S}{V} \log_{10} \left(\frac{1}{1 - \overline{\alpha}}\right) \text{dB/sec}.$$

$$[5.8]$$

Therefore, the time to decay 60 dB is:

$$T_{60} = \frac{1}{2825 \dfrac{S}{V} \log_{10}(1 - \overline{\alpha})} \times 60$$

$$= \frac{60}{1230 \dfrac{S}{V}\left[-2.30 \log_{10}(1 - \overline{\alpha}) \right]}$$

$$= \frac{0.049V}{S\left[-2.30 \log_{10}(1 - \overline{\alpha}) \right]} \text{ sec. } \quad [5.9(a)]$$

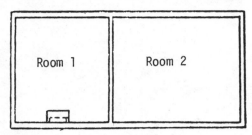

Fig. 5.34. Sound transmission from machine in room.

This equation is referred to as the *Eyring equation* for reverberation time.

Note that the T_{60} equation has been arranged such that, when $\overline{\alpha}$ is small compared to unity (i.e., the room is relatively "hard"), the term $-2.30 \times \log_{10}(1 - \overline{\alpha})$ is approximately equal to $\overline{\alpha}$, and the equation for T_{60} reduces to:

$$T_{60} = \frac{0.049V}{S\overline{\alpha}}, \quad [5.9(b)]$$

the familiar *Sabine equation* for reverberation time.

The above equations are in English units. The comparable metric unit equations are:

$$T_{60} = \frac{0.16V}{S\left[-2.30 \log_{10}(1 - \overline{\alpha}) \right]}$$

$$\text{for any value of } \overline{\alpha} \quad [5.10(a)]$$

$$T_{60} = \frac{0.161V}{S\overline{\alpha}} \text{ for } \overline{\alpha} \ll 1. \quad [5.10(b)]$$

Example: Use of Reverberation Time and Absorption in Treating Rooms. A simple worked example on the use of absorptive material (together with the use of reverberation time measurements) to reduce the noise from a machine in a room (and in an adjoining room) follows. A machine is installed in Room 1 (see Fig. 5.34). The SPL is measured in Room 1 and it is found that, except close to the machine, the sound field is very diffuse and, on average, the SPL in the 1000 Hz octave band is 95 dB (see Fig. 5.35). The SPL in Room 2 in the same band is 75 dB. Can anything be done to reduce the noise in the 1000 Hz band in both rooms?

It is decided that it will be difficult to reduce the noise produced by the machine. Hence it is decided to put some absorption tiles in Room 1 and also in Room 2 to reduce the SPLs.

By using the decay method (producing a steady-state or impulsive noise and using a microphone, amplifier, filter, and level recorder) the reverberation time can be measured against frequency in each room [see Figs. 5.32(a) and (b).]

Assume that the reverberation time T_{60} is measured in both rooms and is given by Fig. 5.36. Then if the volume V of each room is 4000 ft³,

$$A_0 = \frac{0.049V}{T_{60}} = \frac{0.049 \times 4000}{2}$$

$$= 98 \text{ sabin (ft}^2)$$

where A_0 is the room absorption before adding absorptive material.

If absorption tiles are added to Room 1 then the reduction in SPL achieved is given by:

$$\text{SPL}_R = 10 \log_{10} \frac{A_0 + A}{A_0} \text{ dB}, \quad [5.11]$$

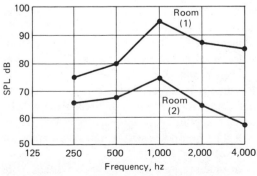

Fig. 5.35. Sound-pressure level in rooms 1 and 2.

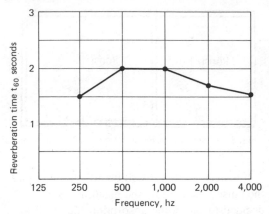

Fig. 5.36. Reverberation times in both rooms.

where A is added absorption in sabin (ft^2). This can be derived from Eq. [4.15]:

$$SPL = PWL + 10 \log_{10} \left[\frac{Q_\theta}{4\pi r^2} + \frac{4}{R} \right]$$

where $SPL_R = SPL_A - SPL_B$, where SPL_A is the level without absorptive material and SPL_B is the level with absorptive material added. (Note that direct-field contribution $Q_\theta/4\pi r^2$ is negligible—field in the rooms is diffuse—and $R = S\alpha/(1 - \alpha) \approx S\alpha$ for small value of α.) If the ceiling of Room 1 is covered with $20 \times 20 = 400$ ft^2 of absorbent tile with $\alpha = 0.9$, then $A = 360$ ft^2 and then the reverberant field SPL will be reduced by $10 \log_{10} [(360 + 98)/98] = 5.6$ dB. If the walls of Room 1 are covered with an additional area of 400 ft^2 of the tile with $\alpha = 0.9$, then $A_{total} = 720$ ft^2 and the reverberant field SPL will be reduced by about $10 \log_{10} [(720 + 98)/98] = 9.2$ dB.

The reverberation time in Room 1 should now be $T_{60} = 0.049(4000)/818 = 0.24$ sec and the SPL should now be 86 dB at 1000 Hz. The SPLs in other frequency bands should be reduced by similar amounts, of course, provided that α has similar values in these frequency bands. Since the sound absorption in Room 1 is now very high, it must be remembered that the SPL in the direct field will be much higher than in the reverberant field. There will now be only a very small reverberant (or diffuse) sound field.

In practice, one cannot hope to attain more than 10 dB reduction in the SPL by adding

sound-absorption material alone. (For instance, Eq. [5.11] indicates that if the total room absorption is doubled, the sound-pressure level in the reverberant field can be reduced 3 dB. A quadrupling of room absorption produces a 6 dB reduction, but a tenfold increase in room absorption is needed for a 10 dB reduction. Hence there is a limit as to how much reduction can be achieved in a reverberant field). The SPL in the 1000-Hz band in Room 2 should also be reduced by 9 dB, and should now be 66 dB. In principle, it should also be possible to reduce the SPL in Room 2 a further 9 dB to 57 dB, by adding 400 ft^2 of absorbent tile to Room 2.

Any further noise reductions in Room 1 would have to be achieved by reducing the noise output of the machine. In Room 2, further noise reduction could be achieved by increasing the transmission loss of the walls (transmission loss is discussed in Section 5.4 of this chapter).

Example: Acoustical Design of Room for Speech. A large rectangular hall with a stage measures 120 ft long, 60 ft wide, and 35 ft high, as shown in Fig. 5.37. The stage is centered on one of the 60 ft walls. The floor has a heavy carpet covering laid on concrete. Unoccupied leather-covered upholstered seating is in place, occupying one-half of the floor area. The side walls and the wall behind the stage are gypsum plaster, smooth finish, on lath. A medium velour 14-oz fabric, draped (i.e., "bunched") to half its area, hangs over the whole of the end wall opposite the stage. The stage surface area can be considered insignificant in comparison with other areas of the room.

If the room is to be used for speaking purposes, what should be the absorption coefficient in the 1000-Hz band of an acoustic-tile ceiling which is planned for the room, as calculated by the Eyring method?

Room volume = $120' \times 60' \times 35' = 252,000$ ft^3 (neglect stage volume).

From Fig. 5.33, optimum reverberation time for speech is

$$T_{60} = 1.02 \text{ sec.}$$

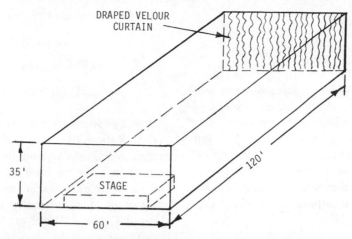

Fig. 5.37. Schematic of hall.

Now from Eq. [5.9(a)], the Eyring equation for reverberation time,

$$T_{60} = \frac{0.049V}{S\left[-2.30 \log_{10}(1 - \bar{\alpha})\right]}.$$

Room surface area is

$$S = (120 \times 60 \times 2) + (120 \times 35 \times 2)$$
$$+ (60 \times 35 \times 2)$$
$$= 27{,}000 \text{ ft}^2.$$

Therefore

$$1.02 = \frac{0.049(252{,}000)}{27{,}000\left[-2.30 \log_{10}(1 - \bar{\alpha})\right]}.$$

Hence

$$-2.30 \log_{10}(1 - \bar{\alpha}) = 0.448$$

and

$$\log_{10}(1 - \bar{\alpha}) = -0.195$$

so that

$$1 - \bar{\alpha} = 0.638$$

and $\bar{\alpha} = 0.362$ (the average room absorption coefficient).

Now from equation [5.6],

$$S_{\text{room}}\bar{\alpha} = S\alpha_{\text{floor}} + S\alpha_{\text{seats}} + S\alpha_{\text{walls}}$$
$$+ S\alpha_{\text{curtain}} + S\alpha_{\text{ceiling}}.$$

From Table 5.1,

$$\alpha_{\text{floor}} = 0.37$$

$$\alpha_{\text{seats}} = 0.61$$

$$\alpha_{\text{walls}} = 0.04$$

$$\alpha_{\text{curtain}} = 0.75$$

Therefore

$$(2700)(0.362)$$
$$= (60 \times 120)(0.37)$$
$$+ (60 \times 120 \times 1/2)(0.61)$$
$$+ (120 \times 35 \times 2)$$
$$+ (60 \times 35)(0.04)$$
$$+ (60 \times 35)(0.75)$$
$$+ (60 \times 120)\alpha_{\text{ceiling}}.$$

In the above, it has been assumed that the curtain covers the one end wall, and therefore the wall contributes no absorption.

Hence

$$9774 = 2664 + 2196 + 420$$
$$+ 1575 + 7200\alpha_{\text{ceiling}}.$$

and

$$\alpha_{\text{ceiling}} = \frac{2919}{7200} = 0.405.$$

Note that, from Table 5.1, this absorption coefficient at 1000 Hz can be obtained by $\frac{1}{2}$ in., $\frac{1}{2}$-lb/ft^3 fiberglass board on #6 mounting.

Of course, a complete solution to the acoustics of this hall would be to make similar calculations for the other five octave-band center frequencies, and then select materials that best fit the preferred absorption coefficient at these frequencies.

5.3.3. Effect of Air Absorption on Decay Rate and Reverberation Time

The pressure of a *plane* sound wave in air decreases with distance as a result of attenuation (due to molecular friction), and this absorption can be considerable at high frequencies.

In a homogeneous medium, it can be shown that pressure at distance x is (2, 4)

$$p = p_0 e^{-mx/2} \qquad [5.12]$$

where p_0 is pressure amplitude at $x = 0$,

m is attenuation coefficient due to molecular friction.

Therefore, the average ratio of reflected to incident pressure now becomes

$$(1 - \overline{\alpha})^{1/2} e^{-mx/2}$$

where x is the distance per reflection (mean free path).

Therefore, using the same method of derivation as in Section 5.3.2, the total average dB decrease in SPL per reflection is (3, 22)

$$10 \log_{10} \left(\frac{1}{1 - \overline{\alpha}} \right) + 4.34 \frac{4V}{S} \frac{\text{dB}}{\text{reflection}}.$$

Hence the average decrease in SPL/sec is

$$-2825 \frac{S}{V} \log_{10} (1 - \overline{\alpha}) + 4900m \frac{\text{dB}}{\text{sec}}$$

$$[5.13]$$

and the reverberation time formula becomes

$$T_{60} = \frac{0.049V}{-2.30S \log_{10} (1 - \overline{\alpha}) + 4mV}$$

$$\text{(English units)}. \qquad [5.14]$$

Values of attenuation coefficient are given in Fig. 5.38 (from Ref. 4) as a function of sound frequency and relative humidity of air at 68°F. At frequencies below 1000 Hz for large rooms, and for all frequencies in smaller rooms, the effect of air on reverberation time can almost always be ignored. However, in large rooms at high frequencies, the absorption of sound by air will be a major factor in determining reverberation time.

Example: Effect of Air Absorption on Reverberation Time. Consider a swimming pool of volume 600,000 ft^3 with a total boundary absorption of 9000 sabins at 4000 Hz and a relative humidity of 30% at 20°C. From Fig. 5.38, $m = 0.0038$. Neglecting air absorption gives

$$T_{60} = \frac{0.049 \times 600,000}{9000} = 3.27 \text{ sec.}$$

Including air absorption gives

$$T_{60} = \frac{0.049V}{S\overline{\alpha} + 4mV}$$

$$T_{60} = \frac{0.049 \times 600,000}{9000 + (0.0038 \times 4 \times 600,000)}$$

$$= 1.69 \text{ sec.}$$

An error of this magnitude could have a large effect on the prediction of speech intelligibility and the design of a public address system for pool areas.

5.3.4. Speech Intelligibility and Reverberation Time

Intelligibility decreases as reverberation time increases, i.e., individual syllables reverberate too long and partially mask following syllables. Also, if reverberation time is too short, loudness level falls sharply—and intelligibility

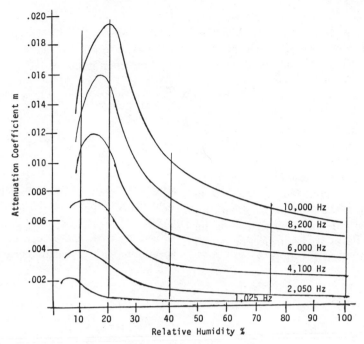

Fig. 5.38. Values of attenuation coefficient m.

suffers. Intelligibility is at a maximum when speaker and listener are close together (small room, short distance in open air) and reverberation time is zero. If the distance between the speaker and the listener increases, then the loudness level falls (and intelligibility decreases), in particular if a disturbing noise is superimposed. By reducing room absorption (increasing reverberation time), the loudness level can be increased and intelligibility will also rise again. It follows that for every size of

room there is an optimum reverberation time for maximum intelligibility, as shown in Fig. 5.39.

5.3.5. Measurement of Random-Incidence Absorption Coefficient

The absorption coefficient α for a material where random-incidence sound is present is calculated from sound decay measurements taken in a reverberation room where the sound

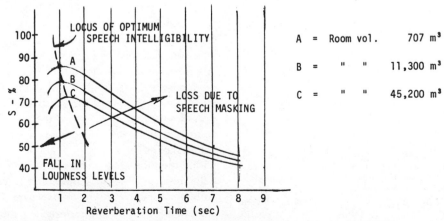

Fig. 5.39. Speech intelligibility versus reverberation time and room volume (8).

field in the room is completely diffuse. The American Society for Testing Materials (ASTM) has set a standard for such measurements (5). The decay of sound is recorded in one-third octave levels of random noise centered at specified frequencies, first with the room empty and then with a test sample of the acoustical material in place. The sample of material is installed exactly as it would be in actual use.

Because the specified test space is a reverberation room with absorption coefficients of the bare room less than 0.05 in all frequency bands of interest (5), the Sabine equation for decay rate can be used. From Section 5.3.2, the average decay rate d is given by

$$d = 2825 \frac{S}{V} \log_{10} \frac{1}{1 - \overline{\alpha}}$$

$$= 1230 \frac{S}{V} - 2.30 \log_{10} (1 - \overline{\alpha})$$

$$= 1230 \frac{S\overline{\alpha}}{V} \text{ dB/sec (for small values of } \overline{\alpha})$$

Therefore

$$S\overline{\alpha} = \text{absorption } A = \frac{Vd}{1230}.$$

But acoustic velocity is

$$c = 1130 \text{ ft/sec}$$

$$= 1230 \times 0.9210$$

or

$$1230 = c/0.9210$$

Therefore*

$$A = \frac{Vd}{c/0.9210} = 0.9210 \frac{Vd}{c}.$$

---

*Note that acoustic velocity is now included in the equation for A to account for variations due to temperature, i.e., $c = \sqrt{kgRT}$, where k = adiabatic exponent, g = acceleration due to gravity, R = gas constant, and T = Rankine temperature.

If

A_1 = absorption of the bare room, and
A_2 = absorption of the room with the specimen in place,

then $A_2 - A_1$ = absorption added to the room by the specimen, i.e.,

$$A_2 - A_1 = \frac{0.9210V}{c} (d_2 - d_1), \quad [5.15]$$

where d_2 and d_1 are the decay rates for the room with the specimen and the bare room, respectively.

The terms d_2 and d_1 can be measured with appropriate instrumentation (see Section 5.3.1), and hence

$$S\alpha \text{ (the absorption for the specimen)}$$

$$= A_2 - A_1.$$

Hence

$$\alpha_{\text{spec}} = \frac{A_2 - A_1}{S_{\text{spec}}} = \frac{\frac{0.9210V}{c} (d_2 - d_1)}{S_{\text{spec}}}.$$

$$[5.16]$$

5.3.6. Normal-Incidence Absorption Coefficient

Normal-incidence absorption coefficients can be obtained for sample materials by use of a standing-wave (or impedance) tube apparatus (Fig. 5.40). The apparatus consists of a long circular tube with a loud speaker (to generate pure tones) at one end, a microphone and microphone probe capable of travelling along the tube axis, and the other end of the tube terminated with a sample of the acoustical material.

As long as the diameter of the tube is small compared with the sound wavelength, transverse modes of vibration cannot be set up within the tube. The plane waves are then partially reflected by the sample and travel back along the tube toward the loudspeaker. This results in a longitudinal interference pattern con-

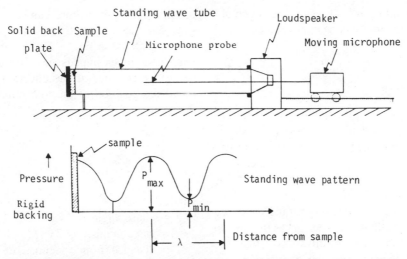

Fig. 5.40. Standing wave apparatus used to determine both the normal incidence absorption coefficient and complex impedance of a sample of material placed at the end of a tube.

sisting of standing waves set up within the tube. The microphone probe is moved along the axis of the tube to measure the variation in sound pressure within the tube. From measurements of the ratio of maximum to minimum sound pressure within the tube, the absorption coefficient of the sample at normal incidence can be calculated. By measuring the distance between the sample and the first standing-wave minimum sound pressure, then the complex acoustic impedance of the specimen can also be calculated.

At any point the incident sound pressure p_i of a plane wave traveling down the tube from the loudspeaker to the specimen can be put in the form

$$p_i = A \cos \omega t,$$

where A is the amplitude and ω the angular frequency ($= 2\pi f$). The reflected wave, at the same point in the tube, has a pressure p_r given in terms of its amplitude B and distance x from the sample surface as

$$p_r = B \cos \omega \left(t - \frac{2x}{c} \right).$$

The total sound pressure p_T at a distance x from the sample is therefore

$$p_T = p_i + p_r$$

$$= A \cos \omega t + B \cos \omega \left(t - \frac{2x}{c} \right).$$

$$[5.17]$$

The maximum pressure will therefore occur when $x = \lambda/2$ and will be equal to

$$p_{max} = (A + B) \cos \omega t,$$

and the minimum pressure will occur when $x = \lambda/4$, to give

$$p_{min} = (A - B) \cos \omega t.$$

The standing-wave ratio n is defined as the ratio of maximum to minimum sound pressures within the tube, so that

$$n = \frac{A + B}{A - B}. \qquad [5.18]$$

Since by definition the reflection coefficient $r = B/A$, we have

$$r = \frac{n - 1}{n + 1}. \qquad [5.19]$$

Using the relation $\alpha = 1 - r^2$, the sound absorption coefficient can be expressed as

$$\alpha = \frac{4n}{(n + 1)^2}. \qquad [5.20]$$

Hence by measuring the standing wave ratio n, one can directly calculate the normal-incidence absorption coefficient. It can be seen from Eq. [5.20] that if $n = \infty$ (i.e., $A = B$) then $\alpha = 0$, and if $n = 1$ (i.e., $B = 0$) then $\alpha = 1$—as would be expected.

It has been found that the normal-incidence absorption coefficient values measured in an impedance tube are generally lower than the random-incidence values obtained from the reverberation-room method (3, 28). At low frequencies the difference is slight, while at high frequencies the tube values are generally 50% lower than those measured in a room.

In general, the length of the tube should exceed $\lambda/4$ and the diameter should not exceed 0.58λ (to ensure only plane, and not transverse, waves in the tube). Hence, a 4-in. (10-cm) diameter, 3-ft-long tube would have a useful range of 90–1800 Hz. In order to make measurements up to, say, 6000 Hz, a smaller-diameter tube would be required to replace the larger-diameter one.

The use of the *complex acoustic impedance*, rather than absorption coefficient, allows a much more rigorous treatment of room reverberation-time analysis. The complex acoustic impedance is defined as the ratio of the sound pressure to the particle velocity at the surface of the material, for a sample of infinite depth, when plane waves strike the face normally, i.e.,

$$Z_0 = \frac{p}{u} \text{ mks Rayls.}$$

This is a complex quantity (4), and it is customary to write Z_0 as

$$Z_0 = r_n + jx_n, \qquad [5.21]$$

where

r_n = normal specific acoustic resistance,
x_n = normal specific acoustic reactance,
$j = \sqrt{-1}$.

The standing-wave tube can be used to measure the parts of the complex acoustic impedance as defined in Eq. [5.21]. It can be shown that the complex impedance of the sample can be written in terms of the reflection coefficient r and the position of the first minimum from the sample as (3)

$$Z_0 = \rho c \left| \frac{1 + re^{j\theta}}{1 - re^{j\theta}} \right|, \qquad [5.22]$$

where

θ = phase angle = $2kd - \pi$,
k = wave number = $2\pi/\lambda$,
d = distance from specimen to first minimum,
r = reflection coefficient given by Eq. [5.19].

Although the use of the complex acoustic impedance involves greater difficulty than does measurement of the absorption coefficient, the impedance technique provides more accurate values of reverberation times in rooms containing uneven distribution of absorption material, even if either pair of opposite walls are highly absorbing or if opposite walls are composed of one very soft and one very hard wall (23).

5.4. SOUND-TRANSMISSION LOSS AND SOUND-TRANSMISSION CLASS

5.4.1. Transmission Loss

The noise-control engineer is often faced with the task of specifying wall materials to be used in the construction of rooms and other enclosures. Structural configurations of materials used for this purpose are classified in terms of their effectiveness for the prevention of sound transmission (a quantity called *sound-transmission loss*, STL).

When a sound wave impinges on a wall, the wall responds to the sound pressure by moving. This in turn excites the air (particle motion) on the other side of the wall, and a sound wave is then propagated in the adjoining air space. The wall motion is frequency dependent, and is influenced by the material used, the type of wall mounting, and the construction of the wall.

A lightweight wall, such as a thin sheet of plywood, glass, or steel, is easily set into motion by an incident sound wave and responds with relatively large amplitude of vibration. Thus it becomes a relatively strong secondary radiator of sound. In contrast, a thick, massive wall, such as an 8-in.-thick solid concrete or masonry wall, resists the vibration with its additional mass and stiffness; it will still vibrate, but at much lower amplitude than the thin surfaces. Because of the lower amplitude of vibration and hence less sound energy transmitted by the wall, the wall is said to have a high transmission loss.

When the wall has natural resonant frequencies, it is more easily set into motion by sound at these frequencies, so the sound-transmission loss decreases. Walls with little internal damping may have many sharp, pronounced resonances, resulting in many low values of transmission loss (see Fig. 5.42), while walls with considerable internal damping have broader, less pronounced resonances and maintain higher values of transmission loss over the frequency spectrum.

The mass law related to transmission loss of walls states that transmission loss of a mass-controlled "limp" wall (i.e., one which has no stiffness) is related to the product of the frequency of the sound and the surface weight of the wall. An appropriate relation for sound transmission loss is (4)

$$STL = 17 \log (f \times m) - 26,$$

where f is the frequency of the sound in Hz, and m is the average surface weight of the wall or panel in lb/ft^2. The slope of the curve of this equation is such that the STL increases about 5 dB for each doubling of either the frequency or the surface weight or the product of the two. This generalization does not apply if the wall is not limp but has stiffness (as most structures do), and it does not apply well in the frequency region where panel resonances occur (such as below 400 Hz). Usually, the wall stiffness and these resonances influence the low-frequency STL of any particular wall.

Airborne sound-transmission loss STL is defined as

$$STL = 10 \log_{10} \frac{1}{\Gamma}, \qquad [5.23]$$

where Γ is the *transmission coefficient*, defined as the fraction of randomly incident intensity that is transmitted by a barrier element, equal to the transmitted intensity/incident intensity.

Measurement of sound-transmission loss (often shortened to "transmission loss") requires mounting the test panel between two reverberation rooms. An example of model apparatus and instrumentation used to measure STL of relatively thin barrier material (e.g., $\frac{1}{2}''$ plywood, $\frac{5}{8}''$ wallboard, etc.) is shown in Fig. 5.41 (see also Ref. 7).

The rooms are isolated from the surroundings and each other, so that all the acoustic energy in the receiving room is transmitted from the source room through the test panel. The source of sound is usually $\frac{1}{3}$-octave bands of random noise in the preferred frequencies extending from 125 to 4000 Hz (7). The $\frac{1}{3}$-octave filter in the microphone system of the receiving room are to reduce the effect of extreme ambient noise that may be present in the room; this filter is tuned to the same center frequency as that generated in the transmitting room. Note that, in actual full-scale measurements, construction and installation of panels are the same as in actual use, and that unwanted transmission paths (for example, at the edges of the panel) are eliminated as much as possible.

Sound transmission loss is calculated from

$$STL = SPL_{TR} - SPL_{RR} + 10 \log_{10} \frac{S}{A}$$

$$[5.24]$$

where $SPL_{TR} - SPL_{RR}$ is the noise reduction between the transmitting room and the receiving room,

S is the area of the panel,

A is the absorption in the receiving room.

This equation is obtained as follows. The sound power impinging on the test panel from the source side is, from Section 4.8,

$$W_1 = \frac{E_1 c}{4} S,$$

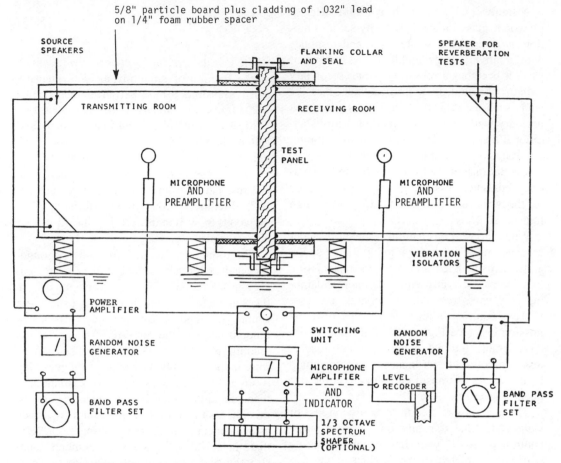

Fig. 5.41. Schematic of apparatus for determining sound-transmission loss of barrier material.

where E_1 is the energy density in the transmitting room,

 S is the panel area,

 c is the acoustic velocity.

The power W_2 transmitted from the wall partition is, by definition,

$$W_2 = W_1 \Gamma = \frac{E_1 cS\Gamma}{4} \qquad [5.25]$$

(Γ is the transmission coefficient). Now, in the receiving room, and applying the ΔS and power-flow concept developed in Section 4.8, the power loss from its reverberant field is

$$\text{power loss}_{RR} = \frac{E_2 c}{4} S_2 \overline{\alpha}, \qquad [5.26]$$

where E_2 is the energy density in the receiving room,

 $\overline{\alpha}$ is the average room absorption coefficient,

 S_2 is the receiving room area.

In the steady state, the power supplied to the reverberant field of the receiving room (from the test wall) must be exactly balanced by the power loss at the walls of the receiving room, i.e., equating Eqs. [5.25] and [5.26],

$$\frac{E_1 cS\Gamma}{4} = \frac{E_2 cS_2 \overline{\alpha}}{4}.$$

Hence

$$\frac{E_1}{E_2} = \frac{S_2 \overline{\alpha}}{\Gamma S} = \frac{A_2}{\Gamma S}$$

Expressed in dB,

$$10 \log_{10} \frac{E_1}{E_2} = \text{STL} + 10 \log_{10} \frac{A_2}{S} \text{ dB.}$$

Since as energy density is proportional to the square of the pressure (Section 4.8),

$$10 \log_{10} \frac{p_1^2}{p_2^2} = \text{STL} + 10 \log_{10} \frac{A_2}{S}$$

and

$$\text{STL} = 20 \log_{10} \frac{p_1}{p_2} - 10 \log_{10} \frac{A_2}{S}$$

$$= 20 \log_{10} \frac{p_1}{p_2} + 10 \log_{10} \frac{S}{A_2}.$$

Now p_1/p_2 is the ratio of pressures in the transmitting room and the receiving room, which can be expressed with reference to the standard of $2 \times 10^{-5} \text{N}/\text{m}^2$, or

$$\text{STL} = 20 \log_{10} \frac{p_1}{p_{\text{ref}}} - 20 \log_{10} \frac{p_2}{p_{\text{ref}}}$$

$$+ 10 \log_{10} \frac{S}{A_2}$$

Therefore

$$\text{STL} = \text{SPL}_{\text{TR}} - \text{SPL}_{\text{RR}} + 10 \log_{10} \frac{S}{A_2}$$

$$[5.27]$$

where

$\text{SPL}_{\text{TR}} - \text{SPL}_{\text{RR}}$ = noise reduction,
S = barrier panel area,
A_2 = absorption in the receiving room.

Figure 5.42 gives a typical panel response curve, showing transmission loss STL plotted as a function of frequency in which the panel response is divided into five frequency regions.

In Region I, there is a 6 dB decrease in STL with doubling of frequency. This region corresponds to the lower frequency response of the panel which moves with the pressure fluctuation to transmit sound. In this region we say that the panel is *stiffness controlled*. The motion is reduced by stiffening the panel, thus causing an increase in transmission loss STL.

In Region II, sympathetic vibrations of the panel, or *panel resonances*, occur. The response of the panel as a function of frequency in this region is erratic because of the many natural (resonant) modes of vibration of the

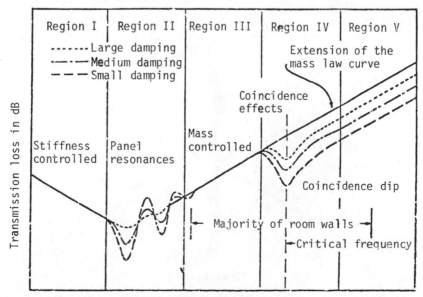

Fig. 5.42. Transmission loss of panels.

panel. In this region the damping properties of the panel are very important. The greater the damping, the less the magnitude of panel resonance and, depending on the frequency, the higher the transmission loss. It is difficult to insulate against the low-frequency sounds of regions I and II, as is indicated by the STL plot in Fig. 5.42.

There is a 6-dB increase in transmission loss with doubling of frequency as we move into Region III. This region is *mass controlled*. The motion of the wall is reduced by increasing its mass, thus resulting in a higher value for transmission loss of the panel. A doubling of mass at any one frequency in this region results in a 6 dB increase in transmission loss (3, 4, 7). A majority of the walls in conventional structures fall into this region.

There is a noticeable dip, called the *coincidence dip*, in the transmission loss curve in Region IV. There is a critical frequency at which longitudinal bending waves will propagate in any panel (4). There is a specific wavelength associated with these bending waves. Whenever the compression and rarefaction of an incident sound wave are such that they coincide (coincidence effect) with the wavelength of bending waves in the panel, then these bending waves will be generated (Fig. 5.43). This increased motion of the panel is transmitted as

sound on the far side of the panel and, hence, the dip in the transmission loss curve.

The coincidence dip is more pronounced in light, stiff panels, such as plywood. Material damping decreases the magnitude of the decrease in transmission loss, as shown in Fig. 5.42. Thicker materials have lower critical frequencies. However, most walls have critical frequencies which fall in the main part of the audible range.

Region V again follows the mass law. However, in this region the increase in transmission loss with doubling of frequency is approximately 10 dB.

Examples of the transmission loss through various materials are shown in Fig. 5.44. All of the curves shown in Fig. 5.44 are derived theoretically and are based on a constant surface weight of 1 lb per sq ft of surface. Fir plywood, being quite stiff in this weight relative to the other materials (the stiffness of a partition increases much more rapidly than its weight as its thickness is increased), shows a plateau beginning at 20 dB. Aluminum, nearly three times as thick as steel in the same weight, is much stiffer than steel and thus starts its plateau at about 30 dB, while steel starts at 40 dB. Limpest of all, lead shows a plateau starting at 56 dB. Of note, therefore, is the fact that the less the stiffness of the partition, the higher will

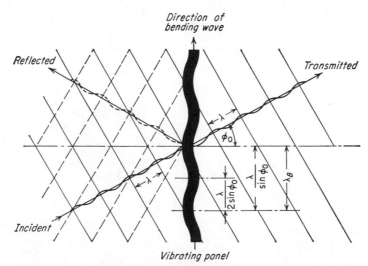

Fig. 5.43. Coincidence effect in panels. Wave coincidence occurs when the projected wavelength of an incident sound wave $\lambda/\sin \phi_0$ equals the bending wavelength λ_B in the panel.

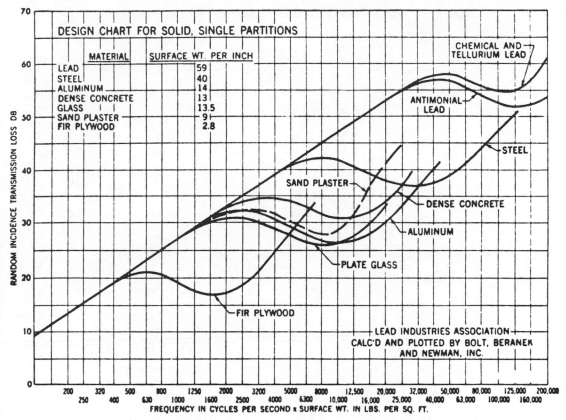

Fig. 5.44. Examples of sound-transmission loss through various building materials.

be the frequency at which the plateau occurs and, in general, the greater will be the overall transmission loss.

Typical sound transmission losses for some common barrier and building materials and constructions are given in Table 5.5. These have been obtained from various literature sources and manufacturers' publications (1, 2, 9, 11, 15, 16, 24, 25, 26, 30, 31, 32).

An example of the effect of baffles, enclosures and vibration isolation in reducing the transmission of noise from a source (a machine) to a receiver (a sound analyzer) is shown in Fig. 5.45. Vibration isolation (and hence reduction of noise transmission due to structure-borne vibration) is discussed in Chapter 6.

5.4.2. Sound-Transmission Class, STC

This is a single number rating used primarily to rate the speech privacy provided by a barrier or other structure. It is obtained from a plot of STL data measured in $\frac{1}{3}$-octave bands from 125 to 4000 Hz. An *STC contour* (7) shown in Figure 5.46 is fitted to the STL data, subject to:

(a) The sum of the deficiencies (i.e., STL data less than contour data) at all frequencies cannot exceed 32 dB;

(b) The STL value at *any* one frequency cannot be more than 8 dB below the STC contour.

The *STC value* is defined as the STL value where the STC contour intersects the 500-Hz line. The sound transmission class number is used to rate the acoustic performance of walls, floors, ceilings, doors, partitions, panels, and other architectural structures.

Relative values of average sound-transmission loss for typical building materials and construction are given in Fig. 5.47 (30). (Note that these are only relative values and are not meant to be definitive; for typical values, see Table 5.5 and/or manufacturers' literature.)

Table 5.5. Transmission Losses of Common Barrier Materials.*

Material	Surface Weight, lb/ft²	Frequency, Hz								
		31	63	125	250	500	1000	2000	4000	8000
Lead										
$\frac{1}{64}$ in. thick	1	—	—	19	20	24	27	33	39	43
$\frac{1}{32}$ in. thick	2	—	—	22	24	29	33	40	43	49
$\frac{1}{8}$ in. thick	$7\frac{1}{2}$	13	19	25	31	37	43	49	53	55
Lead-vinyl	0.5	—	—	11	12	15	20	26	32	37
Lead-vinyl	1.5	—	—	15	17	21	28	33	37	43
Steel										
18-gauge	2.0	3	9	15	19	31	32	35	48	53
16-gauge	2.5	8	14	21	30	34	37	40	47	52
Sheet aluminum										
$\frac{1}{8}$ in. thick	2	1	7	13	19	23	25	26	27	28
$\frac{1}{4}$ in. thick	3.5	7	13	19	23	25	26	27	28	32
Plywood										
$\frac{1}{4}$ in. thick	0.7	—	—	17	15	20	24	28	27	25
$\frac{3}{4}$ in. thick	2.0	—	—	24	22	27	28	25	27	35
Sheet metal laminate, viscoelastic core	2.0	—	—	15	25	28	32	39	42	47
Plexiglass										
$\frac{1}{4}$ in. thick	1.45	—	—	16	17	22	28	33	35	35
$\frac{1}{2}$ in. thick	2.90	—	—	21	23	26	32	32	37	37
1 in. thick	5.80	—	—	25	28	32	32	34	46	46
Glass										
$\frac{1}{8}$ in. thick	1.5	—	—	11	17	23	25	26	27	28
$\frac{1}{4}$ in. thick	3.0	—	—	17	23	25	27	28	29	30
laminated $\frac{1}{2}$ in.	—	—	—	23	31	38	40	47	52	—
Double glass										
$\frac{1}{4} \times \frac{1}{2} \times \frac{1}{4}$ in.	—	—	—	23	24	24	27	28	30	36
$\frac{1}{4} \times 1\frac{1}{2} \times \frac{1}{4}$ in.	—	—	—	23	25	27	31	34	37	42
$\frac{1}{4} \times 6 \times \frac{1}{4}$ in.	—	—	—	25	28	31	37	40	43	47
Thermopane (sealed)										
$\frac{1}{4} \times 1 \times \frac{1}{4}$ in.	—	14	19	23	25	27	31	34	37	42
Concrete										
4 in. thick	48	—	—	29	35	37	43	44	50	55
Concrete block										
6 in. thick	36	—	—	33	34	35	38	46	52	55
Cinderblock, hollow										
6 in. thick	36	26	30	33	33	33	39	45	51	—
Brick, 4 in.	—	—	—	30	36	37	37	37	43	—
Fibertile, filled mineral, $\frac{5}{8}$ in.	—	—	—	30	32	39	43	53	60	—
Door, hardwood, $2\frac{5}{8}$ in.	—	—	—	26	33	40	43	48	51	—
Panels, 16-gauge perforated steel, 4 in. absorbent	—	—	—	25	35	43	48	52	55	56
Curtain, lead-vinyl	1.5	—	—	22	23	25	31	35	42	—
Single-wood-stud wall, 2 in. × 4 in. wood studs, $\frac{5}{8}$ in. wallboard each side, filled mineral wool	—	—	—	20	36	39	46	42	53	—
Staggered wood stud wall, 2 in. × 4 in. studs on 2 in. × 6 in. wood plate, $\frac{5}{8}$ in. wallboard each side, 3 in. fiber in cavity, edges filled and taped	—	—	—	28	36	41	44	47	48	—

*Although most of the above sound-transmission loss values are reasonably accurate (through measurements), some authors indicated that certain STL values have been approximated. Others, even though for presumably the same material, may vary between manufacturers. The reader is encouraged to consult current manufacturers'/suppliers'/technical literature, and to use the values in Table 5.5 as a guide.

5.4.3. Transmission Loss of Composite Walls and the Effect of Flanking

Frequently, a structure will present several transmission paths to arriving sound waves; an example is a wall with glass panels and/or doors (with, possibly, leaks around the doors). The effect of any opening, or viewing panel or flanking, can be estimated from the transmission coefficient and surface area S of each element.

By definition

$$STL = 10 \log_{10} \frac{1}{\Gamma} = 10 \log_{10} \frac{W_1}{W_2}, \quad [5.28]$$

where W_1 or W_2 are the powers in the transmitting and receiving rooms, respectively, each proportional to intensity.

For example, assume a barrier wall with STL = 40 dB and area $S = 10$ ft^2, with a viewing window or panel with STL = 20 dB and area $S = 1.75$ ft^2, and an opening with STL = 0 and area $S = 3$ in^2.

The transmission coefficient Γ_{eq} of an equivalent uniform structure is obtained by equating the sound power for each of the elements, i.e.,

$$\Gamma_{eq} S = \Sigma_i \, \Gamma_i S_i \quad [5.29]$$

were Γ_i and S_i are transmission coefficients and areas of each respective element of the composite wall, and S is the total surface area of the wall.

Hence, for the example, the transmission loss of the wall alone is

$$STL = 10 \log_{10} \frac{1}{\Gamma},$$

i.e.,

$$40 = 10 \log_{10} \frac{1}{\Gamma}$$

and

$$\Gamma = 10^{-4}.$$

Similarly, transmission coefficients for the viewing panel and opening are 10^{-2} and 1.0 respectively. Therefore

$$\Gamma_{eq} \times S = (\Gamma \times S)_{wall} + (\Gamma \times S)_{viewing \ panel}$$
$$+ (\Gamma \times S)_{opening}.$$

Hence

$$\Gamma_{eq} \times 10 = 10^{-4}(8) + 10^{-2}(1.75)$$
$$+ 1.0(0.25),$$

(note that the transmission coefficient for the opening is assumed unity), so that

$$\Gamma_{eq} = 0.0048.$$

The equivalent transmission loss of the composite wall (i.e., original wall + viewing panel and opening) is

$$STL = 10 \log_{10} \frac{1}{\Gamma_{eq}}$$

$$= 10 \log_{10} \frac{1}{0.0048} = 24 \text{ dB}.$$

A graphical method for determining the effect of openings on the transmission loss of a wall with no openings is shown in Fig. 5.48 (29). The effect of airborne flanking on the transmission loss or frequency of a typical wall with a crack under the base $\frac{1}{16}''$ wide by 10' to 12' long, as measured by a laboratory, is shown in Fig. 5.49. In the case of Fig. 5.49, the curves of transmission loss are labelled FSTC, for *field sound-transmission class*, and are obtained from a barrier wall already mounted in a building (such as an apartment), and the transmission loss tests were carried out on the wall in the field (as opposed to having been carried out under controlled conditions in a laboratory) (10). It is easy to see why the STC values, as measured in the field, are often less than laboratory determined STC values, due of course to improper construction procedures which often result in considerable flanking paths.

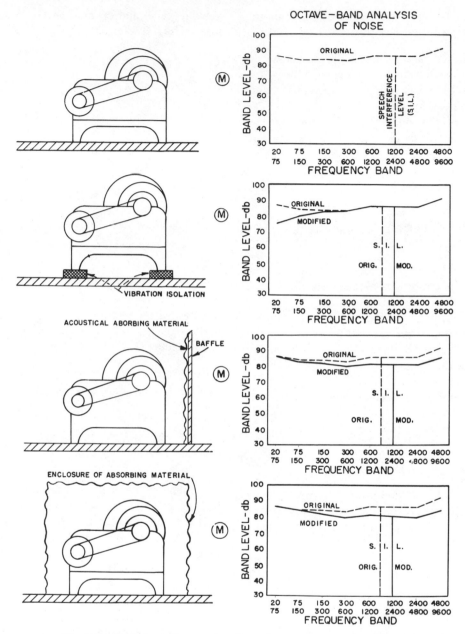

Fig. 5.45. Examples to illustrate the effects of some noise-control measures.

Example: Sound Propagation in a Room and Sound-Transmission Loss through a Barrier. A computing center has several offices separated by barrier walls. One room (dimensions 40′ × 30′ × 12′) has a computer and line-printer installed at the midpoint in the wall at one end of the room. At the other end of the room, on the 30′ wall, is a wall with a glass partition, with a supervisor's office adjoining this wall. The supervisor's office measures 30′ × 15′ × 12′ (Fig. 5.50).

The acoustical center for the combined computer/line-printer is approximately 35 ft from the dividing wall between the computer room and the supervisor's office. The wall consists of 2″ × 4″ wood studs staggered on 2″ × 6″

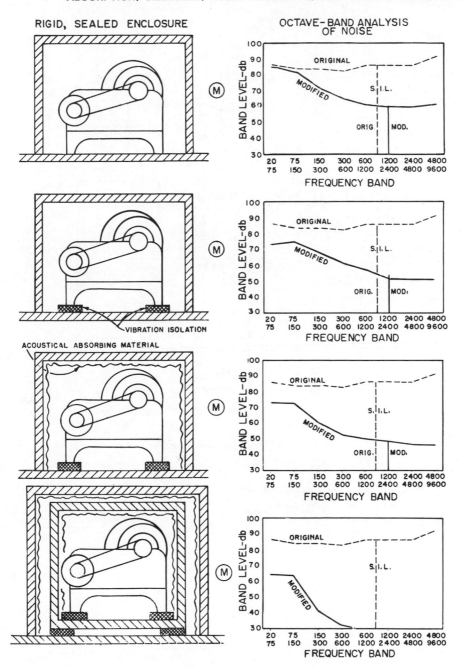

Fig. 5.45. (*Continued*)

wood plates, alternate studs supporting separate walls of $\frac{5}{8}''$ gypsum board, $3''$ fiber in cavity, joints and edges filled and taped. The window partition is double glass, with glass–air space–glass thickness (in inches) of $\frac{1}{4}$–$1\frac{1}{2}$–$\frac{1}{4}$. The window measures $8'$ long $\times$ $4\frac{1}{2}'$ high.

Acoustic panels of 2-in.-thick absorbent material are used to cover the full ceiling area and a pile carpet covers the full floor area in the computer room. The material absorption coefficients in the 1000-Hz octave band are, respectively, 0.9 and 0.5. The untreated room absorption coefficient in the 1000-Hz octave band is 0.05.

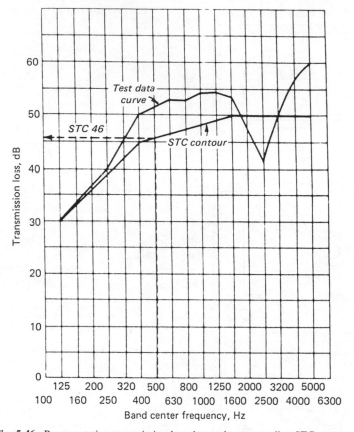

Fig. 5.46. Representative transmission-loss data and corresponding STC contour.

The supervisor's room has an acoustical tile ceiling which is calculated to give an absorption coefficient in the 1000-Hz octave band of 0.85. The remaining surfaces in this room can be considered as untreated, i.e., absorption coefficient = 0.05. Assume the sound distribution in the supervisor's room is diffuse.

The computer manufacturer has submitted PWL (power levels) for the combined computer/line-printer for frequencies from 31 Hz to 8000 Hz. The power level at 1000 Hz is 83 dB re 10^{-12} watts.

What is the sound-pressure level at 1000 Hz in the supervisor's office?

Let the computer room be designated as #1, the supervisor's room #2. For room #1

$$S_{\text{total}}\overline{\alpha_1} = \Sigma\, S_i\alpha_i \quad \text{from Eq. [5.6]}$$

$$[(2 \times 12 \times 40) + (2 \times 40 \times 30)$$

$$+ (2 \times 30 \times 12)]\overline{\alpha_1}$$

$$= (\alpha_{\text{ceil}}S_{\text{ceil}}) + (\alpha_{\text{floor}}S_{\text{floor}})$$

$$+ (\alpha_{\text{untreat}}S_{\text{untreat}})$$

$$4080\overline{\alpha_1} = (0.9)(40 \times 30) + (0.5)$$

$$\cdot (40 \times 30) + (0.05)$$

$$\cdot [(2 \times 12 \times 30)$$

$$+ (2 \times 12 \times 30)]$$

$$4080\overline{\alpha_1} = 1080 + 600 + 84$$

$$\overline{\alpha_1} = 0.432$$

$$R_1 = \frac{S\overline{\alpha_1}}{1 - \overline{\alpha_1}} \quad \text{from Eq. [4.9]}$$

$$= \frac{4080(0.432)}{1 - 0.432} = 3103 \text{ ft}^2.$$

Now

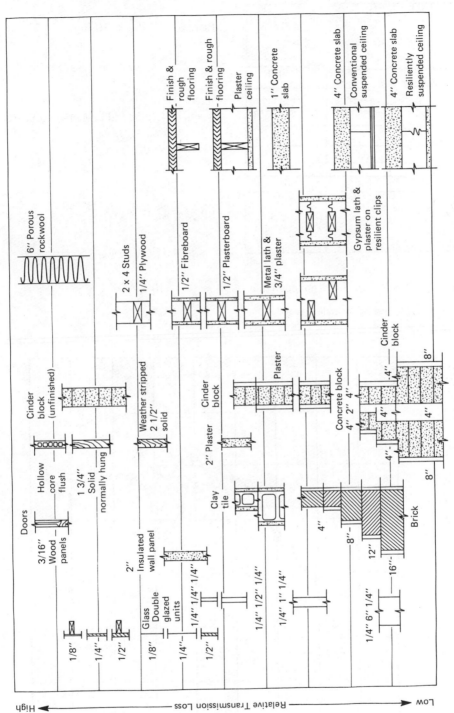

Fig. 5.47. Relative transmission losses for typical construction.

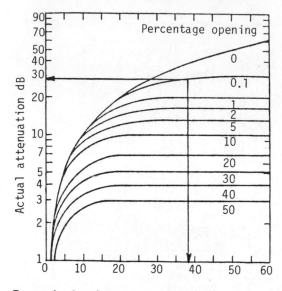

Transmission loss potential (no openings)

Fig. 5.48. Effect of openings on transmission loss.

$$SPL_{1\,wall} = PWL + 10 \log_{10} \left(\frac{Q_\theta}{4\pi r^2} + \frac{4}{R_1} \right)$$

$$+ 10.5\ dB \qquad \text{from Eq. } [4.15b]$$

where

$Q_\theta = 4$ (dihedral corner-section, see Section 4.8),

$PWL = 83$ dB (given).

Therefore,

SPL_1 at wall

$$= 83 + 10 \log_{10} \left[\frac{4}{4\pi(35)^2} + \frac{4}{3103} \right]$$

$$+ 10.5$$

$$= 65.4\ dB.$$

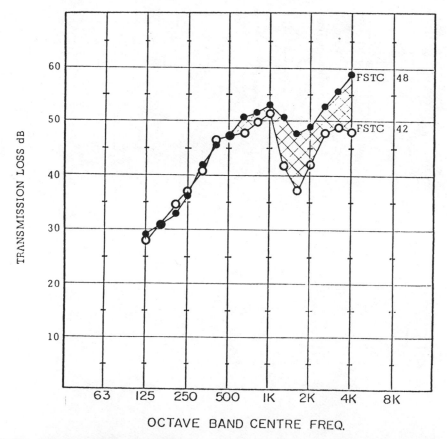

Fig. 5.49. The effect of airborne flanking. *Wall*: 9′0″ × 36′0″ × $\frac{5}{8}$″ gypsum wallboard on each side of insulated metal channel studs, installed in reinforced concrete building. *Crack under base*: Estimated $\frac{1}{16}$″ × 10′ to 12′ long.

COMPUTER/LINE PRINTER

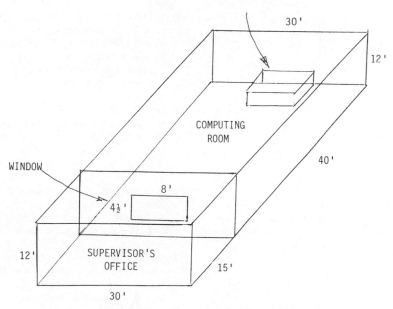

Fig. 5.50. Schematic of computer supervisor rooms.

In order to find the SPL in room #2, it will be necessary to know the sound-transmission loss of the composite wall. Transmission loss STL is defined as $10 \log_{10}(1/\Gamma_{eq})$ from Eq. [5.28], and $\Gamma_{eq}S_{tot} = \Sigma\Gamma_i S_i$ from Eq. [5.29], hence

$$\Gamma_{eq}(30 \times 12) = \Gamma_{wall}S_{wall} + \Gamma_{window}S_{window}.$$

Now

$$STL_{wall} = 44 \text{ dB at } 1000 \text{ Hz} \quad \text{(Table 5.5)}$$
$$STL_{window} = 31 \text{ dB at } 1000 \text{ Hz} \quad \text{(Table 5.5)}.$$

Hence, for the wall

$$44 = 10 \log_{10}\frac{1}{\Gamma_{wall}}; \quad \Gamma_{wall} = 3.98 \times 10^{-5},$$

and for the window,

$$31 = 10 \log_{10}\frac{1}{\Gamma_{window}};$$
$$\Gamma_{window} = 2.94 \times 10^{-4}.$$

Therefore

$$\Gamma_{eq}(360) = (3.98 \times 10^{-5})[(30 \times 12)$$
$$- (8 \times 4.5)] + (7.94 \times 10^{-4})$$
$$\cdot (8 \times 4.5)$$

or

$$\Gamma_{eq} = 1.15 \times 10^{-4}.$$

Hence

$$STL_{total\,wall} = 10 \log_{10}\frac{1}{1.15 \times 10^{-4}}$$
$$= 39.4 \text{ dB}.$$

Now

$$STL_{total\,wall} = SPL_1 - SPL_2 + 10 \log_{10}\frac{S}{A_2}$$

$$(\text{Eq. } [5.27]),$$

where S = barrier wall area = 360 ft². To find A_2 (absorption in room #2),

$$S_2\overline{\alpha}_2 = \alpha_{ceil}S_{ceil} + \alpha_{untreated}S_{untreated},$$

therefore

$$[(2)(12)30 + (2)(30)(15)$$
$$+ (2)(12)(15)]\overline{\alpha_2}$$
$$= (.85)(30)(15) + (.05)[(30)(15)$$
$$+ (2)(12)(30) + (2)(12)(15)]$$

$$980\overline{\alpha_2} = 382.5 + 76.5$$

$$\overline{\alpha_2} = 0.232$$

and hence

$$A_2 = S_2\overline{\alpha_2} = 1980 \times 0.232 = 459 \text{ sabins.}$$

Substituting in the above equation for $STL_{total\,wall}$,

$$39.4 = 65.4 - SPL_2 + 10 \log_{10} \frac{360}{459},$$

from which $SPL_2 = 24.9$ dB, i.e., the sound-pressure level in the supervisor's room in the 1-kHz octave band.

Note that, for a complete analysis, calculations should be made for all frequencies of interest, and not just 1 kHz. This usually covers octave-band center frequencies from 125 Hz to 4 kHz.

REFERENCES

(1) NIOSH. 1975. "Compendium of Materials for Noise Control." Cincinnati, OH: National Institute for Occupational Safety and Health.

(2) Lord, H., W. S. Gatley, and H. A. Evensen. 1980. *Noise Control for Engineers*. New York: McGraw-Hill.

(3) Crocker, M. J., and A. J. Price. 1971. *Noise and Noise Control*. Cleveland, OH: CRC Press.

(4) Beranek, L. L. 1971. *Noise and Vibration Control*. New York: McGraw-Hill Book.

(5) 1966. "Sound Absorption of Acoustical Materials in Reverberation Rooms." ASTM C423-66. Philadelphia, PA: American Society for Testing and Materials ASTM.

(6) ASTM. 1958. "Acoustic Impedance and Absorption by the Tube Method." ASTM C384-58. Philadel-

phia, PA: American Society for Testing and Materials.

(7) ASTM. 1966. "Laboratory Measurement of Airborne Sound Transmission Loss of Building Partitions." ASTM E90-70. Philadelphia, PA: American Society for Testing and Materials.

(8) Knudsen, V. O., and C. M. Harris. 1950. *Acoustical Designing in Architecture*. New York: John Wiley and Sons.

(9) Parkin, P. H., and H. R. Humphreys. 1969. *Acoustics, Noise, and Buildings*. London: Faber and Faber.

(10) ASTM. 1967. "Measurement of Airborne Sound Insulation in Buildings." ASTM E336-67T. Philadelphia, PA: American Society for Testing and Materials.

(11) Doelle, L. L. 1972. *Environmental Acoustics*. New York: McGraw-Hill.

(12) Sabine and Ramer. 1948. "Absorption-frequency Characteristics of Plywood Panels." *J. Acoust. Soc. Amer.*, **20**, 267.

(13) Sacerdote and Gigli. 1951. "Absorption of Sound by Resonant Panels." *J. Acoust. Soc. Amer.*, **23**, 349.

(14) Bolt, R. H. 1947. "On the Design of Perforated Facings for Acoustical Materials." *J. Acoust. Soc. Amer.*, **19**, 917.

(15) Miller, L. N. 1981. "Noise Control for Buildings and Manufacturing Plants." Cambridge, MA: Bolt, Beranek and Newman, Inc.

(16) Irwin, J. D., and L. K. Graf. 1979. *Industrial Noise and Vibration Control*. Englewood Cliffs, N.J.: Prentice-Hall.

(17) Bruel and Kjaer. 1982. *Noise Control: Principles and Practice*. Naerum, Denmark: Bruel and Kjaer.

(18) Davis, D. D., Jr. 1957. "Acoustical Filters and Mufflers." In C. M. Harris (ed.), *Handbook of Noise Control*, Chapter 21. New York: McGraw-Hill.

(19) Mason, W. P. 1946. *Electromechanical Transducers and Wave Filters*. New York: Van Nostrand.

(20) Gosele, K. 1965. "The Damping Behavior of Reaction-Mufflers With Air Flow" (in German). *VDI-Berichte*, **88**, 123–130.

(21) Davis, Stokes, Moore, and Stevens. 1954. "Theoretical and Experimental Investigation of Mufflers with Comments on Engine Exhaust Muffler Design." NACA Report 1192.

(22) Beranek, L. L. 1954. *Acoustics*. New York: McGraw-Hill.

(23) Riva, L., and A. Price. 1974. "Reverberation Time of a Room Having Non-Uniform Distribution of Absorption." University of British Columbia Int. Rept. 7401.

(24) Berendt, R. D., and G. E. Winzer. 1964. "Sound Insulation of Wall, Floor and Door Constructions." Monograph 77 (49 pp). Washington, D.C.: National Bureau of Standards.

(25) Berendt, R. D., G. E. Winzer, and C. B. Burroughs. 1967. "A Guide to Airborne, Impact and Structure-Borne Noise Control in Multi-Family Dwellings." Washington, D.C.: U.S. Dept. of Housing and Urban Development.

(26) Owens-Corning. 1964. "Solutions to Noise Control

Problems in Construction of Houses, Apartments, Motels and Hotels.'' (63 pp.) Toledo, OH: Owens-Corning Fibreglas Corp.

(27) Thumann, A., and R. K. Miller, 1974. *Secrets of Noise Control*. Atlanta, Ga: The Fairmount Press.

(28) ASHRAE. 1987. *HVAC Systems and Applications*. Atlanta, GA: American Society of Heating, Refrigeration and Air-conditioning Engineers.

(29) NIOSH. 1975. *Industrial Noise Control Manual*. Cincinnati, OH: National Institute for Occupational Safety and Health.

(30) Kingsbury, H. F., and G. H. Albright. 1965. *An Introduction to Architectural Acoustics*. Pennsylvania State University.

(31) Warnock, A. C. C. 1985. ''Field Sound Transmission Loss Measurements.'' Building Research Note

No. 232. Ottawa: Division of Building Research, National Research Council of Canada.

(32) Quirt, J. D. ''Measurement of the Sound Transmission Loss of Windows.'' Building Research Note. Ottawa: Division of Building Research, National Research Council of Canada.

(33) Harris, C. M. (ed.), ''Handbook of Noise Control.'' McGraw-Hill Book Co., 1957.

(34) ASHRAE. 1989. *ASHRAE Systems Handbook*. Atlanta, GA: American Society of Heating, Refrigeration and Air-conditioning Engineers.

(35) Erikson, L. J., and M. C. Allie. 1988. ''A Practical System for Active Attenuation in Ducts.'' Bay Village, OH: *Sound and Vibration*, February 1988.

(36) Jones, R. S. 1984. *Noise and Vibration Control in Buildings*. New York: McGraw-Hill.

Chapter 6

VIBRATION AND VIBRATION CONTROL

6.1. INTRODUCTION

Vibration in machinery and structures is caused by the disturbing forces from, for example, reciprocating masses, unbalanced rotating masses, variable fluid pressure in hydraulic systems, turbulent wind loadings, wave action on structures, ground effects on vehicles, etc. Frequently the disturbing forces can be removed or balanced; if not, their harmful effects can be reduced by proper design of the parts and their mountings.

The simplest vibrating system has one degree of freedom, i.e., the system's position can be specified by one number. This is illustrated by a weight suspended by a spring so that it can move only vertically; the weight can be located at any instant by its distance (one number) from a reference point. If the weight could rotate about a vertical axis at the same time that it moves up and down, it would require an additional number to specify its angular position at any instant, and would represent two degrees of freedom. In general, if it requires n numbers to specify the position of a vibrating system, the system has n degrees of freedom.

The single-degree-of-freedom system will be considered here. Many mechanical systems are or can be reduced to this system, and the principles for the simple system can be extended to apply to more complicated ones (1, 16).

In our discussion, two frequencies are involved; one is *the frequency of the disturbing force*. This frequency is equal to or related to the operating speed of the mechanism, which is generally specified and which cannot be altered by the designer. Another frequency, independent of the operating speed, is *the natural frequency of vibration of the system*. This frequency depends on the mass of the system and the stiffness of its support. When the frequency of the disturbing force on a system coincides with its natural frequency, resonance occurs, and the system may vibrate with large amplitudes and develop forces larger than the disturbing forces. These forces may cause annoyance to personnel and failure of parts.

Vibration control consists of removing or balancing the disturbing forces, as mentioned, or of changing the natural frequency of vibration so that it is sufficiently remote from the operating frequency to avoid unsatisfactory operating conditions, or introducing frictional forces that will cause damping, or isolating the vibration by means of elastic supports.

6.2. EQUATION OF MOTION FOR A SYSTEM HAVING A SINGLE DEGREE OF FREEDOM

Consider a spring-supported weight arranged so that the weight can move only in the vertical direction, as shown in Fig. 6.1. The disturbing force is assumed to be $P_0 \sin \omega t$, which is impressed vertically at a time t by an eccentric mass m^1 rotating* with an angular velocity ω.

---

*A convenient way of producing such a force is to rotate two eccentric masses at the same speed but in opposite direction, as shown in the sketch. The masses may be mounted on mating gears. The vertical components of the centrifugal forces equal $P_0 \sin \omega t$. The horizontal components are balanced. This system is used in an experimental vibrator. It will be recalled, however, that such a system as this is used in various forms for offsetting the unbalances that occur in reciprocating slider crank mechanisms (e.g., automobile engines).

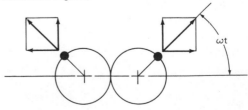

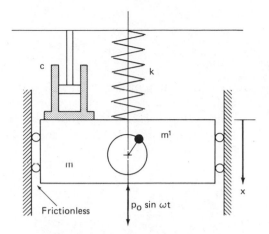

Fig. 6.1. Weight supported by spring.

The spring constant k is equal to the force required to extend the spring a unit distance. The motion of the weight is restrained by damping, which is assumed to vary directly with the velocity of the weight, so that the damping force is the damping constant c multiplied by the velocity of the weight. This is known as *viscous damping*.

Let

m = mass of the weight

x = displacement of weight measured from its position of rest, positive downward

dx/dt = velocity, also represented by $\dot{x}$

d^2x/dt^2 = acceleration, also represented by $\ddot{x}$.

The forces on the weight are:

Spring force = $-kx$

Damping force = $-c\dot{x}$

Impressed force = $P_0 \sin \omega t$.

If the weight is displaced downward, the force on the weight equals $-kx - c\dot{x} + P_0 \sin \omega t$. This force must equal the accelerating force. Therefore

$$m\ddot{x} = -kx - c\dot{x} + P_0 \sin \omega t$$

or

$$m\ddot{x} + c\dot{x} + kx = P_0 \sin \omega t. \qquad [6.1]$$

The above equation is the differential equation of motion for a system having a single degree of freedom.

6.3. FREE VIBRATION WITHOUT DAMPING

For this case, Eq. [6.1] becomes

$$m\ddot{x} + kx = 0$$

$$\ddot{x} = -\frac{k}{m}x \qquad [6.2]$$

The general solution of this equation is (2)

$$x = C_1 \sin t\sqrt{\frac{k}{m}} + C_2 \cos t\sqrt{\frac{k}{m}}. \qquad [6.3]$$

To evaluate the constants C_1 and C_2, let

(a) $x = x_0$ when $t = 0$

(b) $\dot{x} = 0$ when $t = 0$.

By substituting condition (a) in Eq. [6.3],

$$x_0 = C_1 \times 0 + C_2 \times 1$$

or

$$C_2 = x_0.$$

Differentiating the general solution, Eq. [6.3], gives

$$\dot{x} = \sqrt{\frac{k}{m}} C_1 \cos t\sqrt{\frac{k}{m}} - \sqrt{\frac{k}{m}} C_2 \sin t\sqrt{\frac{k}{m}}.$$

By substituting condition (b) in the last equation,

$$0 = \sqrt{\frac{k}{m}} C_1 \times 1 - \sqrt{\frac{k}{m}} C_2 \times 0$$

or

$$C_1 = 0.$$

Therefore Eq. [6.3] becomes

$$x = x_0 \cos t\sqrt{\frac{k}{m}}. \qquad [6.4]$$

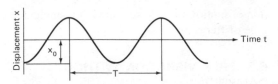

Fig. 6.2. Motion of a single-degree-of-freedom system without damping.

This is the equation of motion for a system having a single degree of freedom vibrating freely without damping. Fig. 6.2 is a graphical representation of the motion.

Inspection of this solution shows that the displacement of the mass is simple harmonic, in which x_0 is a vector rotating at an angular velocity of ω_n, where

$$\omega_n = \sqrt{\frac{k}{m}} = \text{natural frequency,} \quad [6.5]$$

depending on mass and spring constant k. Since there are 2π radians in a cycle of vibration, the frequency in cycles per unit time is

$$f = \frac{\omega}{2\pi}.$$

If ω is in radians/sec, the natural frequency of vibration in cycles per second is

$$f = \frac{\omega_n}{2\pi} = \frac{1}{2\pi}\sqrt{\frac{k}{m}}. \quad [6.5a]$$

In machines, forced vibrations rather than natural vibrations occur. However, natural vibrations are of interest primarily because the natural circular frequency influences the amplitudes of forced vibrations. This will be discussed in a later section.

6.4. FREE VIBRATION WITH DAMPING

For this case, Eq. [6.1] becomes

$$m\ddot{x} + c\dot{x} + kx = 0. \quad [6.6]$$

This may be written

$$\ddot{x} + \frac{c}{m}\dot{x} + \frac{k}{m}x = 0$$

or

$$\frac{d^2x}{dt^2} + \frac{cg}{W}\frac{dx}{dt} + \frac{kg}{W}x = 0. \quad [6.7]$$

Assume a solution of this differential equation in the form $x = Ae^{st}$ (2), where A and s are constants $\neq 0$. Substitute this value into the equation (noting that $dx/dt = Ase^{st}$ and $d^2x/dt^2 = As^2e^{st}$) to obtain

$$As^2 e^{st} + A\frac{cg}{W}se^{st} + A\frac{kg}{W}e^{st} = 0$$

or $\left(s^2 + \frac{cg}{W}s + \frac{kg}{W}\right)e^{st} = 0.$

The desired solution must be such that the above equations be zero. Since e^{st} cannot be zero, then its coefficient must be zero, i.e.,

$$s^2 + \frac{cgs}{W} + \frac{kg}{W} = 0.$$

Using the quadratic formula, the two solutions of s are

$$s = \frac{-\dfrac{cg}{W} \pm \sqrt{\left(\dfrac{cg}{W}\right)^2 - 4\dfrac{kg}{W}}}{2}.$$

The general solution is of the form

$$x = A\exp\left(\left[\frac{-cg}{W} + \sqrt{\left(\frac{cg}{W}\right)^2 - 4\frac{kg}{W}}\right]t\right)$$

$$+ B\exp\left(\left[\frac{-cg}{W} - \sqrt{\left(\frac{cg}{W}\right)^2 - 4\frac{kg}{W}}\right]t\right).$$

The radical may be real, imaginary, or zero depending on the magnitude of the damping coefficient c. The value of c which makes the radical zero is called the *critical damping coefficient* c_c. Its value is obtained by equating the radical to zero and is

$$c_c = \frac{2W}{g}\sqrt{\frac{kg}{W}} = \frac{2W}{g}\omega_n.$$

Note that ω_n is the undamped natural frequency of the system. The ratio of the damping coefficient c in any system to the critical damping coefficient c_c is called the *damping factor d* (often designated as ξ). Its use simplifies the analysis of the problem.

Multiply cg/W by c_c/c_c and substitute $\xi = c/c_c$ and $c_c = (2W/g)\omega_n$ to get

$$\frac{cg}{2W} = \frac{c}{c_c}\frac{g}{2W}c_c = \xi\frac{g}{2W}\frac{2W}{g}\omega_n = \xi\omega_n.$$

The solution for x obtained above may be written

$$x = A\exp\left(\left[-\xi\omega_n + \sqrt{\xi^2\omega_n^2 - \omega_n^2}\right]t\right)$$

$$+ B\exp\left(\left[-\xi\omega_n - \sqrt{\xi^2\omega_n^2 - \omega_n^2}\right]t\right)$$

$$= A\exp\left[(-\xi + \sqrt{\xi^2 - 1})\omega_n t\right]$$

$$+ B\exp\left[(-\xi - \sqrt{\xi^2 - 1})\omega_n t\right].$$

$$[6.8]$$

Three cases arise, depending on the value of ξ.

Case A. Large damping ($\xi > 1$) means that the radical is real and less than 0. Hence both exponents are negative. The value of x is then equal to the sum of two decreasing exponentials. When $t = 0$, $x = Ae^0 + Be^0 = A + B$. The plot of either part of the solution of such a periodic motion indicates the frictional resistance is so large that the weight, after its initial displacement, creeps back to equilibrium without vibrating. Since there is no period to the motion, it is called *aperiodic* (Fig. 6.3).

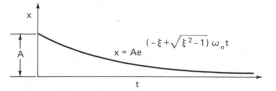

Fig. 6.3. Decay of motion for large damping.

Case B. Light damping ($\xi < 1$) means that the radical is imaginary. Using $i = \sqrt{-1}$, the solution may be rewritten

$$x = A\exp\left(\left[-\xi + i\sqrt{1 - \xi^2}\right]\omega_n t\right)$$

$$+ B\exp\left(\left[-\xi - i\sqrt{1 - \xi^2}\right]\omega_n t\right)$$

$$= e^{-\xi\omega_n t}\left[Ae^{i\sqrt{1-\xi^2}\omega_n t} + Be^{-i\sqrt{1-\xi^2}\omega_n t}\right].$$

$$[6.9]$$

The term in brackets may be expressed in terms of a sine or cosine function. When this is done, we obtain

$$x = X_0 e^{-\xi\omega_n t}\sin\left(\sqrt{1 - \xi^2}\,\omega_n t + \phi\right)$$

where $X_0\sin\phi = $ displacement at $t = 0$ and $\phi = $ phase angle. Note that ω_n, ξ, X_0, and ϕ are all constants. A plot of this solution shows the sine curve with its height continuously decreasing because $e^{-\xi\omega_n t}$ decays with time (Fig. 6.4).

Case C. Critical damping ($\xi = 1$) means that the solution must be written

$$x = (A + Bt)e^{-\omega_n t} \qquad [6.10]$$

The t term is inserted with B because otherwise only one of the two solutions would be found.

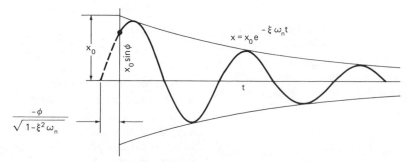

Fig. 6.4. Decay of motion for light damping.

This method is developed in mathematics text-books (2). The graph here is similar to Case A (Fig. 6.3). (See Ref. 2.) The motion is aperiodic, but the time of return to equilibrium is a minimum when the damping is critical.

6.5. FORCED VIBRATION WITHOUT DAMPING

Equation [6.1] now becomes $m\ddot{x} + kx = P_0 \sin \omega t$ or

$$\frac{d^2x}{dt^2} + \frac{kgx}{W} = \frac{P_0 g}{W} \sin \omega t.$$

According to the theory of differential equations, the solution of this equation consists of the sum of two parts: (1) the solution previously determined in free vibrations without damping for the equation when the right side is set equal to zero (the *transient part*), and (2) a solution to make

$$\frac{d^2x}{dt^2} + \frac{kg}{W} x = \frac{P_0 g}{W} \sin \omega t. \quad [6.11]$$

Assume that the second solution, which is called the *steady-state solution*, is of the form $x = x_0 \sin \omega t$. Then $d^2x/dt^2 = -x_0^2 \sin \omega t$. Substituting,

$$-x_0 \omega^2 \sin \omega t + \frac{kg}{W} x_0 \sin \omega t = \frac{P_0 g}{W} \sin \omega t.$$

Hence,

$$x_0 = \frac{P_0/k}{1 - \omega^2 W/gk}. \quad [6.12]$$

Let δ be the deflection which the force P_0 would impart to the spring if acting on it statically, i.e., $\delta = P_0/k$. Also note that $\omega_n^2 = kg/W$, where ω_n is the natural frequency when the disturbing force is absent. Then x_0 may be written

$$x_0 = \frac{\delta}{1 - (\omega/\omega_n)^2}. \quad [6.13]$$

For convenience in analysis let $\omega/\omega_n = r$; then the steady-state solution is

$$x = \frac{\delta}{1 - r^2} \sin \omega t.$$

Note that its frequency is the same as the disturbing frequency. The entire solution is then

$$x = A \sin \sqrt{\frac{kg}{W}} t + B \cos \sqrt{\frac{kg}{W}} t$$
$$+ \frac{1}{1 - r^2} \delta \sin \omega t. \quad [6.14]$$

The first two terms represent the free vibrations and are transient in character because some damping is always present to cause those vibrations to decay. We are interested in the steady-state (or particular) solution. Hence

$$x = \delta \frac{1}{1 - r^2} \sin \omega t.$$

Its maximum value, which occurs when $\sin \omega t = 1$, is $\delta\, 1/(1 - r^2)$ and is called the *amplitude*. The ratio of the amplitude of the steady-state solution to the static deflection δ which P_0 would cause is called the *magnification factor*. Its value is

$$\delta \frac{\dfrac{1}{1 - r^2}}{\delta} = \frac{1}{1 - r^2}.$$

Since

$$r = \frac{\omega}{\omega_n} = \frac{f}{f_n},$$

then

$$\frac{x_0}{\delta} = \frac{1}{1 - (f/f_n)^2}. \quad [6.15]$$

Its value can be positive or negative depending on whether f is less than f_n or not. When $f = f_n$, resonance occurs and the amplitude is theoretically infinite. Actually, the damping which is always present holds the amplitude to a finite amount.

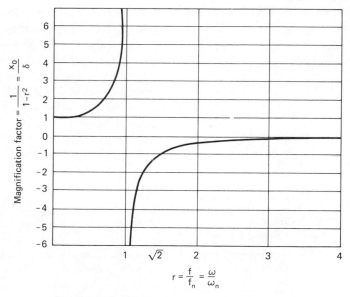

Fig. 6.5. Magnification factor versus frequency ratio.

A plot of magnification factor versus the frequency ratio r is given in Fig. 6.5. This graphical representation is useful in interpreting characteristics of vibrating systems. In this figure, the ordinate x_0/δ is the ratio of the maximum deflection of the mass from its position of rest to the deflection due to a statically applied force P_0. The abscissa ω/ω_n is the ratio of the frequency of the applied force P_0 to the natural frequency of the system. Since δ and ω_n are constants of the system, it may be seen that Fig. 6.5 is a representation of the variation of the maximum deflection x_0 of a vibrating body with the frequency of application ω of the applied force. The figure shows that when ω is zero, the deflection of the body is equal to the static deflection δ, and when ω is equal to ω_n, the deflection in the absence of damping is infinite. For operating frequencies greater than $\sqrt{2}\omega_n$, the deflection of the body becomes less than δ. In using Eq. [6.15] in the region where $(\omega/\omega_n) > 1$, it is necessary to change the sign of the right-hand side of the equation. The physical interpretation of this change is that the deflection and the impressed force are out of phase when ω/ω_n is greater than unity.

Amplitude of Forced Vibrations. Eq. [6.12] can be reproduced as:

$$x_0 = \frac{P_0/k}{1 - \dfrac{\omega^2 W}{gk}}$$

Multiply through by k/m:

$$x_0 = \frac{P_0/m}{\dfrac{k}{m} - \omega^2}. \qquad [6.16]$$

Eq. [6.15] is in nondimensional form showing amplitude as a function of ω and applies where the mass m in Fig. 6.1 is under the action of a periodically reversing force $P_0 \sin \omega t$. The ratio $x_0/(P_0/k)$ is the magnification factor (Eq. [6.15]) since it shows the magnitude of the amplitude in terms of the static displacement caused by P_0. However, it should be noted that for Fig. 6.1, if P_0 is a function of ω in $P_0 = m'r\omega^2$, Eq. [6.16] may be written in comparable nondimensional form:

$$\frac{x_0 m}{m'r} = \frac{(\omega/\omega_n)^2}{1 - (\omega/\omega_n)^2}. \qquad [6.17]$$

Equations [6.15] and [6.17] are different functions of ω. These equations are shown plotted in Figs. 6.6(a) and 6.6(b). As may be

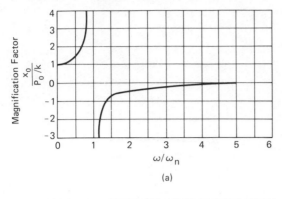

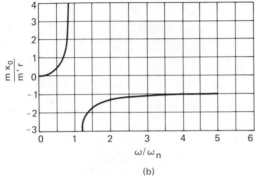

Fig. 6.6. Vibrating amplitude versus frequency ratio.

seen, the equations are somewhat different in form although both curves show a critical condition at $\omega/\omega_n = 1.0$ where amplitude becomes infinite. For the case of the mass m being acted on by a periodically or harmonic reversing force $P_0 \sin \omega t$, the vibrating amplitude becomes smallest at high ratios of ω/ω_n [Fig. 6.6(a)], whereas for the case of a rotating unbalanced mass m', the amplitude x_0 becomes smallest at $\omega/\omega_n = 0$ [Fig. 6.6(b)].

It should be noted that the values in the ordinates of Figs. 6.6(a) and 6.6(b) are negative for $\omega > \omega_n$, indicating that the displacement x_0 is 180° out of phase with P_0. However, since only the magnitude of the amplitude is of importance, the curve may also be shown with positive ordinates for values of $\omega/\omega_n > 1.0$.

6.6. TRANSMISSIBILITY

In a machine which is supported from the floor structure by springs, an unbalanced rotating weight would be transmitted to the floor through the springs. Thus the floor is subject to a forced vibration in which the forcing function

is the periodically reversing spring force $kx = x_0 \sin \omega t$. The floor vibration in turn causes discomfort or the shaking of other objects on the floor such as cameras or instruments. The maximum force transmitted through the springs during a cycle of vibration is $F_{tr} = kx_0$. It is possible to design the springs such that F_{tr} is near zero, in which case the vibrating machine is said to be isolated.

Since x_0 is given by Eq. [6.12], the transmitted spring force is

$$F_{tr} = kx_0 = \frac{P_0}{1 - (\omega/\omega_n)^2}. \quad [6.18]$$

If the machine were rigidly fixed to its foundation, the force transmitted to the floor would be P_0. However, with soft springs, the denominator of Eq. [6.18] may be made large so that the transmitted force F_{tr} approaches zero. The ratio of the force F_{tr} with springs to the force P_0 for rigid attachment without springs is termed the *transmissibility* TR.

$$\text{TR} = \frac{F_{tr}}{P_0} = \frac{1}{1 - (\omega/\omega_n)^2}. \quad [6.19]$$

As shown in Fig. 6.7, the curve of transmissibility versus ω/ω_n given by Eq. [6.19] is exactly the same as that of Fig. 6.6(a). At $\omega/\omega_n = 1.0$ the transmitted force is infinite, but at values of ω/ω_n greater than $\sqrt{2}$ there is a considerable reduction in the transmitted force. To

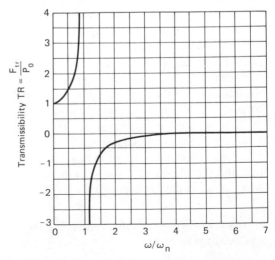

Fig. 6.7. Transmissibility versus frequency ratio.

make ω/ω_n large, either the design speed ω may be increased or the spring constant may be made small. If ω is low, the spring will be soft but may be made harder by increasing ω. Even though the transmitted force is low, the amplitude of vibration x_0 as given in Fig. 6.6(b) does not approach zero but approaches a value given by $mx_0/m'r = 1.0$, which may be appreciable. It will be observed that, even though the transmitted vibratory force F_{tr} may be made to approach zero, the spring transmits a steady force equal to the weight of the machine.

6.7. FORCED VIBRATION WITH DAMPING

In the foregoing, cases of forced vibration are discussed in which the force of friction is neglected. Since friction opposes motion in the sense that it is opposite to velocity, it serves to reduce amplitudes of vibration and is therefore said to *dampen* vibrations. Friction may appear as the viscous resistance of fluids, as the sliding resistance of dry materials in contact, or as internal shearing resistance of the plastic flow of materials, as is evident in the hysteresis loop of stress-strain diagrams. Shock absorbers used in automobiles and aircraft are devices utilizing the frictional resistance of fluids to damp vibrations; commercial spring mounts are made of materials such as rubber, fiber, and cork to utilize the large internal frictional resistance.

Figure 6.1 shows the vibrating mass with a dashpot used to produce damping by the resistance of a fluid. The force of resistance of the viscous fluid in Fig. 6.1 is proportional to the velocity of the mass m but is opposite in sign to the velocity. Thus, the damping force is $-cV = -c\,dx/dt$, in which c is the constant of proportionality having the units lb-sec/in.

The equation of motion for the free body of Fig. 6.1 may be written as follows:

$$P_0 \sin \omega t - cV - kx = mA$$

$$P_0 \sin \omega t - c\frac{dx}{dt} - kx = m\frac{d^2x}{dt^2}$$

$$m\frac{d^2x}{dt^2} + c\frac{dx}{dt} + kx = P_0 \sin \omega t. \qquad [6.20]$$

The steady-state solution to Eq. [6.20] is $x = x_0 \sin(\omega t - \phi)$, in which the amplitude is

$$x_0 = \frac{P_0/k}{\sqrt{\left(1 - \frac{m\omega^2}{k}\right)^2 + \left(\frac{c\omega}{k}\right)^2}} \qquad [6.21]$$

and the angle ϕ is given by

$$\tan \phi = \frac{c\omega}{k - m\omega^2}. \qquad [6.22]$$

ϕ is the angle by which the displacement of the vibration lags the force producing the vibration. Viscosity causes the lag, and as indicated by Eq. [6.22], ϕ is zero in the absence of viscosity.

Of interest is the effect of friction, or damping, on the amplitude of forced vibration given by Eq. [6.21], which may be written in the following nondimensional form:

$$\frac{x_0}{P_0/k} = \frac{1}{\sqrt{\left[1 - \left(\frac{\omega}{\omega_n}\right)^2\right]^2 + \left(\frac{c}{m\omega_n} \cdot \frac{\omega}{\omega_n}\right)^2}}$$

$$[6.23]$$

In Eq. [6.23], $c/m\omega_n$ containing the damping coefficient c is a nondimensional term which may be regarded as a damping factor. When the damping factor is zero, Eq. [6.23] then becomes equal to Eq. [6.15]. Eq. [6.23] is plotted in Fig. 6.8(a).

For the case of an unbalanced rotating mass, in which $P_0 = m'r\omega^2$, Eq. [6.23] may be written in the following form for plotting as shown in Fig. 6.8(b):

$$\frac{mx_0}{m'r} = \frac{(\omega/\omega_n)^2}{\sqrt{\left[1 - \left(\frac{\omega}{\omega_n}\right)^2\right]^2 + \left(\frac{c}{m\omega_n} \cdot \frac{\omega}{\omega_n}\right)^2}}$$

Transmissibility is also affected by damping. Referring to Fig. 6.1, since both the spring and the dashpot are connected to the foundation, the force of these elements is transmitted to the foundation. The force of the spring is kx and that of the dashpot is $cV = c\,dx/dt$. The resultant of the two forces is the transmitted force F_{tr}:

$$F_{tr} = kx + c\frac{dx}{dt}. \qquad [6.24]$$

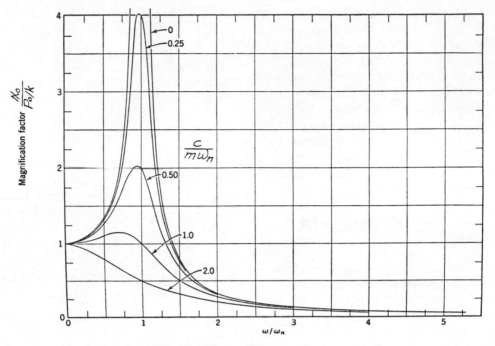

Fig. 6.8(a). Magnification factor versus frequency ratio.

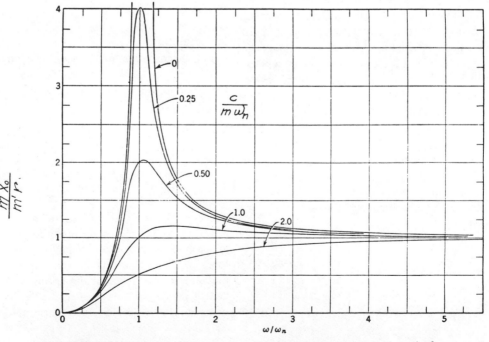

Fig. 6.8(b). Eq. [6.23] plotted for case of unbalanced rotating mass, where $P_0 = m^1 r \omega^2$.

The displacement at any time t is given by $x = x_0 \sin(\omega t - \phi)$, and the velocity is $V = dx/dt = x_0\omega \cos(\omega t - \phi)$. Thus,

$$F_{tr} = kx_0 \sin(\omega t - \phi)$$
$$+ cx_0\omega \cos(\omega t - \phi). \quad [6.25]$$

To determine the maximum value of the transmitted force in a cycle of vibration, Eq. [6.25] may be maximized by differentiating and finding the maxima:

$$\frac{d(F_{tr})}{dt} = kx_0\omega \cos(\omega t - \phi)$$
$$- cx_0\omega^2 \sin(\omega t - \phi) = 0$$

$$\frac{\sin(\omega t - \phi)}{\cos(\omega t - \phi)} = \tan(\omega t - \phi)$$

$$= \frac{kx_0}{cx_0\omega}. \quad [6.26]$$

Eq. 6.26 is represented in the triangle of Fig. 6.9 at the phase for maximum F_{tr}. As shown,

$$\sin(\omega t - \phi) = \frac{kx_0}{\sqrt{(kx_0)^2 + (cx_0\omega)^2}}$$

$$\cos(\omega t - \phi) = \frac{cx_0}{\sqrt{(kx_0)^2 + (cx_0\omega)^2}}$$

Substituting in Eq. [6.25],

$$(F_{tr})_{max} = \frac{(kx_0)^2 + (cx_0\omega)^2}{\sqrt{(kx_0)^2 + (cx_0\omega)^2}}$$

$$= \sqrt{(kx_0)^2 + (cx_0\omega)^2}. \quad [6.27]$$

Transmissibility is the ratio of $(F_{tr})_{max}$ to the

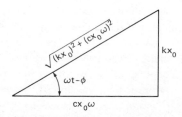

Fig. 6.9. Graphical representation of Eq. [6.26].

forcing function P_0 from Eq. [6.23]:

$$TR = \frac{(F_{tr})_{max}}{P_0}$$

$$= \frac{\sqrt{(kx_0)^2 + (cx_0\omega)^2}}{kx_0\sqrt{\left[1 - \left(\frac{\omega}{\omega_n}\right)^2\right]^2 + \left(\frac{c}{m\omega_n} \cdot \frac{\omega}{\omega_n}\right)^2}}$$

$$TR = \frac{\sqrt{1 + \left(\frac{c}{m\omega_n} \cdot \frac{\omega}{\omega_n}\right)^2}}{\sqrt{\left[1 - \left(\frac{\omega}{\omega_n}\right)^2\right]^2 + \left(\frac{c}{m\omega_n} \cdot \frac{\omega}{\omega_n}\right)^2}}.$$

$$[6.28]$$

The effect of damping on transmissibility is shown by the curves of Fig. 6.10 plotted from Eq. [6.28]. As shown, high damping reduces the transmitted force for values of $\omega/\omega_n < \sqrt{2}$, but it increases the transmitted force when ω/ω_n exceeds $\sqrt{2}$.

When metallic springs, such as coil or leaf springs, are used to suspend a machine, some damping is present because of the internal friction in the material, but the amount is small and the undamped case may be assumed. As Fig. 6.8 and Fig. 6.10 show, k should be chosen such that ω/ω_n at the design speed is greater than 2, in which case both the amplitude of vibration and the transmissibility will be low. However, in the absence of much damping in metallic springs, both the amplitude and transmissibility will be high at some low critical speed ($\omega/\omega_n = 1.0$) through which the vibrating system must pass. In automotive vehicles, dashpots in the form of shock absorbers are added to the system to prevent high motion transmissibility [Fig. 6.8(a)] of the vehicle body for a wide range of speeds; however, transmissibility of force may be increased to some extent at the high speeds as indicated in Fig. 6.10. Commercial isolators of rubber and other nonmetallic materials offer greater amounts of damping than metallic springs ($c/m\omega_n$ is usually not above 0.2).

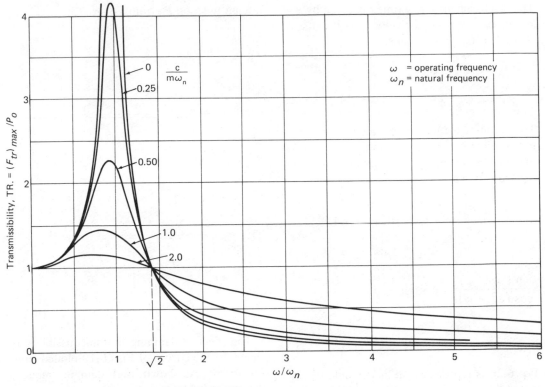

Fig. 6.10. Transmissibility as a function of frequency ratio.

Thus, Fig. 6.10 indicates a number of important concepts: (a) isolators should be chosen so as not to excite the natural frequencies of the system; (b) damping is important in the range of resonance whether the dynamic system is operating near resonance, or must pass through resonance during startup; (c) in the isolation region, the larger the ratio of ω/ω_n (i.e., the smaller the value of ω_n), the smaller will be the transmissibility. Since $\omega_n = \sqrt{k/m}$ (Section 6.3), ω_n can be made small by selecting soft springs. Soft springs used in conjunction with light damping provides good isolation. Note also that ω_n can be reduced by increasing the mass, such as by use of an "inertia base." This is a platform, usually concrete reinforced with steel, upon which a machine (such as a reciprocating pump or compressor, or a large fan) is mounted. The inertia base is in turn supported by isolation springs (usually steel); its weight is usually 1.5–2 times the machine weight.

By referring to Eqs. [6.18] and [6.19], for a given TR, the spring constant would be in-

creased with an inertia base, and the operating amplitude would be reduced. This improves lateral or rocking stability, and simplifies the attachment of ductwork, electrical conduit etc. to the machine. Also the moment of inertia of the system is increased with an inertia base, and oscillatory motion (produced by rotating unbalance) about the center of the mass is reduced accordingly (see later in this chapter). Finally, the transmissibility of the system can be reduced without reducing stability.

6.8. MOTION DISTURBANCE—SINGLE DEGREE OF FREEDOM

The support may be the source of undesirable vibration. Motion of delicate laboratory instruments induced by machinery outside the laboratory area or vibration of the frame of an airplane may make instrument readings unreliable; cams or eccentric and worn rollers may be the source of vibration in machinery; rails and highways give motion disturbances to vehicles;

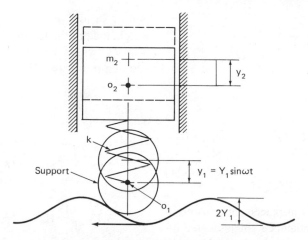

Fig. 6.11. Displacement disturbance.

turbulence produces disturbances in aircraft. Support vibration is measurable as a motion or displacement disturbance; proper flexible suspension between the support and connected masses may decrease the transmission of displacement so that the mass is nearly motionless.

In Fig. 6.11, a schematic representation shows the support receiving a displacement disturbance $Y_1 \sin \omega t$, with a consequent displacement y_2 of spring-supported mass m_2. The force acting upon m_2 is the spring constant times the net displacement of the spring, namely, $-k(y_2 - y_1)$, and the equation of motion is

$$-k(y_2 - y_1) = m_2 \ddot{y}_2$$

or

$$\ddot{y}_2 + \frac{k}{m_2} y_2 = \frac{kY_1}{m_2} \sin \omega t. \quad [6.29]$$

Eq. [6.29] is essentially the same as Eq. [6.11], and its steady-state solution is

$$y_2 = Y_2 \sin \omega t,$$

where the amplitude Y_2 is

$$Y_2 = Y_1 \frac{1}{1 - \left(\dfrac{\omega}{\omega_n}\right)^2}, \quad [6.30]$$

with $\omega_n = \sqrt{k/m_2}$.

The *displacement* or motion transmissibility is defined as the ratio of maximum displacement of suspended mass m_2 to the maximum displacement it would have if it were rigidly connected to the support, namely,

$$\gamma = \frac{Y_2}{Y_1} = \frac{1}{1 - (\omega/\omega_n)^2}. \quad [6.31]$$

This is identical to the expression obtained for force transmissibility, Eq. [6.19]. It may also be of interest to obtain the *force* transmissibility for the case of motion disturbance. Rigidly connected to the support, m_2 received the displacement $y_1 = Y_1 \sin \omega t$, the acceleration $\ddot{y}_1 = -\omega^2 Y_1 \sin \omega t$, and the maximum force $m_2 (\ddot{y}_1)_{max} = m_2 \omega^2 Y_1$. Spring connected, m_2 feels the spring force of

$$-k(y_1 - y_2) = -k(Y_2 \sin \omega t - Y_1 \sin \omega t)$$

with a maximum value

$$k(Y_2 - Y_1) = kY_1 \left[\frac{1}{1 - (\omega/\omega_n)^2} - 1 \right]$$

$$= kY_1 \frac{(\omega/\omega_n)^2}{1 - (\omega/\omega_n)^2}.$$

Hence, the force transmissibility is

$$\text{TR} = \frac{kY_1 \dfrac{(\omega/\omega_n)^2}{1 - (\omega/\omega_n)^2}}{m_2 \omega^2 Y_1}$$

$$= \frac{\dfrac{(\omega/\omega_n)^2}{1 - (\omega/\omega_n)^2}}{(\omega/\omega_n)^2}$$

$$= \frac{1}{1 - (\omega/\omega_n)^2}. \qquad [6.32]$$

Thus, both force and displacement transmissibilities for all the undamped single-degree-of-freedom cases are determined by the same equation of Fig. 6.7.

6.9. ISOLATOR SELECTION

As mentioned previously in Section 6.3, the undamped natural frequency of a single-degree-of-freedom system is given by the equation

$$f_n = \frac{1}{2\pi} \sqrt{\frac{k}{m}} = \frac{1}{2\pi} \sqrt{\frac{kg}{W}}. \qquad [6.33]$$

Be definition, the static deflection is the deflec-tion of an isolator that occurs due to the dead weight load of the mounted equipment, i.e., δ_{st} $= W/k$, where k is the spring constant. Sub-stituting in Eq. [6.33], then,

$$f_n = \frac{1}{2\pi} \sqrt{\frac{980.7}{\delta_{st}}}$$

$$= 4.98 \sqrt{\frac{1}{\delta_{st}}} \text{ Hz} \qquad (\delta_{st} \text{ in cm}) \qquad [6.34a]$$

$$= 3.13 \sqrt{\frac{1}{\delta_{st}}} \text{ Hz} \qquad (\delta_{st} \text{ in in.}). \qquad [6.34b]$$

A graphical representation of this equation is given in Fig. 6.12.

Thus we can determine the natural frequency of a system by measuring the static deflection providing that the spring is linear and that the isolator material possesses the same elasticity characteristics under both static and dynamic conditions (which is usually the case).

Substitution of Eq. [6.34] into Eq. [6.19] (f versus δ_{st} for various TR values) on log-log co-ordinates produces a family of straight lines as shown in Fig. 6.13. Various regions are la-beled according to the application listed.

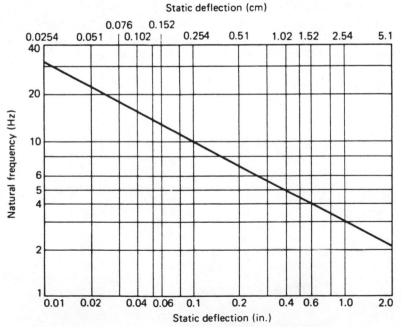

Fig. 6.12. Natural frequency as a function of static deflection.

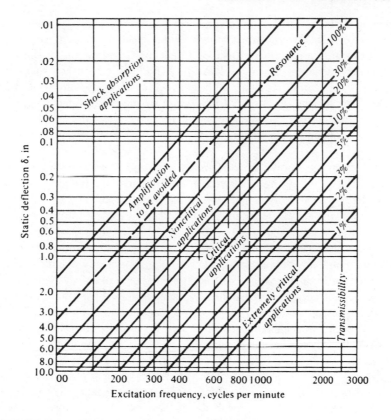

Fig. 6.13. Recommended isolation efficiencies and static deflections for various applications.*

		CRITICAL AREAS: CHURCHES, RESTAURANTS, STORES, OFFICE BLDGS., SCHOOLS, HOSPITALS, BROADCASTING STUDIOS	NONCRITICAL AREAS: LAUNDRIES, FACTORIES, SUBBASEMENTS GARAGES, WAREHOUSES
Recommended minimum isolation efficiencies			
TYPE OF EQUIPMENT			
Air-conditioners (self-contained)		90%	70%
Air handling units		80%	70%
Compressors (centrifugal)		99%	80%
Compressors (reciprocating)	UP TO 10 hp	85%	70%
	15 hp TO 50 hp	90%	75%
	60 hp TO 150 hp	95%	80%
Heating and ventilating		80%	70%
Cooling towers		80%	70%
Condensers	AIR COOLED EVAPORATIVE	80%	70%
Piping		90%	70%
Pumps	UP TO 3 hp	80%	70%
	5 hp OR OVER	95%	80%

*Although the table in Fig. 6.13 has been used in former years (and can still be used as a rough guide for equipment which is mounted on a comparatively rigid base or floor span), a more up-to-date guide to vibration isolator selection is given in the 1987 ASHRAE Handbook, *HVAC Systems and Applications*, Table 27 (15). For a comprehensive coverage of the rationale behind the values given in ASHRAE, the reader should consult Ref. 16.

The forcing or excitation frequency on the abscissa of Fig. 6.13 and the transmissibility lines in the graph locate a point, the ordinate of which is the static deflection of the isolator (and, consequently, the natural frequency of the system) necessary to achieve the required isolation. The system parameters may then be selected or adjusted to obtain this desired natural frequency.

The static-deflection data show that pads are effective only at relatively high frequency (because static deflection is usually small). Although springs, with larger static deflection, are theoretically effective at both high or low frequencies, the spring mass (which is usually neglected in calculations) can reduce the transmission loss at higher frequencies. As will be seen in the next section on types of isolators, placement of an isolation pad (such as neoprene) under each spring provides the isolation at higher frequencies where the spring alone may be inadequate.

Example. A 40 hp centrifugal air-handling unit is to be located in a ground-floor mechanical equipment room of an office building. The unit weighs 1000 lb and operates at 400 rpm. What should be the spring constant of each of four vibration isolators for a transmissibility of 10%?

Excitation frequency is given as 400 cycles/min. From Fig. 6.13, for a transmissibility of 10%, the system static deflection is 2.4 in. and the spring constant of the isolators in the system is

$$k^1 = \frac{W}{\delta_{st}} = \frac{1000}{2.4} = 416 \text{ lb/in.}$$

Since four isolators are to be used, the spring constant of each isolator is

$$k = \frac{k^1}{4} = \frac{416}{4} = 104 \text{ lb/in.}$$

6.10. TYPES OF ISOLATORS

The materials used for isolation are principally steel, rubber and synthetic rubber, polyurethane foam, fiberglass, cork, and felt. They may be applied as simple springs and pads or may be housed to give special characteristics

and convenience in application. Considerations in the selection are not only spring rate (or static deflection) and damping, but also load capacity, sidewise stiffness, snubbing of possible shock overloads, method of attachment, location of mounting brackets, and environment. The need for and determination of sidewise flexibility also has to be considered.

Of the several isolators, steel helical springs are the most versatile for they may be designed for various combinations of load, spring rate, and space. They may be nested or grouped side by side, used in compression if properly recessed and guided, or in tension with hook ends. For the isolation of low-frequency disturbances in heavy machinery they may constitute the only practical method of obtaining sufficient static deflection. Oil, heat, cold, and age do not affect their properties, and lack of damping make their selection reasonably accurate. Stainless steel may be used for corrosive or oxidizing applications.

Steel springs may transmit the high-frequency disturbances associated with high-pitch noise. Rubber pads may then be used between the spring and its supporting surfaces to damp out this noise, and some such damping medium is generally incorporated in commercial isolator units. The isolator of Fig. 6.14 uses a wire-mesh cushion to give damping characteristics. In addition, top and bottom wire-mesh buffers prevent excessive motion in case of shock-loads or temporary overloads. This particular isolator is made entirely of aluminum and stainless steel to meet all environmental conditions. Other helical spring units may have rubber or other organic buffers. Leaf springs give some damping because of rubbing between leaves. Additional damping and deflection limitations may

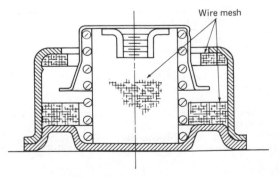

Fig. 6.14. All-metal isolator.

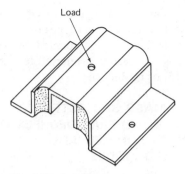

Fig. 6.15. Goodrich channel isolator.

be obtained by separate shock absorbers and snubbers.

When natural and synthetic rubber deflects under compressive or tensile stress, there is no appreciable change in total volume, i.e., it behaves like a liquid. In compressed pads of appreciable area, its deflection is too small to be useful against any but the highest-frequency disturbances. In tension, as a rod or beam, it is easily cut and thereby greatly weakened. However, stressed in shear, it requires no change in thickness, and its flexibility rate is satisfactory.

Inexpensive commercial isolators designed to operate primarily in shear are illustrated in Figs. 6.15 and 6.16. In Fig. 6.15, the rubber is bonded to the steel shapes, and the whole may be purchased in lengths. For a particular application the proper length may be cut off with a hack saw and the metal parts drilled for clamp-down bolts. The isolators of Fig. 6.16 feature snubbing washers. Sidewise motion, because of the tension-compression characteristics, is decidedly restricted in Figs. 6.15 and

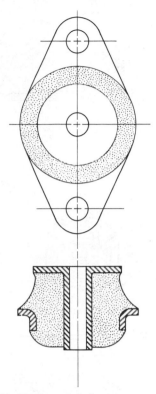

Fig. 6.17. MB Isomode isolator, top and section.

6.16. This may be necessary, as when sidewise location and forces require it, and also is sufficient when the disturbance is primarily vertical and it and the center of gravity are centrally located above the several isolators. For sidewise or noncentral disturbances, an isolator with more or less equal flexibility in the several directions may be desired, as discussed in the following section. Such types are illustrated by Figs. 6.17 and 6.18. The rubber is stressed partially in shear and partially in tension-compression.

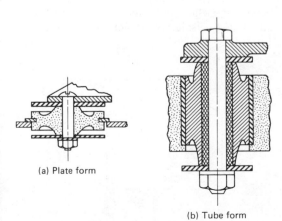

(a) Plate form

(b) Tube form

Fig. 6.16. Lord isolators.

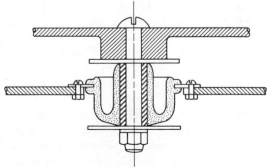

Fig. 6.18. Lord multiplane isolator.

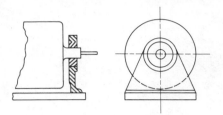

Fig. 6.19. Rubber ring torsional isolator.

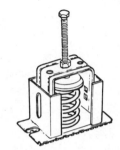

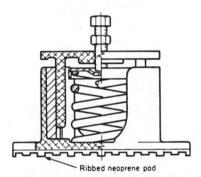

Fig. 6.20. Typical housed vibration isolator.

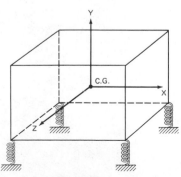

Ribbed neoprene pad

Fig. 6.21. Sectional view of housed vibration isolator.

Purely torsional or torque disturbances may be isolated by such methods as illustrated in Fig. 6.19. This shows a popular method for fractional-horsepower motors. Rubber rings between the stator and its supporting brackets allow rotational flexibility and isolation because they are stressed in shear, but are relatively stiff in the transverse directions, so that belts, gears, couplings, etc. may operate connected to the motor shaft. A "beam spring" is also used under the brackets of large single-phase generator stators to take the torque reaction and isolate 120 cps disturbances. The torque variations of automotive engines are isolated in a somewhat similar fashion but using rubber isolators under the several support brackets. Typical housed isolators are shown in Figs. 6.20 and 6.21.

For a comprehensive guide to vibration isolator selection for different types of equipment on different bases, and mounted on various floor spans, the reader is referred to Table 27 in Ref. 15, or to Ref. 16.

6.11. A MASS SUBJECTED TO MOTION IN SEVERAL DEGREES OF FREEDOM—COUPLED AND DECOUPLED MODES

Figures [6.1] and [6.11] show the mass restricted to linear motion by a schematically represented guide. Such guides do exist in many cases through mechanical linkage or through elastic restrictions. On the other hand, many machines can have no such restrictions. Moreover, they may be subjected to rotating couples and forces which incite angular and transverse motions. Isolation for these couples and forces is desirable, and rigid restriction should not be used. An example is a small engine-generator set where the engine has one or two cylinders, and isolation is desired by supporting the frame of the set on four isolators.

A spring-suspended body may have six degrees of freedom. In Fig. 6.22(a), shown with axes through the center of gravity, there may be translation in the X, Y, and Z directions and rotations about the X, Y, and Z axes, as indicated in the example of Fig. 6.22(b). If the springs are located at random and a variable force is applied at any point on the body, a general motion will result due to the noncolinearity of the resultant elastic force, the applied force, and the inertia force acting at the center of gravity. Fortunately, many machines have planes of symmetry so that the necessary calculations for isolators are not too difficult.

Fig. 6.22(a).

$$\frac{ky_1}{ky_2} = \frac{x_2}{x_1}.$$

The procedure in the analysis (3, 4, 5) is to write the equations of equilibrium for the plane, namely $\Sigma F_y = m\ddot{y}$, $\Sigma F_x = m\ddot{x}$, and $\Sigma M_z = I_z\ddot{\theta}$. Because of the symmetry it is found that the first differential equation contains only the variable y, and is identical to Eq. [6.11] for a single-degree of freedom, whereas the second and third equations each contain the variables x and angular displacement θ. The solutions for y and transmissibility are the same as before, namely, Eqs. [6.15] and [6.19], respectively. Motion in the translatory mode along the Y-axis is said to be *decoupled*. On the other hand, the second and third differential equations must be solved simultaneously for x and θ, and the x-translation and angular modes are said to be *coupled*. Motion in the x-direction induces angular motion and vice versa. Hence, the magnitudes of the motions cannot be independently controlled and the three values of transmissibility will differ, perhaps widely. Only the largest value of transmissibility may determine the effectiveness of the isolators.

A definite improvement is obtained if the modes are decoupled by a proper orientation of the isolators. The necessary condition for decoupling is found from the theory (3, 4) to be as follows: The resultant of the force applied to the mounted body *by the isolators* when the mounted body is displaced in translation must be a force directed through the center of gravity. If we place the isolators in the horizontal plane through the center of gravity (Fig. 6.24) this condition will result. Any disturbance in translation will not induce angular motion; also the flexibilities of the isolators may be arranged so that the transmissibilities in all but the mode of rotation about the Y-axis are approximately equal, and the latter transmissibility not far from the others. Thus the mount provides for efficient isolation from disturbance coming from any direction.

Companies specializing in the manufacture of isolators will recommend center-of-gravity plane mounting for decoupling and most satisfactory results. Unfortunately, many designers of machinery consider the mounting a detail which can be provided at the bottom of the

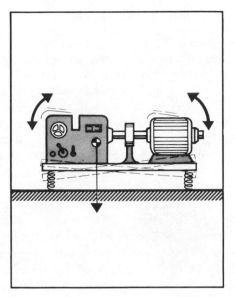

Fig. 6.22(b).

If there are two planes of symmetry, the vertical planes XY and YZ, i.e., if the Y-axis contains the center of gravity, the point of force application, and the center of elastic reaction from the isolators, an analysis may be made in each of the two planes XY and YZ (as shown in Fig. 6.23(a) for the XY-plane only). When the isolators must be unsymmetrically placed, the relationship indicated by Fig. 6.23(b) should be held between the spring constants for a resultant elastic force along the Y-axis, i.e.,

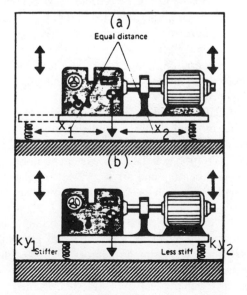

Fig. 6.23.

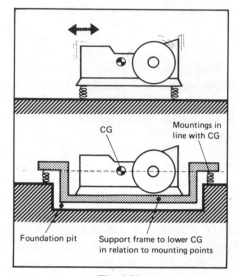

Fig. 6.24.

housing, whether rigid or flexible attachment is required, and center-of-gravity plane brackets are not provided. However, the trend is toward higher mountings.

For slow-speed machinery to obtain a necessary low natural frequency, it has long been common to add weight in the form of a concrete block to which the machine is rigidly assembled (see Fig. 6.25). Recently these blocks have been so shaped that spring mounting of the assembly can be made in its center-of-gravity plane.

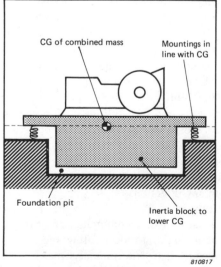

Fig. 6.25.

6.12. VIBRATION CRITERIA

Vibration and vibration-included noise, major sources of occupant complaint, are increasing in modern buildings. Lighter weight construction and equipment located in penthouses or intermediate level mechanical rooms increase structure-borne vibration and noise transmission. Not only is the physical vibration disturbing, but the regenerated noise from structural movement can be heard in other remote sections.

Rigidly mounted pieces of equipment transmit their full vibrational force to the building structure. This can result in disturbing physical vibration felt by the occupant, damaging structural vibration, and annoying noise when the vibration excites the structure at frequencies within the normal hearing range.

Criteria should be established for equipment vibration to determine the excessive forces that must be isolated or that adversely affect the performance or life of the equipment. Figs. 6.26(a) and 6.26(b) show the significance and interrelationship of equipment vibration levels and vibration isolation systems where the isolators provide a fixed efficiency as determined by Eq. [6.32], and where the magnitude of transmission to the building is a function of the magnitude of the vibration force. Theoretically, an isolation system could be selected to isolate forces of extreme magnitude; however, isolators should not be used to mask a condition that should be corrected before it damages the equipment. Rather, isolators should be selected to isolate the vibratory forces of equipment operation, and if transmission occurs, to indicate a faulty operating condition that should be corrected.

Ideally, vibration criteria should: (1) measure rotor unbalance as a function of type, size, mass, and stiffness of equipment; (2) consider the vibration generated by system components such as bearing and drives, as well as installation factors such as alignment; and (3) be verifiable by field measurement. Fig. 6.27 shows some commonly used criteria that do not meet all the requirements above, but are generally satisfactory. A simpler approach is to use the criteria in Table 6.1, which have been developed by individuals and firms experienced in

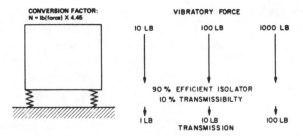

Fig. 6.26(a). Transmission to structure varies as function of magnitude of vibratory forces.

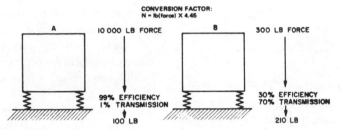

Fig. 6.26(b). Interrelationship of equipment vibration, isolation efficiency, and transmission.

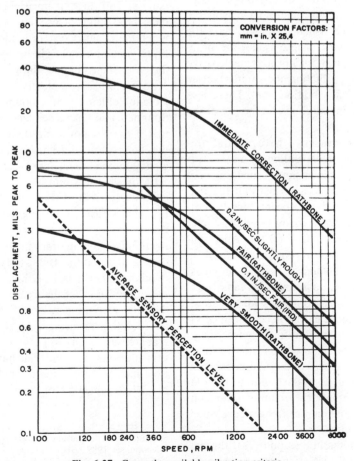

Fig. 6.27. Currently available vibration criteria.

Table 6.1 Equipment Vibration Criteria

Equipment	Maximum Allowable Vibration, Peak-to-Peak Displacement	
	mil	mm
Pumps:		
1800 rpm	2	0.05
3600 rpm	1	0.025
Centrifugal compressors	1	0.025
Fans (vent, centrifugal, axial):		
under 600 rpm	4	0.1
600–1000 rpm	3	0.075
1000–2000 rpm	2	0.05
over 2000 rpm	1	0.025

vibration testing of HVAC (heating, ventilating, and air conditioning) equipment. Table 6.1 shows the maximum allowable vibration levels for steady-state movement taken on the bearing or machine structure if it is sufficiently rigid. These criteria can be met by any properly operating equipment, will determine reasonable vibration levels to be isolated and will make the equipment acceptable. These values are practical levels that allow for misalignment, drive eccentricities, belt vibration and similar factors affecting the overall vibration level. These levels can be maintained throughout the life of the equipment.

Practically all structures have natural frequencies of vibration depending on mass, stiffness, damping, dimensions, methods of mounting, etc. In buildings, many of the natural frequencies of floors, beams, walls, columns, girders, doors, windows, ceilings, etc. fall in the range from 10 to 60 Hz. And typically, much mechanical equipment used in buildings (compressors, elevating devices, cooling tower fans, air circulating fans, etc.) operate at speeds that coincide with the structural natural frequencies and hence produce vibration (and noise) at these frequencies.

Thus, it is important to separate or "isolate" these driving frequencies of the equipment from the natural frequencies of the building, in order the reduce structure-born vibration (and consequent noise radiations) at frequencies where they are disturbing.

As has been noted in Section 6.7, if the natural frequency of the isolation mount just equals the driving frequency of the vibrating source, the system will go into violent oscillation at that frequency—limited only by the damping within the system (see Section 6.13 on damping materials). For a vibration mount to be effective, the driving frequency of the source (which is to be isolated) should be considerably higher than the natural frequency of the isolation mount (by, say, a factor of 4–5). What determines this ratio is the degree of force (or motion) transmissibility which is acceptable. This has been discussed in Section 6.9 and recommended minimum isolation efficiencies for certain areas involving certain equipment, and subsequent static deflection of isolation mounts for different excitation frequencies, were shown in Fig. 6.13.*

In most vibration isolator literature there is usually a curve or a table that shows the vibration isolation efficiency of a mount (such as 80%, 90%, 95%, etc.). However, the accompanying information invariably omits to state that these efficiencies can be achieved only when the isolated system is mounted on an infinitely massive and rigid base—and such is not the case when an upper floor slab deflects up to 1 in. when it is completely loaded. The actual isolation efficiency of a mount decreases as the floor deflection increases, and Miller (11) states that if the isolator static deflection just equals the floor static deflection, the isolation efficiency for the mount approaches approximately 50% of what it would be if the floor had been completely rigid. This means the high-deflection floors (usually comparatively lightweight floors or large-span floors) require larger deflection isolators in order to achieve the desired degree of isolation. By specifying isolator deflection, rather than isolation efficiency, transmissibility or other theoretical parameters, the designer can compensate for floor deflections and building resonances by selecting isolators that are satisfactory to provide minimum vibration transmission, and that have more deflection than the supporting floor. Ref. 16 and Table 27, Chapter 52 of Refs. 15 and 16 give

---

*Again, for a more thorough coverage of vibration isolator selection, see 1987 ASHRAE Handbook "HVAC Systems and Applications," Table 27 (Reference 15).

guidelines for this for a wide range of operating equipment and conditions.

6.13. VIBRATION DAMPING

6.13.1. General

The dynamic responses and sound-barrier properties of a structure are governed primarily by mass, stiffness, and damping. The first two to these have already been mentioned. The last one, damping, is perhaps the most elusive and seemingly complex property, as well as the least understood and most difficult to predict. The word *damping* has been used loosely for many years to denote any number of noise-abatement procedures. The frequent appearance of almost self-contradictory phrases such as "damp out the sound in the room" and "use sound-damping mounts under the machine" indicates the state of popular confusion among the mechanisms of vibration damping, sound absorption, and vibration isolation.

Part of the reason for the difficulty with predicting damping is that there are so many physical mechanisms involved, and any or all may be important factors in a particular problem. These physical mechanisms include viscosity, friction (including internal), acoustic radiation, turbulence, and mechanical and magnetic hysteresis.

While the analysis or prediction of damping may be an elusive thing, the benefits to be obtained from the use of added damping to structures sometimes are worth the effort. Damping reduces the amplitude of vibrations (essentially only at resonant frequencies). These reduced vibrations in turn result in less sound radiation, as well as a lowering of structural stresses and attendant fatigue problems. Reduction of vibrations with damping is particularly helpful when dealing with the noise radiated from the free vibrations associated with the impacts between parts of a structure; e.g., loose parts striking metal chutes. Structure-borne sound is also reduced (i.e., transmission of vibrational energy along the structure is reduced). In addition, the transmission loss performance of a barrier is increased in the resonance region but more particularly at and above the critical coincidence frequency.

While all materials exhibit some natural damping, the amount in most metals, for example, is too low to be of any practical significance. Even lead, which is often thought of as having high damping, only has a loss factor of approximately 0.001. To have any significance the loss factor should be at least about 0.1, and some of the high-damping synthetic rubbers have loss factors as high as 5.0. (Loss factor is a measure of the degree of effectiveness of applied damping material in reducing lateral movement of a radiating surface.)

6.13.2. Types of Treatments

The two major types of surface treatment are the *free-layer* and the *constrained-layer*, as shown in Fig. 6.28. The free-layer treatment consists of a single layer of viscoelastic material applied to the base structure. Bending vibration of the base structure induces extensional deformation of the viscoelastic layer. The constrained-layer treatment consists of two layers: a layer of viscoelastic material and a layer of relatively stiff material, usually sheet metal. Bending vibration of the base structure in this case induces shear deformation of the viscoelastic material due to the constraint provided on the outer boundary of the viscoelastic layer by the added layer of stiff material.

The effectiveness of homogeneous damping can be increased by placing a stiff, rigidly bonded spacer between the panel and damping material, as shown in Fig. 6.28. This construction places the material at a greater distance from the neutral axis of the structure; extension of the material and the composite loss factor increase accordingly.

With either type of treatment the vibrational energy loss is due to the generation of heat by the cyclic deformation of the viscoelastic material. But, because of the different mechanisms of inducing deformation of the material of the two treatments, a damping material that provides good results in one treatment will usually not be effective in the other type of treatment. In general, effective free-layer materials are comparatively stiff and must be applied in thick layers, while effective constrained-layer materials are relatively compliant and may be quite efficient in very thin layers.

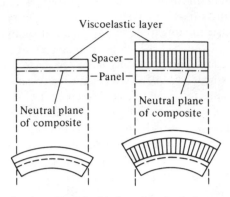

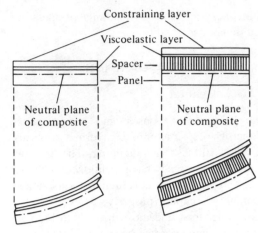

Section of panel with free viscoelastic layer, without and with spacing layer

Section of panel with constrained viscoelastic layer, without and with spacing layer

(*a*) Homogeneous damping.

(*b*) Constrained layer damping.

Fig. 6.28. Homogeneous and constrained layer damping.

Homogeneous and constrained-layer damping can be compared as follows:

- The loss factor with homogeneous damping depends approximately on the square of the weight of the damping treatment. With constrained-layer damping, the dependence of the loss factor on the treatment weight is approximately linear (13).
- The system loss factors provided by either type are approximately equal for treatment surface weights equal to 10–20% of the panel surface weight. For treatment weights less than 10% of panel weight, constrained-layer damping is probably more effective; for treatment weights greater than 20% of panel weight, homogeneous damping is probably more effective (13).

6.13.3 Performance Characteristics

The performance of damping treatments is measured by the *loss factor* η of the damped structure:

$$\eta = \frac{\text{energy dissipated per radian of motion at a given excitation frequency}}{\text{maximum strain energy stored in system at the same frequency}}.$$

The relationship between η and the equivalent viscous damping factor c/c_c, as discussed earlier in this chapter, is (17)

$$\eta \simeq 2\frac{c}{c_c}. \qquad [6.35]$$

The minimum loss factor for satisfactory performance is about 0.1; typical values lie between 0.1 and 0.3. As mentioned earlier, high-damping synthetic rubbers have loss factors as high as 5.0. If the panel surface is an effective radiator, a tenfold increase in η can reduce the radiated sound level by up to 10 dB (13). Loss factor is usually measured in one of two ways (13):

1. The damping treatment is attached to a 20-in.-square, $\frac{1}{4}$-in.-thick steel plate (called a *Geiger plate*) which is supported at the midpoint of each side, as shown in Fig. 6.29(a). A magnetic-force transducer drives the plate at its lowest flexural resonance (160 Hz); when the force is removed, the plate decay is measured by a capacitive-displacement transducer (located near one corner) and recorded on an oscilloscope screen or high-speed chart recorder. The loss factor is simply related to the logarithmic decrement:

$$\eta \simeq \frac{\delta}{\pi} \quad \text{for} \quad \eta \le 0.25, \qquad [6.36]$$

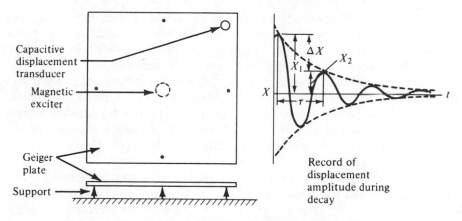

Capacitive displacement transducer

Magnetic exciter

Geiger plate

Support

Record of displacement amplitude during decay

(*a*) Decay of resonant vibrations using Geiger plate.

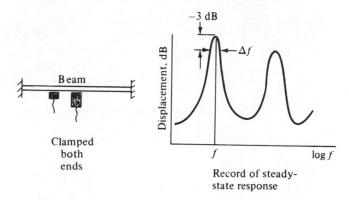

Beam

Clamped both ends

Record of steady-state response

(*b*) Steady-state response of beam near resonance.

Fig. 6.29. Methods for measuring the loss factor of a damped structure.

where $\delta = \ln(x_1/x_2)$ and x_1/x_2 is the amplitude ratio of any two consecutive cycles of plate vibration. The loss factor obtained in this way must be corrected for different plate and treatment thicknesses encountered in actual applications.

2. The damping treatment is attached to a supported beam [Fig. 6.29(b)]; typical dimensions are $\frac{1}{2}$–$1\frac{1}{2}$ in. wide, 5–12 in. long, and 0.02–0.06 in. thick. The beam length is adjusted so that a magnetic-force transducer located at its midpoint will excite one or more flexural resonances. The response of the beam at and near resonance is measured by a capacitive-displacement transducer. For values of η < 0.25,

$$\eta \simeq \frac{\Delta f}{f} \qquad [6.37]$$

where Δf is the bandwidth of response and f is the resonant frequency, both in Hertz. This method has two advantages over the Geiger-plate method: the beam and damping-treatment thicknesses can be similar or equal to those of the actual application, and the variation of η with excitation frequency can be determined.

The effectiveness of layered damping treatments depends on a number of variables. The base structure properties of importance are its thickness, its inherent damping, the resonance frequency, the wavelength of vibration at resonance, and the modulus of elasticity and density of the base structure material. The properties of the constraining layer of importance are its thickness, its modulus of elasticity, and its density. The important properties of the viscoelastic material used in the treatment are its thickness, its density, its modulus of elasticity, and its damping loss factor. The latter two properties of the viscoelastic material are functions of both temperature and frequency. Therefore, it is important to have an accurate description of the modulus and loss factor as functions of temperature and frequency in order accurately to evaluate the effectiveness of the damping treatment made with the material (18). Because the properties of the viscoelastic material are temperature and frequency sensitive, the amount of damping provided by a layered viscoclastic damping treatment will also vary with temperature and frequency. Thus it is essential that a treatment be properly designed if it is to be effective.

The effects of temperature, thickness ratio, and frequency on effectiveness of homogeneous damping are shown in Fig. 6.30(a), (b), and (c) respectively.

The circumstances under which the energy-dissipation characteristics of damping treatment are effective for reducing airborne noise generated by panel vibrations can be summarized as follows (14, 17, 18):

- The panel must be vibrating at or near one or more of its resonant frequencies.
- The resonant vibration of the panel surface must be an effective radiator of

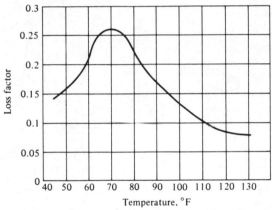

(a) The effect of temperature on the loss factor of a plastic-based vibration damping material (material/metal thickness ratio = 2, frequency = 200 Hz).

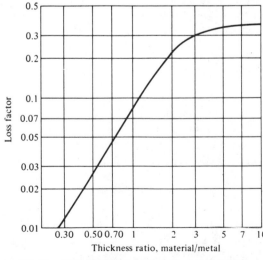

(b) The effect of thickness ratio on the loss factor of a plastic-based vibration damping material (temperature = 72°F, frequency = 200 Hz).

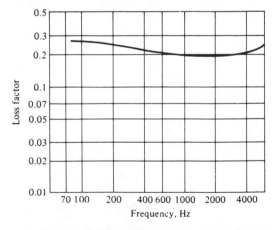

(c) Loss factor as a function of frequency of a plastic-based vibration damping material (temperature = 72°F, material/metal thickness ratio = 2).

Fig. 6.30. Effects of temperature, thickness ratio, and frequency on effectiveness of homogeneous damping (14).

acoustic energy (i.e., the surface area must be relatively large).

- For frequencies above one-half the critical frequency ($f_c/2$), the damping treatment should be attached to the panel surface and concentrated in regions of maximum resonant amplitude. (A damping treatment applied over an entire panel surface can decrease radiation of sound by the panel. However, improvement is primarily the result of increased surface weight rather than energy dissipation by the damping treatment.)
- For frequencies less than about $f_c/2$ for airborne sound excitation, damping treatment should be concentrated near panel mounting and structural attachment points.
- The application of one thick layer of damping material on a panel surface is more effective than two layers (of same total thickness), one on each side of panel.

6.13.4. Frictional Damping

Frictional damping is proportional to the normal force and displacement between the damping material and the panel upon which it is placed. However, it can be approximated by an equivalent viscous-damping coefficient at or near panel resonances; equivalence is determined by equating the average energies dissipated per cycle. Frictional damping requires relatively large values of both the normal force and the coefficient of friction between the surfaces. Three commonly used materials are jute, urethane foam, and cork.

6.13.5. Common Damping Materials

High damping has been designed into some special metals, but vibration damping usually involves the application of viscoelastic materials such as rubbers, or plastics, etc. to the vibrating member. The most suitable substances are the high-molecular-weight polymers.

Materials intended solely for the purpose of vibration damping, are available in semifluid form for application with a spray gun or towel. A few are soft enough after drying to indent with the thumbnail, but most of the highly ef-

fective compositions are hard and seemingly brittle at room temperatures. Damping materials are also available in the form of precured sheets or rolls which can be glued on while others can be supplied already laminated to the metal part. Some even come with a magnetic backing so they can be fastened simply by laying pieces on the part to be damped (Appendix VI-4).

6.13.6. Bonding

In all of the types of damping that are available it is essential that a good transfer of energy between the vibrating panel and the damping material can take place. For this to occur the surfaces must be securely bonded together. And since the bonding agent becomes part of the system its physical properties are also important. Generally, however, the bonding agent is either built into the material that is to be sprayed or trowelled on or can be purchased from the supplier of the damping material. Obviously a secure application of damper will include a thorough cleansing of all grease and dirt from the surface to be damped.

6.13.7. Laminates

Because of the many applications where damping can provide an economical as well as effective solution to noise and vibration problems, the producers of damping products include ready-made laminated panels and sheet materials. There are now available on the market many different combinations of layer-laminated damped materials. These materials offer high damping for manufacturers in particular, but the retrofit noise control engineer should not overlook the possible use of these products in his applications.

REFERENCES

(1) DenHartog, J. P. 1947. *Mechanical Vibrations*, 3rd ed. New York: McGraw-Hill.
(2) Reddick and Miller, 1949. *Advanced Mathematics for Engineers*, New York, Wiley & Sons.
(3) Crede, C. E., and C. M. Harris, eds. 1961. *Shock and Vibration Handbook*, Vol. 1. New York: McGraw-Hill.

(4) Crede, C. E. 1951. *Vibration and Shock Isolation*. New York: John Wiley and Sons.

(5) Thomson, W. T. 1985. *Theory of Vibration with Applications*. Heightstown, NJ: Prentice Hall.

(6) Phelan, R. 1970. *Fundamentals of Mechanical Design*, New York: McGraw-Hill.

(7) Burr, A. H. 1981. *Mechanical Analysis and Design*. New York: Elsevier.

(8) Mabie, H. H., and F. W. Ocvirk. 1957. *Mechanism and Dynamics of Machinery*. New York: John Wiley and Sons.

(9) Catalogues of Barry Corporation, Korfund Co. Inc., Lord Manufacturing Co., M-B Manufacturing Co., Vibro-Acoustics Co.

(10) Irwin, J. D., and L. R. Graf. 1979. "Industrial Noise and Vibration Control," Prentice-Hall.

(11) Miller, L. N. 1981. *Noise Control for Buildings and Manufacturing Plants*. Cambridge, MA: Bolt, Beranek and Newman, Inc.

(12) *Sound and Vibration Magazine*. 1981. "Materials for Noise Control." Bay Village, OH.

(13) Lord, H., W. S. Gately, and H. A. Evensen. 1981. *Noise Control for Engineers*. New York: McGraw-Hill.

(14) Emme, J. 1970. "Composite Materials for Noise and Vibration Control." *Sound and Vibration*, July.

(15) ASHRAE. 1987. *HVAC Systems and Applications*. Atlanta, GA: American Society of Heating Refrigerating and Air Conditioning Engineers, Inc.

(16) Jones, R. S. 1984. *Noise and Vibration Control in Buildings*. New York: McGraw-Hill.

(17) Nashef, A., D. Jones, and J. Henderson. 1985. *Vibration Damping*. New York: John Wiley and Sons.

(18) Beranek, L. L. 1971. *Noise and Vibration Control*. New York: McGraw-Hill.

(19) ANSI. 1983. "Guide to the Evaluation of Human Exposure to Vibration in Buildings. ANSI S3.29-1983. New York: American National Standards Institute.

Chapter 7

NOISE CRITERIA AND REGULATIONS

7.1. INTRODUCTION

The best definition of noise, as many experts will agree, is *unwanted sound*. What is most indisputably unwanted about noise is its capacity to cause hearing loss and the manner in which it invades our everyday lives by causing annoyance. And these seriously limiting disabilities are far more insidiously acquired than most people realize.

Never, insist the worriers, have the hairlike sensors of the inner ear twitched to such a range of roaring, buzzing, beeping, grinding, howling, jangling, blaring, booming, screeching, whining, gushing, banging, and crashing. And even with certain controls and regulations which have been introduced, many people— particularly in large urban areas—feel it is not getting any better. The more militant antipollutionists blame racket for such woes as heart disease, high blood pressure, stomach ulcers, and sexual impotency.

Noise is not new. Juvenal, in imperial Rome, bemoaned the all-night cacophony of the city, observing that "most sick people perish for want of sleep." (Chariot racing was eventually prohibited on village streets at night in Rome because of the adverse effects on the community—the first example of noise control.) To Germany's Schophenhauer, it was clear that "the amount of noise which one can bear undisturbed stands in inverse proportion to his mental capacity, and may therefore be regarded as a pretty fair measure of it."

Moreover, much of what irritates modern man is simply new noise traded in for old. The ear that flinches at the blat of an air horn from a tractor/trailer rig might recoil as much from the clang-rattle-crash of the old trolley. The whine of rubber tires replaces the screech of unsprung cart wheels on cobble-stones; the backfire of unmuffled trucks and motorcycles supplants the ringing hooves of dray horses.

Over the past three or four decades, considerable progress has been made to develop and implement criteria and regulations for the control of today's environmental noise. When interest in, for example, community noise annoyance first developed, the relationship between physical noise measurements and the response of an "average" population was based on the tolerance to nominally steady noise levels which could readily be measured. Controls, where they were instituted, were of a simple type, often only stipulating that a noise criterion level should not be exceeded, with perhaps a lower value to be used at night. More recently, however, much work has been done to determine parameters which adequately describe not only a steady-state noise level, but also the variability of fluctuating or intermittent noise levels, their duration, the time of their occurrence, the location of the noise occurrence (i.e., industrial, commercial, residential, or suburban area), and the level of the background noise in the absence of the annoying noise.

This chapter includes data and discussions on criteria and, where applicable, regulations for (a) acceptable living and working environment *indoors*, (b) conservation of hearing in high-noise-level *workplaces*, and (c) community exposure to noise *outdoors*.

Most of the discussions on criteria and regulations, as they pertain to Canada, relate to situations which evolved in the U.S.A. In the U.S.A., noise control comes under federal legislation, where three primary agencies are designated to issue noise control regulations under the Occupational Safety and Health Act

(OSHA) (1970) and the Noise Control Act (1972). These agencies, and their responsibilities, are as follows:

- *The Department of Labor* (DOL) has the responsibility for implementing the OSHA, which was enacted to ensure safe and healthful conditions at the work place.
- *The Environmental Protection Agency* (EPA) has the responsibility for controlling environmental noise to protect the health and welfare of the general public under the Noise Control Act. Its primary responsibility is to establish national, uniform (federal) noise standards for major products and sources which are involved in interstate commerce; state and local governments still retain the right to establish and then enforce controls on environmental noise by regulating the use, operation or movement of noise sources and by establishing the levels of noise permitted in specific environments.
- *The Department of Transport* (DOT), through the Federal Aviation Administration (FAA) and the Federal Highway Administration (FHWA), is required to work closely with the EPA in establishing aircraft regulations and in undertaking programs to control truck and highway noise.

Other federal agencies concerned with noise regulations include the Department of Defense (DOD), the Department of Housing and Urban Development (HUD), the Department of the Interior, and the General Service Administration. However, the regulations issued by these agencies are usually applicable in limited areas (for example, coal mines) or for special purposes (such as federal procurements or federally insured housing).

As can be seen from the above, noise criteria regulations are well advanced in the U.S.A., both on a federal and state level. This is not the case for Canada, where there has been little effort to regulate noise on a federal level, with, perhaps, the exception of the use of aircraft around airports and the planning and siting of airports, and the regulations under the Motor Vehicle Safety Act of noise emitted by trucks, buses, and passenger vehicles imported or manufactured for sale in another province. (For a very thorough coverage of regulations and the "law relating to noise," refer to the article by J. P. S. McLaren in Ref. 38.) In Canada, noise criteria and regulations vary from province to province (where they exist), and with agreement with the provinces, they may also vary between municipalities within a province (38). Most legislation which has been passed has been at the municipal level, in the form of antinoise bylaws. This noise legislation is quite varied among communities.

However, most of the criteria and regulations that are applicable in the U.S.A. are also applicable in Canada; many have been adopted or are being considered for adoption. Where there are specific exceptions, such as in the Ontario Ministry of the Environment's Model Municipal Noise Control By-Law (1), the Province of Alberta Planning Act (2), the federal Department of Fisheries and the Environment's "environmental assessment policy" (3), and indirect control by the federal government re approval of spending power by Central Mortgage and Housing Corporation where guidelines establishing restriction on housing near airports may be broached, then these will be identified as required.

Currently in Canada, a draft National Guidelines for Environmental Noise Control has been prepared to provide information to municipalities, consultants, provincial planners, industries, designers, and legislators at all levels of government (32). It has been prepared by the Working Group on Environmental Noise for the Federal/Provincial Advisory Committee on Environmental and Occupational Health. (This Committee is comprised of senior officials from the health, labor, and environmental departments of the federal, ten provincial, and two territorial governments in Canada. It is charged with advising Ministers and Deputy Ministers of Health on all matters of environmental and occupational health.)

This document presents methods for the assessment, measurement, and legislative control of environmental noise in such a way as to provide a common basis for this work across Canada, while at the same time providing options to allow flexibility of choice for specific needs. This document may be adopted or modified, in whole or in part, for inclusion into provincial or municipal legislation or into codes of practice or guidelines.

Although noise control can be implemented at any level of government, the intent is to encourage municipal control based on provincial and federal policy direction. This procedure has been taken as it has been found that environmental noise problems are local, and without this municipal involvement, noise problems are difficult to solve.

Municipalities have to carry the burden in environmental noise abatement, while the planning processes are usually conducted through a coordinated approach between the various levels of government.

Although no other country has a similar guideline at this time, the Organization for Economic Co-Operation and Development (OECD) has been active in providing guidance to strengthen noise abatement policies in the area of environmental noise in all member countries (39).

A more complete review and summary of the Guideline is given in Ref. 32. Although the format of the current Guideline has changed somewhat from that discussed in Ref. 32, the basic premise of the Guideline and its intended implementation are still relevant.

7.2. INDICES FOR ENVIRONMENTAL NOISE

Typical criteria employed to describe time-varying noise take both level and duration into consideration. For simplicity, these performance indices are single-number criteria. There are certain measures for quantifying noise exposures that have good correlation with human response to these noises.

7.2.1. Direct Rating

Direct rating of sound levels is probably the most widely-used in community noise studies. As was discussed in Chapter 3, there are:

- *Overall sound-pressure level.* The overall sound-pressure level is a value obtained with the linear response of a sound-level meter. Although it does not correlate with human response to noise, it is useful when determining the total noise power of a source.
- *A-weighted sound-pressure level.* The A-level provides a single-number rating that has been found to correlate well with people's subjective assessment of the loudness or noisiness of many types of sounds and with hearing-damage risks due to continuous noise.
- *C-weighted sound-pressure level.* The C-level is used as an overall measure of sound with equal weighting given to all frequencies from 31.5 Hz to 8000 Hz, the average range of human hearing.
- *D-weighted sound-pressure level.* The D-level was developed as a simple approximation of perceived noise level (PNL—see next section), used mainly for assessing the disturbance likely to be caused by flyover of jet aircraft.

Although the dB(A) was derived from the 40-phon loudness contour (see Section 3.4), it has been found to be a reasonable measure of loudness for all levels of sound pressure, both from the viewpoint of hearing damage and of annoyance. Despite its "loudness" origin, it is overwhelmingly accepted as the easiest common sense way to measure sound for its acceptability.*

7.2.2. Perceived Noise Level (PNL)

Personal experience and environment play an important role with regard to how much noise individuals will tolerate before they deem a particular noisiness unacceptable. Extensive studies have arrived at certain characteristics of noise which are annoying for most individuals.

For example, it has been found that if the noise is concentrated in a narrow bandwidth, it

---

*For a discussion regarding the use of A-weighting in assessment of annoyance or aversiveness, refer to Section 7.5.3.

is considered noisier than if the same energy is spread over a somewhat wider band. A noise with a rapid rise time is found to be noisier than the same noise with a slower rise time. An intermittent or irregular noise is deemed noisier than if it remained steady. And, as we have seen (Section 2.3), high-frequency noise (about 1.5 kHz and up) is judged to be noisier than low-frequency noise, even though both have the same sound-pressure level.

The perceived noise level is a single-number rating of the *noisiness* of a sound computed from octave or $\frac{1}{3}$-octave band sound levels. The PNL is patterned after the loudness level, except that equal noisiness contours are employed instead of equal loudness contours (4, 5). The PNL is used mainly in assessing noise disturbance due to aircraft flyover.

7.2.3. Equivalent Sound Level (L_{eq})

As was discussed in Section 3.8.1 under instrumentation, the equivalent sound level is the level, usually in dB(A), of a hypothetical constant sound which, if substituted for the actual sound history over the same period of time, would represent the same amount of acoustic energy arriving at a given reception point. The purpose of L_{eq} is to provide a single-number measure of time-varying sound for a predetermined time period. The designation of *equivalent* signifies that the numerical value of the fluctuating sound is equivalent in level to a steady-state sound with the same total energy. For definitions of L_{eq}, see Section 3.9.1.

Ambient sounds in a community, highway traffic noise containing variable mixtures and quantities of autos and trucks, airport noise, railroad noise, and variable industrial noise are representative noises that can be analyzed statistically for L_{eq}. Studies are being carried out intensively in many countries to construct rating systems for estimating and predicting community reaction to noise, using statistical descriptors like L_{eq} (see also the later discussion on exceedance levels, Section 7.2.5).

The U.S. Environmental Protection Agency (6) has selected L_{eq} as a preferred environmental noise descriptor for the following reasons:

- It is applicable to evaluation of long-term noise.

- It correlates well with known effects of noise on people.
- It is simple, accurate, and practical.
- It can be used for both planning and enforcement.
- The required measurement equipment is standardized and commercially available.
- It is closely related to existing methods already in use.
- It lends itself to the use of monitors that can be left unattended for long periods of time.

7.2.4. Day-Night Level (L_{dn})

In order to represent the greater annoyance caused by a sound intrusion at night (e.g., when watching TV, radio, or sleeping), the day-night sound level has been developed. (This index has been adopted mainly in the U.S.A. and some European countries; it has not been adopted, other than as a guide, in Canada.)

The L_{dn} suggests that the equivalent sound level L_{eq} occurring between 10:00 p.m. and 7:00 a.m. should be augmented by 10 dB(A) before being combined with the equivalent sound level for the period 7:00 a.m. to 10:00 p.m., to give the day-night level. If L_d is the equivalent sound level for the 15-hour daytime period, and L_n is the equivalent sound level for the 9 hour nighttime period, the day-night L_{dn} is

$$L_{dn} = 10 \log_{10} \frac{1}{24} [15 \times 10^{L_d/10}$$
$$+ 9 \times 10^{(L_n + 10)/10}]. \qquad [7.1]$$

7.2.5. Exceedance Levels (L_N)

With the advances in instrumentation sophistication and statistical analysis procedures, other methods (in addition to L_{eq} and L_{dn}) have evolved for measuring time-varying sound levels such as from highway and airport activities. The acoustic literature now carries such terms as L_1, L_{10}, L_{50}, L_{90}, and others. L_{10} means the sound level that is equaled or exceeded 10% of the time; it can be obtained directly from some types of statistical noise analyzers, or can be derived from the combination of probability distribution plot and cumulative distribution curve, plotted from the sampling of a time-varying noise (see Section 3.9.2.2).

7.2.6. Noise Pollution Level (NPL)

The noise pollution level was developed in an effort to improve upon other single-number noise rating systems which consider only intensity. The NPL attempts to account for the effect of fluctuations of the noise environment (11). NPL is defined as

n
th
er
y-
(3)
any
as

the use of space, ... tion

---

*Standard deviation can be computed from

$$\sigma^2 = \frac{\sum (L_{Ai} - \mu)^2 \Delta t_i}{24}$$

where

$$\mu = \frac{\sum L_{Ai} \Delta t_i}{24} = \bar{L}_A,$$

24 is duration of the survey in hours, and L_{Ai} is the A-level during time Δt_i.

---

of exposure to the noise, the type of noise (impulsive, pure tone, broadband, continuous), the functions which must be performed by the people in the noise environment, etc. As a consequence, some noise criteria are specified in standards related to all three aspects listed above, for both indoor and outdoor space and activities. Hence, some criteria listed in this section for indoor spaces will also be listed in a following section for outdoor environments. (For an extensive listing of criteria used in assessing noise environments, see Refs. 4 and 6).

3.1. NC, PNC, and RC Criteria

From many case histories and studies of people's judgments of noise environments, families of sound-pressure-level curves have evolved to describe acceptable noise environments for a variety of activity or functional indoor areas. *Noise criteria* (NC) or *preferred noise criteria* (PNC) curves are two widely used sets. Intrusion or disturbance from noise and interference with speech sounds by noise (such as in conversation, radio and television reception, and telephone communication) are the basis for these NC and PNC curves. *Room criteria* (RC) curves are preferred where HVAC (heating, ventilating, and air conditioning) systems are involved.

7.3.1.1. NC Curves.
The NC curves (12) are shown in Fig. 7.1. The NC number which is assigned to each curve is the *speech-interference level* (SIL) value, obtained by averaging the values in the octave bands of 500, 1000, and 2000 Hz (see Section 7.3.3 on speech interference). These values originated with studies by Beranek in 1957 (12) and later adapted by ASHRAE (the American Society of Heating, Refrigeration, and Air Conditioning Engineers). The NC rating of a given noise environment is determined by superimposing its octave-band sound spectrum on the family of NC curves, and noting the highest penetration of an NC curve. The next highest NC curve is the *rating*. This rating is then compared with the range of indoor design goals recommended by ASHRAE (see Table 7.1. Section 7.3.1.3 for acceptability).

Generally where a range of NC values is

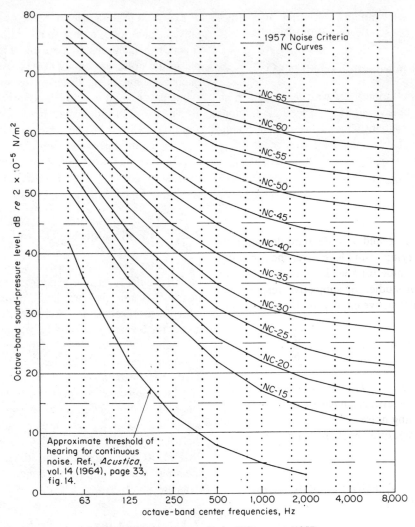

Fig. 7.1. Indoor noise criteria (NC) curves (1957).

given, the lower value should be used for the more critical spaces in the category as for situations where it is known or suspected that receivers of the noise will be somewhat hostile to the noise-making facility; alternatively, the higher NC value may be used for the less critical spaces in the category.

7.3.1.2. PNC Curves. The PNC (preferred noise criteria) curves (13) are shown in Figure 7.2. The application and category classifications of Table 7.1 are applicable to the PNC as well as the NC curves. The term ''preferred'' was first introduced to encourage the use of the new octave frequency bands when North America joined the International Standards Organization (ISO) in acoustic standardization. Also, as a result of noise-control progress over the years, some objection to the NC curves led to their modification in 1971 (13). The PNC curves are lower than the NC curves by a few decibels for low-frequency ''rumbling'' noises and high-frequency ''hissing'' noises, and hence are ''preferred'' also for these reasons. For critical spaces requiring very low background sound levels (such as in studios and auditoria), the PNC curves rather than the NC curves, should be the design target.

7.3.1.3. RC Curves. The Room Criteria (RC) curves (Figure 7.3) are preferred for meeting HVAC system design goals. The shape

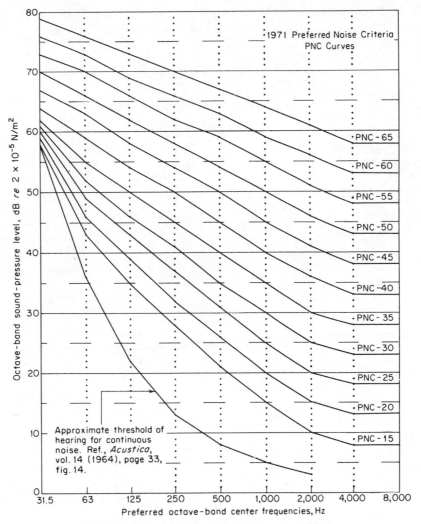

Fig. 7.2. Preferred noise criteria (PNC) curves (1971).

of these curves differs from the NC and PNC curves at low and high frequencies (14).

The RC curve represents a close approximation of a well balanced, bland-sounding spectrum and can be used to maintain background sound levels for masking or other purposes. An optimum balance in sound quality is achieved by approximating the shape of the curve to within $\pm$ 2 dB over the entire frequency range. If low frequency levels (31.5 to 250 Hz) exceed the design curve by 5 dB, the sound is likely to be rumbly; exceeding the design by 5 dB at high frequencies (2000 to 4000 Hz) will cause the sound to be hissy.

Recommended NC and RC design levels are given in Table 7.1. The values for each type of room are justifiable for buildings in quiet locations or where sound transmission is reduced through exterior walls. An increase of 5 dB in NC or RC levels is permissible for buildings in relatively noisy locations without adequate exterior wall sound transmission loss, or where the owner's requirements and budget make lower costs desirable.

The ranges of Table 7.1 are based on the fact that sound radiated from properly designed and maintained air-conditioning equipment is typically steady and broadband in character.

More stringent limits, 5 to 10 dB lower, should be specified for impulsive sounds or sounds containing prominent pure tones. Table

Table 7.1. Recommended Indoor Design Goals for Air-Conditioning System Sound Control[a] (Note: These are for unoccupied spaces, with all systems operating.)

Type of Area	Recommended RC or NC Criteria Range	Typical A-Levels Associated With Criteria Range
1. Private residences	25 to 30	30 to 35
2. Apartments	30 to 35	35 to 40
3. Hotels/motels		
a. Individual rooms or suites	30 to 35	35 to 40
b. Meeting/banquet rooms	30 to 35	35 to 40
c. Halls, corridors, lobbies	35 to 40	40 to 45
d. Service/support areas	40 to 45	45 to 50
4. Offices		
a. Executive	25 to 30	30 to 35
b. Conference rooms	25 to 30	30 to 35
c. Private	30 to 35	35 to 40
d. Open-plan areas	35 to 40	40 to 45
e. Computer/business machine areas	40 to 45	45 to 50
f. Public circulation	40 to 45	45 to 50
5. Hospitals and clinics		
a. Private rooms	25 to 30	30 to 35
b. Wards	30 to 35	35 to 40
c. Operating rooms	25 to 30	30 to 35
d. Laboratories	30 to 35	35 to 40
e. Corridors	30 to 35	35 to 40
f. Public areas	35 to 40	40 to 45
6. Churches	25 to 30[b]	30 to 35
7. Schools		
a. Lecture and classrooms	25 to 30	30 to 35
b. Open-plan classrooms	30 to 35[b]	35 to 40
8. Libraries	30 to 35	35 to 40
9. Concert halls	[b]	
10. Legitimate theaters	[b]	
11. Recording studios	[b]	
12. Movie theaters	30 to 35	35 to 40

[a]Design goals can be increased by 5 dB when dictated by budget constraints or when noise intrusion from other sources represents a limiting condition.
[b]An acoustical expert should be consulted for guidance on these critical spaces.

7.1 is *not* applicable to appliances such as dishwashers and room air conditioners, which operate only when required for short periods under the owner's control. Experience indicates that under these conditions most people seem to tolerate higher sound levels than those listed in the table. Similarly, they are not intended to rate the level in noisy spaces, such as computer rooms or word processing centers.

To determine if the recommended RC design levels listed in Table 7.1 have been met, measurements of at least three of the four octave band levels between 250 and 2000 Hz should lie within the indicated 5 dB RC range. Should the levels in the octave bands below 250 Hz be greater than the RC range established at higher frequencies, a potential rumble problem is indicated.

7.3.2. Direct Rating Guides and Standards

The cumulative noise exposure level using in the EPA levels document (6) relies upon measurement of either L_{eq} (for 8 hr. or 24 hr. periods) and L_{dn}, and therefore is somewhat difficult to measure. Moreover, it has been found that if the sound-frequency spectrum is continuous, contains no sharp peaks or dips, and extends over a wide frequency range, the A-level

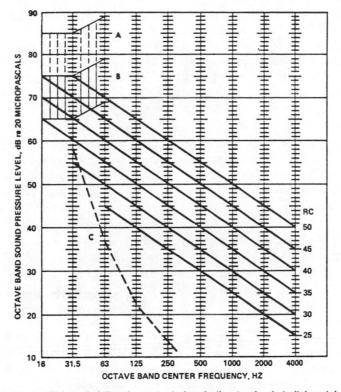

Region A: High probability that noise-induced vibration levels in lightweight wall and ceiling constructions will be clearly feelable; anticipate audible rattles in light fixtures, doors, windows, etc.

Region B: Noise-induced vibration levels in lightweight wall and ceiling constructions may be moderately feelable; slight possibility of rattles in light fixtures, doors, windows, etc.

Region C: Below threshold of hearing for continuous noise.

Fig. 7.3. RC (room criterion) curves for specifying design level in terms of a balanced spectrum shape.

correlates as well with human response as any of the more complicated rating schemes. (Note that many indoor and outdoor noises conform to the above description.) For this reason, the A-level has been adopted in many standards and codes for rating both indoor and outdoor environments.

Although there is no single widely accepted standard for indoor noise environments (with the exception of standards which are generally followed in occupational noise situations), there have been numerous studies which provide recommended A-levels for specific situations. For example, Table 7.1 lists A-level indoor design goals recommended in the *ASHRAE Systems Guide* for steady background noise in various indoor spaces (14). Other recommended A-

level criteria are reported in the EPA levels document (6).

7.3.3. Speech Interference (PSIL)

Extensive data on the intelligibility of speech and the effects of masking have led to procedures whereby one can, from physical measurements correlated with observations of human responses, calculate an index of speech intelligibility. This index is called the *articulation index* AI, and is discussed in a later section of this text. (It is also covered in ANSI S3.5-1969.)

A simplified version of AI is proposed by Beranek (15) for predicting effectiveness of person-to-person speech communication in the presence of noise. In this procedure, the arithmetic average of the 500, 1000, and 2000 Hz octave-band levels of the interfering noise is determined. This average, called the *preferred speech interference level* (PSIL)—"preferred" because of the use of the new ISO standardized octave bands rather than the earlier "old" bands—is compared with subjective data to determine whether speech communication will be easy or difficult in the environment being considered. It can also be used to determine the changes in the band levels needed to improve the ease of communications.

Example. The PSIL of a factory noise environment was determined by measuring the octave-band levels at the three critical center frequencies to obtain (81 + 80 + 80) ÷ 3 = 80 dB. Examination of the rating chart in Fig. 7.4 shows that 80 dB is a high figure for PSIL; verbal communication is possible only at less than 2 ft, and then only if a special effort is made to communicate.

As Fig. 7.4 indicates, the A-weighted sound-pressure level can be used as a simplified alternative index when octave band readings cannot be made. Since PSIL and A-weighted level L either ignore or de-emphasize sound outside the 500—2500 Hz range (the lower and upper cutoff frequencies of the 3-band range), neither should be used when very low-frequency (below 500 Hz) or high-frequency (above 2500 Hz) noise is present. However, most practical situations requiring evaluation of speech intelligibility are covered within this frequency range (see Section 7.3.5 on articulation index and speech privacy).

The levels noted in Fig. 7.4 are for persons with average voice speaking strengths face to face, with the spoken material not familiar to the listener. (A value of 5 dB should be subtracted from these levels for "formal" or experienced speakers). The chart assumes that speaker and listener are in a free field, i.e., the room is not reverberant with reflecting surfaces nearby which could add to the existing background level.

7.3.4. Other Criteria (or Guides) for Indoor Environments

The Noise Pollution Control Section of the Ontario Ministry of the Environment has, in all probability, been the most active group in Canada with regard to establishing acceptable community noise limits and integrating noise considerations and standards into land-use planning. Initially, the section derived a set of noise-level limits for new residential development from a publication of the U.S. Department of Housing and Development (16). Through a process of development, these limits

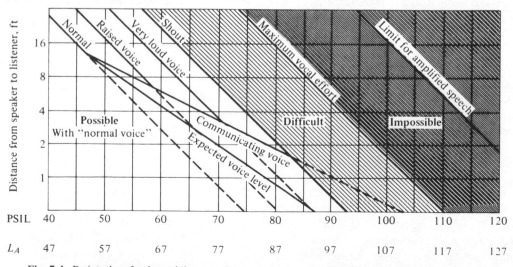

Fig. 7.4. Rating chart for determining speech communication capability from speech-interference levels.

Table 7.2. Indoor Sound Level Limits (Ontario Ministry of the Environment).

	L_{eq}
Bedrooms, sleeping quarters, hospitals, etc. (Time period 2300–0700 hrs)	40 dB(A)
Livingrooms, hotels, motels, etc. (Time period 0700–2300 hrs)	45 dB(A)
Individual or semi-private offices, small conference-rooms, reading room, classrooms (Time period 0700–2300 hrs)	45 dB(A)
General Offices, reception areas, retail shops, stores (Time period 0700–2300 hrs)	50 dB(A)

emerged in the form of Table 7.2, which is extracted from the Ontario Ministry of the Environment's Model Municipal Noise Control By-Law (1).

Another set of sound-level limits currently in use in Canada are those contained in "Road and Rail Noise: Effects on Housing," a Canada Central Mortgage and Housing document produced in cooperation with the Division of Building Research of the National Research Council of Canada (17). Methods are suggested for determining the noise level at a building site caused by road and/or rail activity, and where levels are too high, reducing them to acceptable limits within the various parts of the building where the noise occurs. The relevant criteria (or limits) are outlined in Table 7.3.

To improve the design of new housing projects, the corporation has, for a number of years, reviewed proposed subdivision plans prior to their acceptance for mortgage financing or insurance under the National Housing Act. It therefore has considerable influence with re-

Table 7.3. Maximum Acceptable Levels of Road and Rail Traffic Noise in Dwellings and Outdoor Recreation Areas (Canada Central Mortgage and Housing Corp.).

	24-hr L_{eq}
Bedrooms	35 dB(A)
Living, dining, recreation rooms	40 dB(A)
Kitchens, bathrooms, hallways, utility rooms	45 dB(A)
Outdoor recreation areas	55 dB(A)

gard to seeing that the above-mentioned limits are implemented.

7.3.5. Speech Communication and Privacy

Over the years of offering a consulting service to communities and industry, the writer has often been asked to assist in improving speech communication and privacy in open office planning, open school room concepts, etc. Paramount in the study of these designs is the *articulation index*. Although there are a number of methods of arriving at articulation index (ANSI S3.5-1969 being one), a simplified method has been recorded in Ref. 18.

In Fig. 7.5, the area between the upper and lower curves is the area in which our normal speech most often falls. For example, the part of our speech which a frequency analysis reveals to be in the $\frac{1}{3}$-octave band centered at 500 Hz is found to have SPLs in the range of 43–73 dB for the great majority of people nearly all the time. (This is a 30 dB range.)

Furthermore, studies have shown that those frequencies are not all equally important in speech intelligibility. Speech intelligibility depends on our accurately perceiving speech in approximately the 1500–3000 Hz range much more than in the frequency ranges to either side of this. The number of dots at each center frequency in Fig. 7.5 reflects the relative importances to speech intelligibility of the various frequencies. There are 182 dots in Fig. 7.5.

To calculate AI, one plots on Fig. 7.5 the $\frac{1}{3}$-octave frequency spectrum of the background noise and simply counts the number of dots lying between the spectrum and the upper curve. Dividing this number by 182 gives the AI.

Thus, an intense background sound having a spectrum lying on or above the upper curve will have an AI of zero—implying a large interference with speech. A sound of such low intensity as to have a spectrum lying on or below the lower curve will have an AI of at least unity—implying little or no interference with speech.

Generally, an AI of less than 0.3 is taken to be unsatisfactory for good communication and

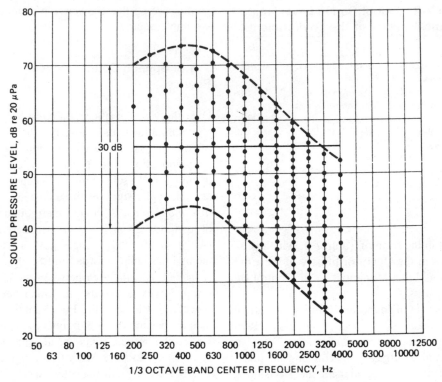

Fig. 7.5. Curves for the computation of the articulation index, (AI). The straight line is the spectrum given in the example.

greater than 0.7 is considered very good. Between 0.5 and 0.7 may be called good, while the range from 0.3 to 0.5 is marginal; but perhaps acceptable. The figures given above have been derived empirically from everyday experience—based upon experiments which have been performed to relate these figures to the percentage of standard syllables, words, or sentences uttered by various voices and perceived correctly by various subjects in various situations. It should, however, be borne in mind that in special situations (such as in aircraft communication, where trained individuals use oft-repeated, phonetically efficient phrases), an acceptable AI may be less than 0.3.

Example of AI. A background noise in an office is expected to have a flat $\frac{1}{3}$-octave band spectrum of 55 dB between 200 and 4000 Hz. Calculate the articulation index.

The spectrum of the noise is plotted on Fig. 7.5 as shown by the horizontal 55 dB line. When the dots between this line and the upper curve are counted, it is found that there are 59.

Dividing this by 182 gives an AI of 0.32. Hence, barely marginal communication may be expected in this case.

7.3.6. Open-Plan Offices

Open-plan offices are an indoor environment requiring particular attention to the assessment of acoustical suitability. The ideal office environment is quiet enough to permit any occupant to talk easily with a visitor or on a telephone, yet not so amenable to speech communication that his/her speech intrudes on the acoustic privacy of other occupants of the same office. Thus for proper speech communication between subjects, a high articulation index (AI) is required; on the other hand, for proper speech privacy conditions, a low articulation index is desirable.

It is a common experience that for effective speech communication in a noisy environment the speaker must reduce the distance between himself and the listener or increase the level of his voice to maintain a sufficient signal-to-noise

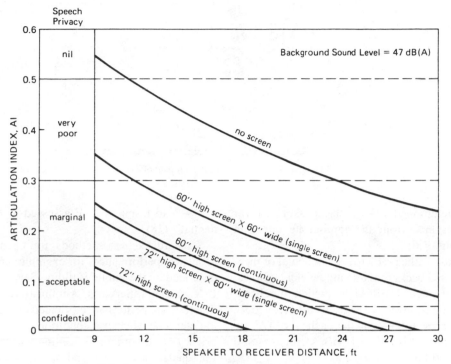

Fig. 7.6. Speech communication with various screen configurations.

ratio* and thus the necessary degree of speech intelligibility. Obviously there is an acceptable limit beyond which the increased levels of speech add to the overall noise level. This culminates in the well known cocktail party effect: with only a few people present, communication using normal voice levels is easy, but as the crowd increases the background noise also increases until speech communication is virtually impossible even with raised voice levels at close range. It is this problem that underlies the success or failure of the open office concept.

It has been shown that most people are unaware of broadband steady sound levels below about 35 dB(A) and will accept levels up to approximately 47 dB(A) (19). Levels above 47 dB(A) will cause the speaker to raise his voice to compensate for the higher background level; consequently, 47 dB(A) represents the practical upper limit to effective sound masking.

Figure 7.6 gives a series of curves that can be used for predicting the privacy conditions in an open plan at any receiver position with re-

spect to any source position. The curves are for a well designed open-plan office environment with a ceiling system having good sound absorptivity (a noise reduction coefficient greater than 0.8) and an electronic masking sound system** set at 47 dB(A), and utilizing either 72- or 60-in.-high screens. To be effective, the screens should have an NRC value in excess of 0.8 and a sound transmission class of 15 (20).

The curves in Fig. 7.6 are based on normal speech levels and assume the optimum speaker orientation, i.e., speaker facing receiver. In order to determine the speech privacy condition prevailing at any workplace the following procedure should be used:

a. Determine speaker-to-receiver distance;
b. Establish screen height and configuration;

---

*Signal-to-noise ratio is the extent to which the speech sound signal exceeds the background noise.

**An electronic masking sound system consists of a broadband random noise generator, amplifier, and suitable speakers. The speakers are usually placed above a drop ceiling, directed at the subflooring in order to diffuse the sound. For a discussion on the preferred random noise signal spectrum, and for further examples in speech privacy, see Refs. 18, 21, and 23.

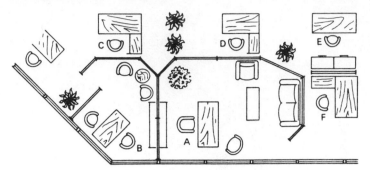

Fig. 7.7. Segment of typical open-plan office.

c. Determine the AI for the speaker-receiver distance from the appropriate curve in Fig. 7.6;

d. Adjust the predicted AI for actual or measured background sound level as follows: for each 1 dB(A) by which the background level is above 47 dB(A), reduce AI by 0.03, and for each 1 dB(A) by which it is below 47 dB(A), increase AI by 0.03;

e. The curves in Fig. 7.6 being for speaker facing receiver, reduce AI by 0.08 and 0.15 for orientations of 90 and 180 degrees, respectively.

An example of the use of Fig. 7.6 is given above. Fig. 7.7 shows a segment of a typical open-plan office and Table 7.4 gives the calculations of speech privacy conditions for a number of workplaces in the office. Note the effect of the continuous screens on speech privacy—A to B and A to F assessed as "confidential" (18).

Standardized test methods for open office acoustical performance, for example, are those described by the Public Building Service of the U.S. General Services Administration (22). The tests are performance-oriented and are used to determine whether a condition of acceptable privacy exists for specific ceiling constructions for standardized screen and background sound conditions. Useful descriptions of open-plan office testing are given in Refs. 19, 20, 21, and 23.

7.4. REGULATIONS OF HEARING-DAMAGE RISK IN INDUSTRY—THE U.S. OCCUPATIONAL SAFETY AND HEALTH ACT STANDARDS

The U.S.A. was the first western country to introduce industrial noise regulations. Under the

Table 7.4. Assessment of Speech Privacy in an Open-Plan Office (see Fig. 7.7).

Step	Reference			Example		
1. Location and interzone distance	Design drawings (Fig. 7.6)	A to B 9 ft	B to C 12 ft	C to D 15 ft	E to F 9 ft	A to F 20 ft
2. Screen height and configuration	Design drawings	72″ continuous	72″ continuous	No screen	60″ single	72″ contin
3. Determine basic AI	Fig. 7.5	0.13	0.08	0.43	0.35	0
4. Adjust for actual background sound level	(19)[a]	+0.03	+0.03	+0.03	+0.03	+0.03
5. Adjust for speaker-to-receiver orientation	(19)	−0.15	0	−0.08	−0.15	0
6. Prevailing AI	Sum of steps 3, 4, 5	0.01	0.11	0.38	0.23	0.03
7. Privacy condition	Fig. 7.5	confidential	acceptable	very poor	marginal	confidential

[a]Assume for the sake of the example that the background level is 46 dB(A)

Walsh-Healey Public Contracts Act of 1969, occupational noise levels were established on a national basis, applicable to industries which had government contracts in excess of $10,000. However, those standards were extended to the employees of all U.S. companies engaged in interstate commerce when the Occupational Safety and Health Act (OSHA) became law (7).

It should be noted that in other countries, industrial noise regulations or codes of practice similar to those enforced by OSHA now exist, or are in the processes of being adopted. However, in most cases these regulations are based on the equal energy principle (with the exception of Canada, at the time of writing*), which stipulates a 3 dB increase in permissible level for halving of the exposure time (the equal energy principle) rather than the 5 dB increase for halving of exposure time allowed by OSHA (see next sections). Both the British Code of Practice (8) published in 1972 and the ISO Recommendation R1999 (9) published in 1971, for instance, were based on the equal-energy principle. The ISO standard 1999 (1984) also stipulates the equal energy principle involving a 3 dB exchange rate.

Canada has chosen, in some provinces, to

*Proposed regulations under the Occupational Health and Safety Act, issued by the Ontario Ministry of Labour (1986), *do* incorporate the equal energy principle—see later section on regulations.

use the OSHA regulation as a guide. At the point of writing, the Ontario Ministry of Labour, through its Occupational Health and Safety Act of 1978, is moving toward identifying noise in the industrial environment as a designated substance—in which case, compliance with regulations so noted would be enforced and a systematic monitoring, control and hearing conservation program would be required (see Section 7.4.4).

7.4.1. Background to OSHA Regulations

For several decades, it has been recognized that high noise levels can cause varying degrees of hearing loss in people exposed for long periods of time. Early studies supported by government agencies had confirmed this finding, and various forms of noise exposure limits were recommended by the military branches as early as 1956 through 1965. In 1965, the document "Hazardous Exposure to Intermittent and Steady-State Noise" was prepared by the Committee on Hearing, Bioacoustics, and Biomechanics (CHABA) of the National Academy of Science and the National Research Council (10). It included hearing damage studies of some 20,000 people whose case histories had been studied on a statistical basis. Fig. 7.8 summarizes one portion of the study. The Part B curves show the influence of noise levels and

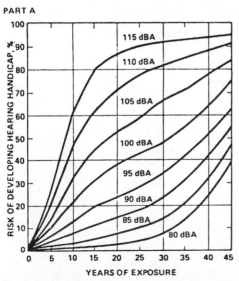

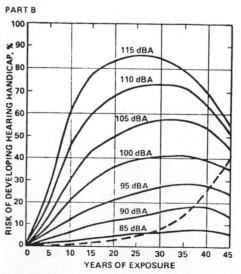

Fig. 7.8. Risk of noise-induced hearing damage versus years of exposure to noise of various A-weighted levels.

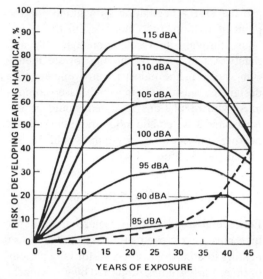

Fig. 7.9. Data from a U.S. Public Health Service study of 2000 people presented in the same format as Part B of Fig. 7.8. Dashed curve shows "risk" versus age (assume 20–65 years of age) for non–noise exposed group.

exposure time on hearing loss for various percentage of exposed people. Fig. 7.9 summarizes a similar study carried out later by the U.S. Public Health Service on approximately 2000 people.

In Fig. 7.8, Part A is for exposure to both occupational and nonoccupational noise and includes nonnoise factors of presbycusis and aging. The high rise of 80 dB(A) curve in the 35–45 year period is attributable to the nonnoise factors. Part B shows effect of noise only. To obtain Part B curves, the 80 dB(A) curve of Part A (assumed to contain negligible occupational noise-induced hearing loss) is subtracted from all the other curves of Part A. "Risk" is the percentage of a normal population expected to receive noticeable hearing loss (25 dB, ISO) (10).

The CHABA Study also included effects of intermittent noise. Table 7.5 lists the CHABA recommendations for maximum sound-pressure levels, in octave bands, for various single daily periods of exposure. Part A is for broadband noise and Part B is for narrowband noise or pure tones. The report also emphasizes the value of rest or recovery periods of relative quietness intermixed with periods of high noise levels. During these periods (which must be at least 10 dB quieter in all bands than the levels shown in the 8-hr column of Table 7.5), it is thought that the ear begins a recovery process from the previous noise exposure that helps prepare for the next noise exposure. In effect, for situations where the steady-state noise lev-

Table 7.5 Maximum Sound-Pressure Levels (in dB) Recommended for Hearing Conservation for Various Periods of Daily Exposure to Noise. From 1965 Chaba Study (10).

Octave Frequency Band (Hz)	Duration of Single Daily Exposure, in Hours					
	8	4	2	1	1/2	1/4
Part A. For Broad-band Noise						
125	97	103	111	119	127	135
250	92	96	101	107	115	123
500	89	90	94	99	105	112
1000	86	88	91	95	100	106
2000	85	86	88	91	95	100
4000	85	85	87	90	93	98
8000	86	87	90	95	100	105
Part B. For Narrow-band Noise or Pure Tones						
125	92	93	95	98	105	112
250	87	88	90	94	100	106
500	84	83	86	91	96	101
1000	81	82	85	89	93	97
2000	80	81	83	86	90	94
4000	80	80	82	85	88	92
8000	81	82	85	90	94	99

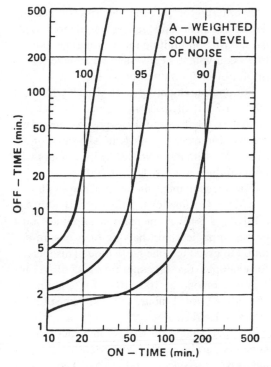

Fig. 7.10. Recommended time intervals of noise *on* and *off* for recovery from preceding noise and preparation for next noise exposure (10).

els are just marginally above the recommended 8-hr noise levels of Table 7.5, it is possible to reduce the effect of the higher noise levels by intentionally providing some scheduled quiet periods. Or, if the nature of the operator's work in the machinery room is somewhat intermittent, it would be possible to permit these higher noise-level exposures, provided that intermittent periods of relative quiet are also assured. Fig. 7.10 shows recommended combinations of *on* and *off* times for noise exposure that permit this recovery process.

The CHABA findings and recommendations have no legislative stature, but they provided much of the technical support for the 1969 noise regulations of the Walsh-Healey Public Contracts Act and the OSHA legislation that followed in 1970. CHABA studies are continuing. Although applications and interpretations of the 1965 CHABA report may vary, and may, at this point, be open to question, the study clearly showed that high noise exposures cause hearing loss in humans.

7.4.2. U.S. Occupational Safety and Health Act of 1970

The Occupational Safety and Health Administration (OSHA) was organized in the United States under the Department of Labor and was given authority to set up and enforce regulations aimed at providing greater safety to people who are normally exposed to somewhat hazardous working conditions and environments, including noise exposures. The OSHA noise regulations are summarized briefly below.

7.4.2.1. Continuous Noise.
The permitted daily noise exposure of the 1970 act are given in Table 7.6 for continuous noise, which is defined as noise of essentially steady-state character whose cyclical time variations, if any, occur at time intervals of less than 1 sec. When these noise levels change very little at the workplace during the day, and when the time spent by the worker at the worksite is fairly well defined, simple sound-level-meter measurements will suffice to establish employee exposure. A-weighted sound levels are used and should be measured with a Type 1 or Type 2 sound-level meter (ANSI Standard S1.4-1971) with the meter movement at the "slow" setting.

7.4.2.2. Mixed Exposures.
When sound levels vary from time to time or place to place in a worker's schedule, the effective "dose" of the mixed exposure is calculated as follows:

$$D = \frac{C_1}{T_1} + \frac{C_2}{T_2} + \frac{C_3}{T_3} + \cdots, \quad [7.4]$$

Table 7.6. Daily Maximum Noise Exposure Permitted by the Occupational Safety and Health Act of 1970.

Duration per Day, hours	Sound Level, dBA
8	90
6	92
4	95
2	97
$1\frac{1}{2}$	100
1	105
$\frac{1}{2}$	110
$\frac{1}{4}$ or less	115

where the C values are the durations in minutes or hours of exposure to the actual sound levels and the T values are the durations permitted under Table 7.6 for those sound levels. The value of D should not exceed 1.

Example. Suppose a worker is exposed for 2 hours at 100 dB(A), 2 hours at 95 dB(A), and 4 hours at 90 dB(A). Then, $C_1 = 2$ hr, $C_2 = 2$ hr, $C_3 = 4$ hr, and $T_1 = 1\frac{1}{2}$ hr, $T_2 = 4$ hr, and $T_3 = 8$ hr. Therefore

$$D = \frac{2}{1.5} + \frac{2}{4} + \frac{4}{8} = 2.33.$$

This would be an unacceptable exposure.

Next, suppose a 5-dB noise reduction is achieved, so that the worker is exposed for 2 hours at 95 dB(A), 2 hours at 90 dB(A), and 4 hours at 85 dB(A). Here, $C_1 = 2$ hr, $C_2 = 2$ hr, $C_3 = 4$ hr, and $T_1 = 4$ hr, $T_2 = 8$ hr, and T_3 has no limit, so it becomes infinity in the calculation. Therefore

$$D = \frac{2}{4} + \frac{2}{8} + \frac{4}{\infty} = 0.75.$$

This would be an acceptable exposure.

When the noise levels fluctuate widely and rapidly, or if the worker moves frequently, measuring the noise levels and determining exposure times can be difficult with simple sound-level-meter measurements. In such cases, a personal noise dosimeter with a lapel- or ear-mounted microphone is used. For such, the noise exposure or noise-dose has been defined. The noise-dose takes into account both the time-varying noise level and its duration. Most dosimeters read out the percentage of allowable daily noise dose, i.e., $100 \times D$. Hence, a readout of 75 corresponds to a value of $D = 0.75$. See Figs. 3.12(a) and 3.12(b) re example of use of personal noise dosimeter.

International Standard ISO 1999-1984 allows a 3-dB increase in noise level per halving of exposure duration, while OSHA Standard (USA) permits a 5 dB increase in noise level. (See Section 7.4.4 on recent studies in Canada in connection with 5-dB tradeoff versus 3-dB.)

The criteria for limiting the risk of hearing damage to an acceptable level is, in most countries, based on limiting the maximum exposure to an equivalent steady state noise level (L_{eq}) of 90 dB(A) for a duration of 8 hours.

For measurement in accordance with ISO 1999, a 3-dB increase in L_{eq} must be accompanied by a halving of the exposure time. This means, for example, that an exposure of 93 dB(A) for 4 hours or 96 dB(A) for 2 hours will both produce a noise dose of 100%.

For measurement in accordance with the American Occupational Safety and Health Act (OSHA), a 5-dB(A) increase in amplitude must be accompanied by a halving of exposure duration to give the same noise dose. This means, for example, that an exposure of 95 dB(A) for 4 hours, or one of 100 dB(A) for 2 hours, or 105 dB(A) for 1 hour, etc., all constitute a 100% noise dose.

The percentage noise dose indicated on dosimeters can be readily converted to L_{eq} averaged over a typical working day. The American National Standard on noise dosimeters (ANSI S1.25.1978) provides descriptions, definitions for use, and tolerances for the instrument.

Most noise dosimeters feature an accelerated measuring mode which permits the user to quickly evaluate the noise emitted by cyclical machines. They also allow the operator to use the meters as an integrating sound level meter to make an estimate of the dose to be encountered in a noisy environment before committing the instrument to, say, a full day's measurement.

Hazardous impulsive noise peaks can be correctly measured by these noise dosimeters, which respond to peaks as short as 100 μs duration. In addition, they provide latched warning indications for exposure to peaks exceeding 140 dB(A). (See next section on impulsive sounds.) The meters also flag a warning of noise levels exceeding 115 dB(A) (slow), according to OSHA noise-exposure regulations.

7.4.2.3. Impulsive Sound. Maximum peak sound levels of 140 dB (overall or flat response, not A-weighted) are permitted for impulsive sound sources, such as punch presses, stamping presses, forging hammers, etc. Impulse or impact noise according to OSHA standards is defined as sound with a rise time of not more than 35 ms to the peak intensity and a duration of not more than 500 ms ($\frac{1}{2}$ s) to the time when the level is 20 dB below the peak. If the impulses recur at intervals of less than 1 s, they are considered to be continuous sound. Peak sound levels are read from a peak-reading sound-level meter, where the true peak level is "captured" and held on the meter for reading. For repeated impacts, the total accumulated exposure should either meet the Table 7.6 permitted limits or the dosage limit of 1 when calculated according to Eq. [7.4].

As an example of an impact noise evaluation, suppose a stamping press operated at 4-second intervals throughout the day and produces 112 dB(A) noise peaks at each stroke. Suppose that a detailed analysis of the impact noise reveals that the peak noise energy has a duration of 50 milliseconds ($\frac{1}{20}$ second). A total of 7200 peaks occur in an 8-hr period. The total exposure time is then 7200/20 = 360 sec = 6 min. According to Table 7.6, a sound level of 112 dB(A) for 6 minutes is acceptable. However, this illustration is misleading in its simplicity. There is no simple field instrument that measures the effective duration of the noise peak. Instead, a rather elaborate collection of laboratory equipment would be required (26). In a critical test, the measurement could be performed but it would be beyond normal field practice with inexperienced personnel.

Current OSHA noise standards cannot adequately treat impact noise since they do not allow for the cumulative effect of exposure to a large number of impacts, nor does it take into account the diminishing effect of decreasing peak levels. (See Section 7.4.4 re Canadian proposal for dealing with repetitive impulses vis-a-vis ISO/DIS 1999-1984.)

A new proposed OSHA noise regulation would prohibit exposures to impulse or impact noise at peak sound-pressure levels above 140 dB (linear weighting) and, in addition, it would limit the number of impacts allowed at or below peak levels of 140 dB. The proposed regulation states that exposure to impulses of 140 dB shall not exceed 100 such impulses per day.

Moreover, for each decrease of 10 dB in peak sound-pressure level of the impulse, the number of impulses to which employees are exposed may be increased by a factor of 10. Hence, the number N of impulses or impacts allowed at a peak level L can be determined by the equation

$$N = \frac{10^2}{10^{0.1(L - 140)}} \qquad [7.5]$$

7.4.2.4. Noise Control Approaches. As of August 1981, OSHA noise exposure standards consist of a two-stage program where hearing conservation measures become mandatory at 85 dB(A) for an 8-hour day and feasible engineering or administrative noise controls are required when exposures exceed 90 dB(A).

Engineering Controls. OSHA requires that engineering controls be applied to solve the noise problem and bring the sound levels into conformance with Table 7.6 values. This includes reducing machinery noise through redesign, replacement with quieter equipment, and reducing the transmission of noise along the path from source to receiver (including enclosing the noise source or receiver).

Administrative Controls. These are to be used to rotate personnel and limit the time an employee is exposed to given noise levels.

If these two control approaches fail to achieve the specified conditions, then personal protective equipment (ear plugs and/or ear muffs) shall be provided and used to meet the specified sound levels. It is strongly suggested that serious and intensive effort be made to the application of engineering and administrative controls.

Hearing Conservation Program. The OSHA noise regulation requires a continuous, effective hearing conservation program to include the following (in addition to engineering and administrative controls): noise surveys, use of effective hearing protectors, employee training and education, dosimeter monitoring, and an effective audiometric testing program. These

Table 7.7. Expansion of Table 7.6 to Include Durations for Each Decibel Level Between 90 and 115 dB(A), and Extension to Sound Levels Between 85 and 90 dB(A) for Noise Dose Calculation (as per NIOSH recommendation).

Permitted Daily Duration (T, hrs)	A-wtd Sound Level (L, dBA)	Permitted Daily Duration (T, hrs)	A-wtd Sound Level (L, dBA)
16.0	85	2.00	100
13.9	86	1.74	101
12.1	87	1.52	102
10.6	88	1.32	103
9.2	89	1.15	104
8.0	90	1.00	105
7.0	91	0.87	106
6.1	92	0.76	107
5.3	93	0.66	108
4.6	94	0.57	109
4.0	95	0.50	110
3.5	96	0.44	111
3.0	97	0.38	112
2.6	98	0.33	113
2.3	99	0.29	114
		0.25	115

(The bracket spanning rows 85–89 in the first column is labelled "NIOSH proposal".)

$$T = 16 \div 2^{[0.2(L-85)]}$$

steps should apply to operator areas where noise levels exceed 85 dB(A) full time or equivalent.

Criteria for Engineering or Administrative Controls. The criteria for these controls are as outlined in Sections 7.4.2.1 and 7.4.2.2.

Criteria for Hearing Conservation Programs. The National Institute of Occupational Safety and Health (NIOSH) is a group that studies and proposes various changes for maintaining employee safety and health. As a result of such proposals, OSHA has extended the dosage Eq. [7.4] to include sound levels in the range of 85–90 dB(A) in the calculation of daily noise dose. In effect, this means that the Table 7.6 duration values are extended upward to 16 hours using the relationship now followed in Table 7.6. Table 7.7 is a modified table that includes this change, and shows the currently permitted duration for each decibel increment between 90 and 115 dB(A) and also the extension sound levels from 89 to 85 dB(A). This means that exposure to sound levels in the range 85 to 89 dB(A) are now counted in the calculation of noise dosage by Eq. [7.4] (62).

Thus, employers shall provide annual audiometric testing of all workers exposed to sound levels equal to or exceeding an 8-hour time-weighted average of 85 dBA, or, equivalently, a daily dose of 50 per cent as measured by the following equation (62):

$$D = 100 \left(\frac{C_1}{T_1} + \frac{C_2}{T_2} + \frac{C_3}{T_3} + \cdots \frac{C_N}{T_N} \right),$$

$$[7.6]$$

where

D = workday dose, %

$1,2,3$ = periods of exposure to different dB(A) levels,

C = actual exposure time at different levels

T = permissible exposure time at a given level in accordance with Table 7.7.

Example Assume exposure of: 100 dB(A) for 1 hour, 90 dB(A) for 4 hours, 85 dB(A) for 3 hours.

$$D = 100[(1/2) + (4/8)$$
$$+ (3/16)] = 118.75\%.$$

Unacceptable, since D exceeds 50%.

Note: The exposure in the Example, when evaluated in reference to OSHA criteria for engineering or administrative controls, can be shown to be acceptable, since levels below 90 dB(A) do not enter into those criteria. However, exposures exceeding a 50% dose still require implementation of hearing conservation programs.

The features of the above criteria, etc. are contained in the 1979 version of the OSHA

"Industrial Hygiene Field Operations Manual" (27).

7.4.3. Ear Protectors

While the merits and costs of the 90 dB(A) versus 85 dB(A) sound-level limits are being debated, people continue to be exposed to high noise levels because it takes time to implement even the 90 dB(A) limit. Ear protectors are an obvious interim measure to protect against these high noise levels (28, 29, 37).

Ear protectors have three problems:

1. They are considered uncomfortable.
2. Sounds seem to "sound different" when heard with ear protectors.
3. There seems to be no urgency to wear them.

Regarding comfort, many ear protectors of 10–20 years ago were uncomfortable. However, there are light-weight plugs and muffs that are much more comfortable than some of the earlier models. They may not be as effective as the heavier, tight-fitting, more uncomfortable ones, but they can still provide at least 15–25 dB(A) reduction to the ear when properly worn, and this can help many noise-exposure problems in industry.

Second, regarding the different quality of the noise when heard with ear protectors, it does indeed sound different. If listeners truly want to wear hearing protectors, however, they can accommodate to the new sound in a relatively short time and recognize its special signatures and clues. Actually, tests show that a listener can hear sounds more clearly and distinctly within the whole mixture of noise when all the sounds are at a lower level.

Third, hearing loss comes on so slowly and so insidiously, that there appears to be no urgency to start right away to do anything about it. Hearing protectors have to be worn every day to provide lifetime protection. Each daily high noise dose causes a small, undetectable amount of hearing loss, and an accumulation of these small amounts brings on a noticeable amount of permanent hearing loss. By the time people realize that they have begun to lose some

hearing, it is too late to recover it. People should be encouraged to use ear protectors voluntarily in their own personal hearing protection program.

7.4.4. Recent Studies in Canada Regarding Noise-Induced Hearing Loss and Damage Risk Criteria

There has been considerable debate over the last 20 years as to: the efficacy of the CHABA studies regarding permissible "noise dose" exposure (10); the 3-dB exchange rate of the equal-energy theory versus the 5-dB exchange rate of the CHABA studies; the question of whether steady, intermittent, varying-level, and impulsive noises should all be included in a comprehensive measurement of "the time integral of the squared A-weighted sound pressure" in accordance with ISO/DIS 1999-1984 (24); whether temporary threshold shift (TTS) is a reliable measure of permanent threshold shift (PTS) when exposed for several years to the same noise level for the TTS; the acceptance or otherwise of the 40-hour work week as the integration period in the measurement and specification of sound exposure defined as the time integral of the squared A-weighted sound pressure.

A special advisory committee was appointed by the Ontario Ministry of Labour in February 1983 to study and report to the Minister on issues pertaining to the formulation of a Noise Regulation under the Occupational Health and Safety Act, R.S.O. 1980 c. 321. Dr. Edgar A.G. Shaw, Principal Scientific Officer, Division of Physics, National Research Council of Canada was appointed scientific advisor to the committee. After an in-depth study of existing literature and after consulting with many authorities in Europe and North America re noise-induced hearing loss, Dr. Shaw submitted a final report to the advisory committee on October 30, 1985 (33). In this report, the scientific evidence shows that there are many deficiencies in the assumptions made in the CHABA studies and report of 1965. The following are comments extracted from Dr. Shaw's report:

- "The definition of A-weighted sound exposure given in ISO/DIS 1999-1984 contains no allowance for the temporal pattern of the noise exposure over the working day or working week. Indeed, this document makes no mention of possible benefits associated with intermittency. As we have seen, such benefits were confidently expected when CHABA Working Group 46 published its landmark report in 1966, and the expectation of benefits was strongly reflected in the formulation of the '5 dB rule' in 1969. This expectation rested mainly on the assumption that the median temporary threshold shift measured two minutes after the end of the working day (TTS_2) accurately foreshadowed the median permanent threshold shift that would accrue over ten years of near-daily exposure to the same pattern of noise."

- "Not long after the publication of the CHABA report, it became apparent that there were significant deficiencies in the procedures used to calculate the contours for intermittent noise exposure. The assumption that all exposure to noise below approximately 89 dBA could be ignored was also shown to be untenable. Much lower background levels were needed to achieve 'effective quiet.' Equally disturbing was the discovery that intermittent exposure to high frequency high intensity noise often produced delayed recovery from temporary threshold shift even though the amount of TTS_2 was moderate. Hence, the magnitude of TTS_2 could no longer be accepted as a reliable indicator of potential long-term hazard."

- "Epidemiological studies of noise-induced hearing loss . . . have failed to confirm that intermittent noise exposures are less hazardous than steady exposures with the same daily A-weighted sound energy."

- "Whatever benefit, if any, there may be from intermittency in the real world, it is clear that the CHABA Damage Risk Contours of 1966 can no longer be accepted as a valid means of assessing that benefit. As regards the 5 dB exchange rate adopted by the U.S. Government in 1969 it was . . .

an 'obviously inadequate standard.' Although supposedly based on the CHABA contours for intermittent noise, it took into account only the total duration of noise exposure without regard to the temporal distribution of the noise."

- ". . . there are cogent reasons for treating industrial exposure to impulse noise in precisely the same way as steady, intermittent and varying noise by including all types of noise, up to a certain instantaneous peak sound pressure level, in a comprehensive measurement or estimation of equivalent continuous A-weighted sound pressure level or, more precisely, the time integral of the squared A-weighted sound pressure as advocated in ISO/DIS 1999-1984."

After due consideration of Dr. Shaw's report, of briefs received from companies, labour unions and other organizations (69 in all), and of a submission of the Canadian Centre for Occupational Health and Safety, the Special Advisory Committee unanimously approved the following conclusions and recommendations (34):

Conclusion I: Measurement of Exposure to Steady, Varying and Intermittent Noise
It is concluded (a) that, for steady, intermittent and varying noise, there is adequate scientific support for the acceptance of the equivalent continuous A-weighted sound pressure level or, in the terminology of ISO/R1999-1984, the "time integral of the squared, A-weighted sound pressure," with an appropriate integration period, as the best available measure of sound exposure; (b) that there is at present no scientifically acceptable alternative measure of sound exposure. In other words, the 3 dB exchange rate should be accepted and the 5 dB exchange rate firmly rejected.

Conclusion II: Comprehensive Measurement of Sound Exposure
It is concluded that, in the measurement and specification of sound exposure, no distinction should be made between impulsive noise and other types of noise. Steady, intermittent, varying and impulsive noise should all be included in a comprehensive measurement of

"the time integral of the squared A-weighted sound pressure," in accordance with ISO/DIS 1999-1984.

Conclusion III: Choice of Integration Period in the Measurement of Sound Exposure
It is concluded that the 40 hour work week is acceptable as the integration period in the measurement and specification of sound exposure, here defined as the time integral of the squared A-weighted sound pressure, provided (a) that a suitable upper limit is placed on the daily duration of exposure to noise and (b) that a suitable lower limit is placed on the duration of the period of "effective quiet" separating successive exposures.

Recommendation 1: Mandatory Hearing Protection
It is therefore recommended (i) that the need for hearing protection be determined from a sound exposure measurement made in accordance with ISO/DIS 1999-1984 and Conclusions I and II with the 8 hours work day as the integration period; (ii) that the use of hearing protectors be mandatory wherever there is exposure to noise at the work place at an equivalent continuous A-weighted sound pressure level of 85 decibels or more over the normal 8-hour work day; and (iii) that a programme of education and instruction be required in parallel with the mandatory use of hearing protectors.

Recommendation 2: Engineering Controls
It is therefore recommended (i) that the need for engineering controls be determined from a sound exposure measurement made in accordance with ISO/DIS 1999-1984 with the 40-hour work week as the integration period; (ii) that engineering controls be required where the equivalent continuous A-weighted sound pressure level at the work place measured in this manner is 90 decibels or more; and (iii) that the acceptance of the 40-hour work week as the integration period be subject to special provision being made for unconventional distributions of hours within the work week.

Recommendation 3: Hearing Conservation
It is therefore recommended (i) that the need for a hearing conservation programme with periodic audiometry be determined from a sound exposure measurement made in accordance with ISO/DIS 1999-1984 with the 40-hour work week as the integration period; (ii) that a hearing conservation programme be required where the equivalent continuous A-weighted sound pressure level at the work place measured in this manner is 85 decibels or more; and (iii) that the acceptance of the 40-hour week as the integration period be subject to special provision being made for unconventional distributions of hours within the work week.

Recommendation 4: Hearing Protector Performance
It is therefore recommended (i) that Table A1, Appendix A, CSA Standard Z94.2-M1984 be recognized as a realistic reflection of the disparity between the "real world" attenuation of personal hearing protectors under typical conditions of use and the laboratory data in current use; (ii) that Appendix A, CSA Standard Z94.2-M1984 be accepted as a starting point for a code of practice on the mandatory use of personal hearing protectors; and (iii) that high priority be given to the development and implementation of realistic and accurate quantitative methods of determining, controlling and maintaining the performance of personal hearing protectors as used at the work place in Canada.

Recommendation 5: Ceiling Level
It is therefore recommended, in the interest of simplicity and in keeping with ISO/DIS 1999-1984, that the use of hearing protectors be mandatory where there is exposure to noise at the work place with instantaneous peak sound pressures exceeding 200 Pa (140 dB relative to 20 Pa).

At the time of this writing, the Occupational Health and Safety Division of the Ontario Ministry of Labour has initiated a "Proposed Regulation under the Occupational Health and Safety Act: Designated Substance—Noise," dated July 7, 1986. This regulation incorporates the conclusions and recommendations of the above-mentioned special advisory committee. The regulation is in various review steps prior to being enacted by the provincial government.

With regard to efforts to control occupational noise on a national basis in Canada, the Federal/Provincial Advisory Committee on Environmental and Occupational Noise established a working group to prepare national guidelines on occupational noise and hearing conservation. (The advisory committee is comprised of senior officials from the health, labor, and environmental departments of the federal, ten provincial, and two territorial governments in Canada. It is charged with advising ministers and deputy ministers of health on all matters of environmental and occupational health.)

The working group prepared a draft "Guideline for Regulatory Control of Occupational Noise Exposure and Hearing Conservation," dated February 1986 (revised August 1986). This guideline is the result of the combined effort of a group of people with expertise in both occupational noise exposure control and in the associated regulations across Canada. Input and comments have been solicited from all the provinces and from acoustics experts working in the area of occupational noise. One of the information bases used in the production of the report was a survey distributed in 1982 by a task force on occupational noise, under the auspices of the Canadian Standards Association Committee on Acoustics and Noise Control, to more than 150 representatives of government, employers, labor, consultants, universities, and hospitals.

It is interesting to note that the major recommendations of the special advisory committee to the Ontario Department of Labour have been incorporated in the above-mentioned guideline. (One exception is that, whereas the advisory committee refers to 8 hours and 40 hours as integration time bases for various provisions, the working group decided on 8 hours as the time basis for all provisions.)

A more complete description of the working group's guideline is given in Appendix V.

7.5. OUTDOOR NOISE CRITERIA

Both physical and social aspects of the environment in which we live have to be taken into consideration when assessing intrusive noise. For instance, physical parameters include the frequency of the sound, its pressure (or energy) amplitude, and the way that each varies with time.

This requires acquiring large amounts of complex and difficult to analyze data. As a result, much effort has gone into the development of rating schemes for outdoor environments which would simplify data acquisition and provide suitable description which would predict community response to noise.

7.5.1. Typical Noise Levels, Statistical Community Noise, and Recommended Outdoor Noise Limits

The A-weighted is most commonly used simplification to combine frequency spectrum information and overall levels (see Section 3.4). The temporal pattern of the A-weighted noise level could then be obtained by a graphic level recorder. An example of two 8-hr samples of outdoor noise in a typical suburban neighborhood is shown in Fig. 7.11.

Note from these two samples that (a) the noise level varies over a range of 33 dB(A), (b) there appears to be a fairly steady lower residual or background noise level on which is superimposed individual higher intrusive levels, and (c) there is a duration factor associated with the single events which could have some effect on the annoyance (e.g., the barking dogs).

However, a further simplification in identifying the noise environment is to eliminate the temporal detail. There are, typically, three ways by which this can be accomplished:

1. The noise level is broken into several periods, i.e.,

Day	7 a.m. to 7 p.m.
Evening	7 p.m. to 11 p.m.
Night	11 p.m. to 7 a.m.

2. The values of maximum noise level of single event intrusive sounds (and after the values of residual noise level) are measured.
3. Statistical properties of the noise level are determined.

Statistical properties most commonly used in rating schemes have been defined earlier in this chapter (Section 7.2). They are L_{max}, L_{min}, L_N (the sound level exceeded n percent of the time

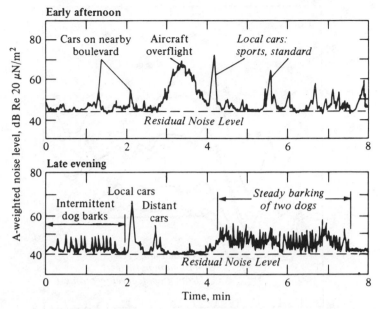

Fig. 7.11. Two samples of outdoor noise in normal suburban neighborhood with microphone located 20 ft from curb (35).

over the sampling period), L_{eq} (the energy equivalent level over the sampling period); and L_{dn} (the day/night noise level).

A statistical portrayal of community noise, for a particular day and location in the community, can be obtained by plotting these data over a 24-hr period. An example is shown in Fig. 7.12, which graphs statistical measures obtained from, say, a statistical noise analyzer, on an hourly basis (the L_N and L_{eq} values), and utilizes a graphic level recorder to plot maximum and minimum values. (Note, however, that statistical noise level meters are becoming available that will record maximum and minimum values in memories, for recall, as well as giving L levels).

All statistical measures exhibit a typical temporal pattern in which the residual noise level drops sharply after midnight, reaching a minimum between 4:00 and 5:00 a.m., there after rising to almost constant daytime values by 8:00 a.m. This correlates well with human activity, particularly vehicle traffic and construction in urban areas.

The result in Fig. 7.12 show that the L_{90}, the level exceeded 90% of the time sampled, is a good example of the residual (or ambient-noise) level. The maximum levels of the intrusive noise can be considerably greater than those predicted by L_1; this is particularly the case if the duration of the event is very short.

The rating schemes which have been developed provide the noise control engineer with the ability to be able to assess the attitude of a community to its noise environment and to predict the community response to a proposed new intrusive noise source (such as an industrial plant or airport). Because of the complexities involved in developing rating schemes for community noise, there has been a proliferation of such schemes, each one developed to satisfy some particular situation. Many studies have been undertaken to relate these different schemes; two of these studies are given in Refs. 4 and 36.

There are several rating scales which are used in community noise assessments. Some of these have been mentioned previously in Section 7.2. Examples of these are: (i) the A-weighted sound-pressure level, which has been defined previously for indoor noise assessment and, as was shown by example in Fig. 7.11, is frequently used for initial noise assessments in outdoor noise; (ii) the noise pollution level NPL, which attempts to account for the effect of fluctuations of the noise environment; (iii)

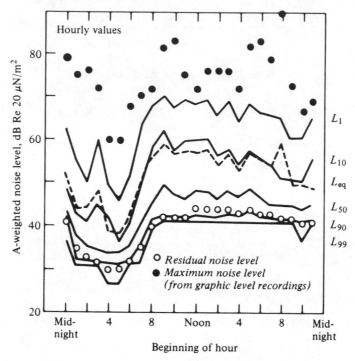

Fig. 7.12. Statistical portrayal of community noise within a 24-hr period. Data include maximum and residual noise levels read from a graphic level recorder, together with hourly and periodic values of the levels which are exceeded 99, 90, 50, 10, and 1 percent of the time, and the equivalent level L_{eq} (35).

A-weighted L_N (percentile levels); for example, the U.S. Federal Highway Administration adopted the A-weighted L_{10} level as the principal indicator for assessing the impact of highway noise (26).

Although, as noted earlier, few criteria for general outdoor noise have been legislated into regulations in Canada, the guidelines for noise control in land-use planning which have been included in the Ontario Ministry of the Environment's "Model Municipal Noise Control By-Law" (1) are worth noting. Table 7.8 gives the sound-level limits for two descriptors (or rating scales), the 50th-percentile sound level L_{50} and the equivalent sound level L_{eq} for *out-*

door recreational areas, where the descriptors are referenced to the 16-hour period from 0700 to 2300.

The 50th-percentile sound level need only be considered for development where the predominant sound is industry or a highway producing near-constant sound levels throughout the 16-hour period. Compliance with these sound-level limits should generally ensure compliance with the appropriate requirements for indoor sound-level limits as outlined in Table 7.2 of Section 7.3.4 for the same time period for any normal building construction nearby.

Table 7.9 gives the sound level limits for two descriptors, the 50th-percentile sound level L_{50} and the equivalent sound level L_{eq}, *for outdoor*

Table 7.8. Sound-Level Limits for Outdoor Recreational Areas (0700–2300).

Sound Descriptor for the Entire Period	Sound-Level Limit, dB(A)
L_{50}	52
L_{eq}	55

Table 7.9. Sound-Level Limits for Outdoor Areas (2300–0700).

Sound Descriptor for the Entire Period	Sound-Level Limit, dB(A)
L_{50}	47
L_{eq}	50

areas, in the vicinity of buildings or proposed buildings containing sleeping quarters, where the descriptors are referenced to the entire 8-hour period from 2300 to 0700. The 50th-percentile sound level need only be considered for development where the predominant sound is industry or a highway producing near-constant sound levels throughout the 8-hour period. Compliance with these sound-level limits should generally ensure compliance with the appropriate requirements of Table 7.2 of Section 7.3.4 for the same time period for any normal building construction nearby. Where the requirements of Table 7.9 cannot be met, special architectural design and construction features will have to be incorporated into the building to ensure compliance with the appropriate requirements of Table 7.2 of Section 7.3.4.

7.5.2. Rating Procedures

Various rating procedures for describing certain aspects of the noise stimulus have been proposed; in addition, these attempt to account for the context in which the noise is experienced. This is accomplished by introducing corrections—both for such peculiarities in the noise as pure tones, impulses, variability, etc. and for such effects as time of day during which the noise occurs, ambient (background) noise level, type of neighborhood (i.e., zoning), whether a previous noise intrusion is experienced, etc.

Two main noise rating procedures will be discussed here. (For a more complete summary of the various procedures proposed, see Ref. 4).

7.5.2.1. EPA Generalized Method. A method for predicting community response to any sound intrusion has been developed in the U.S. from an analysis of 55 community noise situations. About half of these cases were of steady industrial and residential noises, and the other half were of intermittent transportation and industrial noise intrusions. The Environmental Protection Agency has endorsed this method by including it in the levels document (6).

The method first involves calculating the normalized L_{dn} of the intruding noise. This is the actual L_{dn} of the intruding noise normalized by adding or subtracting for the various factors which affect community reaction that are given in Table 7.10. The factors are based on the following reasoning: if the sound occurs only in winter, it will be heard behind closed windows; an already noisy community is less likely to react to a noise intrusion of a given sound level than is a quiet one; a community's previous experience with an intruding noise lessens the impact of further exposure to it; noise control consciousness on the part of the noise-makers (i.e., low misfeasance) lessens community reaction; pure tones and impulsive sounds are specially annoying.

Having normalized the L_{dn} with the corrections, reference is made to Fig. 7.13 which shows the different levels of community reaction observed for the various values of normalized L_{dn} in the 55 cases.

The community reactions may occur at a lower value of normalized L_{dn} than that predicted, should the absolute magnitude of the intruding noise be high enough to have some physical effect, e.g., to interfere with speech.

7.5.2.2. Composite Noise Rating (CNR$_c$). The composite noise rating for assessment of community response uses octave-band sound-level data with appropriate corrections for background noise interference, type of neighborhood, repetitiveness of noise, temporal and seasonal factors, evidence of previous exposure, etc. (4). It has been used (mainly in the U.S.A.) to assess the influence of various intruding noises such as traffic noise, industrial noise, and aircraft noise on the community.

The CRN$_c$ is calculated as follows:

1. Measure the octave-band levels of the intruding noise source. Each band level is assumed averaged (on an energy basis) over a reasonable time interval, and over a reasonable number (or the most critical) of locations in the community. (Note that in some cases, such as where a building or facility is in the planning or design

Table 7.10. Corrections to be Added to the Measured Day-Night Sound Level (L_{dn}) of Intruding Noise to Obtain Normalized L_{dn}.

Type of correction	Description	Amount of Correction to be Added to Measured L_{dn} in dB(A)
Seasonal Correction	Summer (or year-round operation)	0
	Winter only (or windows always closed)	−5
Correction for outdoor noise level measured in absence of intruding noise	Quiet suburban or rural community (remote from large cities and from industrial activity and trucking)	+10
	Normal suburban community (not located near industrial activity)	+5
	Urban residential community (not immediately adjacent to heavily traveled roads and industrial areas)	0
	Noisy urban residential community (near relatively busy roads or industrial areas)	−5
	Very noisy urban residential community	−10
Correction for previous exposure and community attitudes	No prior experience with the intruding noise	+5
	Community has had some previous exposure to intruding noise but little effort is being made to control the noise. This correction may also be applied in a situation where the community has not been exposed to the noise previously, but the people are aware that bona fide efforts are being made to control the noise.	0
	Community has had considerable previous exposure to the intruding noise and the noise maker's relations with the community are good.	−5
	Community aware that operation causing noise is very necessary and it will not continue indefinitely. This correction can be applied for an operation of limited duration and under emergency circumstances.	−10
Pure tone or impulse	No pure tone or impulsive character	0
	Pure tone or impulsive character present	+5

Source: EPA Report 550/9-74-004.

stage, it may be necessary to *estimate* the octave-band levels; see ref. 37).

2. The noise-level rank (expressed as a letter) is determined by plotting the octave levels on the chart of curves given in Fig. 7.14. The highest zone into which the noise spectrum protrudes establishes the noise-level rank.

3. Corrections are made for background noise using either Fig. 7.15 or Table 7.11 (Note that estimates may have to be made as to the octave-band levels of background noise; see Ref. 37).

4. Make corrections for the repetitiveness of the noise using Fig. 7.16.

5. Make final corrections according to the community information and the procedure outlined in Table 7.12. (Note that a total correction of ±3 means that the level rank should be moved up or down by 3 letters, respectively).

6. The final level rank obtained for the intrusive noise is entered in Fig. 7.17 to obtain the response expected from the community.

The CRN$_c$ scheme works fairly well on an absolute basis. The scheme works better in predicting the community response to a change in a noise source which already exists and for which the response is known (36).

7.5.2.3. Noise Exposure Forecast (NEF).

The noise exposure forecast uses the *effective* perceived noise level (EPNL) as its basic measure. EPNL is a single-number mea-

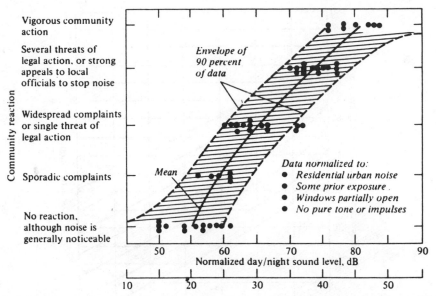

Fig. 7.13. Community reaction to intrusive noise measured in L_{dn} and normalized by factors set out in Table 7.10.

sure of complex aircraft flyover noise which approximates laboratory annoyance response. It includes corrections for both the duration and tonal components of the spectra for different types of non-sonic-boom aircraft flyover noise. EPNL is used in the U.S. by the Federal Aviation Administration and in Canada by the Federal Department of Transport in aircraft certi-

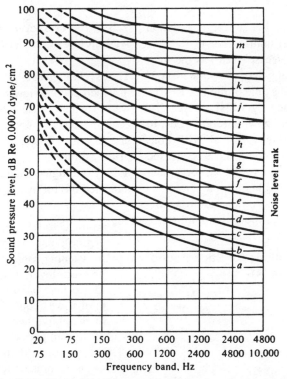

Fig. 7.14. Family of curves used to determine the noise-level rank.

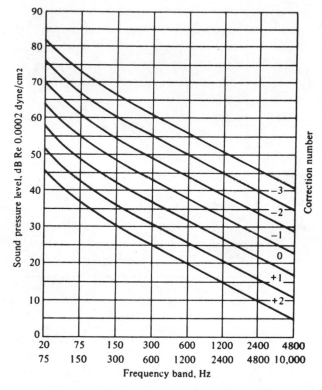

Fig. 7.15. Family of curves used to determine the correction number for background noise.

fication. (See Refs. 4 or 26 for calculation procedures and further information.)

Noise Exposure Forecast (NEF) is the total summation (on an energy basis) over a 24-hour period (weighted for the time of day) of EPNL. Tables and graphs showing EPNL versus distance are available for known aircraft types, to facilitate construction of so-called NEF contour maps (4). NEF contours have been used in evaluating land uses around airports, and they are usually included in environmental impact statements dealing with noise from jet aircraft operations around airport communities.

Table 7.11. Corrections for Background Noise (4).

Neighborhood	Correction Number
Very quiet suburban	+1
Suburban	0
Residential urban	−1
Urban near some industry	−2
Area of heavy industry	−3

7.5.3. Psychoacoustic Studies on Aversiveness to Typical Noises

The psychological effects of noise, i.e., the way in which people view noise as aversive, or the way in which noise interferes with sleep, has been the object of various studies during the last 30 years. Recently, psychoacoustic testing of the attitudinal response of people to corona noise from high-voltage transmission lines has indicated the following (64, 65, 66):

* Corona, traffic, air conditioning, and jet-engine noises were presented to subjects in listening room tests at the same sound pressure level; respondents adjusted a 1-kHz octave-band random-reference sound until it was equal in aversiveness to each of the above test stimuli; the mean level of adjustment was 10 dB higher for the corona noises than for traffic and air conditioning noises; for equal sound pressure level presentation of stimuli, the adjust-

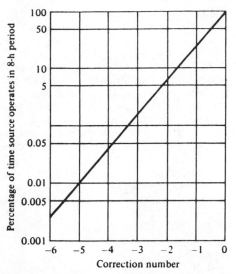

Fig. 7.16. Proposed correction numbers for repetitiveness of the noise when the source operates on a reasonably regular daily schedule (4).

Table 7.12. List of Correction Numbers to be Applied to Noise-Level Rank to Give CNR$_c$ (4).

Influencing Factor	Correction Number
1. Background noise	+2 to −3
2. Temporal and seasonal factors	
(a) Daytime only	−1
Nightime	0
(b) Repetitiveness	0 to −6
(c) Winter	−1
Summer	0
3. Detailed description of the noise	
(a) Continuous spectrum	0
Pure-tone components	+1
(b) Smooth time character	0
Impulsive	+1
4. Previous exposure	
None	0
Some	−1

ment of the reference sound was almost identical for corona and jet engine noises.

• When conventional A-weighting was used in the sound measuring systems, corona noise appeared comparable in aversiveness to most common noises such as from traffic, transformer stations, air conditioners, etc.; it was slightly more aversive than jet engine noise.

• The probability of awakening due to noise interference from transmission lines, transformer stations, traffic and air conditioners is considerably greater for the corona noise from transmission lines than it is for any of the other noise stimuli.

REFERENCES

(1) Ontario Ministry of the Environment. 1978. "Model Municipal Noise Control By-Law," Final Report, August.

(2) Province of Alberta. 1977. "The Planning Act," Bill 15, s. 144.

COMMUNITY REACTION

VIGOROUS ACTION

SEVERAL THREATS OF LEGAL ACTION OR STRONG APPEALS TO LOCAL OFFICIALS TO STOP NOISE

WIDESPREAD COMPLAINTS OR SINGLE THREAT OF LEGAL ACTION

SPORADIC COMPLAINTS

NO REACTION, ALTHOUGH NOISE IS GENERALLY NOTICEABLE

AVERAGE EXPECTED RESPONSE

RANGE OF EXPECTED RESPONSES FROM NORMAL COMMUNITIES

A   B   C   D   E   F   G   H   I

COMPOSITE NOISE RATING

Fig. 7.17. Estimated community response versus composite noise rating (CNR$_c$), modified.

(3) Environment Canada. 1977. "A Guide to the Federal Environmental Assessment and Review Process."

(4) Pearsons, K. S., and R. L. Bennett. 1974. "Handbook of Noise Ratings," NASA CR-2376, Washington, DC: National Aeronautics and Space Administration.

(5) Irwin, J. D., and E. R. Graf. 1979. *Industrial Noise and Vibration Control*. Englewood Cliffs, NJ: Prentice-Hall.

(6) U.S. EPA. 1974. "Information on Levels of Environmental Noise Requisite to Protect Public Health and Welfare with an Adequate Margin of Safety," Environmental Protection Agency Report 550/9-74-004.

(7) The Occupational Safety and Health Act of 1970, S. 2193, *Congressional Record* (U.S.).

(8) U. K. 1972. "Code of Practice for Reducing the Exposure of Employed Persons to Noise." London: Her Majesty's Stationery Office.

(9) ISO. 1971. "Assessment of Occupational Noise Exposure for Hearing Conservation Purposes." International Organization for Standardization, Recommendation R1999.

(10) CHABA. 1975. "Hazardous Exposure to Intermittent and Steady-State Noise," National Academy of Science and National Research Council Committee on Hearing, Bioacoustics and Biomechanics. Also *Journal of the Acoustical Society of America*, **39**, (3), 451–464, (March 1966).

(11) Magrab, E. B. 1975. *Environmental Noise Control*. New York: John Wiley and Sons.

(12) Beranek, L. L. 1960. *Noise Reduction*. New York: McGraw-Hill.

(13) Beranek, L. L. 1971. Noise and Vibration Control. New York: McGraw-Hill.

(14) ASHRAE. 1982. *ASHRAE Handbook and Product Directory System Guide*. New York: American Society of Heating, Refrigeration and Air Conditioning Engineers, Inc.

(15) Beranek, L. L. 1950. "Noise Control in Office and Factory Spaces." *15th Annual Meeting of the Chemical Engineering Conference, Trans. Bull.*, **18**:26–33.

(16) Shultz, T. J. 1972. "Noise Assessment Guidelines—Technical Background." HUD Report No. TE/NA 172 BBN No. 2005R, prepared for USHUD, Office of Research and Technology, by Bolt Beranek and Newman Inc., Washington, DC: Department of Housing and Urban Development.

(17) CMHC. 1986. "Road and Rail Noise: Effects on Housing." Ottawa: Central Mortgage and Housing Corporation.

(18) May, D. N. 1978. *Handbook of Noise Assessment*. Toronto: Van Nostrand Reinhold.

(19) Hegvold, L. W. 1971. "Acoustical Design of Open-Planned Offices." Canadian Building Digest CBD 139. Ottawa: National Research Council.

(20) Warnock, A. C. C. 1974. "Acoustical Effects of Screens in Landscaped Offices." Canadian Building Digest CBD 164. Ottawa: National Research Council.

(21) Edwards, A. T., and J. Kowalewski. 1975. "Open-Office Acoustics—An Engineering Approach." 89th Meeting of Acoustical Society of America, Austin, Tx.

(22) Public Building Service. 1975. "Standard Methods of Test for the Direct Measurement (PBS-PCD:PBS-C1) and for the Sufficient Verification (PBS-PCD:PBS-C2) of Speech Privacy Potential", Washington, DC: U.S. General Services Administration.

(23) Moulder, R. 1975. "Open Plan Office Elements and Testing," *Proceedings of Noisexpo, Atlanta, Georgia, Apr. 30–May 2*.

(24) ISO. 1984. "Acoustics—Determination of Occupational Noise Exposure and Estimation of Noise-Induced Hearing Impairment." International Standard ISO/DIS 1999-1984.

(25) Reynolds, D. D. 1981. *Engineering Principles of Acoustic–Noise and Vibration Control*. Toronto: Allyn and Bacon.

(26) Lord, H., W. S. Gatley, and H. A. Evensen. 1980. *Noise Control for Engineers*. New York: McGraw-Hill.

(27) OSHA. 1979. "Industrial Hygiene Field Operations Manual," U.S. Department of Labor, OSHA Instruction CPL2-2.20, Chap. IV.

(28) Miller, L. N. 1977. Hearing Protection—Start Now." *Sound and Vibration*, May.

(29) NIOSH. 1976. "List of Personal Hearing Protectors and Attenuation Data." U.S. Department of Health, Education and Welfare, NIOSH Publication 76-120.

(30) NIOSH. 1979. "A Field Investigation of Noise Reduction Afforded by Insert-Type Hearing Protectors." U.S. Department of Health, Education and Welfare, NIOSH Publication 79-115.

(31) Alberti, P. W. 1982. *Personal Hearing Protection in Industry*. New York: Reever Press.

(32) Benwell, D.H. 1986. "Draft National Environmental Noise Code." *Proceedings of 12th International Congress on Acoustics*, Toronto, Canada, 24-31 July, Vol. 1, C9-6.

(33) Shaw, Edgar A. G. 1985. "Occupational Noise Exposure and Noise-Induced Hearing Loss: Scientific Issues, Technical Arguments and Practical Recommendations." Report prepared for the Special Advisory Committee on the Ontario Noise Regulation, APS707. Ottawa: National Research Council of Canada, October 30.

(34) Report of Special Advisory Committee on The Noise Regulation to the Ontario Ministry of Labour, December 1985.

(35) Eldred, K. M. 1974. "Assessment of Community Noise." *Noise Control Engineering*, (Sept./Oct.), 88–95.

(36) Shultz, T. J., 1972. *Community Noise Ratings*. London: Applied Science Publications.

(37) Miller, L. N. 1980. "Noise Criteria." Lecture notes on Noise Control for Buildings and Manufacturing Plants prepared in conjunction with Bolt, Beranek and Newman, Inc., Cambridge, MA.

(38) "Noise in the Human Environment," vols. 1 and 2.

Report of the Environment Council of Alberta, 8215-112 St., Edmonton, Alberta.

(39) OECD. 1984. "Noise Abatement: Present Situation and Future Outlook (draft final report)." Environment Committee, Ad Hoc Group on Noise Abatement Policies, OECD, August.

(40) OECD. 1985. "Strengthening Noise Abatement Policies." Report of Environment Committee, OECD, August.

(41) ISO. 1982. "Description and Measurement of Environmental Noise." International Standards Organization 1996 Part 1.

(42) ISO. 1983. "Acquisition of Data Pertinent to Land Use." International Standards Organization 1996 Part 2.

(43) ISO. 1983. "Application to Noise Limits." International Standards Organization 1996 Part 3.

(44) CSA. 1982. "Procedure for Performing a Survey of Sound Due to Industrial Institution or Commercial Activity." CSA Standard Z107.53.

(45) CSA. 1983. "Recommended Practice for the Prediction of Sound Levels Received at a Distance from an Industrial Plant." CSA Standard Z107.55.

(46) CSA. 1984. "Definitions of Common Acoustical Terms Used in CSA Standards." CSA Standard Z107.0.

(47) Gidamy, H. 1984. "Statement of General Principles and Recommendations." CSA Task Force on Community Noise.

(48) CMHC. 1981. "New Housing and Airport Noise." Central Mortgage and Housing Report NHA5185 81/05.

(49) Ontario Environmental Protection Act, SO 1971, c.86 (as amended).

(50) "Acoustic Technology Courses," Vols. 1–3, Ontario Ministry of the Environment, July 1977.

(51) "Summary of Quantitative S and V Criteria in the Model Municipal Noise Control By-Law." Ontario Ministry of the Environment, August 1978.

(52) "Noise Level Guidelines." Ontario Ministry of the Environment, 1984.

(53) "Initial Comments and Procedures for Noise Impact on Proposed Residential Subdivisions." Ontario Ministry of the Environment, 1986.

(54) "Method for Prediction of Road Traffic Noise." Ontario Ministry of the Environment, 1986.

(55) EPA. 1981. "Noise Effects Handbook." U.S. Environmental Protection Agency EPA 550-9-82-106. (Reprinted by Ontario Ministry of the Environment.)

(56) "Land Use Policy Near Airports." Ontario Ministry of Housing, 2M/3-78/PW-43, March 1978.

(57) "Guidelines on Noise and New Residential Development Adjacent to Freeways." Ontario Ministry of Housing, April 1979.

(58) "Land Use Planning for Noise Control in Residential Communities." Ontario Ministry of Housing, August 1980.

(59) Barry, T. M., and J. A. Regan. 1978. "FHWA Highway Traffic Noise Prediction Model." U.S. Federal Highway Administration Report FHWA-RD-77-108.

(60) Lang, J., and G. Jensen. 1970. "The Environmental Health Aspect of Noise Research and Control." World Health Organization Report EVRO2631. Copenhagen, Denmark: WHO Regional Office for Europe.

(61) Transport Canada. 1985. "Land Use in the Vicinity of Airports," 6th edition. TP1247E or TP1247F, Ottawa: Transport Canada.

(62) Hirschorn, M. IAC Noise Control Reference Handbook. Bronx, NY: Industrial Acoustics Company.

(63) ASHRAE. 1987. HVAC Systems and Applications. Atlanta, GA: American Society of Heating, Refrigerating, and Air Conditioning Engineers, Inc.

(64) Foreman, J. E. K. 1986. "Psychoacoustic Study of Human Response to Transmission Line Noise." Proceedings of 12th International Congress on Acoustics, Toronto, Canada. July 1986. Vol. 1, Section C1-7, 23–31.

(65) Pearsons, Bennett, and Fidell. 1979. "Initial Study on the Effects of Transformer and Transmission Line Noise on People, Volume 1: Annoyance." Report to Electric Power Research Institute, Palo Alto, CA.

(66) Horonjeff, Bennett, and Teffeteller. 1979. "Initial Study of the Effects of Transformer and Transmission Line Noise on People, Volume 2: Sleep Interference." Report to Electric Power Research Institute, Palo Alto, CA.

Chapter 8

GENERAL REVIEW OF NOISE CONTROL AND PRACTICAL EXAMPLES

This chapter begins with a general review of noise measurements and noise reduction procedures, together with a discussion of a program for noise control. This is followed by many examples of noise situations and corrections/controls from a publication by Bruel and Kjaer, Naerum, Denmark entitled *Noise Control—Principles and Practice*, 1st edition, 1982. (Some of these examples appeared in earlier chapters.) This book was adapted from a publication of the Swedish Workers' Protection Fund, established by national legislation, with the object of making industry and industrial employees aware of safety and environmental questions and to encourage the provision of an improved and safer working environment throughout Swedish industry. The results were then distributed in a straightforward and understandable form for the benefit on nonexperts in factories who have to deal with the problem at first hand, as well as those with previous and more extensive knowledge. Although these examples are graphic in nature and, consequently, nonrigorous vis-a-vis calculations, it is hoped that the examples will provide an overview as to many of the problems which are encountered in connection with noise, and the preferred methods in the solution of these noise problems.

Chapter 8 is adapted with the permission of Bruel and Kjaer, Ltd., Naerum Denmark. *Note that, consistent with the method of presentation in Bruel and Kjaer's text, there are no numbers associated with the figures in the text of this chapter.*

8.1. GENERAL REVIEW OF NOISE CONTROL

8.1.1. General Noise Control Measures

When attenuating noise, consideration must always be given to the fact that sound is radiated both as airborne and as structure-borne sound, i.e., vibrations. The majority of sound sources produce both airborne and structure-borne noise at the same time. In order to achieve a satisfactory result, a number of different noise control principles must be employed. This section presents a brief description of noise control techniques which have been used with good results in a number of different types of plant.

8.1.1.1. Alteration of Machines and Equipment. In order to be able to carry out noise control measures effectively, many important factors have to be taken into account. Which machine or machines should be quietened? How is the machine tended? Will maintenance and servicing be made more difficult?

Machines. Machines and processes in use can be difficult to alter without adversely influencing production. Attempt, though, to avoid or reduce impact and rattle between machine components. Brake reciprocating movements gently. Exchange metal components with quieter plastic, nylon, or compound components where possible. Enclose, locally, particularly noisy components or processes.

New machines and processes can often be improved by supplying the factory by the same type of techniques but with further possibilities for making more extensive changes. Convince the designer to:

1. Choose power sources and transmissions which give quiet speed regulation, e.g., stepless electrical motors;
2. Isolate vibration sources within the machine;
3. Ensure that cover panels and inspection hatches on machines are stiff and well damped;
4. Provide machines with adequate cooling fins which reduce the need for air flows, and therefore fans.

Equipment. Existing equipment can often be attenuated just as much as new, without complicated operations. Typical sound attenuating measures are to:

1. Provide air exhausts from pneumatic valves with silencers;
2. Change the pump type in hydraulic systems;
3. Change to a quieter type of fan or locate sound attenuators in the ducts of room or process ventilation systems;
4. Replace noisy compressed air nozzles with quieter types.

In a new plant one can go even farther by:

1. Installing quieter electric motors and transmissions;
2. Choosing hydraulic systems with specially stiffened oil tanks;
3. Mounting dampers in the hydraulic lines;
4. Dimensioning these lines for a relatively low flow velocity (a maximum of about 5 m/s);
5. Providing ventilation ducts with sound attenuators to prevent transmission between noisy and quiet rooms via the ductwork.

8.1.1.2. Material Handling. Existing plant can be altered so that impact and shock during manual or mechanical handling and transport of material and items are avoided. This may be done by:

1. Minimizing the fall height for items collected in boxes and containers (see diagram);
2. Stiffening panels which are struck by material and work pieces and damping them with damping panels or materials;
3. Absorbing hard shocks by wear resistant rubber or plastic coatings.

In connection with minimizing fall heights (see diagram), plates which fall from a roller conveyor down on to a collecting table cause very intense impact noise. By making use of a table whose height is controllable, the noise can be reduced.

When obtaining new conveyor equipment consider systems which transport raw materials and products quietly and steadily:

1. Consider choosing conveyor belts rather than rollers. Roller transporters are liable to rattle.
2. Control the speed of conveyor belt transporters, etc. to match the amount of material to be transported. This avoids stops and starts which cause noise from vibrations and impact of the transported material.

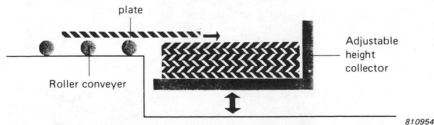

810954

Minimizing fall heights for items collected in containers.

8.1.1.3. Enclosure of Machines. If it is not possible to prevent or reduce the noise at its source, it may be necessary to enclose the entire machine. (See the accompanying figures.) For this enclosure to be satisfactory:

1. Use a sealed material, e.g., panels of metal or plasterboard for the outer surfaces.
2. Provide the inner surface with a sound-absorbent material, e.g., mineral wool, glass wool, foam rubber, or polyurethane material. A relatively simple sealed enclosure of this type can reduce noise by 15–20 dB(A).
3. Mount noise attenuators on any openings for cooling air.
4. Supply the enclosure with inspection

hatches which are easy to open where this is necessary for operation or maintenance.

An example of enclosure of a noisy machine is given in Section 9.3 in Chapter 9.

The enclosure of the hydraulic system shown on page 227 requires sound attenuated ventilation openings. Ventilation is required because heat is radiated by the motor, pump, and oil tank. A sealed inspection cover must be provided.

8.1.1.4. Attenuation of Structure-Borne Sound. A typical cause of vibration in a machine is clatter resulting from wear or from loose bolts and screws. In this case it is rela-

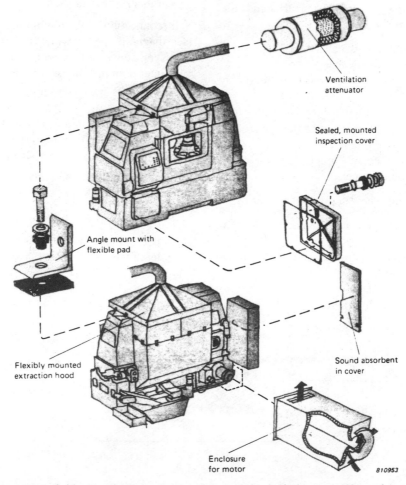

Ventilation attenuator

Sealed, mounted inspection cover

Angle mount with flexible pad

Flexibly mounted extraction hood

Sound absorbent in cover

Enclosure for motor

810953

Different types of airborne and structure-borne sound-prevention methods on a machine tool (a press).

Enclosed hydraulic system with ventilation.

tively easy to reduce the intensity of structure-borne sound by repair and renovation. On the other hand it is more difficult to reduce vibration from a working machine in good condition. It is often possible to reduce structure-borne sound disturbance by preventing transmission of vibration from machines and equipment to the load-bearing structure of the building using the following principles:

1. Vibration-isolate machines with stiff or independent frames. Place the machine on a stable foundation with an elastic separating layer of, e.g., rubber blocks or steel springs. (See Chapter 6.)
2. Place large heavy machines, which cannot be effectively vibration-isolated, on special machine foundations which are otherwise completely separated from the building. (See accompanying diagram.)
3. Vibration isolate machine panels wherever possible in order to minimize radiation of structure-borne noise. Panels should be elastically mounted on the machine frame thus reducing the vibration level transmitted to them. Alternatively, panels can be coated with a special damping material. (See Chapter 6.)

Strongly vibrating machines require separate foundations and isolating joints between floor slabs to prevent the propagation of structure-borne noise. In this case (see diagram on page 228) two joints are used for more effective separation.

(a) Before casting the floor, a thick strip of foamed plastic is placed in all the joints between the floor and the rest of the building structure.

(b) After the floor has been cast, the foam is pulled or burnt out and the joint inspected and cleaned out if necessary. There must be no bridging between the two structures, i.e., mechanical connection by stones, or the isolation will be bypassed.

(c) The joint is then filled with a flexible material, e.g., a synthetic rubber tube and sealed completely with a elastic sealant of high density.

8.1.1.5. Attenuation by Using Absorbents. (See Chapter 5.) In a workshop or in factory premises with hard materials on the ceiling, floor, and walls, nearly all the sound which reaches these surfaces is reflected back into the room.

In a room, the sound level from a machine first falls relatively quickly and then remains approximately unchanged as one moves away from it. This is because close to the machine its noise level falls approximately as if it were in a free field. However, at a certain point the reverberant noise level in the room, i.e., noise coming from all other sources including reflection from the room surfaces, becomes more intense than the direct sound from the single machine and dominates it (see Chapter 4). In such circumstances the noise environment can be improved by:

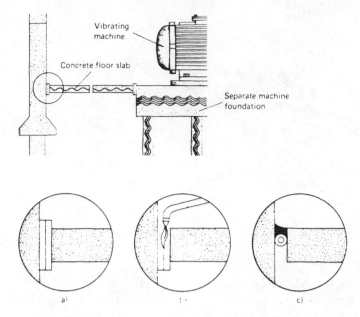

Example of prevention of structure-borne noise.

1. Covering the ceiling with an effective sound-absorbing material, for example, panels of mineral wool or glass fiber, which reduce reflected noise in the room. The reflected or reverberant sound can be reduced by 8–10 dB(A) at distances away from the sound sources.

2. Mounting highly absorbent ceilings and walls, for example, 100 mm of absorbent with perforated panels over it, can reduce the reverberant sound by up to 10 dB in a room with noisy production machinery in one part and relatively quiet work in the other. An attenuation of 10 dB is perceived as a halving of the noise loudness.

3. Using local absorption on the walls and ceiling at the operator's position near a noisy machine to reduce local reflections, and thus lower noise levels by a few dB. This just audible change in sound-pressure level is perceived as an improvement by those who work by the machine.

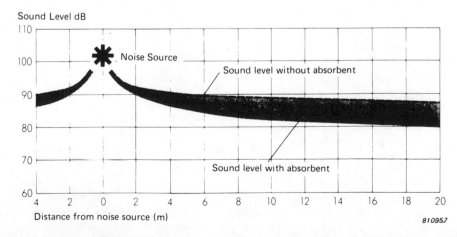

810957

The sound-pressure level at different distances from a source in a room without absorption and in the same room after a large area of absorbent material has been mounted on the ceiling.

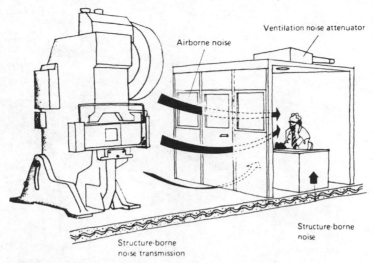

Noise problems in control rooms and workshop offices can be caused by direct airborne sound (because of gaps around doors, windows, etc.), or by the transmission of structure-borne sound, or both.

8.1.1.6. Sound Insulated Rooms.

Current developments within industry are directed toward automating machines and processes, whereby remote control of the process from a control or monitoring room becomes feasible. It is thus possible to limit the noise exposure of machine operators and process controllers to short periods when starting and servicing the machines, and repair and maintenance work. A few important rules of thumb are:

1. Build control and monitoring rooms with good sound insulation properties;
2. Choose door and window designs which are well sealed;
3. Provide ventilation openings with attenuators or acoustic louvers and ensure that cable and pipe cutouts are properly filled with a suitable sealant. Supply all control rooms in machine and process halls with good ventilation and cooling systems, otherwise there is a risk that doors will be opened to obtain sufficient fresh air. This naturally ruins attempts at good sound insulation.

8.1.1.7. Noise Control of New Projects.

There are even better possibilities for achieving good acoustical conditions when planning new projects. By exploiting the best known techniques it is normally possible to reduce noise generation from machinery and pro-

cesses compared with older plants. Acoustic problems should be taken into consideration right from the planning stage of the new building.

When choosing new installations, machinery, and equipment, as well as material handling methods, consideration must be given to the noise disturbance which it could produce. Continual effort should be made to change to quieter processes and working methods by introducing a greater degree of remote control. Personnel can then spend a proportion of the working day in relatively quiet control and operation rooms.

At the project planning stage of a new plant it can often be difficult to obtain an accurate basis for noise calculations. Measurement results from similar workshops and use of manufacturers' noise data for machinery and other equipment to be installed are necessary to make reasoned judgments.

Even though the possibilities for noise control are good when designing new projects, good acoustic conditions cannot always be achieved everywhere. Undesirably high noise levels will still exist in some areas because machine design and production processes cannot immediately be changed.

As a rule, acoustic problems involve a large number of specialist areas, and noise reduction is often difficult to achieve without an extensive knowledge of these. With more compre-

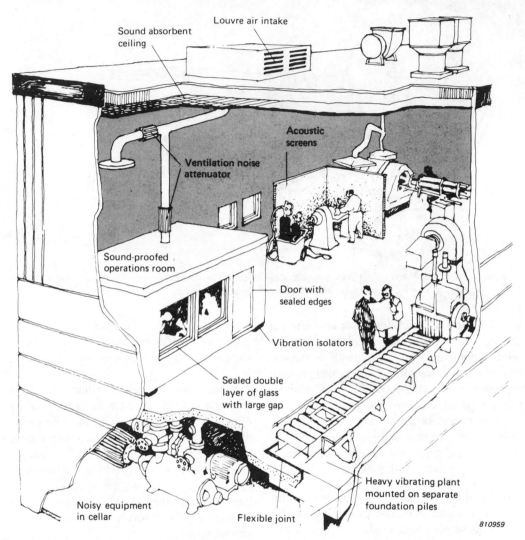

Example showing typical noise-control measures which should be taken in an industrial building in order to minimize the propagation of noise.

hensive projects and more difficult noise problems, it is therefore necessary to seek the advice and help of personnel with knowledge and experience in acoustics and noise.

A target noise climate should be set for all places in the factory where personnel normally operate and should take into consideration the type of work, working conditions, etc.

8.1.1.8. Planning the Building. The acoustically important details of the building's load-bearing structure and work areas should be calculated and fixed early in the planning stage. The need for noise control depends first

and foremost on the way the production plant is designed and laid out. The structural design of the building often depends on where the machinery is placed and the need for insulation against airborne and structure-borne sound.

1. The building's load-bearing structure, floors, and machine foundation should be chosen so that all noise sources can be effectively vibration isolated. Heavy equipment demands stiff and heavy foundations, which must not be in direct contact with other parts of the building structure.

2. Powerful noise sources should be enclosed by constructions which give adequate airborne sound insulation. Doors, inspection windows, and other building elements where there is a risk of sound leakage require special attention.

3. Rooms where there are sound sources or where personnel are present should be provided with ceiling cladding (also wall cladding where high ceilings are concerned) which absorb the incident sound. Sound-absorption characteristics vary widely for different materials which must therefore be chosen with regard to the characteristics of the noise. Good sound-absorption characteristics can often be combined with good thermal insulation.

4. Office areas should be separated from building elements where vibrating equipment is installed by a joint of elastic material.

5. Walls and ceiling construction, windows, doors, etc. should be chosen so as to achieve the required sound insulation.

6. Mounting noisy equipment on light or movable partitions should always be avoided. If ventilation for cooling systems must be mounted on such a light foundation in any case, e.g., a false ceiling, special efforts must be made to obtain sufficient vibration isolation.

7. In open-plan offices and large rooms where there are several office functions carried out in the same room, there must be a ceiling with high sound absorption and soft carpeting on the floor is also beneficial. It should be noted that it is especially important that sound absorption be also effective at low frequencies.

8.1.1.9. Noise Reduction Measures in Rooms.

The shape and size of an industrial workshop is determined to a large extent by the production processes and the flow of materials. The best possibilities for influencing the design and layout of a workshop occur, as mentioned earlier, in the early stages of planning. In the past, environmental questions have far too seldom been discussed in this phase, which often meant being trapped in a program of continuous work to obtain, among other things, satisfactory noise reduction. A few guidelines about the layout of the new plant follow.

1. Work stations and machines should be so placed that the reduction of noise with distance can be exploited, i.e., that there is a certain distance between noisy and quieter activities. Ensure that space is allowed for screens and enclosures.

2. Ensure that separate areas are available for particularly noisy machines, e.g., in cellars.

3. Work which requires a quiet working environment or which does not itself produce noise should be removed to a region with a low noise level. Work areas without noise can, if necessary, be screened from noisy surroundings. Where possible, mount absorbent ceilings in such areas.

4. If noisy work is carried out close to a wall or any other reflecting surface, it should be covered with an absorbent material.

5. Workshop offices, rest rooms, etc., should be provided with sufficient sound insulation and possibly mounted on isolators, or separated from the rest of the building structure by flexible joints, in order to avoid vibration transmission.

6. Fixed installations (ventilation equipment, cooling systems, etc.) should be constructed with sound attenuation in mind, and mounted so that sound from fans is prevented from spreading via ducts, pipes and the building structure itself. Normally a subcontracting ventilation firm will be responsible for attenuation of the system to a preagreed sound level.

7. In open-plan offices and large rooms, work areas must sometimes be located bearing in mind that noise occurs from certain processes while a relatively noise-free environment is required for others so that, for example, conversation is easy.

8. Machine rooms (for compressors, ventilation and cooling systems where service and maintenance staff remain during operation) should be equipped with sound absorbent screens between noise sources.

8.1.1.10. Purchase and Installation of Machinery.

Whenever new machinery is purchased consider the possibility of achieving quieter production and material handling. Before deciding on a purchase, ascertain the sound-pressure levels which the new equipment will cause as well as the feasibility of further noise reduction with possible suppliers. If improvements cannot be carried out within the financial constraints preparations should be made for noise reduction at a later date without expensive alterations.

The following guidelines are given:

1. Machines and equipment which generate vibration should normally be isolated from the building itself, so that disturbing vibration or sound cannot be transmitted. Machines which cannot be vibration-isolated because of their fundamental design or method of operation, e.g., large piston compressors, should be provided with their own foundations on supports which are completely separated from the load-carrying structure of the building.
2. Accessories, for example, hydraulic plant and air compressors, which are located in separate rooms should be provided with attenuators which prevent the propagation of sound and vibration in the installation's connections, pipes, and ducts.
3. Machines and equipment should be designed so that impact and shock are avoided as far as possible when handling raw materials or finished items. These should be slid down chutes rather than dropped into containers, for example.
4. With new purchases of transporting equipment (conveyor belts, roller transporters, traversing cranes, trucks, etc.) be aware of the availability of inherently quieter equipment, e.g., electrically powered fork-lift trucks.
5. When noise problems cannot be solved by other methods, a serious attempt must be made to enclose a whole machine or its particularly noisy parts. Solutions which might make operating or maintenance of the machine more difficult must be discussed with the personnel involved.

8.1.2. A Program for Noise Control

For all noise-control efforts a target noise level must be set. A highest level must be defined for each piece of equipment or room. The examples in the accompanying table should be regarded as guidelines. It is not intended that these should apply rigidly at all times in the future, but should be attainable after a certain time, after which they should be reconsidered, and if necessary lowered. (See also preferred noise criteria outlined in Chapter 7, Section 7.3.1.)

Example of Guideline Noise Levels Within a Factory

Type of Room	Guideline Highest Sound Level, dB(A)
Conference Room	35
Office	40
Workshop office, rest room	45
Laboratory, measurement or inspection room	50
Canteen	50
Changing room	55
Repair workshop	60
Production areas	75
Fan room, compressor room, etc., normally unmanned	90

For work areas with special or particularly severe noise problems it may be advantageous to attack the problem in a number of stages, lowering the noise limit at each successful stage.

By systematically mapping the existing noise situation, a good picture of the noise intensity and distribution within the workshop can be obtained. In order to be able to plan and carry through a noise control program with difficult noise problems it is necessary to carry out as extensive a noise measurement program as possible. Noise often comes from a large number of sources in the workshop (e.g., production machinery and material handling), and as background noise (from ventilation, compressors, circulation pumps, etc.) which may be sited outside the workshop under consideration. When judging the risk of hearing loss in a place of work all the noise sources which would normally be in operation should be in operation during noise measurements.

On the other hand, in order to make decisions about individual noise-control measures on the best possible grounds, each machine and noisy working process should be measured separately. Similarly, check those work processes, machine parts, etc., causing the greatest noise. This forms a good basis on which to judge whether noise control is necessary and possible.

Background noise is often found to contribute significantly to the total noise. Every time a noise source is introduced into the workshop, the noise level is increased to some extent even though the noise level of the new machine may be relatively low.

When a noise map is being made it is important that the people who are to carry out noise control measures, or are responsible for them, discuss the problems with the safety representative or the other employees in the departments affected by the work. They usually have thorough knowledge of the production equipment and can often contribute good practical ideas for improvements. To choose the most cost-effective noise-control measures a table of the different measures needed on various machines and installations is useful. In addition, they should be costed against the amount of attenuation which could be expected. Each project should be described with simple sketches, including:

1. Changes to machines which would reduce noise generation;
2. Alteration of equipment to avoid impact in machinery and when handling materials;
3. Enclosure of noisy machines or machines parts;
4. Mounting of attenuators in gas and air outlets as well as ventilation ducts;
5. Erection of sound absorbing screens, linings, and baffles in work areas.

The likely results for different types of noise control measures are as follows:

1. Mounting an absorbent roof or ceiling in a room will in general give a noise reduction of between 3 and 5 dB(A). Exceptionally, up to 10 dB(A) can be obtained.
2. Damping of vibration of small production machines by applying damping material can give between 3 and 10 dB(A) attenuation.
3. Factory-made screens can reduce noise from between 5 and 15 dB(A).
4. Leakage where pipes go through walls, as well as acoustic leaks between walls, screens, or enclosures, can produce large variations in the attenuation achieved. It is therefore important to seal air gaps carefully when carrying out this type of work.

Always obtain information on materials and costs from the supplier before making a decision on which actions to take. There are several factors which can influence the choice:

1. The intensity of the noise in the workshop (the first priority is to reduce noise which can cause hearing damage);
2. Practical problems in carrying out the work;
3. The number of persons who benefit from the improvements;
4. The costs involved in the measures chosen.

It is often difficult to weigh up all these points, but long-term planning should ensure that the demand for a good working environment is fulfilled in all places of work.

When a large number of projects are planned, a timetable describing each agreed project and stating the order in which they are to be completed is necessary. A plan is also required to determine when machines can be taken out of production to be altered, when absorbents can be mounted, when personnel can be obtained to carry out the assignments, etc.

8.1.3. Methods of Noise Reduction (see also Chapters 4 and 5)

There are three main ways to reduce noise in the factory:

1. Reduce noise at the source.
2. Change to quieter methods of work.
3. Prevent or reduce propagation.

Reduce Noise at the Source. It is often possible to reduce noise radiation from production equipment, material handling, and work in progress, for example, by damping sound radiating panels, quietening power sources and transmissions, and reducing noise from compressed air exhausts.

Sometimes machine alterations or enclosures do not give sufficiently good results, and if it is the work process itself which causes intense noise it can be difficult to predict the results of noise-control measures. In such cases effort might be better aimed at changing the working methods and processes themselves.

Change to Quieter Methods of Work. In many cases changing the method of work is the only way to get to grips with noise generation. This often requires that production equipment or part of it must be replaced and one must be aware of the availability of less noisy equipment for both production and material handling. This requires cooperation between the buyer, supplier, designer, and safety organization.

Prevent Propagation. The noise in a workshop is often dominated by a relatively small number of intense noise sources. Personnel who are working on quieter machinery or with work which does not produce noise are very often unnecessarily exposed to other noise sources in the same room. If these sources are screened or provided with an enclosure the noise level is reduced both close to and far from the source, benefitting everyone in the room.

By setting up sound-absorbing ceiling and wall panels, noise levels within the room far from the noise sources can be reduced. These measures, however, do not significantly reduce the noise exposure of personnel working on these machines.

Alteration and replacement of production equipment may mean that personnel monitoring this machinery need not be in its vicinity if monitoring can be carried out in a sound-insulated control room. However, this should not be exploited in order to avoid or cut down on noise control in areas where maintenance and repair staff spend the greatest proportion of their time.

In order to prevent vibration from noise sources spreading through the building structure and through machinery, it is often necessary to vibration-isolate machines or introduce vibration isolating joints in the building. (For a thorough coverage of this, refer to Chapter 6.)

8.1.4. Noise Measurement

When planning noise-control measures or forming a basis upon which to judge the noise of a projected plant, measurements are the most important starting point. Without existing measurements, or sometimes future predictions from existing measurements, of a noise situation, objective decisions about the need for noise control cannot be made, nor can its effectiveness after installation be judged.

Because of the large variety of noise characteristics and the corresponding large number of measurement and assessment techniques, great care is required in deciding which measurements to make and how to interpret them. The sound-pressure level which is read from a sound-level meter does not always give sufficient information to judge a hearing-noise danger, or for use as a basis for a noise-control program.

Both experience and special training are required to be able to carry out measurements in complicated situations. In many cases, though, a standardized sound-level meter and relatively simple measurement methods are adequate.

8.1.4.1. The Purpose of Measurements.
There are many different reasons for carrying out noise measurements in industry. The most usual are as follows.

1. Determine whether or not noise levels are high enough to lead to permanent hearing damage. Equivalent sound-pressure levels of over 85 dB(A) for an eight-hour working day should be investigated further.
2. Obtain a basis for noise-control measures to be applied to machines and equipment.
3. Determine sound radiation from single machines unambiguously, e.g., to compare with values stated in a noise quarantee or declaration.
4. Ensure that noise levels are not disturbing to third parties, e.g., in residential areas.

Measurement instruments and methods should comply with the standards which apply to the noise measurements to be carried out. The standards include requirements for the measurement instrument, measurement method for noise from different types of machines, and assessment of noise annoyance and damaging effects. The most important international standards are those published by IEC (the International Electrotechnical Commission) and ISO (the International Standards Organization). IEC is concerned primarily with the design and construction of instrumentation, and ISO primarily with the measurement technique, experimental conditions, measurement parameters, and reduction of measured results to a common point of reference. Most are available in English and French and many have been adopted directly or with only small local changes by individual countries as their national standards (e.g., ANSI standards in the U.S.A., CSA standards in Canada). The following sections contain simple rules for the choice of instrument and measuring methods according to the standards relevant to the particular country. On the other hand no attempt is made to describe in detail specialist topics such as impulse noise, infrasound, and ultrasound. (see Sect. 7.4.2.3 for information on impulsive sound).

A sound-level meter is designed to represent the human ear as closely as is possible and practical, while giving a repeatable and objective value. As the human ear responds not only to the level of a sound but also to its frequency, and to some extent duration, these parameters must be built into the sound level meter.

8.1.4.2. Frequency-Weighting Systems.
A number of frequency-weighting systems have been standardized, originally intended to be applied to different sound-pressure level ranges (refer to Chapter 3). These are:

1. The A-weighting system, intended for quiet sounds;
2. The B-weighting system, intended for sounds of medium intensity;
3. The C-weighting system, intended for loud sounds;
4. The D-weighting system, intended for the measurement of jet aircraft noise.

The A-weighting system is by far the most widely used of these weightings and forms the basic of a large number of derived units, e.g., L_{eq} (the equivalent continuous sound pressure level) and L_{10} (the A-weighted sound level exceeded for 10% of the measurement period). Because of its good agreement with subjective response to noise, the A-weighted sound level is now used for assessing noise of all levels, and the D, B, and C weightings are relatively infrequently employed.

Unweighted sound levels are usually only measured in connection with a frequency analysis, e.g., for comparing the frequency spectrum of a machine after applying sound-reduction techniques, with the spectrum produced by the same machine before.

8.1.4.3. Time Constants.
When choosing a suitable time constant for the measurement the characteristics of the noise must be taken into account. The level of a noise always varies to a greater or lesser extent. The display of the measuring instrument, whether a traditional meter needle or a digital display, always has a certain time constant and cannot follow rapid sound-level fluctuations. In addition, the human eye cannot follow the rapid movement, so the display of a sound-level meter is deliberately damped. There is normally a choice between three standardized time constants or dampings:

1. *Slow* has high damping, giving a slow display movement; effective averaging time is approximately 1 s.
2. *Fast* has low damping, giving a more rapid display movement; effective averaging time is approximately 0.125 s.
3. *Impulse* has a very fast rising-time constant and a very slow falling-time constant. This is intended to present a value which represents how loud the human ear judges a short-duration sound, i.e., it is aimed at annoyance rather than hearing-damage risk.
4. *Peak*. In addition, a number of sound-level meters have a further possibility, i.e., measuring the actual peak sound-pressure level of a short-duration sound. This allows the peak values of a sound

Time weighting	Sound level dB
S	93
F	100
I	102
Peak	127

810831/1

When measuring impulse noise, a sound-level meter displays a different value dependent on the time constant used. All will be very much less than the absolute peak value if the impulse is very short, e.g., a hammer impact, gunshot, etc.

whose duration may be as short as 50 microseconds to be accurately recorded.

L_{eq} is a standardized form of long-term average sound level using the A-weighting network described earlier. The sound level is integrated and averaged over the duration of the measurement. This average is carried out using the equal-energy principle. A person would be exposed to the same total sound energy whether he was exposed to the actual noise level including all its fluctuations or the the L_{eq} of that noise exposure for the same duration of time. This is a very useful concept when dealing with typical industrial and environmental noise which fluctuates widely and contains short periods of intense or impact-type noise. The noise exposure allowed for employees in industry is defined as a maximum L_{eq} value for a normal working day. A wide variety of instruments for measuring noise with different characteristics in any situation can be attained today.

To extend the possibilities of measurement it is often practical to connect a tape recorder to a sound-level meter in order to record the source noise. Later analysis of the recorded noise forms a better basis for a more extensive insight into the noise problem and better judgment. This is always to be recommended where a more complete description of the noise source or a better basis for making decisions about sound reduction measures is required.

Noise characteristics classified according to the way they vary with time are shown in the accompanying diagram/table. Constant noise remains within 5 dB for a long time. Constant noise which starts and stops is called intermittent. Fluctuating noise varies significantly but has a constant long-term average (L_{eq}). Impulse noise lasts for less than one second (p. 237).

In a room with closely placed, constantly operating noise sources and many work stations it is usually advisable to draw up a noise map as a first step toward noise control. Measurements of this type are the only way to determine whether or not the working environment is satisfactory. To adequately describe individual noise sources (or work positions), more measurement points than are normally used for a noise map are required. It is a good idea to add a few extra points near particularly noisy sources, as these will tend to be dominant and more likely to require treatment.

8.1.4.4. Practical Noise Measurement Procedure.

The purpose of noise measurement is to make reliable, accurate, and thorough measurements which properly describe the noise situation and which can be depended upon for use in the future. To ensure this the following procedure is recommended:

1. Always calibrate all instrumentation before and also preferably after measurements.
2. Sketch the instrumentation used and note the model and serial numbers.
3. Make a sketch of the measurement situation, position of sources, measurement position, and local reflecting surfaces which may affect measurements.

Characteristics	Type of Source	Type of Measurement	Type of Instrument	Remarks
Constant continous noise	Ventilation systems pumps, electric motors, gearboxes, conveyers,	Direct reading of A-weighted value	Sound level meter	Octave or 1/3 octave analysis if noise is excessive
Constant but intermittent noise	Air compressor during charging, automatic machinery during a work cycle,	dB(A) value and exposure time or L_{eq}	Sound level meter / Integrating sound level meter	"
 Periodically fluctuating noise	Mass production, surface grinding	dB(A) value, L_{eq} or noise dose	Sound level meter / Integrating sound level meter	"
 Fluctuating non periodic noise	Manual work, grinding, welding, component assembly	L_{eq} or noise dose / Statistical analysis	Noise dose meter / Integrating sound level meter	Long term measurement usually required
 Repeated impulses	Automatic press, guillotines, pneumatic drill, rivetting	Measurement of L_{eq} or noise Dose, and "Impulse" noise level Check "Peak" value	Impulse sound level meter	Difficult to assess / More harmful to hearing than it sounds
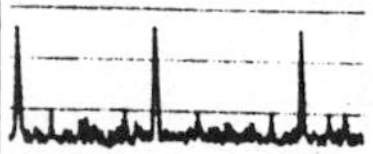 Single Impulse	Hammer blow, material handling,	L_{eq} and "Peak" value	Impulse sound level meter	Difficult to assess. Very harmful to hearing especially close.

Left-margin labels accompanying the waveform diagrams:

0 5 10 15 20

small variation

intermittent noise / background noise

Large fluctuations

large irregular fluctuations

Similar impulses

isolated impulse

810339

Noise characteristics, source, and type of measurement/instrument.

Sound level dB(A)

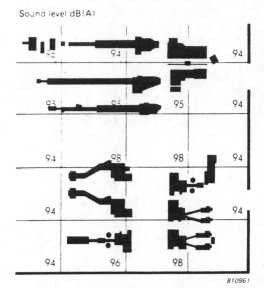

810961

Grid mapping a machine hall with a large number of closely spaced machines. Each grid is 6 × 7.5 m.

4. When working outdoors, note the meteorological conditions, especially wind direction and strength, temperature, and humidity.

5. Check the background noise level to ensure that it is sufficiently below the measurements being taken or correct if necessary; see below.

6. Carry out the measurements, noting down relevant equipment settings such as A-weighting, fast time constant, etc.

7. Keep a log, noting changes made to equipment settings and unusual occurrences, and make notes where relevant.

8. When tape or level recording, always make notes in the field directly on the tape or level recorder paper where they are obvious when you return to the recording at at later date for further analysis.

8.1.4.5. Background Noise. Noise from unwanted sources, so-called *background noise*, must be at least 10 dB below the level of the noise emitted by the source being considered for measurements to be valid. If this is the case, the measurement is accurate to within 0.5 dB. The background noise must therefore always be checked before making measurements. However, if the difference between the source noise measured in the presence of background noise and the background noise alone lies between 3 and 10 dB, a correction may be made as in the

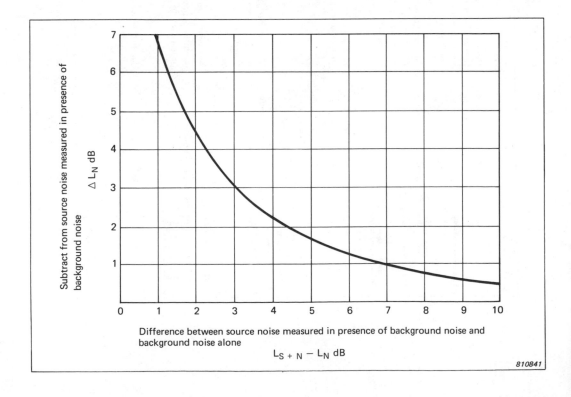

810841

diagram. If this difference is less than 3 dB, the source noise level is less than the background noise level and a reliable value for the source noise alone cannot be obtained. If measurements are made in frequency bands, the background noise measured in each band should be at least 10 dB lower than the source.

8.2. PRACTICAL EXAMPLES OF NOISE-CONTROL TECHNIQUES

In this section a number of noise control principles are presented in the form of general techniques with a practical example of each. These deal both with the factors which influence the generation of noise and its propagation in materials, structures and rooms. First and foremost, noise must be limited at the source. When solving a noise problem, the question, ''What is the caused of the noise disturbance?'' has to be answered. The technique described most often uses that method of noise reduction which directly reduces noise generation. Most techniques have a limited area of use, though, and may not be available for all possible situations.

Noise can be caused by a large number of factors, and extensive measurements may often be required before a decision can be made as to which measure should be tried first. It is not usually sufficient to employ just one of the techniques shown to reduce noise. In most cases several different actions need to be taken to achieve the desired result.

In order to make the drawings easy to understand, different symbols are used, e.g., large arrows to indicate high noise radiation and small ones to indicate less noise, wavy lines to indicate structure-borne noise, etc. In a particular case only those symbols which clarify the particular problem under consideration are used.

These practical noise-control solutions are relatively simple examples—and hence are not accompanied by theoretical derivations. However, appropriate theory, where required, has been covered in preceding chapters.

The following pages are reproduced by permission directly from *Noise Control—Principles and Practice*, published by Bruel and Kjaer, Naerum, Denmark.

Practical examples of noise control techniques

In this chapter a number of noise control principles are presented in the form of general techniques with a practical example for each. These deal both with the factors which influence the generation of noise and its propagation in materials, structures and rooms. First and foremost, noise must be limited at source. When solving a noise problem, the question, "What is the cause of noise disturbance?" has to be answered. The technique described most often uses that method of noise reduction which directly reduces noise generation. Most techniques have a limited area of use, though, and may not be available for all possible situations.

Noise can be caused by a large number of factors, and extensive measurements may often be required before a decision can be made as to which measure should be tried first. It is not usually sufficient to employ just one of the techniques shown to reduce noise. In most cases several different actions need to be taken to achieve the desired result.

In order to make the drawings easy to understand, different symbols are used, e. g. large arrows to indicate high noise radiation and small ones to indicate less noise, wavy lines to indicate structure-borne noise, etc. In a particular case only those symbols which clarify the particular problem under consideration are used.

More rapid changes produce higher dominant frequencies

The dominant frequency of the noise produced by an impact is dependent upon the speed of the force, pressure, or velocity change which gives rise to the noise. A rapid change produces a shorter pulse which has higher dominant frequencies. The speed of this change is often determined by the resilience of the two impacting surfaces: The more they deform, the longer they are in contact and the lower the dominant frequencies are. When bouncing a basketball on the floor, the ball is in contact with the floor for a relatively long time. The dominant frequency is therefore low. When playing table tennis the ball is in contact with the bat or table for only a very short time. The dominant frequencies are therefore much higher.

Principle

Long-lasting impact against floor
— Low frequency noise

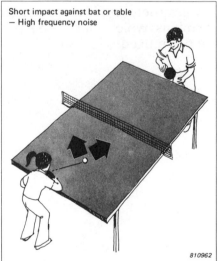

Short impact against bat or table
— High frequency noise

810962

Example

A sonic boom lasts approximately as long as it takes for the aircraft to fly through its own length. This may take e.g. 0,25 s. The dominant frequency is therefore about 4 Hz. A gunshot may last only 1 ms. Its dominant frequencies are therefore much higher, at about 1 kHz.

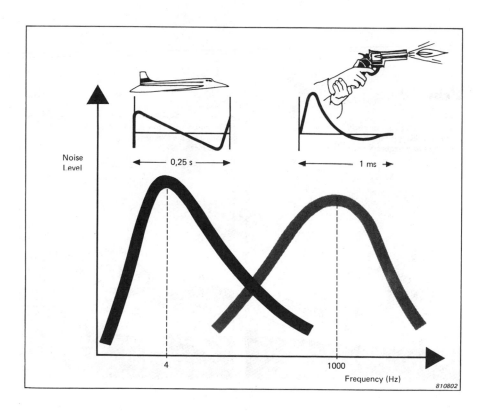

Slow repetitions give low frequencies, fast repetitions give high frequencies

A noise-producing event which repeats, generates frequencies which depend on the time between repetitions. A slowly repeating event gives rise to predominantly low frequencies and a rapidly repeating event gives rise to high frequencies. The level of the sound depends on the magnitude of the change which gave rise to it.

Principle

The distinctly separate beats of a low reving oil engine produce low frequency noise

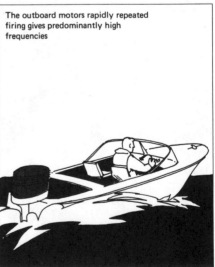

The outboard motors rapidly repeated firing gives predominantly high frequencies

810963

Example

Two gearwheels having the same size but one having twice as many teeth as the other will have predominant frequencies a factor of two apart. The main source of noise is the contact of one tooth on the corresponding tooth on the gear-

wheel in mesh with it. For the same diameter and speed of rotation the gearwheel with twice as many teeth will have twice as many tooth contacts per second and therefore radiate noise at twice the frequency of the other.

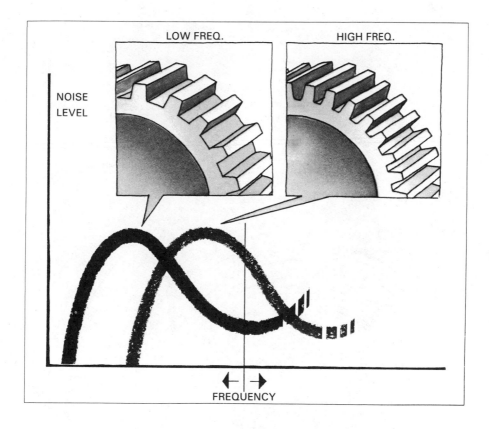

Low frequency sound bends round obstacles and through openings

Low frequency sound radiates approximately equally in all directions. It diffuses round edges and through holes without losing intensity, and reradiates from the edge or from the hole as if it were a new source, again equally in all directions. For this reason screens and barriers are not very effective against it unless they are very large.

Principle

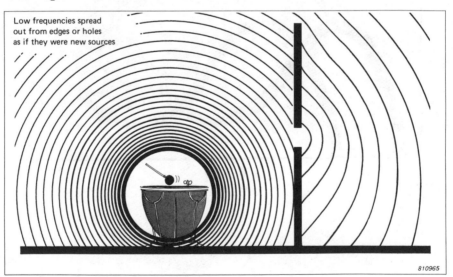

Low frequencies spread out from edges or holes as if they were new sources

810965

Example

Diesel driven compressors produce high levels of low frequency noise, even if they are furnished with efficient intake and exhaust silencers. Partly open or louvred covers for the intake of cooling air are of little use as noise attenuators. Noise easily radiates out through the openings and gaps.

Solution

Effective quietening of a powerful compressor requires a well sealed cover eliminating air and noise leaks. The cover can be constructed as a double wall containing ducts with sound absorbent linings. Air for the compressor, for the engine, and for cooling purposes is carried through these ducts, entering and leaving via acoustic louvres. The exhaust silencer is also enclosed within the outer cover. All inspection hatches and access panels must also be tightly fitting and well sealed.

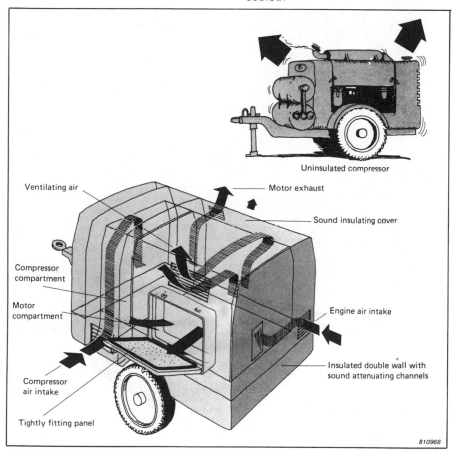

Uninsulated compressor

Ventilating air

Motor exhaust

Sound insulating cover

Compressor compartment

Motor compartment

Engine air intake

Compressor air intake

Insulated double wall with sound attenuating channels

Tightly fitting panel

810966

High frequency sound is highly directional and easy to reflect

High frequency sound is often produced by sources which radiate a high noise level in some directions but low levels in others. It can be reflected from a hard surface just as light is reflected by a mirror, and passes through holes in a panel like a beam without being diffused to the sides. Also it cannot diffuse around edges, so barriers are effective against it.

Principle

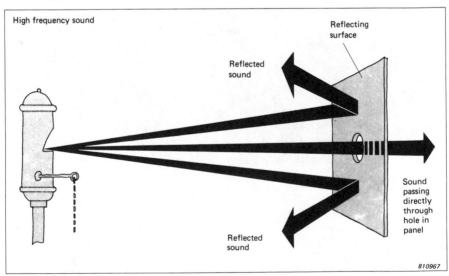

Example

Noise from processes which involve punching, hammering, or other forms of impact give rise to high levels of high frequency noise which can be dangerous to the operator.

Solution

A local enclosure built around the noise source with an opening for access and safety glass to view the work protects the operator's ears from direct sound from the machine. Reflections from the safety glass and most direct sound in other directions are absorbed by the absorbent lining. The small area open for access emits sound only away from the operator's ears.

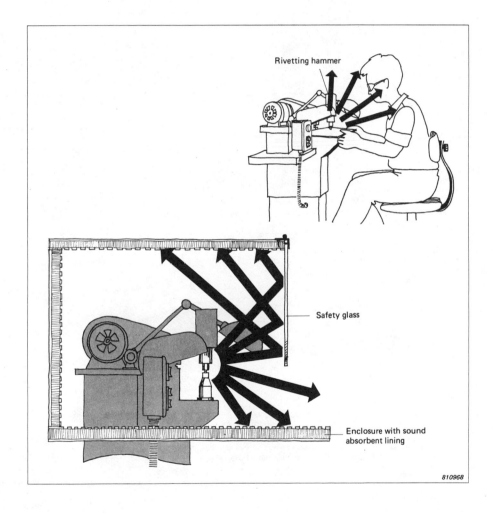

Rivetting hammer

Safety glass

Enclosure with sound absorbent lining

810968

Close to the source high frequency noise annoys more than low frequency noise

The ear is more sensitive to high frequencies than to low frequencies; so to produce the same amount of annoyance, a low frequency noise must have a higher sound level than a high frequency noise. In some circumstances it may be possible to reduce the annoyance of a close noise source by moving the dominant sound energy to lower frequencies.

Principle

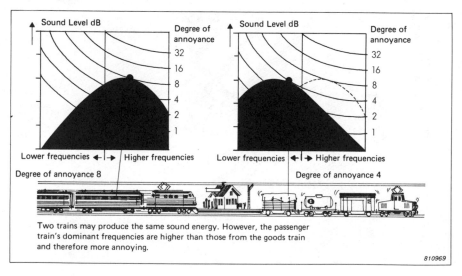

Two trains may produce the same sound energy. However, the passenger train's dominant frequencies are higher than those from the goods train and therefore more annoying.

810969

Example

A ships propeller turns at the same speed as the motor, 125 revolutions per minute, and is the source of most noise on board.

Solution

A design with a larger propeller driven through a gearbox at a lower speed lowers the dominant frequencies and reduces the annoyance caused.

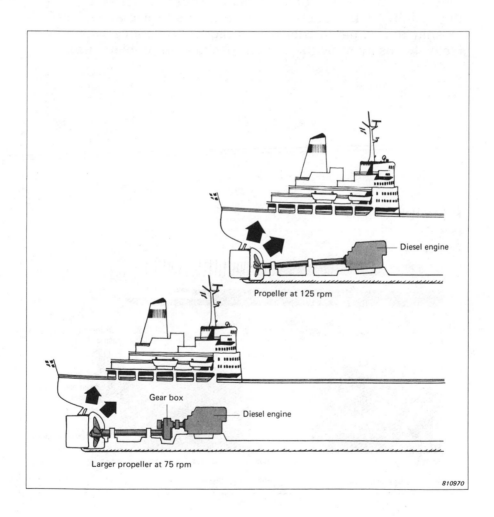

Diesel engine

Propeller at 125 rpm

Gear box

Diesel engine

Larger propeller at 75 rpm

810970

Far from the source high frequency noise annoys less than low frequency noise

High frequency noise is attenuated much more, by absorption in air, than low frequency noise is over large distances. This is because absorption is dependent on the number of cycles, and there are more cycles of a high frequency sound than a low frequency in a given distance. In addition, it is normally easier to reduce or shield a source of high frequency noise. If noise in the vicinity of the source is not a problem, it may be possible to shift the dominant noise to a higher frequency which is effectively absorbed by the time it reaches the problem area.

Principle

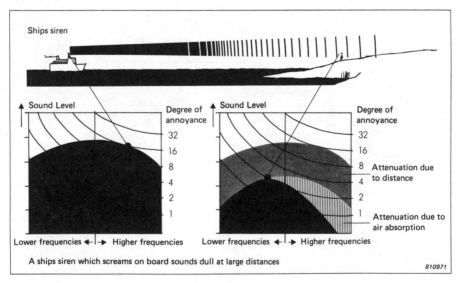

A ships siren which screams on board sounds dull at large distances

810971

Example

Low frequency noise from industrial fans causes noise annoyance in a distant residential area. High frequencies are attenuated on the way.

Solution

The fans can be exchanged with a type with more blades which shift the major sources of noise up in frequency. The higher tones are absorbed sufficiently by the atmosphere so that they are not a source of annoyance in the residential area, and the low frequency tones are no longer produced. The noise is also easier to attenuate at source.

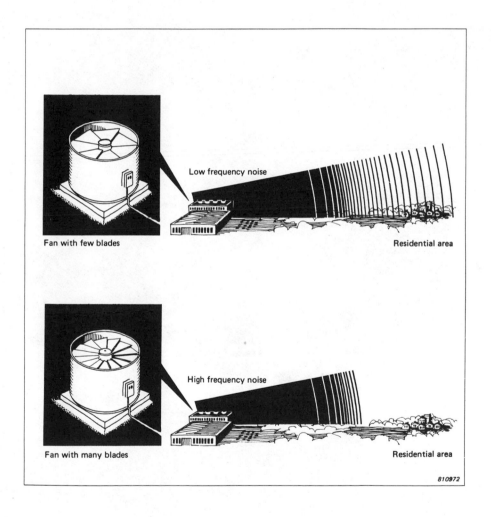

Low frequency noise

Fan with few blades

Residential area

High frequency noise

Fan with many blades

Residential area

810972

Sound sources should be sited away from reflecting surfaces

The closer a sound source is placed to a reflecting surface, the more of the sound radiated is directed back into the room. The worst position is against three surfaces, i. e. in a corner. The best position is free-hanging: away from all reflecting surfaces.

Principle

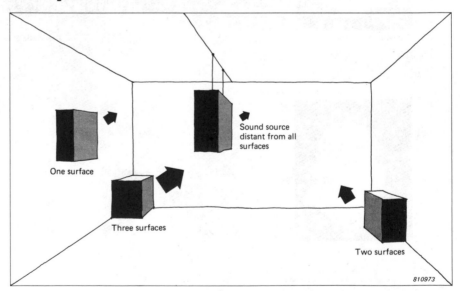

One surface

Sound source
distant from all
surfaces

Three surfaces

Two surfaces

810973

Example

In a machine hall a number of machine tools are placed in four lines, two of them against the walls, with three access lanes between them. This increases the noise from the two lines of machines placed next to the walls.

Solution

The machines along the walls are moved beside the other two lines so there are only two lines. The space along the walls is used as access lanes, of which there are still three, and the overall noise in the hall is reduced.

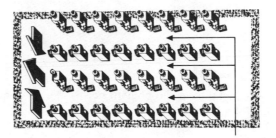

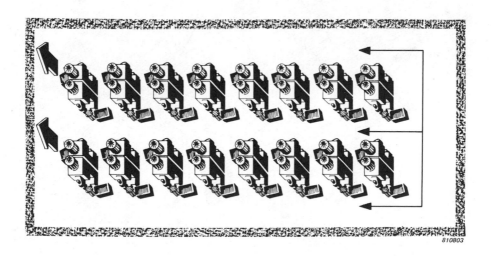

810803

Changes in force, pressure, or speed, lead to noise

Noise always occurs where there is a change of force, pressure, or speed. Large changes produce the greatest noise, small changes produce less. In many cases the same result can be achieved either with the application of high power over a short period or with less power for a longer period. The first case causes high noise levels, the second, where the power required is small, produces much lower noise levels.

Principle

A metal strip can be bent noisily using a hammer

or quietly using pliers

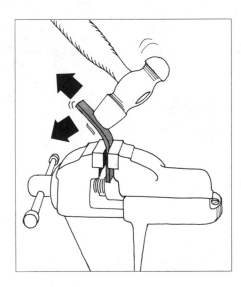

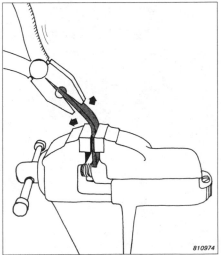

810974

Example

Panels and sheets can be fastened in a number of different ways, some of them very much noisier than others. Those involving impact e.g. nails and rivets are particularly bad from the point of view of hearing damage as they produce very high peak levels of noise.

Solution

In many cases quiet methods such as screws and bolts can be substituted directly without loss of effectiveness or increase in cost, and with the advantage of improved access and ease of dismantling at a later date.

Two panels can be fixed together

using nails — **noisy**

or screws — **quiet**

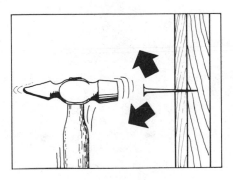

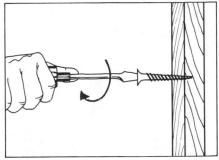

Steel sheet may be rivetted
Very noisy

or bolted — **very quiet**

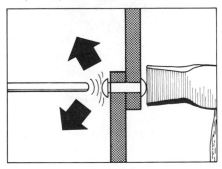

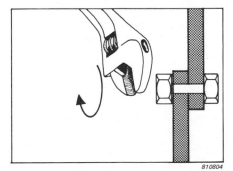

810804

Example

Paving breakers have traditionally been handheld and usually pneumatically powered to reduce weight. High levels of impulse noise are produced both by the chipping process itself and from the exhaust. The operator is exposed to high levels of both noise and vibration.

Solution

A tractor-mounted hydraulic ram driving a hammer can exert a very large static force as well as vibrate. The paving is fractured and the cracked surface can then be levered up by a bucket loader. The noise levels are lower and the operators are further from the source, often in noise protecting cabins.

Pneumatic drill

Hydraulically operated drill

810805

Example

Cardboard in a carton machine is chopped using a guillotine. The knife must fall very quickly using high power in order to cut perpendicular to the production line, causing high noise levels.

Solution

Using a knife which is driven across the production line, the material can be cut with a low force over a longer period, virtually silently. The knife must be set at an angle to the moving line of board to cut perpendicular to the direction of motion.

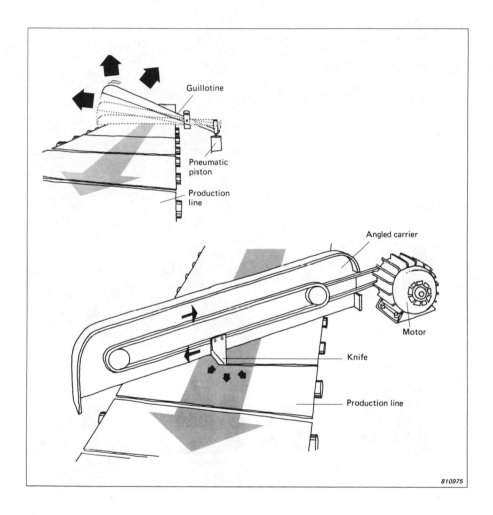

810975

Low mass and low fall heights give least sound

The noise level generated when a panel is struck by a falling object depends primarily on the mass and velocity of the object. The greater the mass and fall height, the louder the noise, because greater energy is available for transfer into the panel via the impact. A reduction in height or in mass by a factor of ten reduces the noise generated by approximately 10 dB.

Principle

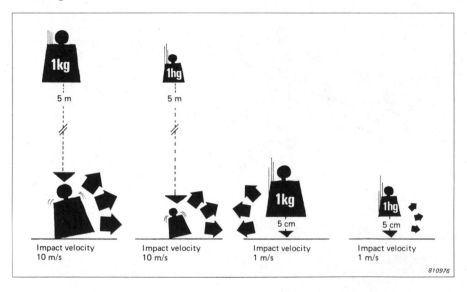

Example

Manufactured items are carried from the producing machine by conveyer and dropped into a collecting bin from a fixed height. When the bin is empty the fall height is large and the noise level is therefore high. There may also be danger of damaging the items.

Solution

The conveyer is constructed so that its height can be adjusted, and is supplied with a case with a number of rubber flaps inside to break the fall of the material. The fall height is therefore never greater than the distance from the collected material to the lowest rubber flap, the conveyer rising automatically as the bin fills.

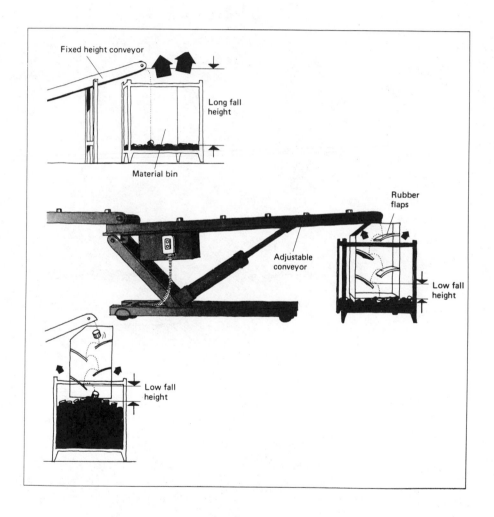

Problem

A material conveyer feeding a hopper deposits the material in the centre of the hopper and the fall height is therefore large. The hopper itself is also a very resonant structure.

Solution

Mount the conveyer so that the material falls on the edge of the hopper so that the free height is minimised. The interior of the hopper can be lined with wear-resistant material to absorb the impact better, and the external surfaces can be mounted with damping sheets to reduce resonances even further.

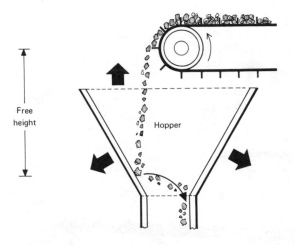

Free height

Hopper

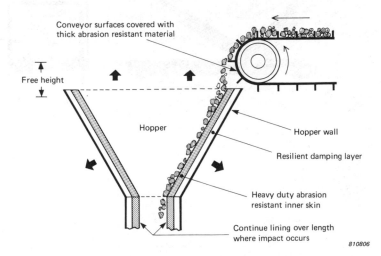

Conveyor surfaces covered with thick abrasion resistant material

Free height

Hopper

Hopper wall

Resilient damping layer

Heavy duty abrasion resistant inner skin

Continue lining over length where impact occurs

810806

Problem

Sheet piles are normally driven via the impact of a heavy mass dropped from a great height, often powered up again by exploding a diesel charge. Dangerous local noise levels are generated both by the impact on the pile and from the explosion in this case, and annoyance may be caused at distances of up to several kilometers.

Solution

In many situations it is possible to use a completely different technique which avoids impact completely. A set of hydraulically operated rams grip a number of sheet piles simultaneously.

One pile is forced down at a time while the machine pulls upwards on all the rest, which anchor it to the ground. Vibration of the ram holding the pile being driven assists its progress. Impact is avoided completely and noise levels are as low as the hydraulic equipment allows.

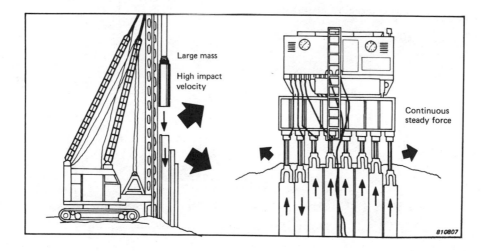

Structure-borne sound travels long distances

Vibration which gets into a structure, especially homogenous structures such as concrete buildings or ships, travels a very long way because of the very low internal damping of the structure. The energy does not reduce and as soon as a large surface, which acts as a loudspeaker, is connected to the vibrating structure, a high noise level is generated. It is best to isolate the structure from the source of vibration as near to the source as possible.

Principle

Vibration from the train is transmitted directly along the rails and can be heard at great distances

810978

Example

Vibration and stop/start shocks from an elevator can be heard throughout a building. The sound is carried for large distances virtually unattenuated via the concrete slabs.

Solution

The winding machinery must be isolated completely from the building structure using a spring support. Further reduction can be achieved by building the lift shaft and driving mechanism separately from the rest of the building structure.

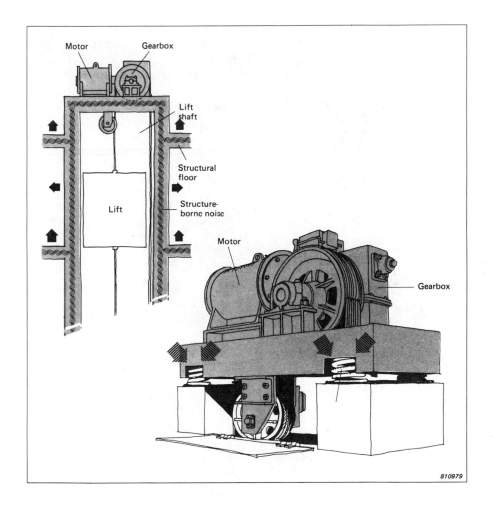

Structure-borne vibration needs large areas to convert it to airborne sound

The vibration of a small object will not generally give a high noise level because the area of air set in motion by the object will also be small. The vibration is thus badly matched to the air. However, connecting a large panel transfers the vibration energy into airborne sound much more effeciently by spreading the vibration over a much greater area which gives a high noise level. A tuning fork generates hardly any noise unless connected to a "sounding board". The circulation pump of a central heating system causes the pipework to vibrate, but little noise is transmitted until a large panel in the form of a radiator is connected; radiating not only heat, but noise as well.

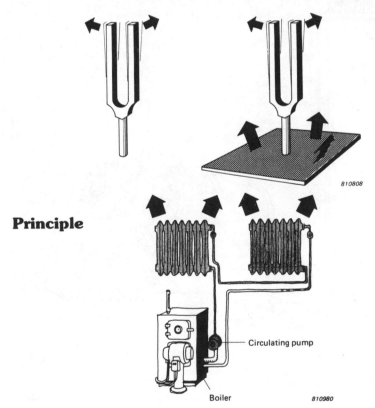

810808

Principle

Circulating pump

Boiler   810980

Example

Structure-borne sound in a pipe, perhaps vibration from the circulation pump or noise from the fluid itself, has little opportunity to develop airborne sound as it is of small area. Fixing the pipe to a wall or panel gives the vibration a chance to excite a large area and therefore generate a high airborne sound level.

Solution

The pipework must be properly mounted and isolated from the wall or panels so that they are not set into vibration. This may be done using one of a number of different types of isolator employing springs, rubber strips, foam rubber washers, etc.

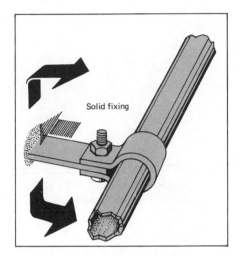

Solid fixing

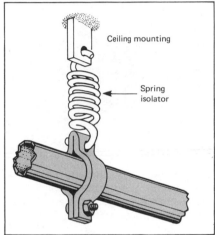

Ceiling mounting

Spring isolator

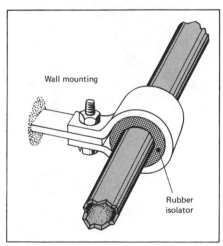

Wall mounting

Rubber isolator

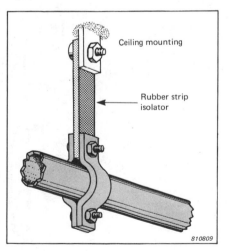

Ceiling mounting

Rubber strip isolator

810809

Small vibrating objects radiate less noise than large

A small object may vibrate without giving rise to high noise levels because the surface cannot transfer the vibration energy into sound energy efficiently. Connecting a large panel to the object increases its ability to convert vibration to sound. As most machines produce some vibration, the size of the machine and its panels should be kept as small as possible.

Principle

The shaver's vibration is transmitted to the glass shelf which vibrates over a large area, amplifying the noise substantially.

The vibration is no longer transmitted and the noise is reduced.

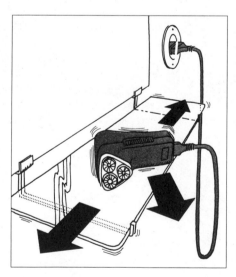

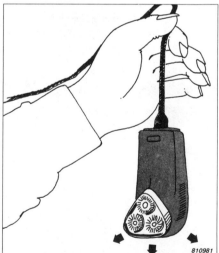

Example

A hydraulic supply system was a significant sound source even though the panels of the oil tank were damped by the oil inside. The chief source of noise was found to be the instrument panel which was set into vibration by the motor.

Solution

Removal of the panel from the machine uncoupled the source of sound from the source of vibration and reduced the sound level.

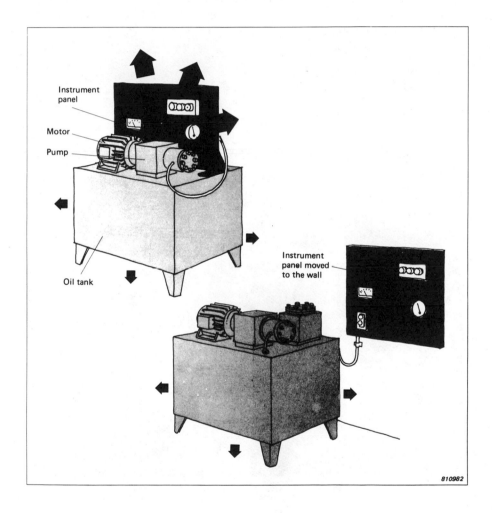

Instrument panel

Motor

Pump

Oil tank

Instrument panel moved to the wall

810982

Vibrating machinery or parts of machinery should be mounted on a heavy foundation wherever possible

Tapping on a light partition generates noise because the partition is easily moved by the force of the tap and therefore transmits the sound. Tapping on a heavy masonary wall produces little noise because the force available is so small it cannot have much effect on the wall. To avoid noise transmission from motors, pumps, etc., they should not be mounted on the relatively flexible equipment which they serve but separately on heavy bases where possible.

Principle

810983

Example

Pumps and motors serving large pieces of equipment such as hydraulic presses, machine tools, and turbines are often mounted directly on structural panels. These are set into vibration, radiating high noise levels from the entire area of the machine.

Solution

The services should be mounted, on isolators, away from the main frame of the equipment, on a solid floor whereever possible. Pipework carrying fluids should be connected via flexible piping and include attenuators to avoid the transfer of vibration via these connections back to the main structure of the equipment.

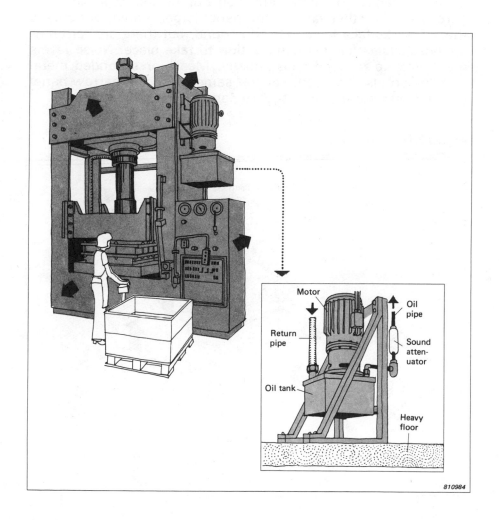

810984

Free edges on panels allow pressure equalization around them and reduce radiated noise levels

It is not always possible to avoid the use of large vibrating panels which give rise to high noise levels. In many cases these may be replaced by a perforated panel or another type with a broken surface. A plain panel radiates noise from all its area efficiently as there are only four sides along which the sound pressure can be partially cancelled out by the negative pressure from the other face. If the panel is perforated, not only is there less surface to radiate the sound, but there are far greater possibilities for this equalization to take place. Noise levels are therefore reduced substantially. Mesh, or expanded metal panels can also be used. For the same reason, a narrow panel radiates less noise than a square panel of the same area.

Principle

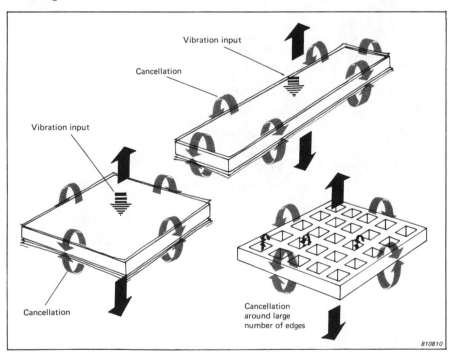

810810

Example

The protective cover over the fly-wheel and belt drive of a punch press radiates noise efficiently.

Solution

A replacement cover of wire mesh reduces the noise radiation.

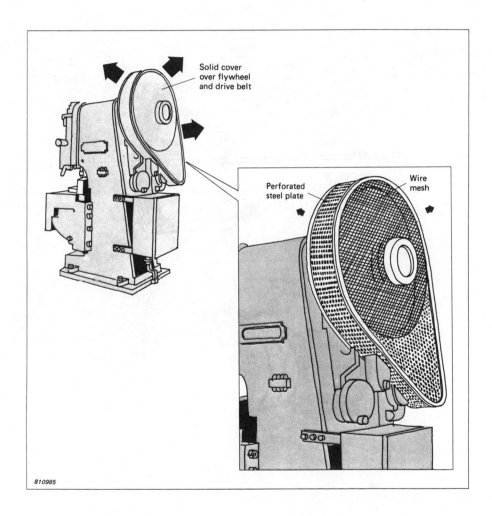

Solid cover over flywheel and drive belt

Perforated steel plate

Wire mesh

810985

Example

Vibration of wide drive belts on industrial drives can lead to high levels of low frequency noise.

Solution

Replacing the single drive belt with a number of narrower drive belts with gaps between them increases the amount of cancellation which is possible between the top and bottom of each belt and between one belt and the next one. The noise level is therefore reduced.

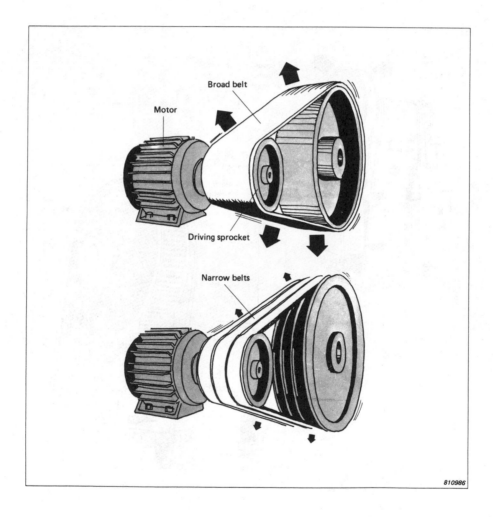

810986

Example

Bins for the transport of material radiate noise when being loaded and emptied and when being transported over uneven surfaces. With this type of construction, pressure equalization can only occur around the upper edges.

Solution

The side panels can be fixed to edge frames with narrow brackets so that there is a much greater length of free edge around which pressure equalization can take place. If the size of the material or components allows, the side panels may be made of wire mesh to reduce radiation further.

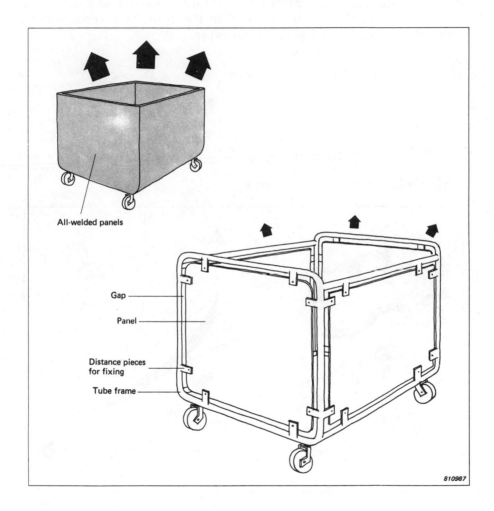

All-welded panels

Gap

Panel

Distance pieces for fixing

Tube frame

810987

Damped structures give rise to less noise

If a panel is set into vibration, the level of vibration, and therefore the noise level will dimminish with time. The speed of this reduction depends on the material's internal damping. The higher the damping the quicker the drop in level. The damping also has an effect on the maximum level that can be generated from a given excitation; a well damped panel cannot be excited as much, as the resonances are reduced. Unfortunately most common metals have very low internal damping and a damping layer must usually be introduced in the form of a ready made laminate or as a spray or stick-on layer.

Principle

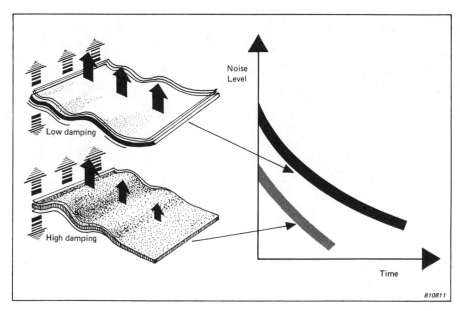

Example

Panels on machinery containing motors or pumps are prone to vibration and are therefore a normal source of radiated noise.

Solution

By using a laminated panel with high damping properties the noise can be reduced signficantly.

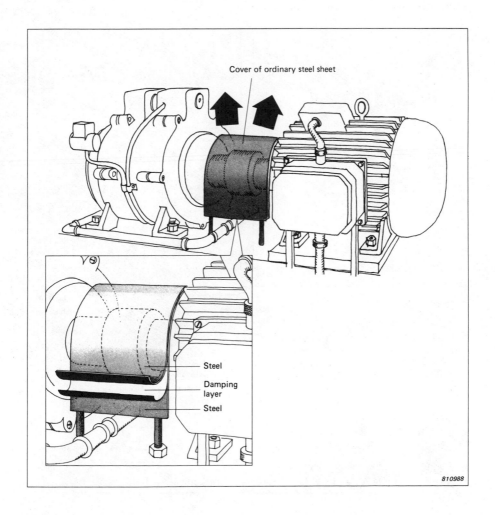

Cover of ordinary steel sheet

Steel

Damping layer

Steel

810988

Resonances amplify noise radiation but can be damped easily

Resonances strongly amplify the noise emitted by vibrating panel and plates, especially in homogenous structures. However, relatively small additions of extra damping can reduce the resonance peaks, and therefore the noise radiated, enormously. Pieces of damping material, fixed to a work piece temporarily, are also very effective.

Principle

Tapping a glass produces a loud resonance

Damping the glass removes the resonance

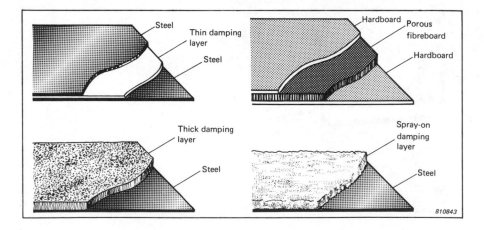

810843

Example

A circular saw blade in a sharpening machine generates a high level of noise because of resonance and very low internal damping.

Solution

A disc of rubber damping material fastened to the blade by a stiff disc during sharpening, adds both mass and damping to the blade and reduces the amplification of the resonances.

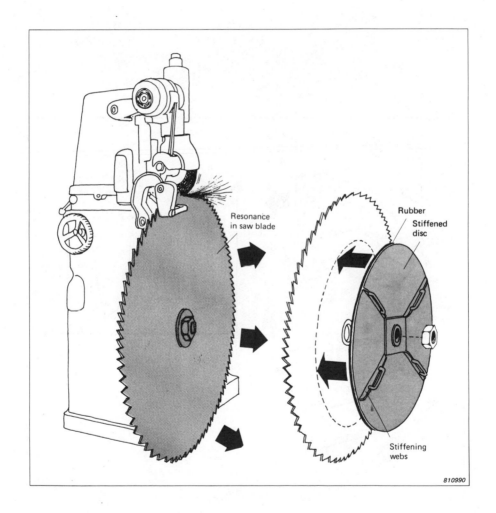

Resonance in saw blade

Rubber

Stiffened disc

Stiffening webs

810990

Example

During sawing operations on steel plate the sawing motion induces strong resonances in the work piece whose large area radiates a high level of screeching unpleasant noise.

Solution

Temporary addition of a magnetically held damping panel reduces the intensity of the resonances and reduces the noise to an acceptable level.

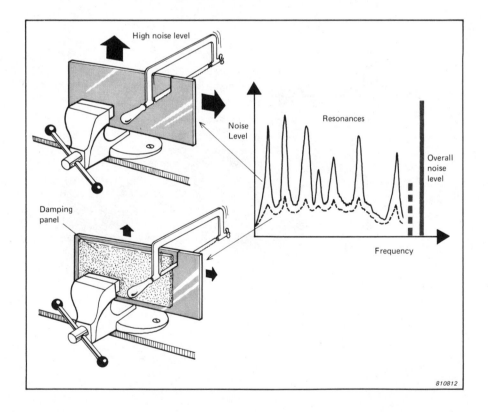

Example

The process of rivetting large structures such as aircraft, ship, or process plant components leads to high noise levels because of the impact caused and the large size of the component which efficiently converts the vibration energy into noise.

Solution

The application of temporary damping pads to the structure as it is rivetted reduces the intensity of the resonances and attenuates the vibration as it travels from the rivetting site to the rest of the panel.

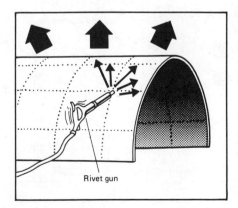

Rivet gun

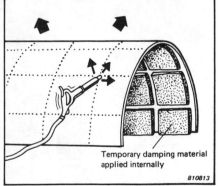

Temporary damping material applied internally

810813

Resonances transferred to a higher frequency are easier to damp

Large vibrating plates and shells often have low frequency resonances which are difficult to damp out. If the plate or shell can be stiffened, the resonance is shifted to a higher frequency which is easier to damp. In some cases it may still be difficult or expensive to apply the damping material to the separate areas, and therefore advantageous to mount a thin damped panel over the stiffening webs.

Principle

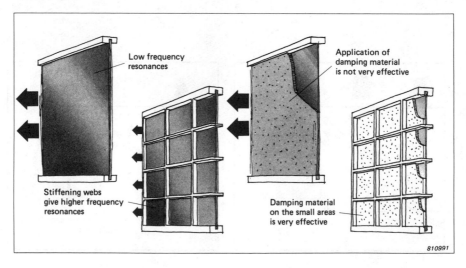

Example

The chief source of noise from a shearing machine proved to be radiation from the support and not from the workpiece as expected.

Solution

Stiffening webs were fitted to the support panels, and damped panels mounted on them.

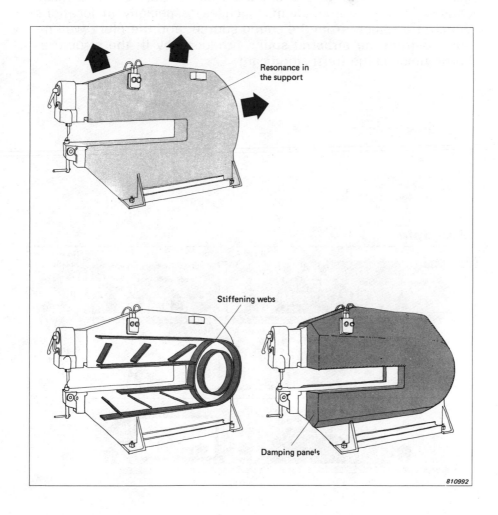

Resonance in the support

Stiffening webs

Damping panels

810992

Flexible mountings isolate machine vibration

Nearly all structure-borne noise can be eliminated or at least significantly reduced by mounting vibration sources on flexible supports. In some cases it may be necessary to mount the receiver room on flexible mountings as well, e. g. where sensitive apparatus is used or for low level acoustic measurements. Normally it is preferable to isolate the source, or at least as near to the source as practically possible. In this way every region of the building is protected from structure-borne noise caused by the source. In many cases, especially in locations which are remote from the sound source, and therefore well insulated from the airborne sound produced by it, the structure-borne noise is the most significant.

Principle

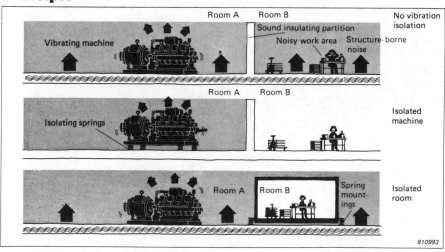

Example

Flexible mounts for vibration isolation can be obtained in a wide variety of types and materials to cope with any load requirement and any practical situation. For mounting heavy machinery, individual springs are used, with or without additional damping; for mounting light structures, pads of cork, expanded polystyrene, foam rubber, or rubber are often employed. Ceilings, ducts, and pipework are normally suspended from spring hangers or artificial rubber straps. Other isolators for special purposes and some individual types are shown in the drawing.

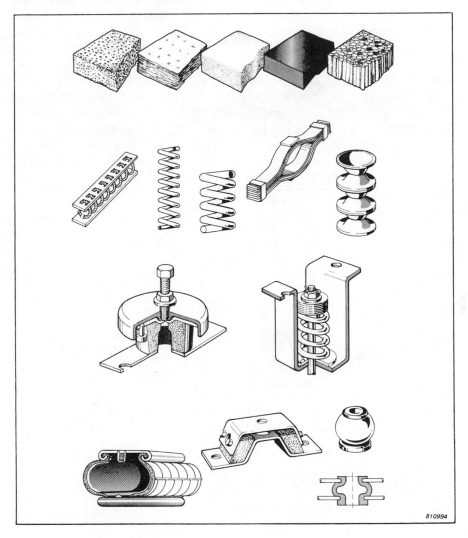

810994

Incorrectly chosen mountings can amplify vibrations

A flexibly mounted machine always has a characteristic resonant frequency on its mountings. This resonance is determined by the weight of the machine and the stiffness of the mountings. A light machine and stiff mountings give a high resonance frequency; a heavy machine and low stiffness of the mountings give a low resonant frequency. Vibration produced by the machine at frequencies lower than its mounted resonant frequency are not isolated. Vibrations well above the resonant frequency are isolated. Vibrations at the resonant frequency may be highly amplified if the internal damping of the mountings is low, and in any case will not be isolated. The natural frequency of the machine on its supports must always be below the normal running speed of the machine or the frequency of the vibrations to be isolated. Where the machine may spend some time at its resonant frequency, e.g. during run-up or run-down, mountings with very high internal damping should be chosen to keep the degree of amplification at resonance as low as possible.

Principle

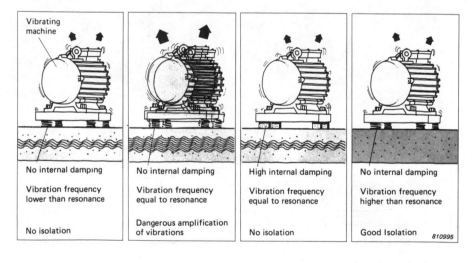

Vibrating machine			
No internal damping	No internal damping	High internal damping	No internal damping
Vibration frequency lower than resonance	Vibration frequency equal to resonance	Vibration frequency equal to resonance	Vibration frequency higher than resonance
No isolation	Dangerous amplification of vibrations	No isolation	Good Isolation

810995

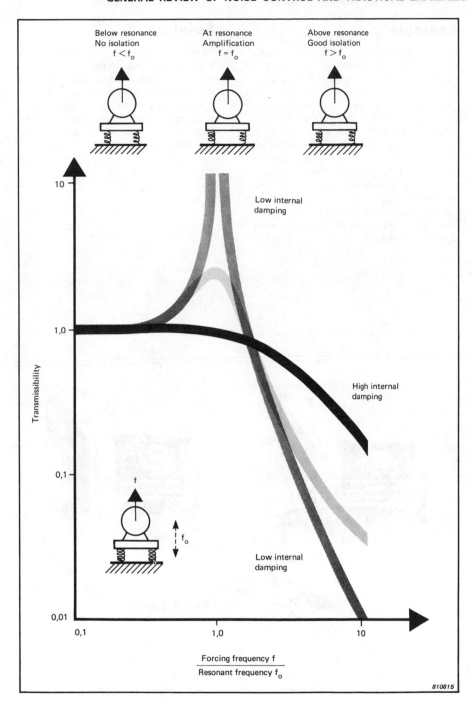

Example

Where a machine is run at a steady speed and continuously, without a large number of stops and starts, the resonant frequency of the suspension can be arranged to be well below the frequency of the machine and with a very low internal damping to give the highest isolation using, for example, simple spring isolators. If the machine spends a significant amount of time at resonance, e.g. because of frequent stops and starts of a compressor, this may lead to damage after a short time.

Solution

Supplying a type of isolator with high internal damping, e.g. pads of artificial rubber laminate, reduces the vibration significantly at the resonant frequency while only reducing the effectiveness of the isolation at the normal running speed by a small amount.

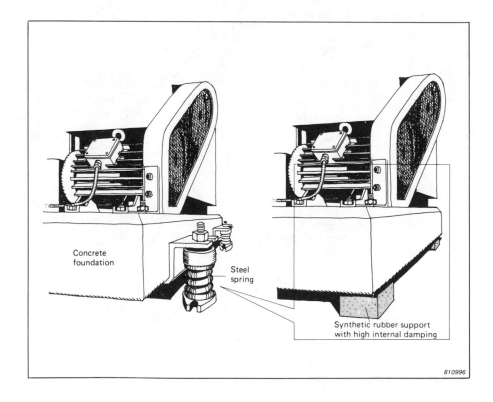

Concrete foundation

Steel spring

Synthetic rubber support with high internal damping

810996

Example

A heavy machine producing low frequency vibration may cause the floor itself to resonate even though isolators of the correct rating are used. This problem is particularly common in concrete buildings, whose floors have low internal damping.

Solution

For the best isolation the natural frequency of the machine on its isolators should not only be well below the exciting frequencies from the machine, but should also be lower than the resonances of the floor. In practice, this may be achieved by reinforcing the floor structure to provide a more stiffer solid base. Alternatively, the machine may be mounted on pillars founded directly in the ground.

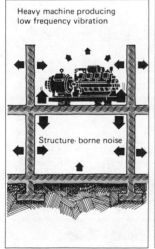

Heavy machine producing low frequency vibration

Structure-borne noise

Highly stiffened floor

Supports under the mountings

810997

Displacement of all mountings must be equal to avoid rocking motions

Flexible mountings for a machine should be chosen so that the static deflection at each mount is the same. If they are different, rocking motions may be forced at higher frequencies than the up-and-down motion of the machine, which are not effectively isolated by the mountings. On a machine with non-uniform weight distribution, mountings nearer the centre of gravity must be stiffer than those more remote from it.

The most efficient isolation is obtained if the mountings are fixed so that lines joining their points of application on the machine pass through the centre of gravity.

Principle

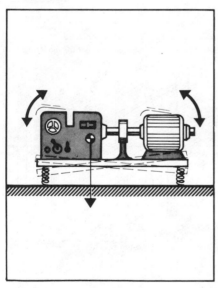
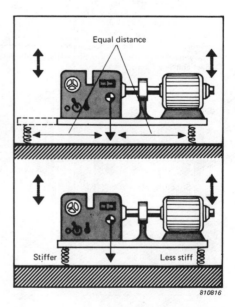

810816

Example

If the centre of gravity of a vibrating machine is above the line of action of the horizontal forces from the mounting, there is the danger of setting up a rocking motion from this source. This is particularly a problem when dealing with out of balance forces on rotating machines where the sideways component is large.

Solution

Mounting the machine in a tray or on an inertia block enables the mounting points to be placed in the same horizontal plane as the centre of gravity.

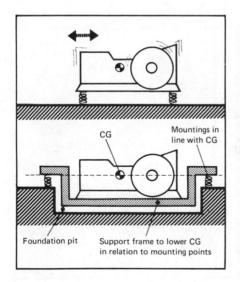

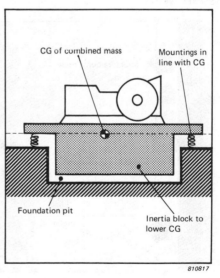

810817

Structure-borne sound via connections must be avoided

The most effective vibration isolation can be made totally ineffective if the vibration is transmitted by connections such as pipes, electrical conduits, supply ducts, etc. These must be flexible or contain flexible sections if vibration transmission is to be avoided.

Principle

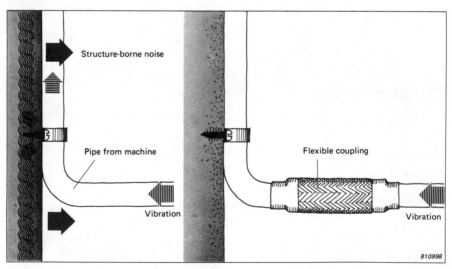

Example

Refrigeration plant can be a serious noise source because of the large pressure changes in the fluid during its passage through the compressor unit. Careful vibration isolation of the entire plant is necessary, and all ingoing and outgoing pipework should be isolated from the plant by flexible couplings.

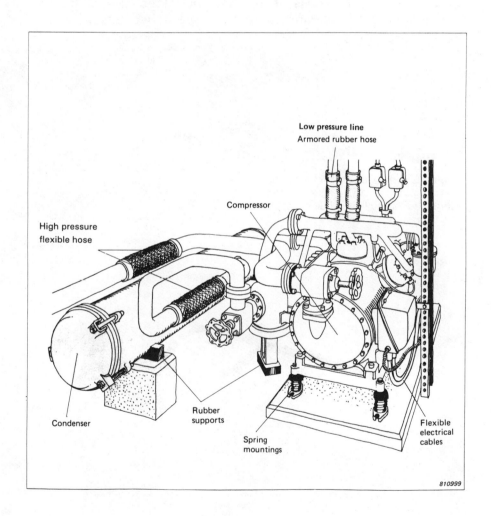

Insulation of single walls and panels depends on the surface density

When sound meets a wall, it sets the wall into vibration and sound is radiated from the other side. How much is radiated is a measure of the insulation of the wall, and is dependent on the surface density of the wall, i.e. its weight per unit area. In general, insulation increases as the frequency of the sound increases or the thickness of the wall increases, until a point is reached where it begins to fall off again because of coincidence.

Example

What insulation does 15 mm chipboard give at 500 Hz? Surface weight is 10 kg/m². 10 × 500 = 5000. Insulation is 26 dB.

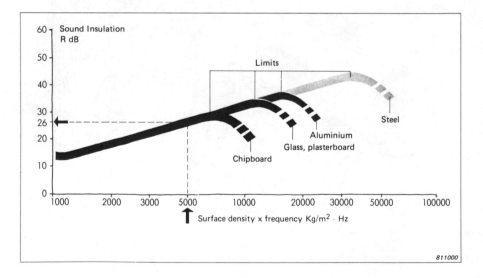

811000

Example

A sand blasting system is the dominant noise source in a workshop, and is only separated from the rest of the area by thin curtains.

Solution

A sound insulated machine room can be built for the sand blasting plant and a partial enclosure erected round the work area. Access to this area is via heavy lead/rubber laminate curtains which have high insulation while being flexible and easy to fold to allow easy access to the work area.

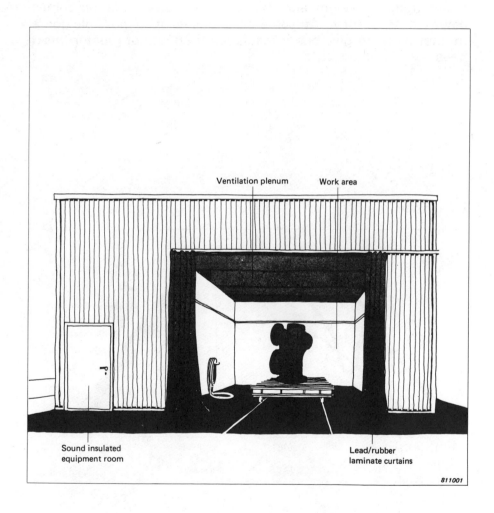

Ventilation plenum   Work area

Sound insulated
equipment room

Lead/rubber
laminate curtains

811001

Single walls have a region of poor insulation — Coincidence

A single panel has natural bending modes which are dependent upon the stiffness and thickness of the panel. When the wavelengths of the incident sound and the panel modes are equal, they vibrate in sympathy and the panel's insulation ability is reduced before increasing again as the resonance condition is passed through. The coincidence dip only disappears if the internal damping of the panel is high. As the thickness of a single wall or panel is increased, its stiffness rises relatively faster than its weight and the region of low insulation comes down in frequency. At some frequencies it is possible for a thinner panel to give better insulation than one of greater thickness.

Principle

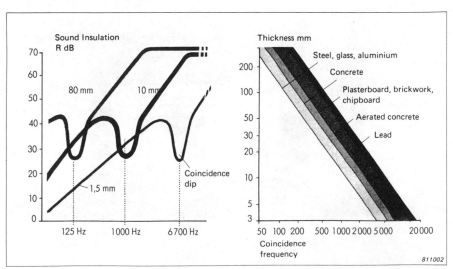

811002

Example

One end of a workshop contained machines with a high noise output at 1000 Hz. The area had been separated from the rest of the workshop by a 25 mm chipboard partition with 6 mm windows. The insulation was not as expected because the partition had a coincidence dip at 1000 Hz, where the noise from the machines was also a maximum. The window's coincidence dip, on the other hand, was up at 2000 Hz.

Solution

Exchanging the partition with one made of two layers of plasterboard improved the situation by 10 dB. Although the weight was about the same, the new partition's stiffness was only a quarter of the previous one. The coincidence dip was therefore much higher at approximately 2500 Hz.

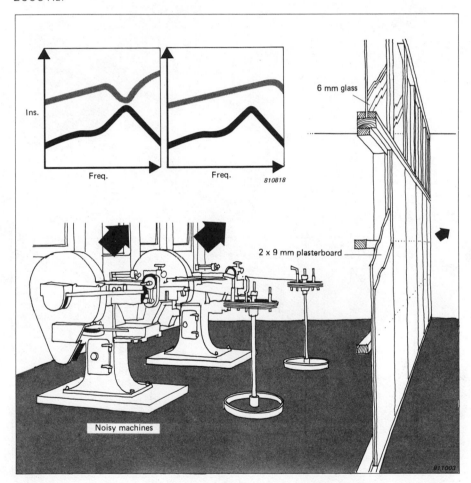

Stiffness and weight are both beneficial in thick walls

Most practical single walls have coincidence dips around 100 Hz. At frequencies above this, the insulation increases both with increased weight and increased stiffness. A cast concrete wall has a greater stiffness than a block one and may give the same insulation value though be lighter.

Principle

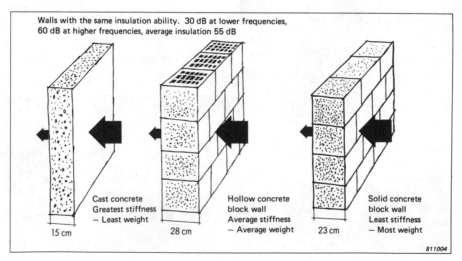

Walls with the same insulation ability. 30 dB at lower frequencies, 60 dB at higher frequencies, average insulation 55 dB

Cast concrete
Greatest stiffness
— Least weight

15 cm

Hollow concrete
block wall
Average stiffness
— Average weight

28 cm

Solid concrete
block wall
Least stiffness
— Most weight

23 cm

811004

Example

The chief source of noise in an industrial building was found to be a fan unit.

Solution

A single block wall was built around the equipment with access via sound attenuating doors. This type of construction was chosen because it had to resist damage from transport trucks and also help to support the heavily loaded floor above.

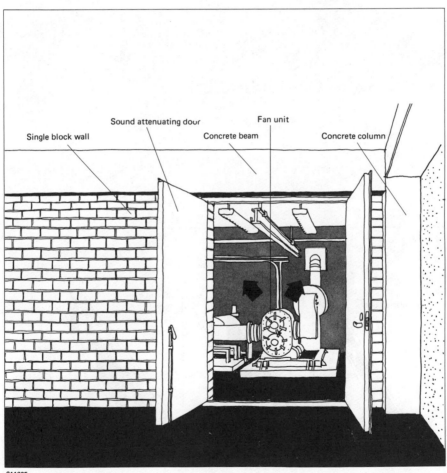

Single block wall   Sound attenuating door   Concrete beam   Fan unit   Concrete column

811005

Lightweight double partitions give good insulation

Two light partitions separated by an air gap give an insulation which increases as the air gap increases. If an absorbent material is placed in the gap, the insulation is further increased. The most insulation is achieved if the two panels are completely disconnected from each other, mounted on separate frameworks. If connection is necessary the highest insulation is achieved if the ties are as small, as few, and as elastic as possible. Insulation can be obtained which would normally require a single partition of 5 or 10 times the weight.

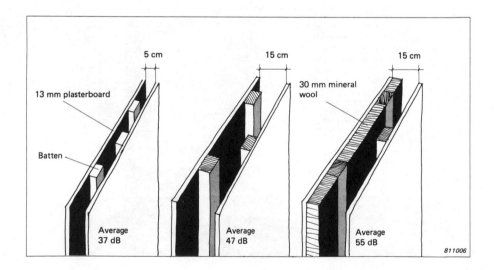

Example

Two connected workshops both contain noisy machinery, which is annoying in both locations.

Solution

The Noisy equipment should be located together at one end of one of the workshops and a double partition constructed with a large air gap in the form of an air lock, a gap large enough to serve as a corridor to the two quiet rooms. The doors are placed at opposite corners so that even if one of them is open an insulation of at least 35 dB is achieved. When closed over 50 dB can be expected. To achieve the same insulation with a single wall would require heavy masonary or concrete construction and the use of a special sealed sound-proof door.

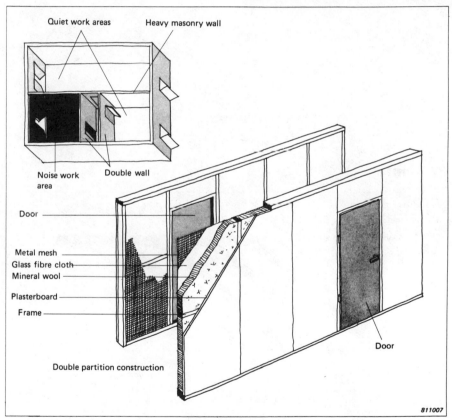

Quiet work areas   Heavy masonry wall

Noise work area   Double wall

Door

Metal mesh
Glass fibre cloth
Mineral wool

Plasterboard
Frame

Door

Double partition construction

811007

Thick porous layers are effective absorbers at both high and low frequencies

Porous materials through which air can pass are often excellent sound absorbers. Some examples are: Glass fibre, mineral wool, foam rubber, woodwool and sintered metal. If the material has closed cells the absorption is low. Thin layers are only capable of absorbing high frequencies, whereas thick layers can absorb over a wide frequency range including both high and low frequencies. To be effective below about 100 Hz a layer has to be impractically thick or mounted over an air space.

Principle

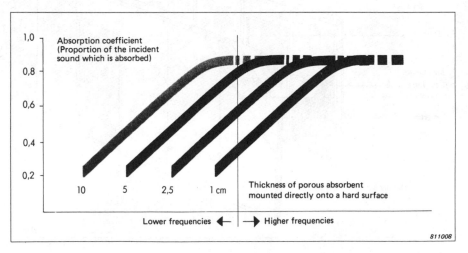

811008

Example

A workshop with a high noise level, especially at low frequencies, has to be treated to reduce the noise levels over the entire frequency range. Hanging panel absorbers can be used in a large part of the workshop where the ceiling is free of obstructions. These are very efficient, having two absorbing sides to each panel. A traversing crane makes it impossible to use these in the other part of the workshop. Instead, horizontal absorbent panels are mounted well below the ceiling to obtain improved low frequency absorption. Except in regions close to a noise source it is possible to reduce the overall noise level by up to 10 dB.

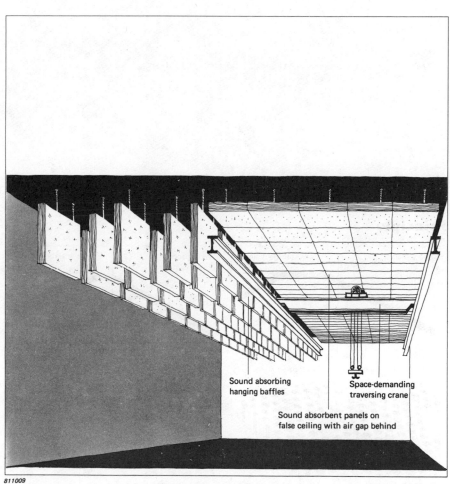

Sound absorbing
hanging baffles

Space-demanding
traversing crane

Sound absorbent panels on
false ceiling with air gap behind

811009

Perforated panels over absorbent need not reduce its effectiveness

In order to protect the absorbent and improve its appearance it is often covered with a perforated material of some kind. This does not significantly alter the characteristics of the absorbent if the perforations are a sufficient proportion of the total area. A perforation ratio of 15% is sufficient for thin panels. The ratio must be greater, the thicker the panel becomes. In addition, it is generally better to perforate with many small holes than with a smaller number of large holse.

Principle

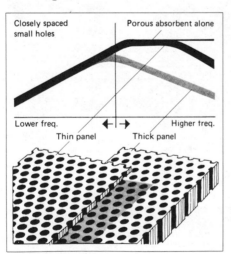

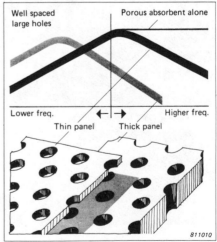

Example

In order to improve the appearance in a building it is possible to choose from a wide variety of materials to cover large areas of absorbent. The only requirement is that the uncovered area is sufficiently large to allow the absorbent material to do its job satisfactorily. Textiles, wood strips, expanded metal, and various types of partially open panel can be used to create the desired appearance.

Porous absorbent

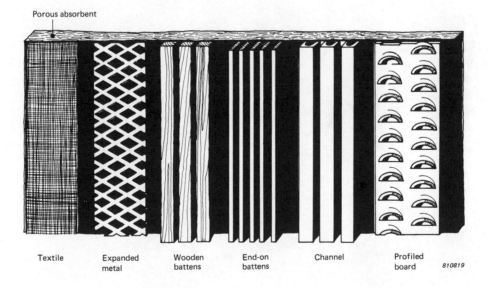

| Textile | Expanded metal | Wooden battens | End-on battens | Channel | Profiled board |

810819

Panel absorbers are effective at low frequencies

Thin panels mounted on a framework absorb low frequencies well in a fairly narrow range, whose frequency depends on the size and thickness of the panel and its distance from the wall. The effective absorption bandwidth depends on the internal damping of the panel; high internal damping giving a wider frequency range. For a porous absorbent to be as good at low frequencies, it would have to be extremely thick.

Principle

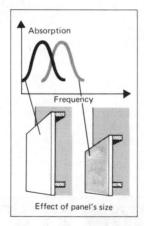

Effect of panel's size

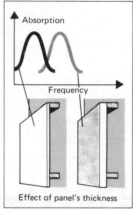

Effect of panel's thickness

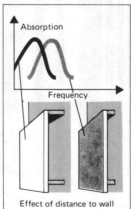

Effect of distance to wall

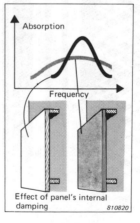

Effect of panel's internal damping

810820

Example

In an engine test room a resonance is excited when the test engine is run at or near its normal speed. At speeds away from the normal running speed the resonance disappears completely.

Solution

The walls may be covered in panels fixed to a wooden frame, the dimensions being chosen so that frequency range includes the unwanted resonance. To make the panel effective at frequencies at either side of the resonance, a panel material with high internal damping should be chosen, e.g. a laminate or fibre board.

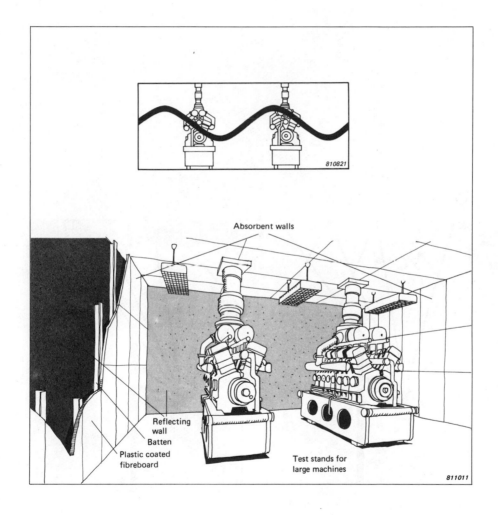

Absorbent walls

Reflecting
wall
Batten
Plastic coated
fibreboard

Test stands for
large machines

810821

811011

Screens should be combined with absorbent ceilings

Where the noise is dominated by high frequencies, a noise screen can be very effective against it. The screen is more effective the higher it is and the nearer to the source it is placed. However, if the ceiling above the source is non-absorbent, it reflects noise from the source into the "quiet area" on the far side of the screen. An absorbent ceiling prevents this reflection taking place, and improves the performance of the screen.

Principle

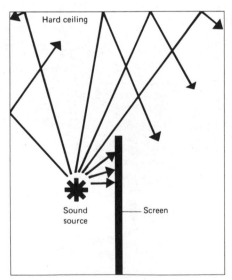
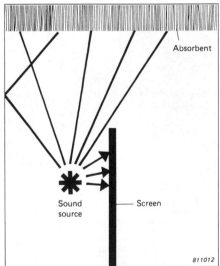

811012

Example

In an assembly hall with a number of production lines running parallel, one of the lines produces much more noise than the others. Noise from body preparation produces high frequency noise which affects everyone in the assembly hall.

Solution

By using lightweight absorbent screens on both sides of the noisy line, and hanging sound absorbing baffles above it, the noise levels are reduced at the quieter lines while not worsening the situation for those actually working on the noisy line. If it had merely been enclosed in a sound attenuating tunnel without sound absorbents, the reverberent sound level would have increased locally.

Relatively quiet line   Noisy line

Sound absorbent baffles

Absorbent screen

811013

Changes in ducts reduce noise transmission

At all changes along the transmission path, part of the sound energy is reflected back toward the source. In a duct this change may be the shape or area of the cross-section, bends, branches and wall material. This fact can often be exploited during design of a duct system to achieve a level of attenuation. Attenuation obtained in this way is termed reactive.

Principle

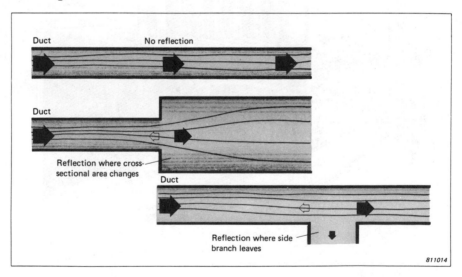

Example

A large landscaped office is to be fitted with a mechanical ventilation system. Space limitations prevent the use of an attenuator on the fan outlet large enough to provide the necessary noise reduction.

Solution

By making use of the sound reduction obtained when the duct contains a number of area changes, direction changes, and distribution branches, it is possible to achieve the required sound level limits at the delivery vents, even though the fan could not be sufficiently silenced at source. Absorbent on the duct walls and smooth bends avoid the generation of more noise and a better distribution of air is achieved around the office.

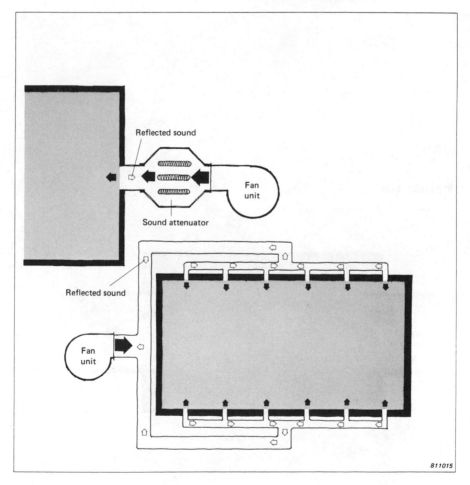

Reactive attenuators are efficient in a narrow frequency range

A reactive attenuator is an effective method of reducing low frequency noise over a limited frequency range, and is relatively compact. By coupling a number of attenuators of different sizes together, most conveniently within the same external casing, it is possible to cover an extended range of frequencies. Perforated tubes are often used within the attenuator to improve gas flow and provide some absorption.

Principle

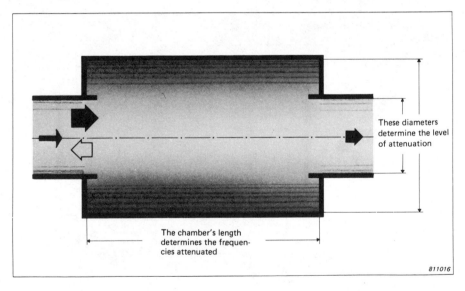

These diameters determine the level of attenuation

The chamber's length determines the frequencies attenuated

811016

Example

Absorbent (resistive) attenuators are simple to design and produce, and provide noise reduction over a wide frequency range. However, the absorbent material can be easily blocked with residues and carbon deposits which make it less effective after a time, and absorption of combustible materials can cause a fire risk.

Solution

A multi-step reactive attenuator can be used in this case. It attenuates over a wide range of frequencies, is less sensitive to deposits, and rugged. The example shown is for a large piston engine, and is especially suitable where the motor speed only varies over a fairly narrow range.

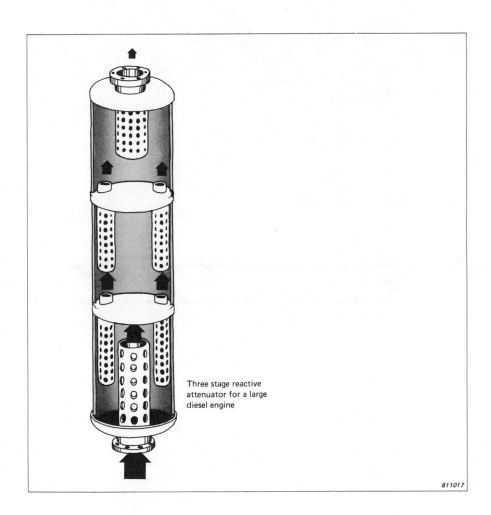

Three stage reactive attenuator for a large diesel engine

811017

Expansion chambers are effective where low frequencies dominate

If a pipe or duct is provided with an enlargement or chamber, then the low frequency fluctuations are evened out. This is very effective where continuous pressure pulses are concerned, e.g. in engine exhausts or compressor outlets. The lower the frequency, the larger the chamber must be to be effective.

Principle

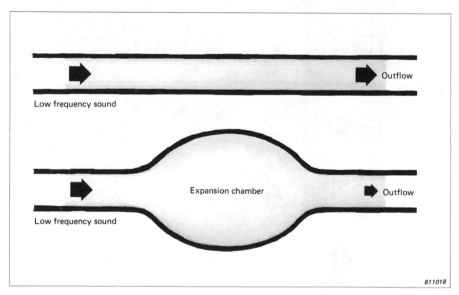

Example

The exhaust of a pneumatic drill produces both high and low frequency noise; low frequency from the repetitive pulses, and high frequency from the jet of gas escaping with each pulse.

Solution

By surrounding the drill body with a jacket acting as an expansion chamber for the low frequencies, and as a noise screen for the high frequencies, the noise level is reduced substantially. Because the energy in the pulse is spread more evenly over a longer time by the jacket, the noise is also reduced in that way. As the impact of the drill on the concrete is an inherently noisy process which cannot be easily quietened, there is a limit to the amount of quietening which is worthwhile.

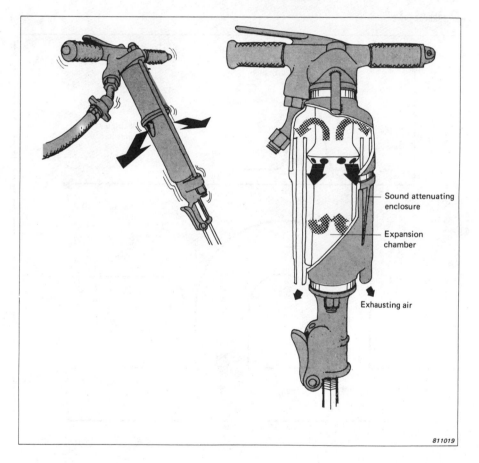

Sound attenuating enclosure

Expansion chamber

Exhausting air

811019

Pure tones can be reduced by interference

When a noise contains a predominant tone or a few tones, it can be substantially reduced by interference, i. e. the presence of two out of phase tones of the same frequency at the same place, which tend to cancel each other out. An interference attenuator consists of a branch which leaves and later returns to the main channel. Sound passing through it travels an odd number of half wavelengths further than in the main channel, so that when the two sounds meet again they are out of phase and cancel each other out.

Principle

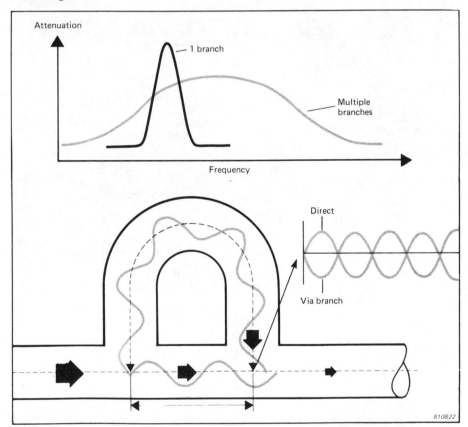

Example

If a tone is not completely fixed, the bandwidth of the interference attenuator can be broadened by having a number of branches of slightly different lengths. The attenuation at a given frequency will be a little lower, however. This type of device is suitable for engines operating at a more or less constant speed, generator sets, for example, and for fan and blower units.

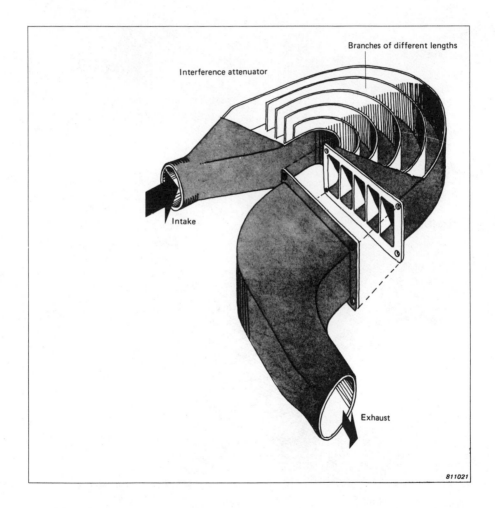

Unused spaces can be used as absorbent plenum chambers

An absorbent room is a simple but effective sound attenuator. Ducts can be fed into a chamber whose walls are lined with sound absorbent material which absorbs the sound energy. To prevent the direct passage of high frequencies, the incoming and outgoing ducts should not be directly opposite each other. The larger the volume of the chamber and the thicker the absorbent lining, the lower the frequencies which can be effectively absorbed.

Principle

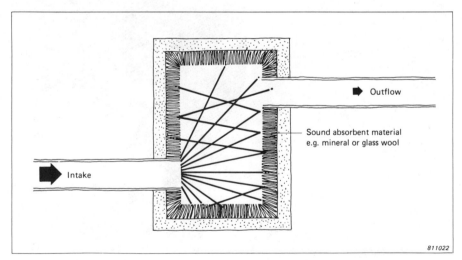

Outflow

Sound absorbent material
e.g. mineral or glass wool

Intake

811022

Example

The shape of the space used as a plenum chamber is not important, as long as it is well lined with absorbent and large enough to deal with the lowest frequencies encountered, so any unused part of the building can be utilised. It should also be well sealed in order to prevent leakage of the air out of the system.

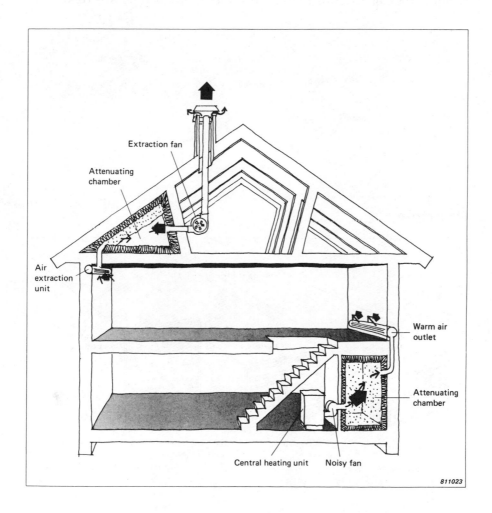

Extraction fan

Attenuating chamber

Air extraction unit

Warm air outlet

Attenuating chamber

Central heating unit

Noisy fan

811023

Absorbent attenuators are effective over a wide frequency range

The simplest type of absorbent attenuator is a duct with sound absorbent material on the walls. The thicker the materials, the lower the frequency which can be absorbed. For higher frequencies though, thinner absorbent layers are effective, but the large gap allows noise to pass directly along. This layers and narrow passages are therefore more effective at high frequencies. For good absorption over the widest frequency range, thick absorbent layers and narrow passages are best.

Principle

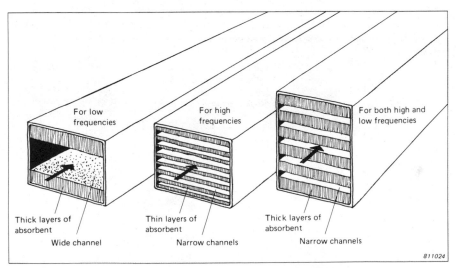

For low frequencies

For high frequencies

For both high and low frequencies

Thick layers of absorbent

Wide channel

Thin layers of absorbent

Narrow channels

Thick layers of absorbent

Narrow channels

811024

Example

Gas turbine powered standby generator sets are used extensively both as emergency power supplies and to complement normal generating plant in peak periods. It can be necessary to quieten the set by anything up to 70 dB, over a wide frequency range. Extensive use is made of absorbent materials in the form of splitters, baffles, and linings on the walls of plenum chambers on both the intake and exhaust sides.

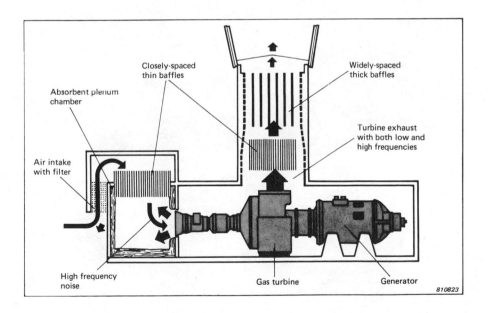

Wind generated tones can be avoided by profile changes or spoilers

When air flows past an object, a powerful pure tone, known as a Karman tone, can be produced at certain wind speeds. The tone is caused by the regular shedding of vortices from alternate edges on the downwind side of the object as the laminar boundary layer separates from the surface. In addition to a loud tone this phenomenon may also lead to severe structural vibration and damage if the modes of vibration of the object are at the same frequency. By lengthening the object in the direction of the air flow, i.e. by "streamlining", it is possible to keep the boundary layer attached right round the object, greatly reducing the tone level. Where the flow may come from any direction this method is not practicable, e.g. where chimneys are concerned. Spoilers can then be added to break up the air flow so that random turbulence is produced which does not generate the disturbing pure tone.

Principle

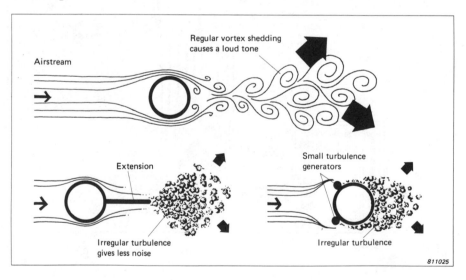

Example

At certain speeds the wind causes a powerful pure tone to be emmitted from the chimney.

Solution

A spiral fin of metal mounted on the chimney causes local tubulence regardless of the wind direction, so regular vortex shedding and the tone which arises from it are avoided. The pitch of the spiral should not be the same all the way up. As well as presenting an irregular shape to a wind coming from any direction, the spiral also increases the strength and stiffness of the chimney.

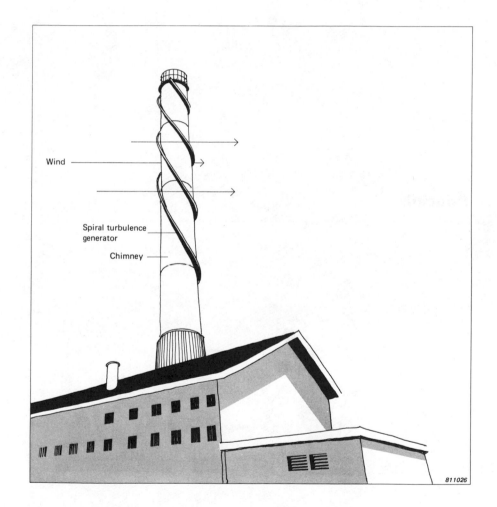

Wind

Spiral turbulence generator

Chimney

811026

Avoid air flows over cavities

When air is blown over openings with a cavity behind them, a loud pure tone is produced via what is called a Helmholz resonance. Wind instruments such as the organ and the flute operate on this principle. The frequency produced depends primarily on the volume of the cavity and the size of the opening to it. The larger the volume and the smaller the opening, the lower the frequency.

Principle

Hollow key

Empty bottle

811027

Example

When the cutter of a planing machine or saw blade screeches even when idling, it is often because the air passing over the blade excites the cavity behind the blade into resonance. In this case it is the cavity which is in motion and the air which is stationary, but the effect is the same. Turbulence formed by the passage of air over the cutter's edge produces a whistle which is amplified at certain frequencies, giving the sound its screeching character.

Solution

Rounding the edge reduces the whistle and partially filling the cavity avoids resonant amplification of the sound. Detail changes to the blade fixing mechanism are necessary to carry out these improvements.

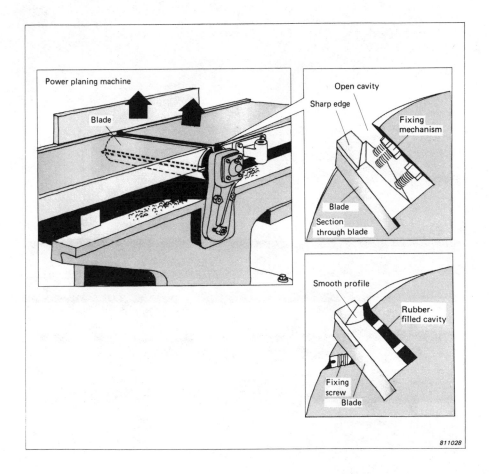

811028

Smooth ducts and pipes create less turbulence noise

Flow through pipes and ducts always gives rise to the generation of some turbulence and noise at the walls. If the flow suddenly has to change direction because of obstacles in the duct, or sharp bends, strong turbulence and high noise levels are formed, which increase with increasing speed. If the obstacles are near each other, the flow does not have a chance to settle down again and the turbulence is made worse by the second obstacle.

Principle

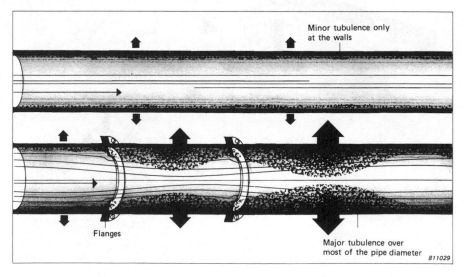

Minor tubulence only at the walls

Flanges

Major tubulence over most of the pipe diameter

811029

Example

Three valves in a branch from a steam system produce an unacceptable screeching noise. The branch has sharp corners and a number of closely spaced valves which produce turbulence noise.

Solution

The bends can be made gentler to avoid the generation of so much turbulence, and the valves placed further apart. Any turbulence generated in the flow by one corner of valve has space to settle down before reaching the next.

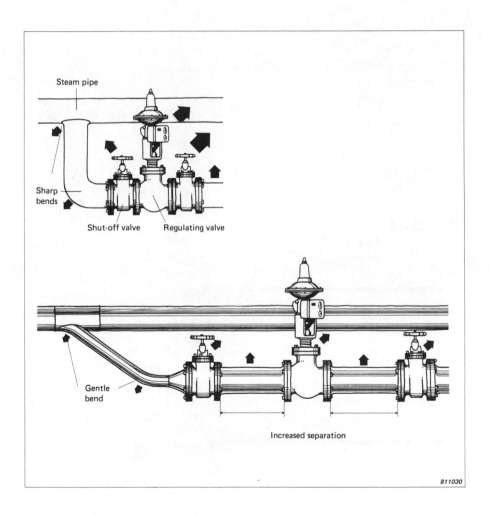

Undisturbed flow gives rise to less exhaust noise

When a fast moving air or gas stream mixes with still air, turbulence is formed which radiates noise. If the stream is disturbed before the exhaust so that there is already turbulence in the flow, the mixing region amplifies the noise level further by up to 20 dB for the same exhaust velocity. A lower exhaust velocity leads to a lower noise level. Halving the speed leads to a noise reduction of about 15 dB.

Principle

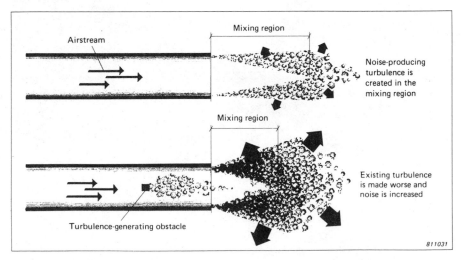

Airstream

Mixing region

Noise-producing turbulence is created in the mixing region

Mixing region

Turbulence-generating obstacle

Existing turbulence is made worse and noise is increased

811031

Example

The exhaust air from a compressed air driven grinder generated high noise levels. The air was already very turbulent on leaving the motor and entering the handle, which was hollow and acted as the exhaust. The turbulence noise was amplified in the mixing region.

Solution

The handle was replaced with a new type with a steel wool packing kept in place by mesh washers at either end to allow easier transition into the air. The turbulence was smoothed by the air's passage through the porous material and the exhaust noise level was therefore reduced.

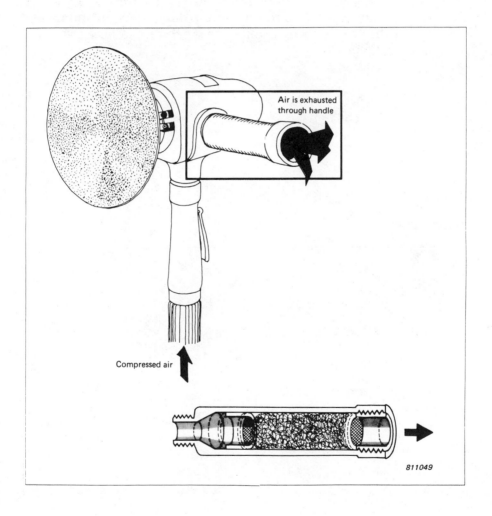

Air is exhausted through handle

Compressed air

811049

Jet noise can be reduced by an extra airstream

At velocities above about 100 m/s jet noise occurs. Because the formation of turbulence in the mixing region outside the exhaust is so violent, the condition of the airstream before the outlet is not important. Halving the outlet velocity reduces the noise level by approximately 20 dB under these conditions. The strength of the turbulence is determined by the relationship between the speed of the jet and the speed of the ambient air. The noise level can therefore often be significantly reduced by introducing an extra airstream with a lower speed alongside the jet, so that the velocity profile across the jet is less steep.

Principle

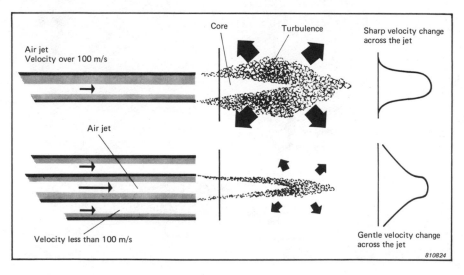

Example

Cleaning machine components with compressed air to remove swarf and dirt is often done using a single jet from a simple nozzle. The high velocity airstream required for this purpose gives rise to unacceptable high frequency noise.

Solution

The simple nozzle can be replaced by a compound nozzle which feeds a lower speed annular airstream around the main high-speed airstream. The transition from the high speed of the central airstream to the still air is much less sudden and the noise level is reduced substantially.

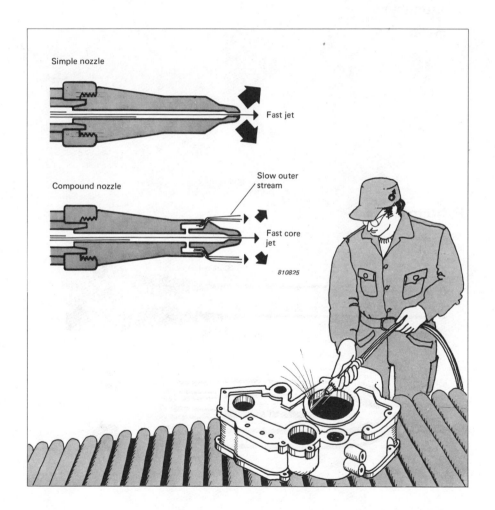

Low frequency exhaust noise transformed to higher frequencies is easier to attenuate

If the outlet of an exhaust pipe is large, the noise which arises is predominantly low frequency; if it is small, high frequencies are dominant. By replacing a single large exhaust by a number of small ones with the same capacity, it is possible to reduce the intensity of the low frequencies. The high frequency noise level is usually increased, of course, but this is relatively easier to attenuate.

Principle

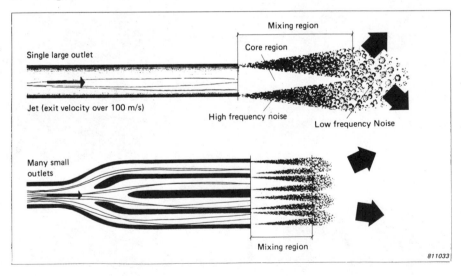

Example

In power stations and in many major industries large quantities of steam are produced, and safety valves may vent excesses to the atmosphere at high pressure several times a day, at the rate of many tons per hour. This requires large diameter pipes producing unacceptable levels of noise, mainly at low frequencies.

Solution

On the existing pipe, a diffuser and frequency transformer, followed by a high frequency attenuator, can be mounted. The diffuser lowers the exhaust velocity by a factor of four which results in a noise reduction of up to 40 dB compared with the original free pipe. The diffuser is made as a perforated cone which splits the single large jet into a large number of small jets generating predominantly high frequency noise which can be significantly attenuated by the highly absorbent spiral insert and walls.

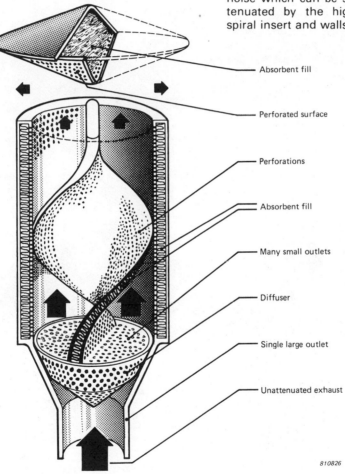

Absorbent fill

Perforated surface

Perforations

Absorbent fill

Many small outlets

Diffuser

Single large outlet

Unattenuated exhaust

810826

Position fans in smooth, undisturbed flow

Fans generate turbulence which radiates noise. Existing turbulence in the incoming air is made worse by the fan and the noise is amplified. If there is sufficient distance from the source of turbulence to the fan, the turbulence has a chance to die down and the noise level is reduced. Fans should therefore be placed well downstream of obstacles, valves, corners, and changes of cross-section. The same principle also applies to liquids, e.g. ships propellors should operate in a smooth flow.

Principle

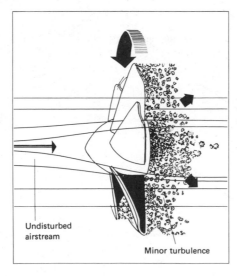

Undisturbed airstream

Minor turbulence

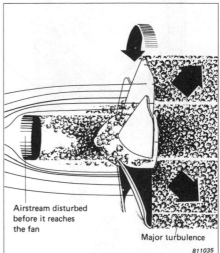

Airstream disturbed before it reaches the fan

Major turbulence

811035

Example

In one case the fan is too close to an obstacle, in the other too close to a bend. In both situations disturbances are formed which cause increased noise after passing through the fan.

Solution

The regulator is moved further from the fan so that the turbulence has a greater distance in which to settle down. The bend should be made gentler to reduce the strength of the turbulence and the fan moved further downstream to increase the settling distance.

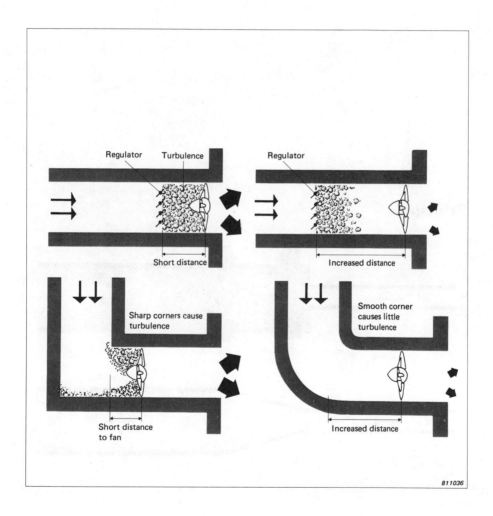

Flow noise in pipes is formed by sudden pressure changes

As in airflow, sudden pressure changes in pipes carrying liquids also give rise to noise. Air or vapour bubbles are released which cause noise, but they disappear again rapidly. The pressure change is most often caused by sudden area changes, and can be avoided by ensuring that all area changes are smooth and gradual.

Principle

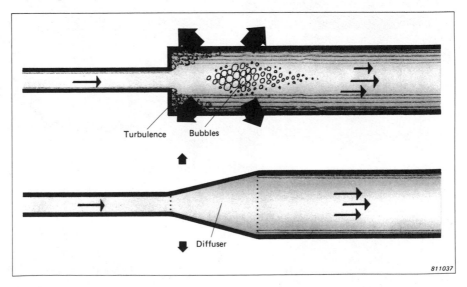

Turbulence   Bubbles

Diffuser

811037

Example

Regulating valves in fluid systems often have small seats which lead to a high local flow velocity around the valve at high pressures. Tortuous paths and sharp edges lead to the formation of strong turbulence. Sound is radiated directly from the valve and vibration is transferred along the pipe to appear elsewhere as structure-borne sound.

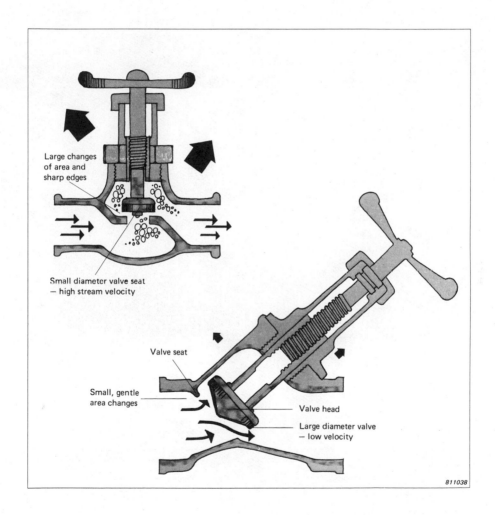

Large changes of area and sharp edges

Small diameter valve seat — high stream velocity

Valve seat

Small, gentle area changes

Valve head

Large diameter valve — low velocity

811038

Sudden large pressure changes cause cavitation

If a pressure change is large and quick enough, vapour bubbles are formed which collapse again almost immediately, causing both intense noise and high levels of vibration. Cavitation, as this phenomenon is called, occurs with regulating valves, pump impellers, and propellers, and is very common in hydraulic systems where pressures are high. By lowering pressures in a number of smaller steps it is possible to avoid cavitation.

Principle

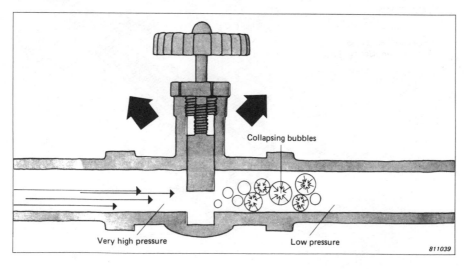

Collapsing bubbles

Very high pressure

Low pressure

811039

Example

The full capacity of a pump in a hydraulic system is only rarely exploited, so the pressure is usually reduced by means of a regulating valve. The sharp pressure change across the valve can give rise to unacceptable noise radiation from the valve, which may be spread as structure-borne sound throughout the building whereever the pump is connected.

Solution

A pressure reducer can be inserted into the pipe beside the valve. The insert contains a number of exchangeable discs with different perforations. Suitable discs are chosen to give a pressure drop no larger than that necessary to prevent cavitation.

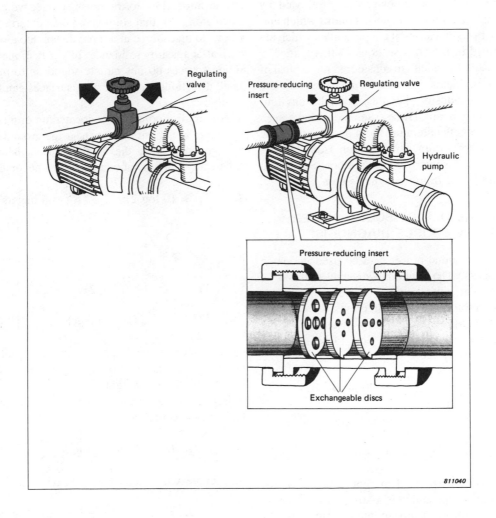

Regulating valve

Pressure-reducing insert

Regulating valve

Hydraulic pump

Pressure-reducing insert

Exchangeable discs

811040

Chapter 9

NOISE SOURCE DIAGNOSIS AND CASE STUDIES

This chapter begins with a discussion on source diagnosis by sound-power data, followed by typical case studies on noise control which amplify the application of the principles which are outlined in previous sections of this text. The case studies which are considered are typical of noise problems and control measures encountered in industry, building, and the environment in general. These case studies have been taken from the literature (with permission of the publishers), from a publication from NIOSH (National Institute for Occupational Safety and Health), and from the writer's own files.

9.1. THE USE OF SOUND-POWER DATA IN SOURCE DIAGNOSIS*

How can sound-power data be used in source diagnosis? The sound-power level radiated by an "ideal point source" (a source radiating sound uniformly in all directions) is related to the sound-pressure level at a distance r by the following equations (see Eqs. [4.4] and [4.6] in Chapter 4):

$$L_w = L_p + 10 \log 4\pi r^2, \qquad [9.1]$$

where r is expressed in meters, or

$$L_w = L_p + 10 \log 4\pi r^2 - 10, \quad [9.2]$$

where r is expressed in feet. For these two equations, the source is assumed to radiate its sound with no nearby reflecting surfaces. As was pointed out in Chapter 4, this would be known as spherical radiation in a free field, a relationship fundamental to source diagnosis.

To review its use, note that if we measure L_p at a location close to the noise source, we can calculate L_w for that source and then determine L_p due to that source at a more distant location, such as at a nearby residence. In practice, many sound sources do not radiate sound uniformly in all directions, and reflecting surfaces can be nearby.

For an ideal point source located on or close to a large-area floor or at or near the ground in a large open area, the sound radiates hemispherically, and the above equations become:

$$L_w = L_p + 10 \log 2\pi r^2 \qquad \text{(for } r \text{ in meters),}$$
$$[9.3]$$

and

$$L_w = L_p + 10 \log 2\pi r^2 - 10$$
$$\text{(for } r \text{ in feet),} \qquad [9.4]$$

In the more general case, the source is not a point source; instead, it has finite values of length, width, and height. In this case, sound power and sound-pressure levels are interrelated by the equations:

$$L_w = L_p + 10 \log S \qquad [9.5]$$

for S expressed in square meters, or

$$L_w = L_p + 10 \log S - 10 \qquad [9.6]$$

for S expressed in square feet. In these last two equations, S is the area of an imaginary shell all around the source, and L_p is the sound pressure level that exists at any point on that imaginary shell.

In a further extension of Eqs. [9.5] and [9.6],

*From *Industrial Noise Control Manual*, U.S. National Institute for Occupational Safety and Health, 1974; reprinted by Canadian Acoustical Association, 1984.

suppose that the source does not radiate its sound uniformly through all portions of the shell. Perhaps one part of a large, complex sound source radiates higher sound-pressure levels (SPLs) than some other part of the source. For such situations, Eqs. [9.5] and [9.6] must be broken down into several parts, where L_{p1} is the SPL at one element of area S_1 on the shell, L_{p2} is a different SPL at another element of area S_2, and so on over the entire range of L_p values over the entire area. Then,

$$L_w = \sum_{i=1}^{n} [L_{pi} + 10 \log S_i] \quad \text{(for S in m}^2\text{)}$$

$$[9.7]$$

or

$$L_w = \sum_{i=1}^{n} [L_{pi} = 10 \log S_i - 10]$$

$$\text{(for S in ft}^2\text{).} \quad [9.8]$$

Example. Figure 9.1 shows an imaginary shell around a sound source of interest, at a 1-m distance. The source dimensions are 2 m × 3 m × 5 m, as shown in the sketch. The north and south surfaces of the imaginary shell each have an area of 21 m², the east and west ends have an area of 15 m² each, and the top of the shell has an area of 35 m². For this simple example, suppose the SPL all over the north surface of the shell is uniform at 98 dB; for the south surface, it is 93 dB; for the east end, it is 88 dB; for the west end, it is 90 dB; and for the top surface, it is 95 dB. The total sound power radiated from this source would be as follows, using Eq. [9.7]:

$$L_w = (98 \times 10 \log 21) \text{ dB (N)}$$

$$+ (93 + 10 \log 21) \text{ dB (S)}$$

$$+ (88 + 10 \log 15) \text{ dB (E)}$$

$$+ (90 + 10 \log 15) \text{ dB (W)}$$

$$+ (95 + 10 \log 35) \text{ dB (Top).}$$

These components are to be added by decibel addition (see Chapter 1). Thus,

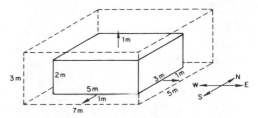

Fig. 9.1. Assumed sound source (solid lines) on a factory floor, surrounded by an imaginary shell (dashed lines) at 1-m distance.

$$L_w = 111.2 + 106.2 + 99.8 + 101.8$$

$$+ 110.4$$

$$= 114.9 \text{ dB} \quad \text{or} \quad 115 \text{ dB re } 10^{-12} \text{ W.}$$

Calculations can be carried out to 0.1 dB values, but the final value should be rounded off to the nearest whole number.

Two practical considerations limit the validity of this example. First, in practice it is unlikely that a uniform sound level would exist over an entire large area of the imaginary shell, so it might be necessary to take several SPL values over each large area of interest and to assign a subdivided area value to each SPL value. Second, when SPL measurements are made close to a relatively extensive or large source, the sound is not radiating as though it were from a point source in a free field. Instead, the SPL value is taken in the near field of the sound source, where the sound field is distorted and is not necessarily representative of the true total sound power that would be radiated to a large distance out in the free field. As a result, errors of a few decibels may be encountered at these close distances from large sources, and it is essentially impossible to predict the amount of error to be expected. Thus, be prepared to have an unknown error (possibly up to 5–8 dB for large sources, but fairly negligible for quite small sources).

In spite of these drawbacks, the concept of sound-power level is very helpful in identifying and diagnosing sound sources. To illustrate this assistance, suppose the microphone of a sound-level meter can be brought up to within 5 cm of a small sound source in a large machine, and the sound-pressure level is found to be 105 dB in the 1000-Hz octave band. Over another, much larger, area of the machine, the close-in sound level is 95 dB in the 1000-Hz octave band. Estimate the sound power levels of these two sources to determine the controlling source at this frequency. Suppose the 105-dB value is found to exist over an area of about 100 cm × 10 cm, or 1000 cm^2 (= 0.1 m^2), whereas the 95-dB value is found to exist over a surface area of about 2.5 × 4 m, or 10 m^2. From Eq. [9.5], the approximate sound-power level of the small-area source is

$$L_w = 105 + 10 \log 0.1$$

$$= 95 \text{ dB re } 10^{-12} \text{ W,}$$

while the approximate sound-power level of the large-area source is

$$L_w = 95 + 10 \log 10$$

$$= 105 \text{ dB re } 10^{-12} \text{ W.}$$

Even if the power-level values are in error

by a few decibels, this comparison indicates that the large-area source radiates more total sound power than the small-area source, even though the small source has a higher localized sound-pressure level. For noise control on that machine, the noise from the large-area source must be reduced by about 10 dB before it is necessary to give serious consideration to the small source.

For another illustration of how sound power level data are used in source diagnosis, look at Fig. 9.2. It shows the noise spectrum found at the property line of a plant and a sound spectrum indicative of a target goal for the situation. Note that the sound-pressure levels are excessive in the 250-Hz and 8000-Hz octave bands.

Close-in data, obtained 1 m from each of the three possible sources (Fig. 9.3) of the property-line noise, were then examined to determine which noise sources should be treated. From Eq. [9.5], the power level of each source is obtained, and from Eq. [9.7], the expected contribution of each source to the property-line

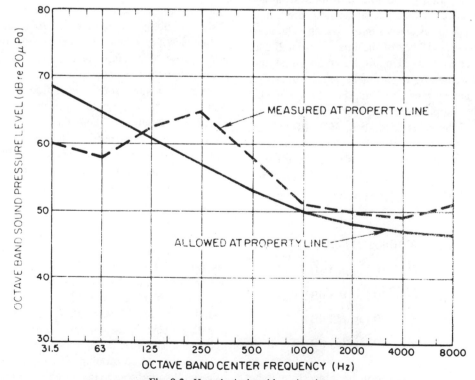

Fig. 9.2. Hypothetical problem situation.

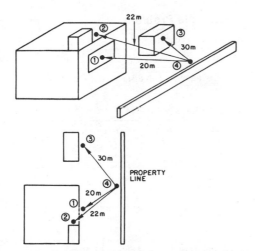

Fig. 9.3. Location of noise sources (1–3) relative to property-line position (4) for use in example.

measurement is estimated (in this example, each noise source is assumed to radiate hemispherically). Fig. 9.4 illustrates the results of the computations shown in Table 9.1. The calculations indicate the vent noise is responsible for the 31.5-Hz and 63-Hz octave-band sound-pressure levels, the compressor noise is responsible for the 125-Hz to 500-Hz octave-band sound-pressure levels and partly responsible for the 1000-Hz and 2000-Hz octave-band sound-pressure levels, and that sound coming through the window contributed to or is responsible for sound-pressure levels in the 1000-Hz to 8000-Hz bands.

Because the 31.5-Hz and 63-Hz octave-band

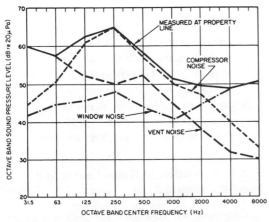

Fig. 9.4. Results of power-level extrapolations for problem shown in Fig. 9.2.

levels are not consequential to the problem, the vent need not be treated. However, both the window and compressor do require treatment, and the amount of treatment required is indicated by the difference between the estimated levels from the window or compressor and the target goal (in those octaves dominated by the individual source).

These examples illustrate the importance of obtaining close-in SPL values near each operating mechanism or component of a source and of estimating the area of that component or the area through which its SPL is radiating. Sound-control work should be directed to those components that yield large values of sound-power level. It is also necessary to investigate the frequency variation of the component sources as measurements are being carried out. Some components may shift from small-valued sound sources in some frequency regions to high-valued sources in other frequency regions.

In the previous examples, the sound source was presumed to be located in a large open area, so that nearby reflecting surfaces (other than the floor or ground) would not alter the free-field radiation of the sound. In most indoor plant situations, the confining walls and ceiling of the work space keep much of the sound from escaping to the outdoors. Instead, each ray of sound from the source strikes a solid surface and is reflected to some other direction inside the room. That same ray may travel 300 m and be reflected a dozen times before its energy is sufficiently dissipated for it to be ignored. In the meantime, other rays of sound are also radiated and reflected all around the room until they are dissipated. In a small room, the sound-pressure levels caused by the confinement of sound can be built up to values as much as 15–30 dB above the values that would exist at comparable distances outdoors. This buildup of sound can influence the sound level at the operator position of a machine. In fact, a machine that might have an 85-dB(A) sound level at a 2-m distance when tested outdoors in a large, open parking lot could produce a sound level of 95 to 100 dB(A) at the same distance when it is moved indoors into a small, highly reverberant room. Note that the sound-power level of the source did not change, but that the

Table 9.1. Calculations for Example Problem.

Description	Octave-Band Center Frequency, Hz								
	31.5	63	125	250	500	1000	2000	4000	8000

Source 1 (window on side of building) 2 m $\times$ 4 m; shell surrounding window at 1 m distance has area of 4 m $\times$ 6 m = 24 m^2.

L_w of window = L_p + 10 log 24 = L_p + 13.8, say L_p + 14

	31.5	63	125	250	500	1000	2000	4000	8000
L_p at 1 m	82	65	66	68	64	61	65	69	71
plus	14	14	14	14	14	14	14	14	14
L_w of window	76	79	80	82	78	75	79	83	85

Window noise at property line (from Eq. [9.3]): $L_p = L_w - 10$ log $2\pi r^2$; $r = 20$ m.

$L_p = L_w - 10$ log $2\pi 20^2 = L_w - 34.0$

	31.5	63	125	250	500	1000	2000	4000	8000
minus	34	34	34	34	34	34	34	34	34
Window noise	42	45	46	48	44	41	45	49	51

Source 2 (small vent on roof) surface area of sphere centered at vent radius of 1 m = $4\pi r^2$ = 12.56 m^2

L_w of vent = L_p + 10 log 12.56 = L_p + 10.99, say L_p + 11

	31.5	63	125	250	500	1000	2000	4000	8000
L_p at 1 m	84	81	76	74	76	69	62	56	54
plus	11	11	11	11	11	11	11	11	11
L_w vent	95	92	87	85	87	80	73	67	65

Vent noise at property line (from Eq. [9.3]): $L_p = L_w - 10$ log $2\pi r^2$; $r = 22$ m.

$L_p = L_w - 10$ log $2\pi 22^2 = L_w - 34.8$, say $L_w - 35$

	31.5	63	125	250	500	1000	2000	4000	8000
minus	35	35	35	35	35	35	35	35	35
Vent noise	60	57	52	50	52	45	38	32	30

Source 3 (compressor) 2 $\times$ 3 m; shell surrounding compressor with 1 m distance = 3 m $\times$ 5 m = 15 m^2.

L_w of compressor = L_p + 10 log 15 = L_p + 11.8 dB, say L_p + 12 dB

	31.5	63	125	250	500	1000	2000	4000	8000
L_p at 1 m	71	77	87	91	84	76	73	66	59
plus	12	12	12	12	12	12	12	12	12
L_w compressor	83	89	99	103	96	88	85	78	71

Compressor noise at property line (from Eq. [9.3]): $L_p = L_w - 10$ log $2\pi r^2$; $r = 30$ m.

$L_p = L_w - 10$ log $2\pi 30^2 = L_w - 37.5$, say $L_w - 38$

	31.5	63	125	250	500	1000	2000	4000	8000
minus	38	38	38	38	38	38	38	38	38
Compressor noise	45	51	61	65	58	50	47	40	33

acoustic environment made a major difference in the sound levels. This aspect of room acoustics was covered in Chapter 5.

9.2. FURTHER CASE STUDIES IN NOISE CONTROL*

In this section the principles of earlier chapters will be applied, as well as the concepts and relationships developed, to a variety of practical problems. From these case studies the reader

should develop an appreciation that noise control is neither a black art nor an exact science. Rather, it is an approach that utilizes physical principles (often in simplified form), engineering judgment, and knowledge of available technology. In the sections that follow, a variety of noise-control problems and the methods employed to solve them are described.

9.2.1. Reduction of Noise Levels in a Food-Packaging Area

Description. An area used for packaging cereals and dry prepared mixes occupies an entire floor of a large building. The area measures

*Reprinted with permission from *Noise Control for Engineers* by Lord, Gatley, and Evensen, New York: McGraw Hill, Inc., 1980.

Table 9.2. Noise Levels at 10 ft from Box-Forming Machine, in dB.

		Octave-Band Center Frequency, Hz						
		31.5	63	125	250	500	1000	2000
Former on, dB(A)	89	84–89	82–86	84–86	83	86	85	82
Former off, dB(A)	88	85–89	82–84	83–85	82	84	84	82

about 180 ft long by 120 ft wide by 20 ft high; all surfaces are poured concrete. Three packaging lines, each with similar equipment, are set up on the floor. In each line, the major noise sources are a paper shear, a box former, and a wrapping/sealing machine. The machines are old but well maintained. Some employees are stationed at specific locations, while others circulate among the major pieces of machinery.

Noise Measurements. A survey of the packaging area showed that A-weighted noise levels varied from 88 to 91 dB(A) at employee locations. Table 9.2 compares sound levels measured at a typical employee location, with the nearest box former operating and shut off. The similarity between levels indicates that other contributions are almost as significant as those from the box former, a nearby major source. Because of the surroundings (poured concrete), and the fact that no other major sources were nearby, it was concluded that reverberant sound was a significant contributor. This conclusion was reinforced by the uniformity of noise levels throughout the packing area.

Analysis. Addition of acoustical absorption to the packaging area, in the form of film-covered hanging baffles, was recommended for several reasons:

- No one type of source was dominant; noise levels were produced by operation of many sources.
- The space was highly reverberant, so that a reasonable amount of additional absorption could reduce noise levels by a significant amount.
- A reduction in noise level of about 7 dB(A)

would satisfy present and proposed OSHA regulations and was attainable by use of absorption.
- Hanging baffles covered with a thin Mylar film would not interfere with lighting, the sprinkler system, or production. In addition, they could resist vermin infestation and withstand a periodic 130°F soak temperature, as required by health regulations.
- The only other alternative, enclosure of dominant sources, would interfere with production and be of only limited value for reducing noise levels.

In the octave bands of greatest interest for reduction of A-weighted sound levels (500, 1000, and 2000 Hz), the average absorption coefficient was estimated to be 0.04. The existing absorption was estimated by calculating $S\bar{\alpha}_{SAB}$.

Design. The amount of additional absorption required to produce a 7-dB reduction in the 500-, 1000-, and 2000-Hz bands was calculated from

$$\Delta L_p = -10 \log_{10} \frac{A_f}{A_i},$$

where A_f = final value of absorption and A_i = initial value of absorption, both in sabins. The 500-, 1000-, and 2000-Hz band levels were dominant and approximately equal; therefore, a 7-dB reduction in each band would provide an overall reduction of about 7 dB(A). From calculations, it was determined that 800 hanging baffles of the type described in Table 9.3

Table 9.3. Absorption Data for Panels.

Frequency, Hz	125	250	500	1K	2K	4K
Sabins	2.1	5.9	9.8	13.3	11.6	7.6

23″ × 48″ × 1½″ glass-fiber board (4.7 lb/ft³) wrapped with Mylar film, hung vertically in rows on 4 ft centers, sabins/unit.

would be needed; 25 rows of baffles hung from cables stretched on 4-ft centers, with 32 baffles in each row, were recommended.

Evaluation. Only a partial coverage of the packaging area was completed initially. Baffles were concentrated over the noisiest areas of two lines. In these areas, a reduction of about 6 dB(A) was measured; this decrease was very noticeable to employees.

Note that the existing absorption must be estimated with some care, either by use of published values or by the measurement of reverberation times.

9.2.2. Reduction of Engine-Generator Cooling-Fan Noise

Description. Three diesel-powered emergency engine-generator units are located on the ground floor of a three-story brick building, in a room with an outside wall. The wall is penetrated by five louvered openings, each approximately 3 ft square. Two of the openings are connected by a 3-ft section of ductwork to a radiator which serves one engine-generator unit. An engine-driven multibladed fan forces cool air through the radiator and discharges it through the ductwork to an alleyway outside the building. During operation of the engine-generator units (about 3 h per week), fan noise is objectionable to residents located about 250 ft away.

Noise Measurements. Table 9.4 lists sound-pressure levels (overall, A-weighted, and in octave bands) corresponding to operation of all three units, and typical daytime ambient noise.

Analysis. The noise produced by the engine-generator units was broadband in nature, without distinct pure-tone components. Engine exhaust passed through silencers and into a vertical discharge stack that terminated above the third-floor roof. Exhaust noise at the discharge was negligible; the cooling fans were clearly the dominant sources of noise.

The exterior wall of the engine-generator room was masonry faced with brick; transmission of noise by the wall was insignificant. Modification of the cooling fans was not feasible, and insufficient space was available in the discharge ducts to permit installation of silencers. Therefore, a decision was made to design exterior silencers for each of the five openings. The possibility of a concrete-block plenum was discussed with the building architect. However, since footings would be required, this proposal was rejected in favour of custom-fabricated silencers designed for attachment to the outside wall of the building.

Design. Each of the five silencers consisted of a $1\frac{1}{2}$-in. welded-steel angle framework to which $\frac{3}{8}$-in.-thick Fiberglas board was attached with screws (see Fig. 9.5). This material was selected because of its relatively high surface weight, its relatively low cost, and its resistance to weather. Each silencer contained parallel baffles, lined on all sides with 1-in.-thick commercial duct liner, and a lined 180° bend. Baffles were spaced 6 in. apart because of the relatively high frequency content of the noise. The outlet of each silencer faced downward to prevent entry of rain or snow; however, individual designs were modified so that cooling-air discharge and inlet-air openings were separated by as great a distance as possible.

Table 9.4. Outdoor Noise Levels Near Engine-Generator Units, in dB.

		Octave-Band Center Frequencies, Hz							
		31.5	63	125	250	500	1000	2000	4000
Alleyway, 15 ft from louvers, dB(A)	82	77	81	87	79	78	78	75	66
Nearest residence, dB(A)	61	57	61	67	59	58	58	53	42
Daytime ambient-noise level, dB(A)	49	56	55	56	45	46	40	35	39

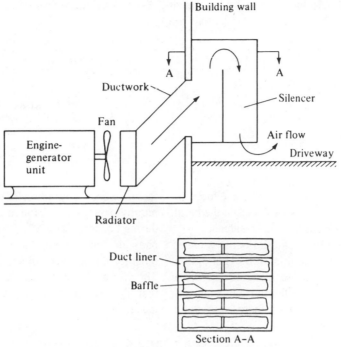

Fig. 9.5. Schematic of typical silencer.

Removal of the existing louvers before installation of the silencers maintained about the same flow resistance; however, the two inlet-air openings in the building wall were undersized, so that a negative pressure existed in the engine-generator room during operation of the units. This problem was largely eliminated by locating a centrifugal fan in the room, with its suction connected to one of the inlet-air openings.

Evaluation. After completion of the installation, operation of the units was inaudible at the nearest residence and caused no measurable increase in ambient-noise levels. An estimated noise reduction of 25 dB(A) was obtained by use of the silencers.

It should be noted that although this project was highly successful, the necessity for custom fabrication of the silencers resulted in a cost of several thousand dollars.

9.2.3. Reduction of Noise from a Mine Ventilation Fan

Description. A large (6000 ft³/s) vaneaxial fan provides ventilation air for a coal mine in

a rural area. The discharge duct measures 12 ft square and is located approximately 300 ft from the nearest residence. The fan operates continuously; residents have complained about excessive noise levels, especially during nighttime hours, and have threatened legal action.

Noise Measurements. Noise levels at the residence are shown in Table 9.5. The A-weighted sound level is approximately 35 dB above the nighttime ambient-noise level. A comparison of close-in noise levels radiated by the fan housing and discharge duct with those radiated from the duct opening showed clearly that the opening was the dominant source of noise.

Analysis. The noise spectrum of the fan was broadband, with a dominant pure tone at 360 Hz; energy was concentrated in the 125-, 250-, 500-, and 1000-Hz octave bands. Two possible noise-control methods were considered:

1. Addition of a 45° bend to the existing discharge duct. The bend would direct sound waves away from the residence and provide some noise reduction, particularly at frequen-

Table 9.5. Measured Sound Levels and Estimated Insertion Loss at Nearest Residence, in dB.

	Octave-Band Center Frequency, Hz						
	31.5	63	125	250	500	1000	2000
Sound level at residence no silencer, dB(A)	68–70	62–66	64–68	70–74	68–72	64–68	57–62	48–51
Estimated insertion loss of silencer, dB(A)	35	15	25	35	40	40	30	20

cies above 250 Hz. However, the effects of wind could reduce the insertion loss of the bend to an unacceptable value at the residence.

2. Addition of a lined parallel-baffle silencer to the discharge-duct termination. This option was recommended because it would provide a predictable attenuation of fan noise in a reasonable length, with a minimum increase in static pressure.

Design. A plan view of the silencer is shown in Fig. 9.6. The floor is a poured-concrete slab, exterior walls are Fiberglas board attached to 6-in. steel studs, and the roof is corrugated Fiberglas sheet. The transmission loss of the walls and roof was selected to exceed the insertion loss of the silencer in each frequency band. Each of the interior wall surfaces and the five internal baffles is fabricated from 6-in. steel studs covered by 30% open perforated-steel sheet. Cavities within the walls are filled with 3-lb/ft^3 glass-fiber insulation. This construction is adequate to protect the glass fiber from erosion by flow; the flow velocity is about 60 ft/s. The cross section of the silencer is larger than the duct opening, to minimize the additional increase in static pressure. Glass-fiber fill is used in the transition duct to provide additional attenuation.

The estimated attenuation of the silencer is shown in the table. The reduction in A-weighted sound level at the residence was estimated to be 35 dB(A); this reduction would make the fan barely audible, even at night.

Evaluation. Although the silencer was designed to minimize construction costs, the only bid received was excessive. Two contributing factors were the remoteness of the area and the requirement that the fan remain in operation during construction. As a result, the coal company decided to purchase the residence (no other residences were affected by the fan noise). The cost of the residence was about $7000 less than the estimated construction cost of the silencer.

9.2.4. Reduction of Railcar Retarder Noise

Description. Retarders are used in switching yards to reduce the speed of railcars on individual spurs. Most retarders consist of steel bars about 30 ft long by 3 in. wide, located just above and on each side of a section of track (see Fig. 9.8). Their purpose is to clamp the sides of the railcar wheels during passage of a railcar through the retarder. The clamping force is usually determined by the weight of the railcar, through a system of springs and levers. Although retarders are effective for control of railcar movements, the clamping action is a source of high-intensity pure-tone noise that is a hazard to employees and a source of annoyance in nearby residential areas. The noise-generating mechanism is friction between the retarder rails and a railcar wheel, which excites a resonance in the wheel at about 2000 Hz. Because the wheel is an effective radiator of acoustic energy at this frequency, a piercing noise is produced. In fact, the level can exceed 130 dB(A) at a distance of 20 ft, and the noise is clearly audible in the surrounding area.

Noise Measurements. A characteristic time history of retarder noise is shown in Fig. 9.7 (upper trace). Typically, each railcar will pro-

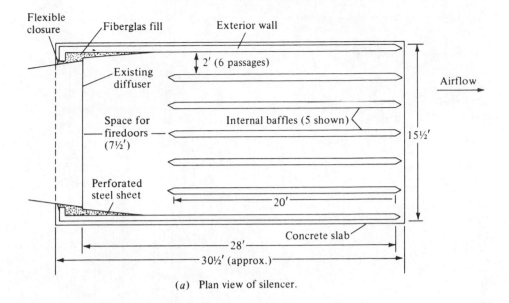

(a) Plan view of silencer.

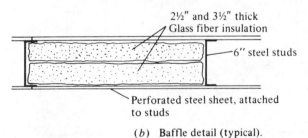

(b) Baffle detail (typical).

Fig. 9.6. Details of lined parallel-baffle silencer.

duce two to four peaks (or "squeals") during its passage through the retarder.

Design. The purpose of this study was to evaluate a new design for the friction surfaces, as shown in Fig. 9.8. Instead of a continuous steel bar, the redesigned surface consisted of a series of cast-iron pads welded to a steel backing strip with random spacing.

Analysis and Evaluation. The new design produced a dramatic reduction in sound level, as shown by the lower trace of Fig. 9.7. Not only was the average level reduced (almost to the yard ambient level), but the character of the noise changed from repeated high-intensity squeals to a low-frequency "grinding" sound punctuated with occasional squeals of reduced intensity. Apparently the random spacing of the pads prevents reinforcement of vibrations at the resonant frequency, so that much less acoustic power is produced.

Although the wear rate of the new design is somewhat greater, the increased maintenance is not excessive. The new design is currently being manufactured and marketed by at least one firm.

Note that the new retarder design is an example of noise reduction at the source, which—when feasible—is the most desirable approach to noise control.

9.2.5. Reduction of Mechanical-Equipment Noise in a Penthouse Apartment

Description. A mechanical equipment room, containing two boilers, a chiller, a 40-hp supply pump for a cooling tower, and several smaller pumps, adjoins a penthouse apartment. The common wall is 8-in. cement block, fin-

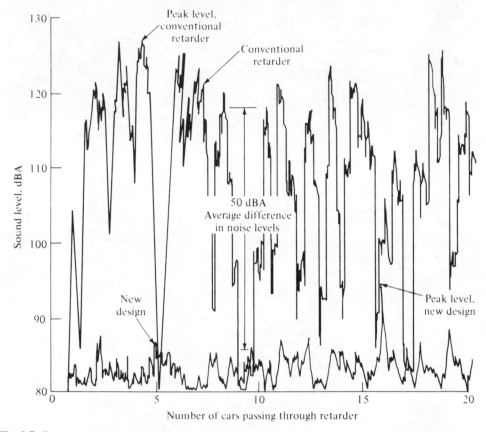

Fig. 9.7. Representative sound levels for conventional and modified retarders, measured to a distance of 20 ft.

ished on one side with gypsum board. In addition to the equipment mentioned, several ventilation fans (for kitchens and bathrooms), are mounted on curbs installed on the roof. The apartment tenant had moved out because of high noise levels.

Noise Measurements. Noise levels were measured in the apartment with all equipment operating and with individual sources shut off (see Table 9.6). From these measurements, the contribution from each source could be identified.

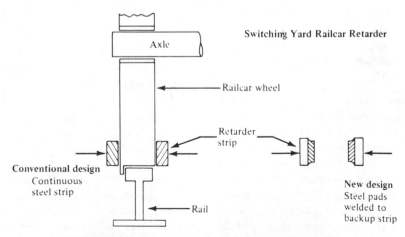

Fig. 9.8. Details of conventional and modified retarder designs.

Table 9.6. Noise Levels with Various Sources Operating, in Decibels.

	Octave-Band Center Frequency, Hz							
	31.5	63	125	250	500	1000	2000	4000
All HVAC sources operating	75	63	62	57	47	38	30	29
Cooling-tower pump off	69	57	56	53	43	34	26	23
Cooling-tower pump and ventilation fans off	63	53	54	51	43	34	26	23
All HVAC sources off	61	50	52	50	43	34	26	23

Analysis. Comparison of band levels shows that the cooling-tower pump is a major noise source, followed by the rooftop ventilation fans. Contributions from the apartment air-conditioning unit were relatively minor. Levels with all sources off resulted from daytime traffic; nighttime levels would be significantly less.

The common-wall transmission loss was adequate, except for an opening where it joined the building exterior wall.

Inspection of the cooling-tower pump showed that its vibration isolators were effectively short-circuited by a pipe support (see Fig. 9.9). The rooftop fans were rigidly attached to the roof and to the supply and return ductwork.

Design. The following steps were recommended, in priority order:

1. Paint the common cement-block wall with two coats of bridging paint; seal all openings with resilient caulk.
2. Remove the rigid discharge-pipe supports; support the pipe with resilient pipe hangers; install a flexible rubber coupling in the discharge pipe near the pump to reduce the load carried by the vibration isolators.
3. Measure the noise reduction obtained in steps 1 and 2. If inadequate, proceed to step 4.
4. Mount the rooftop ventilation fans on

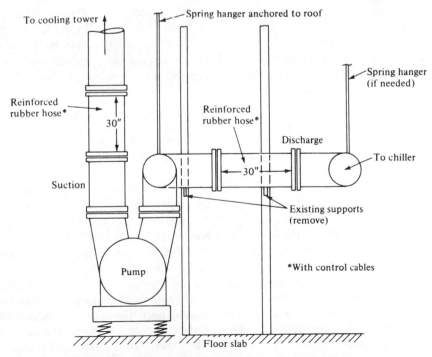

Fig. 9.9. Schematic of pump, showing existing and recommended pipe supports.

spring isolators; install flexible vinyl duct connectors between each fan and the supply and return ductwork.

Evaluation. The owner and the architect would not agree on how the costs for the noise control program should be shared. None of the steps were implemented, and the apartment remained vacant.

9.3. CASE STUDY OF NOISE REDUCTION FROM A CENTRIFUGAL PUMP AND MOTOR DRIVE IN A LABORATORY

This case study concerns a noisy pump and drive motor in the Hydraulics Laboratory at the writer's university. The approach to the control of the noise involves typical measurements and analysis, treatment of the room by absorption material, and finally enclosure of the noise source by a suitable barrier material. A step-by-step outline of the procedure is given first, followed by actual measurement data, calculations relevant to the procedure outline, and their prediction as to the final results. The calculations were performed by a graduate student in the Sound and Vibration Laboratory.

9.3.1. Introduction

The Allis-Chalmers centrifugal pump and motor drive in the Hydraulics Laboratory in the Faculty of Engineering Science at the University of Western Ontario supplies water to a constant-head (pressure) manifold from a sump in the laboratory floor. The manifold supplies a flow at constant head to various experiment stations in the laboratory (see general arrangement diagram of laboratory in Fig. 9.10); this constant head is obtained by means of a diaphragm control valve (air operated) in a bypass line from the manifold supply line to the sump.

The pumping and supply unit has been the center of numerous complaints because of the excessive noise levels in the laboratory when the unit is in operation. This noise also carries to adjacent laboratories, offices, and lecture rooms, with a resultant inconvenience to the occupants of these areas.

A plan and elevation view of the motor/pump unit is shown in Fig. 9.11.

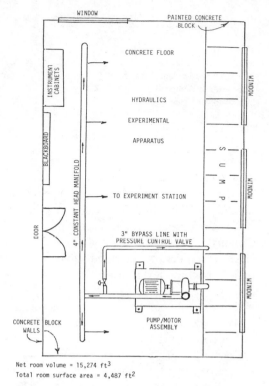

Net room volume = 15,274 ft^3
Total room surface area = 4,487 ft^2

Fig. 9.10. General arrangement—Hydraulics Laboratory.

9.3.2. The Problem

Simply stated, the problem is to carry out an analysis of the noise in the laboratory due to the pump-motor-piping assembly, and report on recommendations to reduce the noise to acceptable levels.

9.3.3. Suggested Procedure

Octave-Band Sound-Pressure Levels. In all noise abatement problems such as this, it is, of course, necessary to establish whether the noise levels are above acceptable or safe levels, and to what extent. This could be determined by preliminary octave-band sound-pressure levels and dB(A) levels and comparing these with preferred level criteria (NC curves, speech interference, etc.).

Narrowband Frequency Correlation of Sound and Vibration. The noise-radiating surfaces would have to be identified. In a case where there appears to be a wide range of sound spectra with discrete frequency components, it would probably be necessary to carry out a nar-

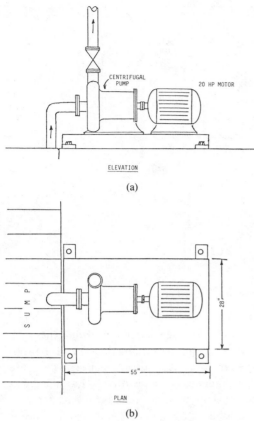

ELEVATION

(a)

PLAN

(b)

Fig. 9.11. Hydraulics Lab pump.

rowband frequency analysis on the sound, and correlate with stroboscopic measurement of speed of the pump/motor and a companion vibration spectrum analysis on the various radiating components; see Figs. 9.12(a) and 9.12(b). This should include a study of the possibility of isolating structure-borne vibration causing noise to radiate from adjacent surfaces in the laboratory.

Source Reduction. The first noise-abatement method should be that which reduces the noise at the source. In the event that such procedures are not feasible, or prove to be too costly, then alternative methods of attenuating the noise by treatment of the noise path should be employed.

Room Absorption and Absorptive Treatment. If the room in which the noise is generated is reverberant, it may be possible to bring about some measure of noise abatement by introduction of absorptive treatment on the walls, ceiling, and/or floor of the room. The existing reverberation time and absorption coefficient characteristics of the room should be determined (see Figs. 9.13 and 9.14 for schematic

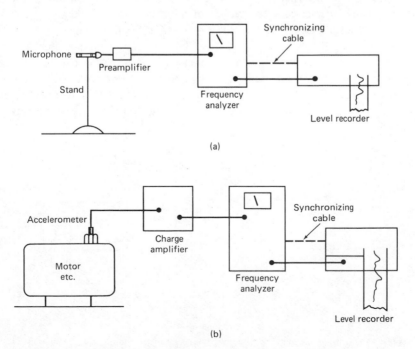

(a)

(b)

Fig. 9.12. (a) Measurement of sound spectrum, (b) measurement of vibration spectrum.

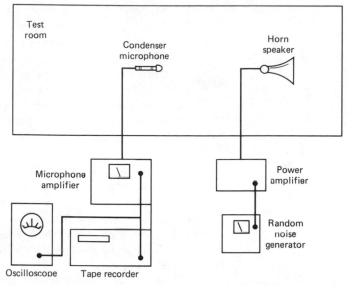

Fig. 9.13. Instrumentation for recording of sound decay.

of instruments), and the effect of adding additional absorptive material (possibly absorptive panels on the walls, acoustic tile ceiling, or hanging absorptive panels or cylinders) should be estimated.

Reduction by Enclosure.   In the event that absorptive treatment does not provide sufficient reduction of noise in the important frequency bands, it will probably be necessary to investigate placing an enclosure around the major noise source.

In order to estimate the effect of an enclosure barrier around the noise source, an indication of the free-field (or direct) SPLs in various frequency bands should be obtained. In the absence of such measurements from free-field conditions, an approximation could be obtained by measuring the SPLs close to the noise source (but not in the near field), i.e., the direct field, as opposed to reverberant or far-field measurements. Selection of appropriate barrier material should allow an estimation of sound transmission loss and an approximation of resulting SPLs in the room (taking the absorption of the room into consideration). This is summarized in the following.

The prediction of the effect of enclosing a noise source by some barrier material is most complex (see *Noise and Vibration Control* by L. L. Beranek, 1971). However, an approximation can be made if the power levels (PWLs)

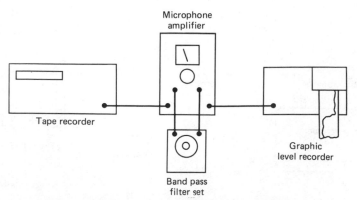

Fig. 9.14. Instrumentation for determining reverberation time at different frequencies.

of the noise source in frequency bands of interest are known.

The sound-power level is related to the sound-pressure level by the relative sound-pressure level versus distance (Eq. [4.15(b)]):

$$SPL = PWL + 10 \log_{10} \left[\frac{Q_\theta}{4\pi r^2} + \frac{4}{R_r} \right]$$

$$+ 10.5 \text{ (English Units)}$$

where

SPL = sound-pressure level at distance r from source,

PWL = sound-power level,

Q_θ = directivity factor,

r = distance from source,

R_r = room constant = $S_r \bar{\alpha}_r / 1 - \bar{\alpha}_r$ (see previous notes),

S_r = floor, ceiling, walls, etc. areas,

$\bar{\alpha}_r$ = average absorption coefficient of room.

Conversely, if the SPL for a particular noise source at a certain distance in an enclosed room is known, the PWL can be obtained as

$$PWL = SPL - 10 \log_{10} \left[\frac{Q_\theta}{4\pi r^2} + \frac{4}{R_r} \right]$$

$$- 10.5.$$

From the reverberation time measurements in the room, $\bar{\alpha}_r$ can be calculated from Eq. [5.9(b)]:

$$T_r = \frac{0.049V}{S_r \bar{\alpha}_r}$$

(assuming $\bar{\alpha}$ small compared to unity, i.e., < 0.20–0.25)

where

T_r = reverberation time,

V = room volume,

S_r = room surface area,

$\bar{\alpha}_r$ = average absorption coefficient.

Hence, R_r and then PWL can be calculated for any particular distance r and directivity factor Q_θ (assume $Q_\theta = 2$ in this case).

Recall the equation for sound-transmission loss (STL) of a barrier material, Eq. [5.27]:

$$STL = NR + 10 \log_{10} \frac{S}{A}$$

where

NR = noise reduction
   = SPL (trans. room) − SPL (receiving room),

S = barrier area,

A = absorption of receiving room.

This relationship is developed (see Section 5.3) from an analysis of the ratio of *power* in the transmitting room and receiving room, where sound-transmission loss is defined as the ratio of power between the two rooms, i.e.,

$$STL = 10 \log_{10} \frac{W_{tr}}{W_{rr}} \text{ dB.}$$

Once the PWLs for the pumping unit are known for the frequency bands of interest (i.e., PWL_{tr}), the values of W_{tr} for each frequency band can be obtained from PWL = $10 \log_{10}$ (W/W_{ref}) where $W_{ref} = 10^{-12}$ watts. Then by selecting a barrier material for the enclosure with certain STL values in the frequency bands of interest, the values of W_{rr} can be calculated from the above equation. Note that W_{rr} is the power in the laboratory space, i.e., on the outside of the intended barrier enclosure around the pump. From W_{rr}, the PWLs in the receiving room can then be calculated.

(The foregoing procedure is strictly applicable only to diffuse-sound-field conditions in the transmitting and receiving rooms. Because of the relatively small size of the pumping unit enclosure and resultant introduction of standing waves at lower frequencies, this diffuse field is not practical. Hence, it must be emphasized that the method is *approximate* only in the lower frequencies.)

Having then obtained the PWLs in the receiving room (i.e., outside the enclosure), the

SPLs at various distances can now be estimated for an existing (or modified) room constant.

Note, again, that lining the enclosure with an absorptive material should reduce the SPLs in the enclosure with a resultant further reduction in the SPLs in the room.

Beranek, in his book entitled *Noise Reduction*, p. 282 (see Ref. 3 of Chapter 4), gives the following relationship which can apply to an enclosure surrounding a noise source; the equation *estimates* the SPL outside the enclosure (as above), but takes into consideration any absorption within the enclosure as well (i.e., through inclusion of term R_1 below):

$$SPL_2 = PWL - TL + 10 \log_{10} \frac{S_w}{R_1}$$
$$+ 10 \log_{10} \left(\frac{1}{S_w} + \frac{4}{R_2} \right) dB$$

where

SPL_2 = is the sound pressure level in room 2 measured in the vicinity of the enclosure partition,

PWL = is the sound power level of the source in room 1 (enclosure) re 10^{-13} watts (*Note:* PWL here is equal to PWL re 10^{-12} watts plus 10 dB),

S_w = is the area of the inter-room partition in square ft,

R_1 = is the room constant for room 1 (enclosure) in square ft,

R_2 = is the room constant for room 2 (treated lab area) in square ft,

TL = is the partition transmission loss in dB.

9.3.4. Instrumentation

The instrumentation which could be used for the above-mentioned tests is as noted in Figs. 9.12, 9.13, and 9.14 and described in Chapter 3.

9.3.5. Comments on and Results of Typical Measurements

Octave-Band Sound-Pressure Levels. These are shown plotted in Fig. 9.15 for measurements taken at 6-ft and 25-ft distances from

the unit. The preferred noise contour PNC-50 (see Section 7.3.1.2) is compared to the octave-band measurements.

Narrowband Frequency Analysis of Sound. A narrowband sound analysis (6% constant-percentage bandwidth filter) showed that the sound spectrum was wideband in nature. The electrical noise floor of the system, obtained with a "dummy"—but similar—microphone substituted for the test microphone, was at least 10 dB below the test signal.

Vibration Measurements. The narrowband sound-frequency analysis showed little in the way of discrete frequency components, and hence a full vibration frequency analysis was not carried out. Instead, spot vibration measurements were taken at a number of points on the pumping/motor system, and the results are shown in Table 9.7. Besides indicating the major noise radiating sections of the system, these measurements aid in determining if isolation of the pump/motor assembly from the piping and/or floor, or isolation of the piping assembly from the ceiling, could reduce adjacent surface vibrations.

Reverberation-Time Measurements. Decay measurements were carried out in the room in accordance with Section 5.3.1. (See Figs. 9.13 and 9.14 for schematic of instrumentation.) Reverberation plots are shown in Fig.

Table 9.7. Vibration Measurements.

Location	Acceleration, in./sec.^{2}
Motor	2,500
Base plate	2,600
Coupling guard	1,100
Bearing housing	2,600
Pump body	2,000
Pump packing	9,000
Discharge valve	6,000
Discharge piping (2 ft from pump)	7,000
Discharge piping (8 ft from pump)	5,000
Discharge piping (10 ft from pump)	4,500
Suction piping at pump entrance	6,000
Motor switch box	700
Laboratory floor adjacent to motor	50
Bypass valve body	1,500
Bypass valve diaphragm	1,100
Piping adjacent to bypass valve	2,800
Bypass line to sump	1,500

OCTAVE BAND ANALYSIS

LOCATION Hydraulics Laboratory EMB 1074   DATE March 20, 1981

PROCESS   Allis Chalmers Centrifugal Pump/Motor   GRAPH NO.

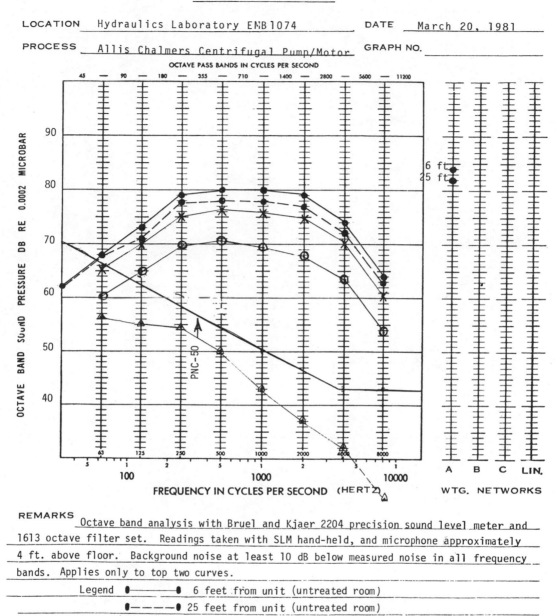

REMARKS Octave band analysis with Bruel and Kjaer 2204 precision sound level meter and 1613 octave filter set. Readings taken with SLM hand-held, and microphone approximately 4 ft. above floor. Background noise at least 10 dB below measured noise in all frequency bands. Applies only to top two curves.

Legend ●————● 6 feet from unit (untreated room)

        ●————● 25 feet from unit (untreated room)

        X————X 6 feet from unit (treated room)

        O————O 30 feet from unit (treated room)

        △————△ 30 feet from unit (treated room & enclosure)

Fig. 9.15. Octave band measurements before and after treatments, relative to PNC50.

9.16, and calculated reverberation times are given in Table 9.8.

9.3.6. Sample Calculations

Reverberation time is defined as the time for a sound in a room to decay 60 dB.

Reverberation time can be calculated from the decay curves of Fig. 9.16. As an example, and referring to the 2000 Hz decay in Fig. 9.16(b), it can be seen that the distance l for 30-dB decay is 27 mm. The paper speed was 30 mm/sec [Fig. 9.16(c)]. Therefore, 1 mm

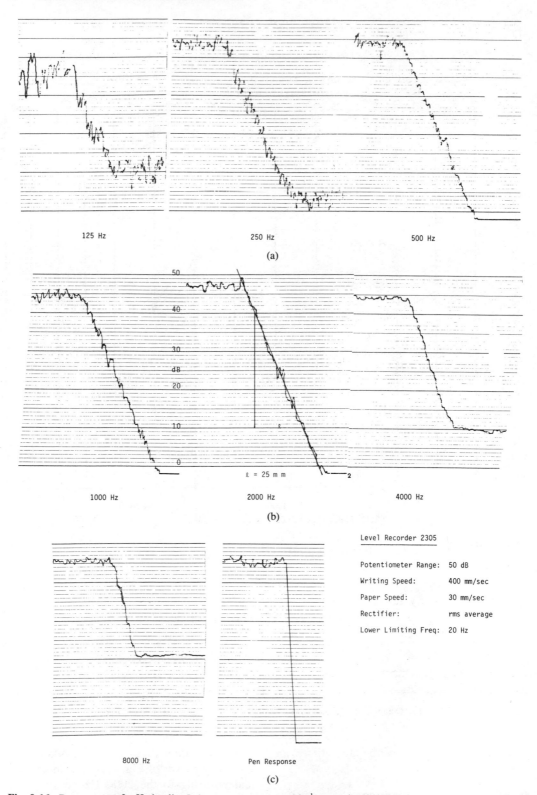

125 Hz   250 Hz   500 Hz

(a)

1000 Hz   2000 Hz   4000 Hz

$\ell = 25 \, mm$

(b)

8000 Hz   Pen Response

(c)

Level Recorder 2305

Potentiometer Range:	50 dB
Writing Speed:	400 mm/sec
Paper Speed:	30 mm/sec
Rectifier:	rms average
Lower Limiting Freq:	20 Hz

Fig. 9.16. Decay curves for Hydraulics Laboratory (as measured in $\frac{1}{3}$-octave bandwidths of random noise). (a) 125 Hz, 250 Hz, 500 Hz. (b) 1000 Hz, 2000 Hz, 4000 Hz. (c) 8000 Hz, pe$N$ response.

Table 9.8. Reverberation Times.	
⅓-Octave Center Frequencies	Reverberation Times T_{60}
63	1.56
125	1.64
250	1.84
500	1.86
1000	2.00
2000	1.80
4000	1.56
8000	1.23

Table 9.9. Absorption of drum sound absorbers.	
Frequency, Hz	Absorption, sabins per unit
63	6
125	6
250	8
500	8
1000	10
2000	11
4000	11
8000	11

paper movement took $\frac{1}{30}$ sec., and hence 27 mm occurred in $\frac{27}{30}$ sec. Therefore, for 60-dB decay (reverberation time T_{60}), the time was

$$T_{60} = \frac{27}{30} \times 2 = 1.8 \text{ sec.}$$

The reverberation times for all the frequency bands are recorded in Table 9.8.

A commercially available sound-absorber product by Insul-Coustic/Berina Corporation of New Jersey, made of mineral wool or glass fibers packed into a drum of 12″ diameter and 24″ long and protected with a glass fiber reinforcing mesh, has the absorption characteristics shown in Table 9.9.

From Table 9.8, the reverberation-time measurement at 63 Hz is 1.56 sec. Using the equation

$$T_{60} = \frac{0.049V}{S\overline{\alpha}},$$

the untreated room absorption is

$$S\overline{\alpha} = \frac{0.049V}{T_{60}} = \frac{0.049(15274)}{1.56}$$

$$= 479.8 \text{ sabins.}$$

Therefore the untreated room absorption coefficient is

$$\overline{\alpha} = \frac{479.8}{S} = \frac{479.8}{4487} = 0.11.$$

From Table 9.9, the absorption of the Insul-Coustic sound absorber drum at 63 Hz is 6 sabins/unit. With 200 units, total added absorption is 1200 sabins. Therefore

Treated room absorption

$$= \text{original} + \text{added}$$

$$= 479.8 + 1200 = 1679.8 \text{ sabins.}$$

Hence, treated room absorption coefficient is

$$\overline{\alpha}_A = \frac{1679.8}{S_A}$$

$$= \frac{1679.8}{4487 + \text{area of units}} = 0.28.$$

The room constant is given as

$$R = \frac{S\alpha}{1 - \overline{\alpha}}$$

For the untreated room,

$$R = \frac{479.8}{1 - 0.11} = 539.1 \text{ ft}^2$$

For the treated room,

$$R_A = \frac{1679.8}{1 - 0.28} = 2333 \text{ ft}^2.$$

From Fig. 9.15, the untreated room SPL at 63 Hz at a distance $r(= 6 \text{ ft}) = 68$ dB. Hence the power level PWL of the noise source can be calculated at this frequency, (Eq. [4.15]), i.e.,

$$PWL = SPL - 10 \log_{10} \left[\frac{Q_\theta}{4\pi r^2} + \frac{4}{R} \right]$$

$$- 10.5 \text{ dB}$$

$$= 68 - 10 \log_{10} \left[\frac{2}{4\pi(6)^2} + \frac{4}{539.1} \right]$$

$$- 10.5 \text{ dB}$$

$$= 76.8 \text{ dB}.$$

The SPL of the untreated room at distances of $r = 20'$ and $r = 30'$ can be found from

$$SPL_{r=20} = PWL + 10 \log_{10} \left[\frac{Q_\theta}{4\pi r^2} + \frac{4}{R} \right]$$

$$+ 10.5$$

$$= 76.8 + 10 \log_{10} \left[\frac{2}{4\pi(20)^2} \right.$$

$$\left. + \frac{4}{539.1} \right] + 10.5$$

$$= 66.2 \text{ dB},$$

$$SPL_{r=30} = 76.8 + 10 \log_{10} \left[\frac{2}{4\pi(30)^2} \right.$$

$$\left. + \frac{4}{539.1} \right] + 10.5$$

$$= 66.1 \text{ dB}.$$

Note that the values of SPL could also have been obtained from the graph of relative sound pressure level of Fig. 4.12 in Chapter 4.

Similarly, the SPL of the treated room at the various distances are as follows:

$$SPL_{r=6} = 76.8 + 10 \log_{10} \left[\frac{2}{4\pi(6)^2} \right.$$

$$\left. + \frac{4}{2333} \right] + 10.5 \text{ dB}$$

$$= 65.1 \text{ dB},$$

$$SPL_{r=20} = 76.8 + 10 \log_{10} \left[\frac{2}{4\pi(20)^2} \right.$$

$$\left. + \frac{4}{2333} \right] + 10.5 \text{ dB}$$

$$= 60.5 \text{ dB},$$

$$SPL_{r=30} = 76.8 + 10 \log_{10} \left[\frac{2}{4\pi(30)^2} \right.$$

$$\left. + \frac{4}{2333} \right] + 10.5 \text{ dB}$$

$$= 60.0 \text{ dB}.$$

Therefore reductions in noise levels at various distances for the 63 Hz frequency are

$$NR_{r=6} = SPL_{untreated} - SPL_{treated}$$

$$= 68 - 65.1 = 2.9 \text{ dB},$$

$$NR_{r=20} = 66.2 - 60.5 = 5.7 \text{ dB},$$

$$NR_{r=30} = 66.1 - 60.0 = 6.1 \text{ dB}.$$

(Note that when SPLs are calculated for each octave frequency for $r = 30$ ft, Table 9.12, and plotted on Fig. 9.15, there is still considerable deficiency in noise reduction vis-a-vis the PNC of 50.)

Hence, it is necessary to consider employing an enclosure around the motor-pump unit to bring about further noise reduction. By definition,

$$TL = 10 \log_{10} \frac{1}{\Gamma_{eq}} \quad (\text{Eq. } [5.28]).$$

In this particular example, it was decided to use 18-gauge sheet steel as a barrier material in order to reduce the SPLs in the treated room so as to reach the PNC 50 of Fig. 9.15. To provide ventilation to the motor, it was also decided to use lined 180°-bend ducts in the barrier.

Continuing with the example at 63 Hz, the transmission loss of 18-gauge sheet steel is 9 dB (Table 5.5.1); that of polyurethane-lined 180°-bend duct (with $L/D = 5$ and absorption coefficient $\alpha = 0.1$) is 5 dB (Fig. 5.12). Therefore

$$TL_{duct} = 10 \log_{10} \frac{1}{\Gamma_{duct}},$$

$$\Gamma_{duct} = [10^{5/10}]^{-1} = 3.16 \times 10^{-1},$$

$$\Gamma_{barrier} = [10^{9/10}]^{-1} = 1.26 \times 10^{-1}.$$

Now, by definition the equivalent transmission loss of a composite barrier is

$$\Gamma_{eq} = \frac{\sum \Gamma_i S_i}{S_{tot}} \quad (\text{Eq. } [5.29])$$

Hence

$$\Gamma_{eq} = \frac{\left[(\Gamma S)_{\text{duct}} + (\Gamma S)_{\text{barrier}}\right]}{S_{\text{total}}}$$

$$= \frac{(3.16 \times 10^{-1} \times 0.25) + (1.26 \times 10^{-1} \times 47.88)}{48.13}$$

$$= 1.27 \times 10^{-1}.$$

(Note that the area of the barrier and the area of the duct openings can be calculated from the proposed barrier/duct configuration of Fig. 9.17).

Therefore

$$\text{STL}_{\text{enclosure}} = 10 \log_{10} \frac{1}{\Gamma_{eq}}$$

$$= 10 \log_{10} \frac{1}{1.27 \times 10^{-1}}$$

$$= 8.96 \text{ dB} \quad (\text{say } 9.0 \text{ dB}).$$

Now

$$\text{PWL}_{\text{tr}}^* = 10 \log_{10} \left(\frac{W_{\text{tr}}}{W_{\text{ref}}}\right).$$

Hence

*Note: tr = transmitting room

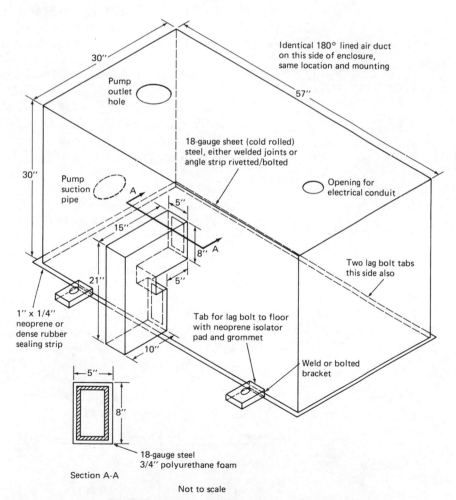

Fig. 9.17. Isometric sketch of pump-motor enclosure.

$$76.8 = 10 \log_{10} \left(\frac{W_{tr}}{10^{-12}} \right)$$

and

$$W_{tr} = (10^{76.8/10}) \, 10^{-12}$$

$$= 4.74 \times 10^{-5} \text{ watt.}$$

Since

$$STL_{enclosure} = 10 \log_{10} \left(\frac{W_{tr}}{W_{rr}} \right)$$

then

$$9.0 = 10 \log_{10} \left(\frac{4.74 \times 10^{-5}}{W_{rr}} \right)$$

or

$$W_{rr}^* = (10^{9.0/10})^{-1} \, 4.74 \times 10^{-5}$$

$$= 6.02 \times 10^{-6} \text{ watts}$$

Now

$$PWL_{rr} = 10 \log_{10} \left(\frac{W_{rr}}{W_{ref}} \right)$$

Therefore

$$PWL_{rr} = 10 \log_{10} \left(\frac{6.02 \times 10^{-6}}{10^{-12}} \right)$$

$$= 67.8 \text{ dB.}$$

From

$$SPL_{rr} = PWL_{rr} + 10 \log_{10} \left[\frac{Q_\theta}{4\pi r^2} + \frac{4}{R_A} \right]$$

$$+ 10.5$$

Note: rr = receiving room

we get

$$SPL_{rr} \text{ (at } r = 6') = 67.8 - 11.6 = 56.2 \text{ dB}$$

$$SPL_{rr} \text{ (at } r = 20') = 67.8 - 16.3 = 51.6 \text{ dB}$$

$$SPL_{rr} \text{ (at } r = 30') = 67.8 - 16.7 = 51.1 \text{ dB}$$

(Note that the room constant R_A used above is that of the treated room.)

9.3.7. Results

Tables 9.10 to 9.15 summarize the results of the analysis, where procedures comparable to that in the sample calculations (Section 9.3.6) have been applied to all of the frequency bands of interest.

9.3.8. Discussion

Because of the cost of replacement of the pump and motor drive unit, it was not deemed possible to reduce the noise at the source (the first line of attack). Hence the noise path was treated, first by increasing the absorption in the room and reducing reverberations, and then by building an enclosure around the noisy unit.

Increased absorption was achieved by the installation of 200 Insul-Coustic Sound Absorber Drums hanging from the ceiling of the laboratory. Calculations showed the sound-pressure levels varied (at different distances from the source) from just under 3 dB to approximately $9\frac{1}{2}$ dB (Table 9.12). The reductions are greatest in the reverberant field.

The above analysis showed that the absorptive treatment alone did not provide sufficient noise reduction to bring the levels in range of the PNC 50 curve (see Fig. 9.15). However, a decided advantage is to be noted in the increase of absorption coefficient $\bar{\alpha}$ of the treated room (Table 9.10) where absorption coefficients increased from a range of 0.08/0.14 to 0.27/0.46. The respective NRCs and reverberation times, calculated from the Sabine equation, are as follows:

	Untreated	Treated
$T_{60} = \dfrac{0.049V}{S\bar{\alpha}}$	$\bar{\alpha} = 0.09$	$\bar{\alpha} = 0.40$
$V = 15{,}274 \text{ ft}^3$	$T_{60} = \dfrac{0.049(15274)}{4487(.09)}$	$T_{60} = \dfrac{0.049(15274)}{4487(.40)}$
$S = 4{,}487 \text{ ft}^2$	$= 1.85 \text{ sec}$	$= 0.42 \text{ sec}$

Table 9.10. Average Absorption Coefficient ($\bar{\alpha}$) and Room Constant (R) Values.

Frequency, Hz	Untreated Room Absorption, Sabins	Untreated Room Absorption Coefficient, $\bar{\alpha}$	Added Absorption per Cylinder, Sabins	Total Added Absorption (200 cylinder), Sabins	Treated Room Absorption, Sabins	Treated Room Absorption Coefficient, $\bar{\alpha}_A$	Room Constant, ft^2 Untreated Room	Room Constant, ft^2 Treated Room
63	479.8	0.11	6	1200	1679.8	0.28	539.1	2333.0
125	456.4	0.10	6	1200	1656.4	0.27	507.1	2269.0
250	406.8	0.09	8	1600	2006.8	0.33	447.0	2995.2
500	402.4	0.09	8	1600	2002.4	0.33	442.2	2988.6
1000	374.2	0.08	10	2000	2374.2	0.39	406.8	3892.2
2000	402.4	0.09	11	2200	2602.4	0.43	442.2	4565.6
4000	479.8	0.11	11	2200	2679.8	0.44	539.1	4785.3
8000	608.5	0.14	11	2200	2808.5	0.46	707.5	5200.9

Table 9.11. Noise Source Power Level (PWL).

Frequency, Hz	Untreated Room SPL, dB, $r = 6'$, as measured	$10 \log \left(\frac{Q}{4\pi r^2} + \frac{4}{R}\right) + 10.5,$* dB $r = 6'$	$r = 20'$	$r = 30'$	PWL, dB
63	68	-8.8	-10.6	-10.7	76.8
125	73	-8.6	-10.3	-10.4	81.6
250	79	-8.2	-9.8	-9.9	87.2
500	80	-8.2	-9.8	-9.9	88.2
1000	80	-8.0	-9.4	-9.5	88.0
2000	79	-8.2	-9.8	-9.9	87.2
4000	74	-8.8	-10.6	-10.7	82.8
8000	64	-9.5	-11.7	-11.8	73.5

*R values used here were the untreated room constants from Table 9.10.

$$PWL = SPL \text{ (untreated, at } r = 6') + 10 \log_{10} \underbrace{\left(\frac{Q}{4\pi r^2} + \frac{4}{R}\right)}_{\text{at } r = 6', R \text{ is untreated room constant}} + 10.5.$$

Table 9.12. Sound-Pressure Level (SPL) of the Untreated and Treated Room and Noise Reduction.

Frequency, Hz	Untreated Room SPL, dB $r = 6'$	$r = 20'$	$r = 30'$	$10 \log \left(\frac{Q}{4\pi r^2} + \frac{4}{R_A}\right) + 10.5,$* dB $r = 6'$	$r = 20'$	$r = 30'$	Treated Room SPL, dB $r = 6'$	$r = 20'$	$r = 30'$	Noise Reduction, dB $r = 6'$	$r = 20'$	$r = 30'$
63	68	66.2	66.1	-11.6	-16.3	-16.7	65.1	60.5	60.0	2.9	5.7	6.1
125	73	71.3	71.2	-11.6	-16.2	-16.6	70.0	65.5	65.0	3.0	5.8	6.2
250	79	77.5	77.3	-11.9	-17.1	-17.7	75.3	70.1	69.5	3.7	7.3	7.8
500	80	78.5	78.4	-11.9	-17.1	-17.7	76.3	71.1	70.5	3.7	7.4	7.9
1000	80	78.6	78.5	-12.1	-18.0	-18.7	75.8	70.0	69.3	4.2	8.6	9.2
2000	79	77.5	77.4	-12.3	-18.5	-19.3	75.0	68.8	67.9	4.1	8.7	9.4
4000	74	72.2	72.1	-12.3	-18.6	-19.5	70.5	64.2	63.3	3.5	8.0	8.8
8000	64	61.8	61.6	-13.0	-18.8	-19.7	60.5	54.6	53.7	3.5	7.2	7.9

*R_A values used here were the treated room constants from Table 9.10.

$$SPL \text{ (treated room)} = PWL + 10 \log_{10} \underbrace{\left(\frac{Q}{4\pi r^2} + \frac{4}{R_A}\right)}_{\text{Table 9.11}} + 10.5 \text{ dB.}$$

Table 9.13. Barrier Sound Transmission Loss and Noise Attenuation of Lined 180° Bend Duct.

Frequency, Hz	Barrier Sound Transmission Loss,* dB	$\bar{\alpha}$ of Acoustic Lining,**	Attenuation of Lined 180° Bend Duct ($L/D = 5$).
63	9	0.1	5
125	15	0.1	5
250	21	0.28	9
500	27	0.49	10
1000	33	0.96	25
2000	38	1.00	27
4000	39	0.95	24.5
8000	39	0.97	26

*Barrier Material: 18-gauge sheet steel, cold rolled (as per Ref. 15, Chapter 5).
**Lining Material: ¾" Scott fine-pore acoustic foam (as per Ref. 1, Chapter 5).

The optimum reverberation time for speech above 500 Hz for a room of this volume is about 0.75 sec (Fig. 5.33); although the reverberation time is now too low, conditions have improved somewhat (over the untreated room condition) as far as speech intelligibility is concerned.

Inasmuch as drum absorbers are relatively expensive, it is the writer's opinion that fewer drums (say 150) at less cost would still produce a noticeable effect in the room. Alternatively, the possibility of procuring 2 in. × 2 ft × 4 ft hanging absorptive panels, at less cost and perhaps with greater absorption, should be explored with material suppliers.

The placement of an enclosure around the unit, with appropriate ventilation ducts, considerably reduced the noise levels in all frequency bands. In fact, as can be observed from Fig. 9.15, the calculated noise levels were below the PNC 50 curve at all frequencies.

Sheet steel, 18-gauge thick, was chosen initially because of (1) its superior sound transmission loss values and (2) durability and ease of fabrication (welding). However, considering the margin of safety (Fig. 9.15), ¾-in. or 1-in. plywood has good STL values (Table 5.5.1) and would undoubtedly be less expensive than the steel.

Three-quarter-inch Scott fine-pore acoustic foam (Ref. 1, Chapter 5) was used as the lining material of the ventilation ducts because of high absorption coefficients in the frequency range of concern (Table 9.13).

Lining the total enclosure with an absorptive material would reduce the SPLs in the enclosure with a resultant further reduction in SPLs in the room. (It is suggested that the reader perform calculations based on, say, ¾-in. plywood material to estimate whether such material, lined with foam or fiberglass, would give as satisfactory results as the more costly steel).*

---

*The procedure followed in the foregoing problem was one of trial and error, i.e., select a barrier material and note the effect on the resultant room sound-pressure levels. However, knowing the sound energy which has to be contained, the treated room constant and the required SPL in the room (commensurate with the PNC curve), it is possible to work backward and find the required sound transmission loss STL, and hence select an appropriate material—a good followup exercise to this problem.

Table 9.14. Enclosure Transmission Coefficient (Γ_{eq})

Frequency, Hz	$(\Gamma_{eq})_{duct}$	$(\Gamma_{eq})_{barrier}$	$(\Gamma_{eq})_{enclosure}$
63	3.16×10^{-1}	1.26×10^{-1}	1.27×10^{-1}
125	3.16×10^{-1}	3.20×10^{-2}	3.35×10^{-2}
250	1.26×10^{-1}	7.94×10^{-3}	8.55×10^{-3}
500	1.00×10^{-1}	2.00×10^{-3}	2.51×10^{-3}
1000	3.16×10^{-3}	5.00×10^{-4}	5.10×10^{-4}
2000	2.0×10^{-3}	1.60×10^{-4}	1.70×10^{-4}
4000	3.55×10^{-3}	1.30×10^{-4}	1.50×10^{-4}
8000	2.51×10^{-3}	1.30×10^{-4}	1.40×10^{-4}

Table 9.15. Sound-Pressure Level of the Treated Room with Enclosure and Air Vents.

Frequency, Hz	STL, dB, equivalent barrier (i.e., enclosure plus vents)	W_{tr}, watts	W_{rr}, watts	PWL_{rr}, dB	$SPL_{rr} = PWL_{rr} + 10 \log \left[\dfrac{Q}{4\pi r^2} + \dfrac{4}{R_A} \right] + 10.5,*$ dB		
					$r = 6'$	$r = 20'$	$r = 30'$
63	9.0	4.74×10^{-5}	6.02×10^{-6}	67.8	56.2	51.6	51.1
125	14.8	1.45×10^{-4}	4.86×10^{-6}	66.9	55.3	50.7	50.3
250	20.7	5.30×10^{-4}	4.53×10^{-6}	66.6	54.7	49.5	48.9
500	26.0	6.62×10^{-4}	1.66×10^{-6}	62.2	50.3	45.1	44.5
1000	32.9	6.25×10^{-4}	3.19×10^{-7}	55.0	42.9	32.1	36.4
2000	37.7	5.26×10^{-4}	8.93×10^{-8}	49.5	37.3	31.1	30.2
4000	38.2	1.89×10^{-4}	2.83×10^{-8}	44.5	32.2	25.9	25.1
8000	38.5	2.22×10^{-5}	3.11×10^{-9}	34.9	22.0	16.1	15.2

$*R_A$ values used here were the treated room constants from Table 9.10.

A schematic diagram of the proposed enclosure and ventilation ducts is shown in Fig. 9.17. It should be noted that acoustic caulking should be used around pump inlet and outlet pipes, electrical conduit piping, etc., in the enclosure.

Finally, a satisfactory "fix" for the noise problems of this pumping unit would be incomplete without recommendations for the following:

(a) Flexible spring hangers supporting the discharge pipe from the pump to the constant pressure manifold (to eliminate pipe vibrations exciting other areas/units in the laboratory through the solid pipe hangers currently used);

(b) The use of one of the recently introduced "quiet" valves at the outlet from the pump in order to reduce turbulence and hence pipe vibration to a minimum;

(c) A flexible coupling in the outlet pipe from the pump, which will minimize structure-borne vibration from the pumping unit to the attached pump discharge piping.

9.4. EXAMPLE OF USE OF SILENCER AND SILENCING MATERIAL IN VENTILATION SYSTEM

A fan used to provide ventilating air to an office (Fig. 9.18) is located in an equipment room 1, and transmits air through ducting into the adjacent private office, room 2. Room 2 has a "drop ceiling" which is perforated, and thus provides only cosmetic covering of the ducting. Room 1 is 60 ft long × 30 ft wide × 12 ft high. Room 2 is 60 ft long × 30 ft wide × 10 ft high. Room 1 has a reverberation time of 3.0 sec (hard and diffuse) and room 2 has a reverberation time of 1 sec, each at 1000 Hz. The fan is hung at the interface between the ceiling and the end wall in room 1, midway in the 30-ft wall. The fan radiates a sound-power level (PWL) of 90 dB at 1000 Hz into the equipment room, and a sound-power level of 90 dB at 1000 Hz into the duct. Assume the ducting is lagged on the outside (with, say, a layer of $\frac{3}{4}$-in. Fiberglas surrounded by $\frac{1}{8}$-in. polyvinylchloride (PVC) septum impregnated with lead powder, and appropriately bound in place). Hence, minimal sound radiates through the duct walls into either of rooms 1 and 2, and vice versa.

A 5-ft, low-pressure-loss commercial silencer, placed in the ducting immediately after the fan, has an insertion loss at 1000 Hz of 20 dB (Table 5.4).* The ducting from the fan to room 2 is 1 ft × 1 ft square section; it is unlined and has an attenuation of 0.04 dB/ft of duct length at 1000 Hz (Table 5.2).

There is an end-reflection loss where the duct terminates in room 2. (This occurs when a relatively small duct opens abruptly into a large-area space.) The "area mismatch" (see section on expansion chambers, 5.1.4.2) at this inter-

---

*Insertion losses as given for each element (i.e. silencer, ducting, elbows, end reflection) are subtracted arithmetically from the power level generated by the fan.

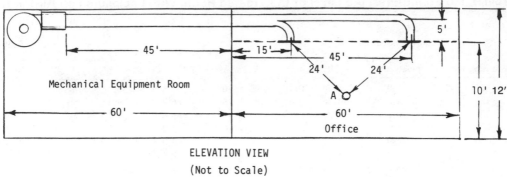

ELEVATION VIEW
(Not to Scale)

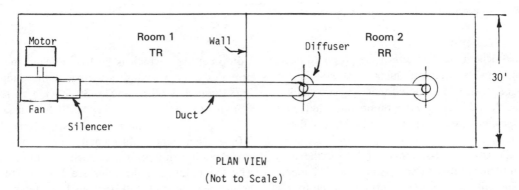

PLAN VIEW
(Not to Scale)

Fig. 9.18. Elevation and plan views of rooms and ducting in example of Sect. 9.4.

face has the effect of reflecting some of the sound back along the path from which it came. This end-reflection loss is 1 dB at 1000 Hz (see reference 27, Chapter 5).*

With the expected flow rate, a sound-power level of 20 dB will be produced by each room exit diffuser (i.e., the existing fan noise PWL will be increased by this amount—see Refs. 15, 27, and 28 of Chapter 5). The sound transmission loss of the common wall between rooms 1 and 2 is 35 dB at 1000 Hz. Assume that, because of poor sealing between the duct and the wall, there is the equivalent of a $\frac{1}{4}$-in. crack between the periphery of the duct and the wall. Assume, initially, that because of a smooth-contoured elbow, there is no flow-induced noise as a result of turbulence generated in flow around bends. However, there is an attenuation of 7 dB due to reflection at the 90° elbow in room 2 at 1000 Hz (see Ref. 27 of Chapter 5).

The total air flow from the fan to the main duct is 1000 CFM. Each outlet to room 2 has a flow of 500 CFM.

Calculate the sound-pressure level (SPL) at 1000 Hz in room 2 at a receiving point A as shown in Fig. 9.18. Assuming that this SPL represents the maximum penetration of all octave bands into the PNC curves of Chapter 7 (Fig. 7.2), what is the PNC value for the office, and is it acceptable according to ASHRAE standards?

What would be the effect on the SPL at the receiving point of inserting turning vanes in the 90° elbows leading to each of the diffusers in room 2?

SPL in transmitting room is

$$\text{SPL}_{\text{TR}} = \text{PWL}_{\text{fan}} + 10 \log_{10}\left(\frac{Q_\theta}{4\pi r^2} + \frac{4}{R_{\text{TR}}}\right)$$

$$+ \ 10.5 \text{ dB} \qquad (\text{Eq. } [4.15\text{b}]).$$

Since sound is diffuse in TR, then there is no distance term and $Q_\theta/4\pi r^2 = 0$. Therefore

---

*Note that end reflection loss is reduced (minimal) if a diffuser terminates the duct.

$$SPL_{TR} = 90 + 10 \log_{10} \frac{4}{R_{TR}} + 10.5 \text{ dB}.$$

Now reverberation time $T_{60TR} = 0.049V/S\bar{\alpha}$ (Eq. [5.9b]). Therefore

$$3.0 = \frac{0.049(21,600)}{S\bar{\alpha}},$$

$$V_{TR} = V_{RR} = 60 \times 30 \times 12$$

$$= 21,600 \text{ ft}^3$$

and $S\bar{\alpha} = 352.8$ sabins.

$$S_{TR} = (2 \times 30 \times 60) + (2 \times 60 \times 12)$$

$$+ (2 \times 30 \times 12)$$

$$= 5,760 \text{ ft}^2$$

Therefore

$$\bar{\alpha} = \frac{352.8}{5760} = 0.061.$$

Hence

$$R_{TR} = \frac{S\bar{\alpha}}{1 - \bar{\alpha}} = \frac{352.8}{1 - 0.061} = 376 \text{ ft}^2$$

and

$$SPL_{TR} = 90 + 10 \log_{10} \frac{4}{376} + 10.5$$

$$= 80.7 \text{ dB}.$$

This is the SPL acting against the wall in the transmitting room (i.e., room 1 in Fig. 9.18).

Now consider the transmission loss TL of the wall between rooms 1 and 2:

$$TL = 10 \log_{10} \frac{1}{\Gamma} \qquad \text{(Eq. [5.28])}$$

$$\Gamma_{eq} S = \Sigma_i \Gamma_i S_i$$

$$= \Gamma_1 S_1 + \Gamma_2 S_2 + \ldots + \Gamma_n S_n$$

$$\text{(Eq. [5.29])}$$

$$S_{hole} = 48 \text{ in.} \times \frac{1}{4} \text{ in.} = 12 \text{ in.}^2 = \frac{1}{12} \text{ ft}^2$$

$$= 0.083 \text{ ft}^2$$

$$S_{wall} = (30 \times 12) - (1 + 0.083)$$

$$= 358.9 \text{ ft}^2.$$

Now $TL_{wall} = 35$ dB and $TL_{hole} = 0$ dB. Hence, for wall,

$$35 = 10 \log_{10} \frac{1}{\Gamma_{wall}}$$

and

$$\Gamma_{wall} = 3.162 \times 10^{-4}$$

and for hole,

$$0 = 10 \log_{10} \frac{1}{\Gamma_{hole}}$$

and $\Gamma_{hole} = 1$. Therefore

$$\Gamma_{eq}(360 \text{ ft}^2) = (3.162 \times 10^{-4})358.9 \text{ ft}^2$$

$$+ 1(.083 \text{ ft}^2)$$

and

$$\Gamma_{eq} = 5.46 \times 10^{-4}.$$

The equivalent transmission loss of the wall plus crack is

$$TL_{eq} = 10 \log_{10} \frac{1}{\Gamma_{eq}}$$

$$= 10 \log_{10} \frac{1}{5.46 \times 10^{-4}} = 32.6 \text{ dB}.$$

Now the SPL in room 2 due to the sound transmitted through the wall can be found from Eq. [5.24]:

$$STL = SPL_{TR} - SPL_{RR} + 10 \log_{10} \frac{S_{wall}}{A}$$

where S_{wall} is area of barrier wall, and
A is the absorption in the receiving room
$= S_{RR}\bar{\alpha}.$

Note that this equation was derived on the basis of a diffuse field in both the transmitting and receiving homes (Section 5.3.1). With a reverberation time in the receiving room of 1 second, it is quite possible that the sound is not diffuse—and hence Eq. [5.24] will be approximate only.

For the receiving room*

$$T_{60} = \frac{0.049V}{S\overline{\alpha}}$$

$$1.0 = \frac{0.049(21,600)}{5760\,\overline{\alpha}}$$

Therefore

$$\overline{\alpha} = 0.184 \text{ and } S\overline{\alpha} = 5760(0.184) = 1060$$

and

$$SPL_{RR} = SPL_{TR} - TL_{eq} + 10\,\log_{10}\frac{S}{A}$$

$$= 80.8 - 32.6 + 10\,\log_{10}\frac{360}{1060}$$

$$= 80.8 - 32.6 + (-4.69)$$

$$= 43.6 \text{ dB}.$$

Now, considering the noise introduced into the room 2 through the ducts:

PWL$_{fan}$ = 90 dB,
Silencer attenuation = 20 dB,
Duct attenuation = 0.04 dB/ft × 80 ft (average to two diffusers)
                 = 3.2 dB,
Elbow attenuation = 7 dB,
End reflection attenuation = 1 dB.

There are two outlets and diffusers in room 2. The sound power is divided into proportion to the area of the duct, or CFM (cubic feet per minute) through the duct, as follows (from Ref. 27, of Chapter 5):

*The volume of the receiving room is assumed to be 60' × 30' × 12' (21,600 ft^2); the drop ceiling is perforated, and hence "acoustically" the ceiling is transparent. The area of room 2 is thus 5,760 ft^2 (same as transmitting room).

PWL$_{outlet}$ = PWL$_{total}$ − 10 log$_{10}$

$$\cdot \left(\frac{\text{total area or CFM}}{\text{area or CFM of outlet}}\right).$$

Therefore the dBs to subtract from total power level to obtain power level per outlet is

$$10\,\log_{10}\frac{1000(\text{CFM})}{500(\text{CFM})} = 3 \text{ dB}$$

Thus

$$PWL_{\substack{\text{diffuser}\\\text{outlet}}} = 90 - 20 - 3.2 - 7 - 1 - 3$$

$$= 55.8 \text{ dB}.$$

Now, the diffuser adds a noise PWL of 20 dB at each outlet (due to turbulence, etc. as a result of air flow past the diffuser grill). This must be added by "dB addition." It can be determined that the addition of 20 dB to 55.8 dB still results in a combined value of 55.8 dB for each diffuser (from Fig. 1.9). Hence, the final noise PWL at each diffuser is 55.8 dB.

Now to the determination of SPL at point A in Fig. 9.18. For the distances shown in the figure between the diffuser outlets and point A (i.e., each is 24 ft), and assuming $Q_\theta = 2$, and $R_{RR} = S\overline{\alpha}/(1 - \overline{\alpha}) = 1060/(1 - 0.184) = 1292$ (from before), then interpolating for R and $r/\sqrt{Q_\theta}$ in Fig. 4.12, relative SPL (or SPL − PWL) is −14 dB. Therefore SPL at point A due to one diffuser is

$$SPL = PWL - 14$$

$$= 55.8 - 14$$

$$= 41.8 \text{ dB}.$$

The combined effect of the noise of two diffusers on point A is

$$SPL_{comb} = 41.8 + 3$$

$$= 44.8 \text{ dB (from Fig. 1.9)}.$$

The noise from the wall must now be combined with the noise from the diffusers at point A in order to get the total noise.

The noise at the wall in the receiving room

was calculated at 43.6 dB. The distance from the wall to point A is 30 ft, and assuming a value of $Q_\theta = 2$ for the wall and $R_{RR} = 1292$ (from before), then interpolating for R and $r/\sqrt{Q_\theta}$ in Fig. 4.12, relative SPL (or SPL − PWL) is −15 dB, and hence the SPL at point A, due to noise through wall, is $43.9 - 15 = 28.9$ dB.

The noise from the wall and the noise from the ducts at point A must be combined logarithmically. Hence, from Fig. 1.9, the combined SPL is still the higher of the two SPLs, i.e., 44.8 dB from the ducts.

From Fig. 7.2, the PNC value for a SPL of 44.8 dB at 1000 Hz is 45—which, according to Table 7.1, indicates that the noise level in the office is borderline for acceptability for a typical office.

We were asked to assess the effect of adding turning vanes in the 90° elbows leading to each of the diffusers in room 2. From Ref. 27 of Chapter 5, the sound power produced by turning vanes at 1000 Hz for a duct of 1 square foot is 53 dB. This is based on a velocity through the duct of 2000 feet per minute. The flow velocity in this case is 1000 CFM ÷ 1 square foot = 1000 ft/min. According to Ref. 27, a correction factor for 1000 ft/min is −18 dB. Hence, the sound power produced by the turning vane is $53 - 18 = 35$ dB.

The end-reflection loss at each diffuser is, as noted earlier, 1 dB at 1000 Hz. Hence, the total noise at each diffuser as a result of the turning vanes in the elbows leading to each diffuser is $35 - 1 = 34$ dB.

If we added this noise source with that of the fan (55.8 dB) at each diffuser, it can be seen that the power generated through the vanes has no effect on the overall level. Many times, the effect of vanes is deleted from the analysis; however, in applications where the flow velocity is high, the effect of vanes should always be checked.

9.5. NIOSH CASE HISTORIES

These case histories are selected from a publication by NIOSH (National Institute for Occupational Safety and Health) of the U.S. Department of Health, Education and Welfare, which was originally published in 1975 and re-vised in 1978. The publication is called *Industrial Noise Control Manual*. The case studies were compiled by P. Jensen, C. R. Jokel, and L. N. Miller of Bolt, Beranek and Newman, Inc. under Contract No. 210-76-0149.

A revised edition of the above-mentioned Manual has been issued by the Canadian Acoustical Association (1984) and can be obtained by writing to:

The Canadian Acoustical Association
P.O. Box 1351, Station F
Toronto, Ontario
M4Y 2V9
Canada.

Case History 1. Gas Turbine Test Station (Hearing Conservation and Speech Communication Noise Problem)

Walt Jezowski
General Electric Company
Gas Turbine Division
Building 53-303
Schenectady, New York 12345
(518) 385-7544

Problem Description. Operations of a gas turbine test stand at the General Electric Company's Schenectady, New York plant involve fabrication and assembly workers on the 128,000-ft^2 workfloor surrounding the test area. In particular, sound between 90 and 95 dBA was at times present in the vicinity of the test stand where some 40 employees work for varying periods of time.

Problem Analysis. The test station responsible for the high sound levels is partially treated; the test stand is surrounded by a 14-ft-high acoustically lined, open-topped barrier. Noise is emitted over the top of the partially enclosed test area, which remains open for crane accessibility. Alternatives for reducing the sound levels in the area surrounding the stand narrowed to treating the room surfaces to reduce the effects of reverberation. Hanging baffles, wall and ceiling blanket linings, and spray-on materials were investigated, the latter eventually being selected for implementation.

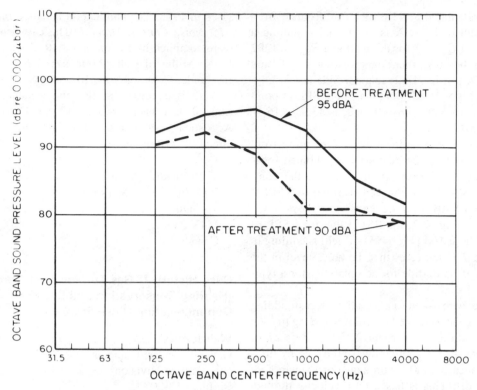

Fig. 9.19. Reduced aisle sound levels, as predicted.

Prior to installation, estimates of the expected acoustical benefit were made on the basis of calculations of the existing and modified room constants.

Control Description. The selected treatment consisted of a 1-in.-thick layer of sprayed-on cellulose-fiber-based material called K-13, available from National Cellulose. The material is applied directly to the surface to be coated, where it forms a permanent thermal and acoustic lining. In this installation, approximately 28,000 ft² of ceiling and wall area were coated at a cost of about $1.10/ft².

Results. Aisle sound levels were reduced, as predicted, from 95 dB(A) to 90 dB(A), as shown in Fig. 9.19. The manned area surrounding the test stand with above-90-dB(A) sound levels has been eliminated.

Comments. In addition to having improved the acoustic environment, General Electric also achieved added thermal insulation. Annual savings of about 13¢/ft² are estimated in heat-

ing costs for the treatment—one of the major reasons for selecting a surface-applied material. Additional benefits include lower maintenance costs (there is no longer the need to paint the 65-ft-high ceiling and wall areas) and improved light reflection and diffusion.

Case History 2. Nail-Making Machine* (OSHA Noise Problem)

Problem Description. A nail-making machine was operating under conditions causing severe impacts. The vibration was solidly transmitted to a weak concrete floor, which radiated considerable noise. There were 10 machines, operating at 300 strokes/min. Operator sound level was 103.5 dB(A).

Control Description. It was decided to use vibration-isolating mounts to reduce floor-radiated noise. Because of the repeated shock sit-

---

*From M. J. Crocker, and J. F. Hamilton, 1971. "Vibration Isolation for Machine Noise Reduction." *Sound and Vibration* **5**(11), 30.

uation, selection of the isolator followed these rules:

1. The natural period of isolator plus machine should be much greater than the shock pulse duration (10 msec).
2. The natural period of isolator plus machine should be less than the time between pulses (200 msec).

Elastomer-type isolators were used, which had a static deflection of 0.1 in. under machine load. This corresponds to a natural period of 100 msec, thus fulfilling the design conditions.

Results. Figure 9.20 shows octave-band spectra at the operator's position after all machines had been vibration-isolated. The sound levels have been reduced about 8.5 dB to 95 dB(A), a level still in excess of permitted levels. Additional noise control is needed.

Comments. To maintain the isolation, maintenance people should be warned not to short-circuit the isolators by any solid connection from machine to floor. This short-circuiting can also occur when dirt and grease are allowed to build up around the pods.

As a reduction to a sound level of 95 dB(A) is not considered satisfactory for full-day operator exposure, additional noise reduction could be obtained by the design of a barrier between the major noise source in the machine and the operator. Depending on the needs for vision through the barrier, plywood, lead-loaded vinyl curtain, or Plexiglas could be used. Such a barrier should yield a reduction of 5–8 dB at the operator position. (For calculated design parameters see Case History 22, and for rule-of-thumb parameters see Case History 7.) This noise reduction should result in lowering of the sound level to 87–90 dB(A).

Where there is a series of machines, additional reduction of several decibels could be obtained by added room absorption, either in the form of spray-on acoustic absorbent on ceilings and walls or in the form of hanging absorbent baffles from the ceilings.

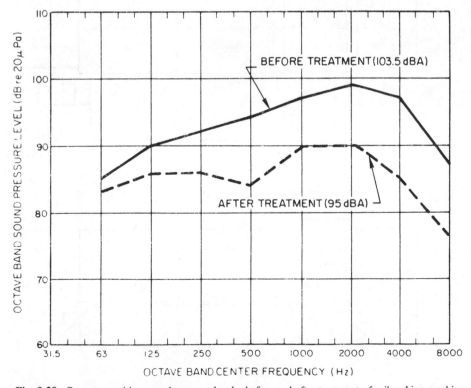

Fig. 9.20. Operator position sound-pressure levels, before and after treatment of nail-making machine.

Case History 3. Pneumatic Scrap Handling (OSHA Noise Problem)

Problem Description.   In the folding carton industry, printed sheets are cut on Bobst and similar cutting presses equipped with automatic strippers for removal of waste material between cartons. When the press is operated and is in good mechanical adjustment, there is no serious noise problem. Often, however, noise from the scrap disposal system results in sound levels above 90 dB(A) on the pressman platform.

This popular scrap disposal system (see Fig. 9.21) uses a horizontal air vane conveyor to move the scrap from under the stripping station to the intake of a centrifugal fan that pushes the scrap to a baler or to bins at a baler in a remote location.

The noise problem arises from the pieces of paper scrap striking the sides of the intake conveyor under the press stripper, the sides of the intake hood to the fan, and the fan and outlet ducts. All these contributed noise that resulted in sound levels of over 90 dB(A) at the pressman station. Depending on amount of scrap and size of pieces, the sound level reached 95 dB(A) on each stroke of the press, normally making the noise almost continuous.

Problem Analysis.   In this type of problem, it was not considered necessary to make octave-band measurements when simple, direct sound-level readings would tell the story of the obvious problem before and the results after damping. Octave-band sound-pressure levels aid in determination of the noise source, but in

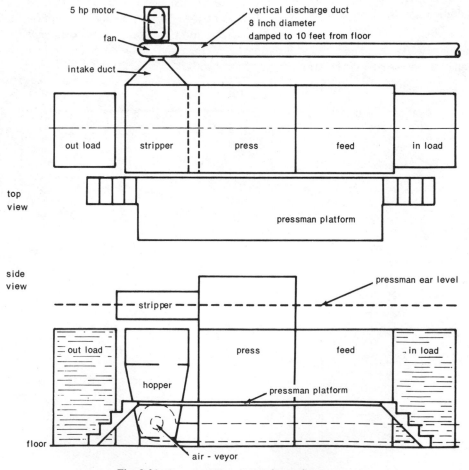

Fig. 9.21. Scrap handling system for cutting press.

this case the noise source was known and before-and-after levels could be expressed in dB(A).

Control Description. The sheet metal of the stripper intake, fan intake from horizontal air vane, the fan, and outlet ducts were all damped (and transmission loss improved) by gluing a layer of lead sheeting to the outside surfaces, using a resin glue recommended by the supplier of the sheeting. Sheeting used was $\frac{1}{32}$ in. thick, 2 lb/ft². Other sheet damping materials that are on the market could have been used as effectively, as discussed below.

Results. The damping of the sheet metal reduced the sound level at the pressman platform to 88–90 dBA.

The concept of using sheet lead to damp the sheet metal ducts came from supplier literature citing successful sheet metal damping on ducts and fans and other surfaces. (Cost is about $0.90/ft².) For less damping, a 1 lb/ft² material may be used at $0.46/ft². For minimum damping, stiff roofing felt may do. For even greater damping, there are many products on the market in sheet form and tape form. Suppliers can be consulted on specific problems; prices range from $1.50 to $3.50/ft².

For very high vibration and sound levels, a further duct treatment step would be lagging, which is a spring-absorber-mass combination of 1 to 3 in. of resilient acoustic absorbing material (glass fiber or polyurethane) with a heavy cover sound barrier of sheet lead or lead-loaded vinyl sheeting over the entire surface.

Case History 4. Electric-Powered Towing Machine (OSHA Noise Problem)

Robert C. Niles
Uniroyal, Inc.
Oxford Management and Research Center
Middlebury, Connecticut 06749
(203) 573-2000

Problem Description. At the Uniroyal tire manufacturing facility in Opelika, Alabama, noise to which a "Green Tire Truck Tugger" operator is exposed was measured and found excessive under OSHA regulations. The employee operates a "standup" electric-powered towing machine that moves green tires from the tire building machines to the spray machine and returns with empty trucks from the curing process.

Problem Analysis. The problem is the noise caused by hauling the empty trucks. The "truck" that carries the tires consists of a metal frame with hollow metal elliptical prongs that hold the green tires. When the truck is empty, the prongs vibrate and act like a sounding drum, emitting a loud noise. The loudest noises occurred on concrete floors because of unevenness caused by globs of rubber on the floor. Metal plate aisles were quieter.

Noise at the operator's ear measured 100 dB(A) when he was towing the empty trucks—a sound level that exceeds the OSHA allowable limit. In addition to the operator exposure, adjacent employees are subjected without warning to a loud intermittent noise, which is motivationally depressant.

Control Description. The prongs were filled with a rigid foam, developed by Rubicon at Naugatuck through the cooperation of Mr. Thomas Haggerty. It is an MDI, polyurethane foam, formula RIA Nos. 553A and 553B. The product is shipped as liquid foam in two parts, which are combined on the job. Cost is estimated at about $1 per kilogram, depending upon quantity, and comes to about $10 per truck.

Results. The original and after-treatment noise data were taken by riding the tugger next to the operator. Both sets of data were taken in a warehouse in order to ensure low ambient noise conditions. The same tugger, same route, and ambient sound levels were used for both the "before" and "after" tests. The noise-abatement program of filling the prongs with a rigid foam resulted in a 10-dB reduction, adequate to alleviate the noise problem as defined by OSHA.

Case History 5. Blanking Press (OSHA Noise Problem)

Problem Description. In forming operations, large blanking presses are used. The ram, which is like a connecting rod in a reciprocating engine, is hollow. The forming die runs in grooves on the side of the press, like a piston in the cylinder of a reciprocating engine, and completely closes off the end of the hollow ram. There are slots in the ram that are used normally when the press is used in blanking operations to extricate the work from the die, similar to removal of a cookie from a cookie cutter. These slots are in the side of the ram (see Fig. 9.22). When the press is being used in the forming mode, these slots are not required, and when the die "snaps through," it cuts off the work. This gives rise to high sound levels.

Problem Analysis. In a vibration-isolated forming press, operator position sound levels were L_A = 94 dB(A), L_C = 100 dB(C) in the "slow" reading position. An octave-band measurement disclosed that sound pressure levels in the 250-, 500-, 1000-, and 2000-Hz bands were much higher than in other bands. This ringing noise, which had a maximum near 2 kHz, was easily discernible by ear. By careful listening, it was determined that the source was radiation from the slots in the ram.

The technical conclusion was that the hollow ram interior, with the slots, was essentially behaving as a shock-excited Helmholtz resonator. A Helmholtz resonator is a closed volume of air connected by a tube to the outside air; it resonates at various frequencies (as when air is blown across a glass jug opening).

The one approach that would obviously work would be to fill the cavity in the ram with rubberlike material. Another approach would be simply to plug the slots, thus keeping the noise inside the ram. The second approach was chosen because it was easy to try, inexpensive to test, and allowed the machine to be reconverted easily to a blanking operation.

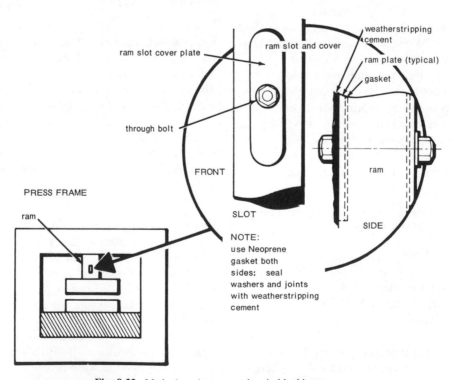

Fig. 9.22. Method used to cover slots in blanking press ram.

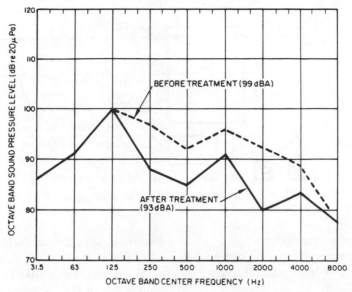

Fig. 9.23. Quasi-peak readings of blanking press after ram ringing was contained.

Control Description. The ram slots were each covered with a plywood plate sealed with a Neoprene gasket, as shown in Fig. 9.22. Weatherstripping (nonhardening sealant) was used to prevent small leaks. These control measures were easily installed.

Results. The first attempt was satisfactory and achieved a 6-dB reduction of quasi-peak sound level from 99 to 93 dBA. See Fig. 9.23. Applying this to the observed slow A-reading of 95 dB(A) yields the observed 88 dB(A).

This case history demonstrates both the simplicity (the solution) and the complexity (the resonator) of noise control. It also demonstrates a more subtle feature: Simple solutions are worth trying if there is a good physical reason for them.

Comments. The obvious pitfall here would be to apply this solution to a press that had not first been vibration-isolated. If the press were on other than piers isolated from the building, or had sheet metal guards, one would probably not have been able to measure any improvement. Filling the ram cavity would have been another pitfall. It would have accomplished the noise reduction, but would have prevented easy reconversion of the press to blanking operation.

Case History 6. Boxboard Sheeter (OSHA Noise Problem)

Problem Description. The sheeter, starting from large rolls of boxboard about 6 ft in diameter, cuts the web to length with a rotary knife that can be adjusted to rotary speed, and therefore sheet length, by means of a variable-speed drive (Reeves Drive). The cut sheets are delivered to a pallet. The speed is about 700 ft/min.

Problem Analysis. At the operator control station near the sheeter (see Figure 9.24), the sound level was found to be 93 dB(A). Close-in probe readings at the variable-speed drive were high, indicating that the drive is a major noise source. Readings were as follows:

96 dB(A) close to front drive guard, in aisle, 98 dB(A) close to front drive guard, in aisle 3, 105–107 dB(A) close to front drive vent openings.

The drive box enclosure was a steel shell 6 ft high, 3.5 ft wide, and 3.5 ft deep, having two vent openings in the side for natural air cooling (see Fig. 9.25).

Other operator locations that were far from the drive were checked:

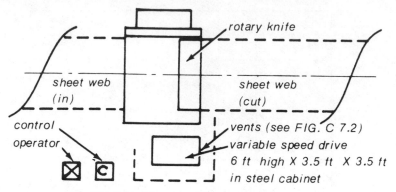

Fig. 9.24. Floorplan of sheeter for boxboard.

90 dB(A): operator at delivery;
88 dB(A): operator at rollstand in feed (see Fig. 9.26 for general layout).

From the close-in readings, the drive was determined to be the major noise source and not the roll unwind stands, rotary cutter, or delivery belts to finished pallet of boxboard.

Control Description. To reduce the drive noise within the steel box enclosure, it was decided to line the interior walls with an acoustic absorbing polyurethane foam with a layer of 0.017-in.-thick sheet lead to provide damping of the steel surface panels. To reduce the noise coming out of the air vents, an acoustic trap was designed to absorb the noise at the vents but allow full normal air circulation. This acoustic trap is shown in Fig. 9.25.

Results. The sound level at the operator control panel near the drive unit was found to be 89 dB(A), reduced from 93 dB(A). In ad-

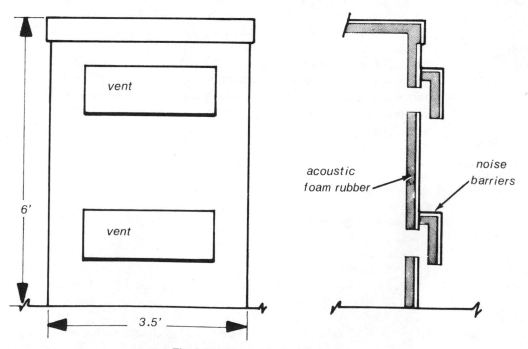

Fig. 9.25. Sheeter drive box enclosure.

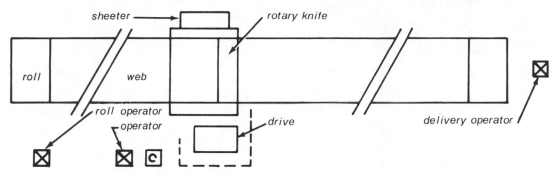

Fig. 9.26. Layout of sheeter and operators.

dition, some reduction was obtained in other operator positions:

86–87 dB(A) from 90 dB(A), operator at delivery;
86 dB(A) from 88 dB(A), operator at roll stand.

Sound levels close in to the vents were reduced to 94 dB(A) from 105 dB(A); this is not an operator position.

Sound-absorbing polyurethane foam with a lead septum designed for combined damping and absorption is available from various suppliers at less than \$4/ft; material cost was about \$400, and in-house labor to glue in place and fabricate a holder for the sound trap was about another \$400; total cost was about \$800.

Comments. Without close-in reading to locate the drive unit as the major noise source, the conclusion could have been that the entire sheeter, including the drive unit, must be installed in an acoustic enclosure, and a great deal more money would have been spent for the solution.

This kind of noise reduction is typically not as satisfactory as one would like. The major problem that can arise is the existence of other direct sound paths from the knives to the operator.

Another pitfall for sheeters is the knife design. Some of the older models have straight knives instead of an angular striking or cutting edge. Straight knife sheeters will probably require an acoustic-absorbent-lined metal or wood hood over the knife assembly and perhaps under the knife assembly.

Case History 7. Folding Carton Packing Stations, Air Hammer Noise (OSHA Noise Problem)

Problem Description. In the manufacture of folding cartons, the individual cartons are cut, and the cut sheets are stacked by the cutting press on a pallet. To deliver the multiple sheets from the press, the cartons are held together with a nick or uncut portion. When stacked, the individual cartons are separated by stripping with an air-driven chisel which breaks the nicks and frees an entire stack. When no additional operations are needed, these stacks are packed in cases for shipment.

Air hammers/chisels produce noise that has not yet been eliminated by equipment manufacturers. Currently available air hammer mufflers do not reduce the noise to an acceptable level. The air hammer operator therefore must wear ear protection. The problem in this case was to protect other workers (packers) from the air hammer noise. A typical production air hammer stripping and packing set-up is shown in Fig. 9.27.

The production sequence for this operation is for the stripper to air hammer a stack of cartons (precounted by the cutting press) and place them on the conveyor at Point C. The packer, at the end of conveyor E, prepares the case, packs the stacks of cartons, seals, labels, and stacks the finished pack on a delivery skid. Two packers are required to handle the output from one stripper. The stripper is actually using the air hammer about 50% of his time, with the balance of the time used in stacking or preparing the load. Thus, he can get some relief from

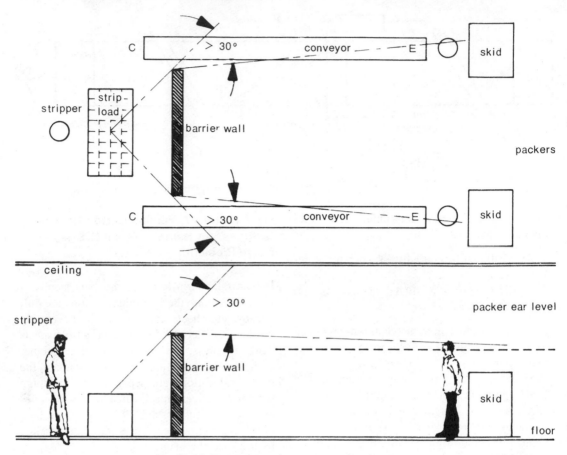

Fig. 9.27. Air hammer stripper and packer line.

continuous use of his ear muffs by hanging them around his neck while not actually using the hammer. It is easier to promote the use of ear muffs when needed if the operator can get some relief when muffs are not needed.

Problem Analysis.   As frequency analysis is not critical in this problem, no octave-band readings were made; all data were based on A- and C-scale readings from an acceptable Type 2 sound-level meter.

Control Description.   It was decided to protect the packers from the air hammer stripping noise by using a barrier wall. A convenient rule of thumb is that useful protection is afforded by the barrier wall beyond 30 degrees into the acoustical shadow. Note that in Fig. 9.27, the packers behind a wall 10 ft long and 6 ft high are within this protected zone in both top view and side view of the operation.

The barrier will need to be no better acoustically than the attenuation afforded around the sides and top of the wall. Therefore, the wall was fabricated with a 2 × 4-in. frame faced on both sides by $\frac{1}{4}$-in. plywood for a simple sturdy barrier wall.

If there had been any reason to reduce noise reflections from the noise source side, this side could have been faced with sound-absorbing acoustic materials.

The rule of thumb of aiming for the packer to be well within the 30-degree line from the acoustic shadow line was used in this case. Other means of estimating the attenuation of barrier walls are covered by Beranek* in *Noise and Vibration Control*, p. 178, and illustrated in Fig. 9.28. The attenuation calculated for this barrier wall ranges from 10 to 15 dB, depend-

---

*Beranek, L. L. 1971. *Noise and Vibration Control*. New York: McGraw-Hill.

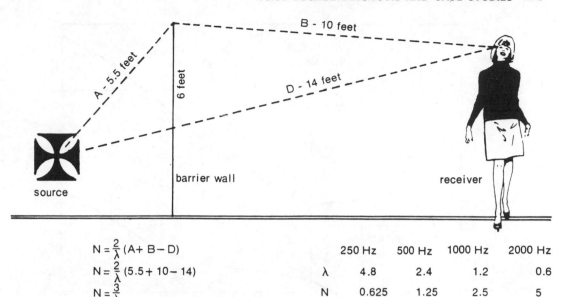

$$N = \frac{2}{\lambda}(A + B - D)$$

$$N = \frac{2}{\lambda}(5.5 + 10 - 14)$$

$$N = \frac{3}{\lambda}$$

	250 Hz	500 Hz	1000 Hz	2000 Hz
λ	4.8	2.4	1.2	0.6
N	0.625	1.25	2.5	5
dB	10	11	13	15

Fig. 9.28. Barrier wall theory.

ing on the wavelength. This agreed with the measured attenuation of 7–12 dB and the noise reduction from the 92- to 97-dBA range to about the 85-dB(A) average measured at the packer's ear level.

Comments.   In this installation, there were, fortunately, no low ceilings, which would have established a serious sound reflection problem and defeated the barrier wall. Barrier walls will not give good results in a highly reverberant, low-ceilinged room. If there had been a low ceiling, useful noise reduction would still have been possible by adding sound-absorbing material at the reflecting portion on the ceiling (about 12 ft over the barrier wall and the noise source). The amount of attenuation gained is easily estimated by using the ratio of absorption of new material to that of the existing ceiling. Ceiling reflection is a major pitfall of the use of barrier walls indoors. The design of the wall alone is based on freefield conditions.

Case History 8. Straight-and-Cut Machines (OSHA Noise Problem)

Problem Description.   The straight-and-cut machine straightens heavy-gauge wire in an in-feed to a cutoff unit set to cut repeat lengths,

resulting in sound levels of 92 dB(A) at the operator position. The client in this case sought to reduce the sound level to a maximum of 85 dB(A) at the operator position.

Problem Analysis.   Figures 9.29 and 9.30 are close-in octave-band analyses of the diagnostic measurements made in front of the clutch mechanism. In Fig. 9.29, curve A shows peak cutting levels, and curve B is the slow response of the same cutting sound pressure levels (wide separation indicates impact noise). Curve C is the idling, noncutting machine sound level. The differences indicate dominance of the total spectrum by the cutting noise. In Fig. 9.30, curves D and E exceed curves A and B, indicating some directionality of the cutting noise.

Figures 9.31 and 9.32 are octave-band analyses made at the operator position. Most of the operator time is represented by Fig. 9.31, with the cutting cycle sound level at 92 dB(A) (idling cycle at only 83 dB(A), indicating that the dominant noise source of the clutch cutter mechanism is the same form as in the close-in diagnostic measurements. Comparison of the measured sound-pressure levels with the 90-dB(A) criterion indicates the required attenuation is between 5 and 11 dB in the 1000- to 8000-Hz octave bands.

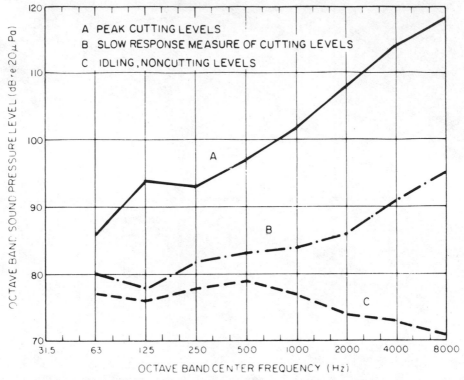

Fig. 9.29. Straight-and-cut machines: close-in measurement near west side of clutch cutter mechanism (1.2 m above floor, 0.5 m from cutter).

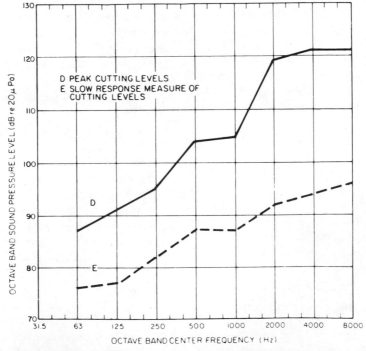

Fig. 9.30. Straight-and-cut machine: close-in measurement near east side of clutch cutter mechanism (1.2 m above floor, 0.5 m from cutter).

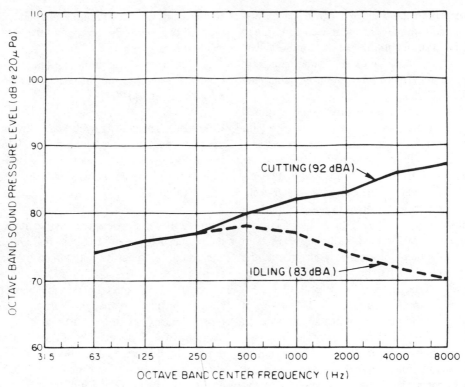

Fig. 9.31. Straight-and-cut machine: operator's nearfield exposure.

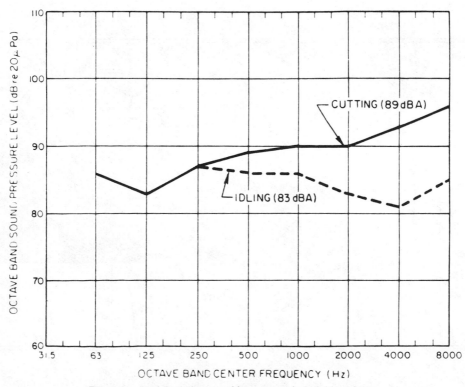

Fig. 9.32. Straight-and-cut machine: operator's farfield exposure.

Control Description. On the basis of discussions with management, it was determined that noise control should take the form of a barrier wall that would block the sound path from the cutting assembly to the operator, rather than machine redesign.

Barrier materials for obtaining the required attenuation were $\frac{1}{4}$-in. plywood, with $\frac{1}{8}$- to $\frac{1}{4}$-in. Plexiglas for viewing ports where necessary. The barrier wall was extended 26 in. past the extremities of the area encompassed by the cutter and was close to the cutter, about 6–8 in. away. The barrier was hung in place, supported by chains from overhead. In addition, an absorbent layer was hooked to the barrier on both sides. To prevent clogging of absorbent, the 1-in. polyurethane foam absorbent was supplied with Mylar facing. See Fig. 9.33.

Normally, the noise absorbent for barriers is used only on the machine noise source side. In this case, however, noise absorbent was used on the operator side of the barrier as well, to reduce sound field build-up in the space between barriers. With the barrier close to the cutter, the operator would be within the safe sound shadow area—the area beyond a line at least 30 degrees from the edge of the acoustical shadow line.

As the barrier was built in-plant, no actual costs are available, but material costs are estimated at about $100.

Results. The cutting cycle sound levels at the operator location were reduced from 92 dB(A) to 85 dB(A), a 7-dB reduction in sound level. Idle cycle sound level was reduced from 83 dB(A) to 76 dB(A).

Comments. Barriers are easy to remove by the operator for many reasons, real and imaginary, and use must be maintained by supervision.

Location of an effective portable barrier must be standardized so that the barrier is not bypassed. Barriers can be bypassed by noise reflections from a low ceiling. If this problem had existed in this case, a section of the ceiling above and about 4 ft on each side of the barrier could have been treated with absorbing material.

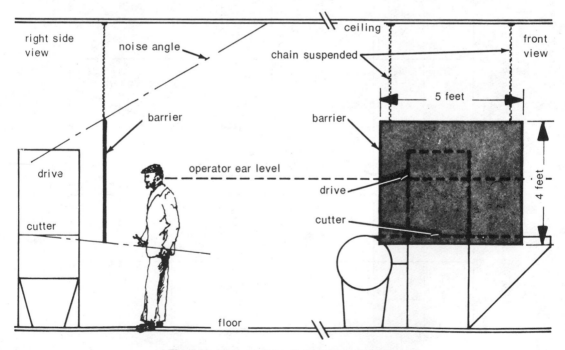

Fig. 9.33. Barrier wall for straight-and-cut machine.

Case History 9. Metal Cutoff Saw*
(OSHA Noise Problem)

Problem Description.   A common problem in industry is that of protecting workers from noise produced by machines that the worker must guide or manipulate directly. An example is a cutoff saw used on metal shapes. Noise comes from two main vibrating sources: the saw blade itself and the workpiece. The saw itself is actuated downward and into the work by a lever attached to the hinged and counter-balanced (or spring-loaded) saw and motor.

The worker must visually monitor the cutting operation. In addition, the vibration and opposing force transmitted to him through the lever arm furnish useful cues on the progress of the cutting operation. The problem is to reduce the noise he receives, without undue interference with work flow, with visibility, and with the use of the lever arm.

*Handley, J. M. 1973. *Noise—The Third Pollution.* IAC Bulletin 6.0011.0.

Control Description.   The solution was an enclosure covering the whole saw. Workpieces pass transversely through slots in the enclosure. Flaps of lead-loaded vinyl close off the opening and reduce to a small amount the unavoidable leakage area when a workpiece is present. The front, above saw bed height, is closed by two doors whose surface is mostly $\frac{1}{4}$-in. clear plastic (polymethylmethacrylate). This plastic provides very good vision. The doors close with a gap the width of the control lever. Each door has a flap of lead-loaded vinyl about 3 in. wide to close the gap. The lever pushes aside the flaps only where it protrudes. Thus, the leakage toward the worker is greatly reduced.

Results.   Figure 9.34 shows the sound pressure levels at the worker position before and after the enclosure was installed. The decrease in sound level is 13 dB. The standard panels used in the enclosure are very much better than indicated by the reduction measure, illustrating

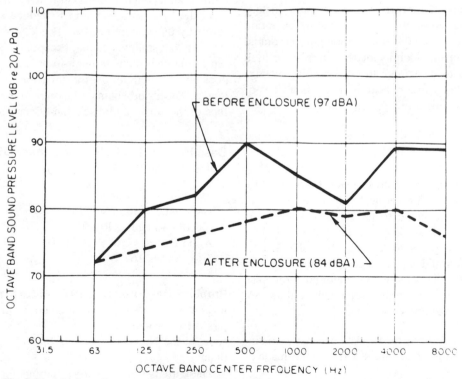

Fig. 9.34. Metal cutoff saw: operator position sound-pressure levels before and after enclosure of saw.

again the importance of leaks in determining the performance of enclosures.

Comments. Several features of the design could be improved. The ears of the workers are very close to the leak at the door flaps. It should be relatively simple to offset the saw feed lever to the right (for the right-handed worker). This change has several advantages: (1) it places his right hand in a more comfortable position, (2) with the door gap and flaps moved to the right, his vision is greatly improved, and (3) the noise leak is moved farther from his ears. A non-acoustical improvement would be to have the doors slide open, rather than open out, which can be a safety hazard.

Case History 10. Punch Press (OSHA Noise Problem)

Keith Walker
U.S. Gypsum Company
P.O. Box 460
Round Lake, Illinois 60073

Problem Description. This case history concerns noise emissions caused by operation of a high-speed 290-ton stamping press. Sound levels in the vicinity of the press were high enough to contribute to OSHA noise overexposures of workers near the press as well as of the press operator.

Problem Analysis. Sound levels were found to be in the 95-dB(A) to 101-dB(A) continuous slow meter response, at distances of 15–25 ft from the operating press when it was the only noise source operating.* The U.S. Gypsum Company decided to install their Acoustisorber™ Industrial Sound Control Panel System around the press, to determine how effective the system would be in reducing sound levels in the shielded positions. (Operator position noise exposures were studied separately and are not discussed in this case history.)

Control Description. The panel system employed consists of 2-ft × 8-ft modules made of

---

*Distances chosen to represent possible nearby worker locations.

hardboard on one face, expanded and flattened metal on the other side, with a mineral fiber absorbent sandwiched in between. The absorbent is fully wrapped with a thin, heat-shrunk plastic film. Individual panels are joined together by light steel framing to form enclosure walls. The two long walls in this example were suspended on an overhead roller track for access to the press. The installation is open-topped and about 24 ft × 32 ft in size. Walls are 16 ft high, except at one short end where the height was dropped to 8 ft to allow for overhead crane clearance. Material feed and discharge are through openings cut into the short sides of the walls. Material costs were approximately $1600.

Results. Sound levels at the original measurement locations were reduced by 7–14 dB to a maximum of 88 dB(A) at those locations. (See Fig. 9.35.) Enclosure systems need not always be elaborate when moderate amounts of noise reduction are needed, and relatively inexpensive materials can be used. The panels provide more than enough transmission loss, mainly from the hardboard backing, to reduce sound levels by the amount needed. The key is making sure that spillover sound, escaping over the top of the enclosure, through joint leaks, etc., does not short-circuit the transmission loss potential. The absorbent material on the inner surface of the walls minimizes that effect here.

Case History 11. Filling Machines (OSHA Noise Problem)

Dr. Walter W. Carey
The Nestlé Enterprises, Inc.
100 Bloomingdale Road
White Plains, New York 10605
(914) 682-6716

Problem Description. Two Nalbach filling machines used to fill freeze-dried coffee in glass jars were located in a 65 ft × 23 ft × 10 ft room at the Nestlé Company's Sunbury, Ohio plant.

There are two fixed worker stations for each machine. An operator station is directly in front of the filling machine, and an inspection station

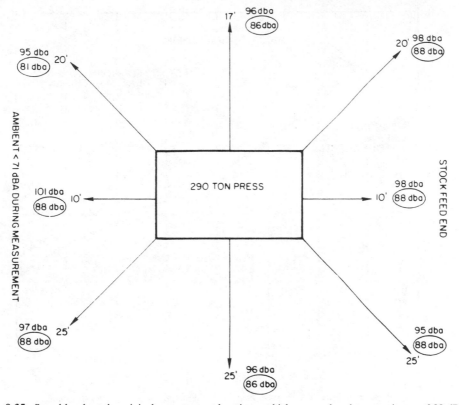

Fig. 9.35. Sound levels at the original measurement locations, which were reduced to a maximum of 88 dB(A).

is located downstream of the machine discharge conveyor. A roving worker also works in this area. The filler operator maintains a steady flow of bottles into the filling machine and checks and adjusts the filled weight of spilled product as required. The inspector's function is to ensure that each jar is properly filled and that lids are securely fastened to the jars. The roving worker fills the lid bins with lids and maintains cleanliness in the area.

Problem Analysis.   The Nestlé Company retained Bolt, Beranek and Newman Inc. as consultants to evaluate the noise environment and recommend controls to ensure that all noise exposures in the area met OSHA limits. The highest worker noise exposure occurred at the filling machine operator location, where the sound level varied between 94 and 96 dB(A). The sound level was at or above 92 dB(A) elsewhere throughout the space, because of the highly reverberant nature of the room (typical for food processing facilities where easy-to-

clean, hard surfaces are required by FDA regulations). The filling machines were most responsible for the above-90-dB(A) sound levels, as the sound level dropped to 74 dB(A) when both filling machines were stopped.

To determine what part of the machines radiated noise, measurements were made close in to suspected important noise sources. Observation of the operation indicated that likely candidates were the constant jar-to-jar contact at the infeed to the filling machine, the vibrations developed by the feed mechanism in the filling machine, and gear noise. Measurements were taken near each of these sources.

The data obtained appeared to confirm the significance of the suspected source. For example, the octave-band spectrum measured 6 in. from the filling machine inlet indicated that the sounds generated in that area were largely responsible for the octave-band sound-pressure levels measured at the operator's ear, at least for those octave bands that penetrated the 90-dB(A) criterion curve appropriate for this situ-

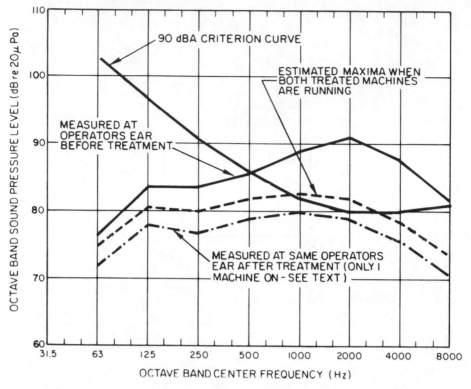

Fig. 9.36. Sound-pressure levels in filling machine room before and after treatment.

ation. Fig. 9.36 summarizes these findings. Note the similarity in spectral shape between the upper two curves. Other close-in measurements indicated that openings in the bottom part of the filler structure were important contributors to the overall noise environment relative to the 90-dB(A) criterion, but were of lesser significance than noise sources on the filler table itself.

The analysis suggested that the most significant noise was generated by jar-to-jar and jar-to-machine impacts. Clearly, a possible remedial solution would be to minimize or eliminate the force of these impacts. However, an equally acceptable acoustical treatment would be to contain the sounds. In view of the problems inherent in redesigning the machine feed mechanism to yield softer impacts, strong consideration was given to noise containment. In fact, the solution attempted was a cover for the infeed and discharge parts of the machine, combined with a closure for the bottom parts of the machine.

Control Description. Because of the intricate design of these machines, the selected noise control was not attempted until after a careful analysis had been made of the possibility of rotating filler-associated personnel with workers in other departments who were exposed to equivalent sound levels lower than 90 dB(A). However, such rotation was discarded as totally infeasible.

The major problem associated with this project was the amount of design work needed. Mr. John Meyer, the design engineer, spent approximately 3 weeks on site before sufficient details were gathered and design concepts fully developed. The design phase was also extended because of the constraints of sanitation, maintenance, and operator access.

Figure 9.37 is an example of the conceptual design drawings that were developed in connection with this project. The treatments were fabricated by the E. A. Kaestner Company of Baltimore, Maryland.

Excluding engineering design costs but in-

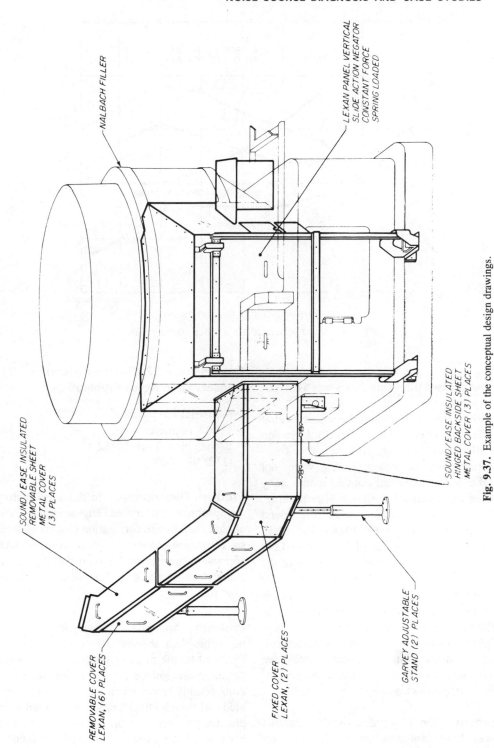

NALBACH FILLER

LEXAN PANEL VERTICAL
SLIDE ACTION NEGATOR
CONSTANT FORCE
SPRING LOADED

SOUND/EASE INSULATED
HINGED BACKSIDE SHEET
METAL COVER (3) PLACES

SOUND/EASE INSULATED
REMOVABLE SHEET
METAL COVER
(3) PLACES

REMOVABLE COVER
LEXAN, (6) PLACES

FIXED COVER
LEXAN, (2) PLACES

GARVEY ADJUSTABLE
STAND (2) PLACES

Fig. 9.37. Example of the conceptual design drawings.

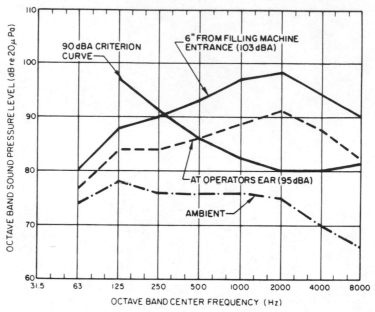

Fig. 9.38. Sound-pressure levels in filling machine room.

cluding material and fabrication cost, the treatment for the two filling machines was $16,300.

Results.   Before treatment, the sound level at the filling machines was 94–96 dB(A), when both fillers were running. Although after-treatment octave-band measurements were not available for the identical running modes, they exist for the condition with one filler running. For the one-filler-running mode, the sound level has decreased to 85 dB(A). Fig. 9.38 shows octave-band spectra of the measured before-and-after situations and an estimate of the maximum expected sound-pressure levels for the two-filler-running mode. All operators are now exposed to sound levels less than the 8-hr 90-dB(A) level allowed by OSHA. Operators and plant management indicate complete satisfaction with the controls, as sound levels have been reduced with no perceptible effect on productivity or product quality.

Comment.   Dr. Carey discusses the conflict between FDA sanitation and OSHA noise reduction requirements in the July 1978 issue of *Sound and Vibration* in an article entitled "The Ramifications of Noise Control in Food Plants."

Case History 12. Gearbox (Hearing Conservation Noise Problem)

Industrial Acoustics Co.
1160 Commerce Ave.
Bronx, New York 10462
(212) 931-8000

Problem Description.   In this case history, the problem concerned engine room noise aboard the Matson Navigation Company's vessel *Hawaiian Queen*. At full power, the 9000-shp steam turbine used aboard the ship causes sound levels exceeding 120 dB(A) in the engine room.

Problem Analysis.   Investigation of the noise problem showed the cause of the high levels to be the primary stage of a nested-type double-reduction gear unit. Sound levels are considerably lower when this unit is not operated. Although consideration was given to replacing gearing, that alternative was rejected because of the expense involved, in favor of enclosing the reduction gear casing. An enclosure design was sought to bring the engine room noise environment down to ambient levels measured when the gear unit was inoperative.

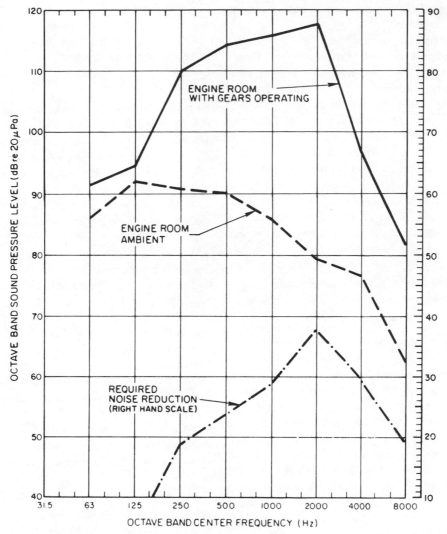

Fig. 9.39. Engine-room sound-pressure levels.

The required noise reduction is indicated in Fig. 9.39, which also compares sound pressure levels measured in the engine room with and without the gear unit in operation. The required noise reduction is the algebraic difference between the two curves.

Control Description. IAC Modular™ acoustic panels were used as the basis for the enclosure because of the high transmission loss properties. A notable feature of this enclosure is the use of a split commercial silencer at the propeller shaft penetration into the enclosure, to attenuate sounds that would otherwise es-

cape around the shaft. Penetrations for thermocouples, lubricating oil lines, and other pipes were cut in the enclosure and provided with seals. Materials for a similar enclosure would cost about $9000 today.

Results. The actual effectiveness of the enclosure is not measurable because after the enclosure was put in place, the engine room noise environment decreased to the ambient levels. However, it is clear that the enclosure met design objectives.

The major problem with the enclosure was rearrangement of piping necessitated by close tolerances between the gearbox casing and the enclosure walls.

The operating temperature of the gearbox did not change as a result of its enclosure.

Note that in most cases of enclosure construction, achieved noise reduction obtained probably will not reach the amount indicated by the given laboratory-determined transmission loss of the enclosure walls. The reason is that when an enclosure is made, noise is confined, resulting in a buildup of sound levels inside the enclosure. This effect is predictable when the principles of room acoustics are used. In this case, however, the use of nonreflective panels for the enclosure walls minimized the effect.

Case History 13. Pneumatic Scrap Handling Ducts (OSHA Noise Problem; Speech Interference Problem)

Problem Description. In a corrugated box factory, slit side-trim is removed from the conveyor by air. Trim blower fans with extra heavy blades cut the trim while conveying it via ducts to bins and balers. The 12-in. ducts are suspended from the ceiling about 10 ft from the floor, crossing a 40-ft-long work room en route to the bins and baler room. The trim is carried along in the ducts by the air, which normally moves about 6000 ft/min. The trim often strikes the duct walls (mainly at bends), causing noise.

In the case described here, ear-level sound level was 93 dB(A). Noise reduction was desired to improve worker communication for operations under ducts and to meet the requirements for OSHA compliance.

Problem Analysis. Octave-bend data were collected because this case required not only a reduction of a few decibels to comply with minimum regulations, but noise reduction for safety reasons, to control speech interference. Octave-band data provide a truer measure of speech interference than a single-number dB(A) reading. For the minimum compliance data, the dB(A) reading would have been adequate.

Control Description. The solution chosen was to wrap the problem duct locations with 2

in. of mineral-wool building insulation to furnish a resilient and absorbing layer. Over this insulation were placed two impervious layers of heavy tar paper, spirally wrapped with 50% overlap.

Result. Noise was reduced considerably in the problem area: The sound level changed from 93 dB(A) to 72 dB(A), a reduction of 21 dB. To the ear, the noise could hardly be heard above other noise. The octave-band comparison is shown in Fig. 9.40. Although the standard materials used were very economical, special acoustic absorber pipe coverings with lead-filled vinyl sheeting could also have been used and may have given even more attenuation. This was not needed here.

There are no detailed costs for this case history. However, since the materials are inexpensive, the major cost must have been labor. The job was probably done for less than $200.

If less attenuation had been required for OSHA compliance only, the sheet metal ducts could have been damped and transmission loss improved by gluing $\frac{1}{32}$-in. sheet lead to the duct outer surface. Comparison with the experience at other installations indicated that a 5-dB attenuation would probably have been attained. (See Case History 3 for other methods.)

Comments. Use the most economical methods to attain the attenuation required. The building insulation plus roofing paper used here is a very economical solution. Note that a large overlap was used; lack of overlap on any wrapping will cause leaks and reduce attenuation.

Case History 14. Blood Plasma Centrifuge (OSHA Noise Problem)

Problem Description. This plasma production room has two parallel banks of centrifuges, 15 to a bank, plus refrigeration units. A sketch of one centrifuge is shown in Fig. 9.41. Centrifuge spinning frequency is 13,000 rpm (217 Hz). Though centrifuges appeared to be the major noise source, refrigeration units were also evaluated. The same refrigeration units without centrifuges are used in a separate reconstituting room.

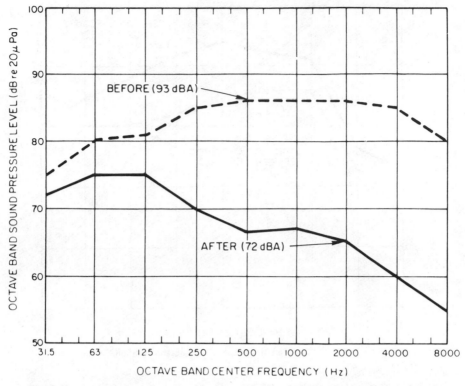

Fig. 9.40. Noise levels in scrap duct for corrugated box industry, before and after covering.

Problem Analysis. Operator sound level was 97 dB(A) with one bank in operation; 100 dB(A) was predicted with both banks operating. Figure 9.42 shows the measured sound-pressure levels at the operator positions in both the centrifuge and refrigeration rooms and also a 90-dB(A) criterion spectrum.

Close-in diagnostic readings were made around the centrifuges at the locations shown in Fig. 9.43.

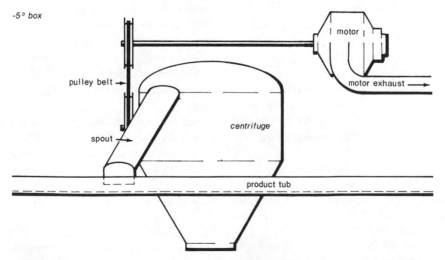

Fig. 9.41. Sketch of one centrifuge (front view).

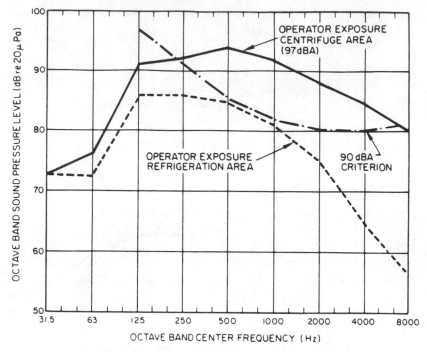

Fig. 9.42. Sound-pressure levels at operator position.

A comparison between the operator position spectrum in the centrifuge room and the 90-dB(A) criterion curve indicates that the 500- to 4000-Hz bands merit the most attention. The fact that the operator position spectrum in the refrigeration room is considerably lower in level in all octave bands suggests that the refrigeration-unit noise is not a significant contributor to the noise exposure in the centrifuge room.

Close-in data show that the maximum sound pressure levels in the 500- to 4000-Hz octave bands occur close to the motor exhaust and near the pulley guard surface. The pulley guard surface was presumed not to be an important noise source; it was reasoned that the sound-pressure levels measured near the pulley resulted from motor exhaust and other sounds being reflected from the highly reflective pulley guard surfaces.

Results of the measurements indicated an unacceptable exposure when the operator was exposed for 4 hr to both centrifuge banks [predicted level at 100 dB(A)]. Under these conditions, the operator's daily dose was 2.0, which thus exceeded the acceptable exposure of 1.0, as specified by OSHA noise regulations.

Control Approaches Considered. There are three general locations for controlling noise: at the source, along the transmission path, or at the receiver. Two could be used in this case. A properly designed and constructed muffler or partial enclosure for the centrifuge motor exhaust would provide the necessary source control, using materials and geometric configuration that have effective attenuation in the octave bands of interest. Such a control measure would provide an expected 3- to 7-dB attenuation.

Noise-control measures can also be used along the path of transmission. The paths of airborne noise transmission were from direct and reverberant fields such as walls, floor, and ceiling, which supply very little sound absorption.

For direct field reduction, barriers with the proper transmission loss, dimensions, and orientation may be used. The reverberant field can be controlled by the addition of absorbent materials. These combined measures would reduce operator exposure 5–15 dB.

Control Description. In this case, after discussion, two of the possible control methods were eliminated. The hard and impervious walls had to remain because of the need for

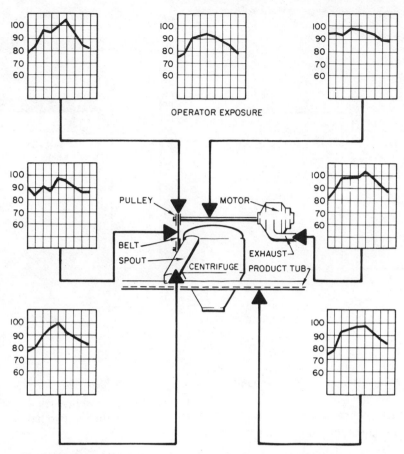

Fig. 9.43. Sketch of centrifuge showing locations for close-in diagnostic readings.

daily sterilization by high-pressure steam and water hosing and hand scrubbing. A porous absorbent surface would provide areas for bacterial growth and would not withstand the rigorous daily cleaning. A barrier to shield the operator from the centrifuge motor noise was eliminated from consideration because it would block the proper flow of refrigerated air. Even a small warming of the plasma would produce an unusable product. It was decided to try the motor exhaust muffler and see how much noise reduction could be achieved.

The muffler was designed with a stainless steel outer skin, lined with acoustical absorbent spaced 2 in. from the inside of the steel shell, with small blocks. The 2-in. air space allows absorption since it reflects from the inner steel surface and back through the absorbent. The muffler has one 90° bend, as shown in the sketch in Fig. 9.44. In-house shop cost was estimated at $300.

Results. Noise at the operator's position was reduced from 97 dB(A) to 92 dB(A), satisfactory for a 4-hr exposure. The motors of both centrifuges must be treated if they operate together.

Comments. An airtight seal at the junction of muffler and motor is very important; resilient caulking compound was used as a sealant. An air leak with an area of no more than 10% of the muffler cross section would produce as much noise as if the muffler were not there at all. An intake muffler would probably also help, but noise from other sources would then become prominent.

Effectiveness of noise control is reduced if the path of vibration transmission is not held to minimum. Accordingly, it was very important to use as few absorbent spacers as possible. In so doing, the steel skin vibration was kept small. The spacers can be made of damping

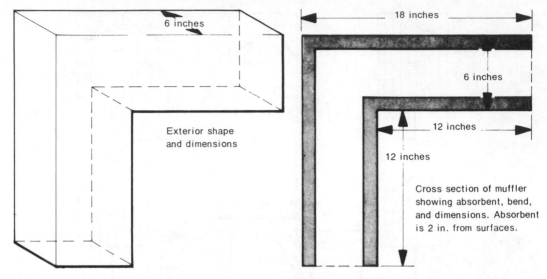

Fig. 9.44. Motor exhaust muffler.

material to reduce resonant vibration of the skin.

Although one main control solution would satisfy the noise control requirements, alternatives should be discussed with company representatives, as they are able to specify important operating, maintenance, and production constraints that may limit the ideal noise-control treatment.

Case History 15. Pneumatic Motors (OSHA Noise Problem)

Problem Description. Air-operated motor hoists are a noise source in many industries that make extensive use of materials-handling systems.

Control Description. As the noise source is the exhaust air, this exhaust can be muffled by using off-the-shelf mufflers selected for the air pressure and delivery of the exhaust.

Results. A typical octave-band analysis, before and after installation of an exhaust muffler, is shown in Fig. 9.45. Note the rising spectrum that is characteristic of freely escaping high-pressure gas. Another case showed the following A-weighted sound levels at the floor for a 1-ton air hoist:

	No Muffler	With Muffler
Up—no load	98	85
Up—600 lb	96	84
Down—no load	102	88
Down—600 lb	100	86

Air exhaust from other tools can be similarly muffled. Newer designs include mufflers, which should be specified at purchase.

Case History 16. Process Steam Boiler Fans (OSHA Noise Problem)

Industrial Acoustics Co.
1160 Commerce Avenue
Bronx, New York 10462
(212) 931-8000

Problem Description. At several of their outdoor process steam boilers in Winston-Salem, N.C., staff members of the R.J. Reynolds Company found that excessive noise was being generated by the fans supplying air to the boilers and the blowers feeding air to the firing units.

Problem Analysis and Control Description. The use of silencers to minimize the fan and blower noise at the inlets to this equipment was considered, as was the effect of the silencer on available pressure head. A careful analysis of the system determined that at the peak operating condition the centrifugal fan could sustain a total additional loss of 0.9 in.

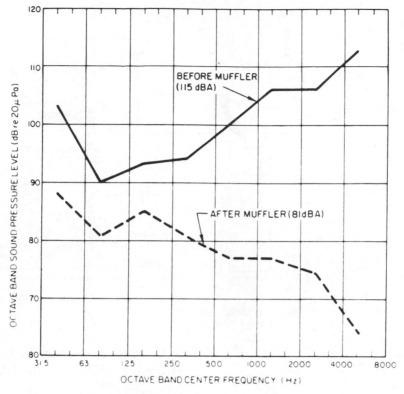

Fig. 9.45. Effect of muffler on air exhaust from hoist.

of water, and the head loss available for the overfire air fan was 0.25 in. of water.

Next, it was necessary to select a silencer configuration that was compatible with the air inlets to the fan and blower as well as the surrounding equipment. An IAC Model 3PL 24-in. × 72-in. rectangular Power-FLOW silencer was chosen for the centrifugal fan. This silencer provides required acoustical performance at a satisfactory pressure loss. The cross section of this particular Power-FLOW silencer is readily mated with the fan inlet duct.

A tubular Power-FLOW silencer, Model 16 PCL 36, was chosen for use with the overfire air fan, as the round shape was easily adapted to match the blower inlet. The acoustical and aerodynamic performance requirements of the silencer were closely examined in selecting the required silencers.

Placement of the silencers is shown in Fig. 9.46. At current prices, the two silencers would cost approximately $3000 (circa 1984).

Results. Silencers for one boiler system were installed and an acoustical test conducted. With the silencers installed, there was no change in the sound levels measured with or without the boiler in operation. As a result of these tests, silencers were installed on three other boiler systems. Fig. 9.47 shows the sound-pressure levels measured 3 ft from the fans before and after the IAC Power-FLOW silencers were installed. Because of extraneous noise sources, it was not possible to measure the full effectiveness of the silencers. The residual sound pressure levels measured during boiler operation are therefore indicative of sounds from both the silenced boiler and the ambient noise sources.

Case History 17. Gas Turbine Generator (Community Noise Problem)

Robert M. Hoover
Eric W. Wood
Bolt Beranek and Newman Inc.
50 Moulton Street
Cambridge, Massachusetts 02138
(617) 491-1850

Gas Turbine. Gas-turbine (also called combustion-turbine) generators are used to supply

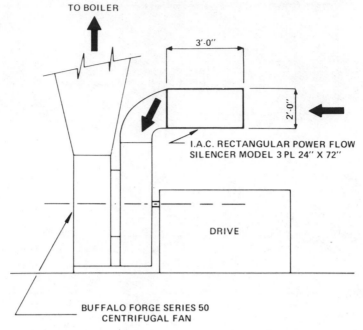

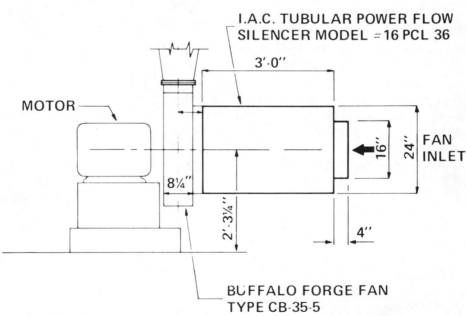

Fig. 9.46. Elevation drawings showing how two fans at an R. J. Reynolds Tobacco Co. plant in Winston-Salem, North Carolina were quieted by IAC tubular and rectangular Power-FLOW silencing units.

emergency reserve capacity and peaking power for electric utility systems. When they are located near residential areas, they can cause community noise complaints unless adequate noise-control treatments are provided. This case history is a discussion of the installation of additional exhaust mufflers at a gas turbine in-

stallation to alleviate community complaints about low-frequency exhaust noise.

Problem Description. Three gas-turbine units capable of generating 60 MW of electricity were installed in a rural/suburban area of New England. Each generating unit had a sin-

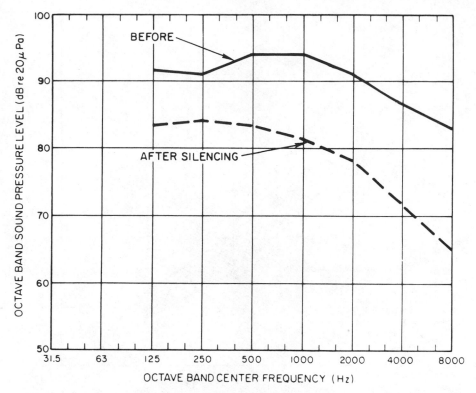

Fig. 9.47. Sound-pressure levels 3 ft from fans.

gle generator driven by four aircraft-type jet engines; each pair of engines shared a common exhaust. Each generating unit was originally installed with two muffled exhaust stacks approximately 4 m in diameter and 15 m tall.

The owner of the generating station received complaints about low-frequency noise from neighbors living about 300 m from the station.

Problem Analysis. The owner's acoustical consultant, Bolt, Beranek and Newman Inc., was asked to investigate the generating station noise problems and to recommend corrective actions. Octave-band sound-pressure level measurements and tape recordings were made at the station, at the nearest residential area, and at various intermediate locations during several station operating conditions. Measurements were also made along the stack wall and at the top of the stack. In addition, ambient measurements were obtained without the station operating.

Measurements obtained outside a neighbor's house are summarized in Fig. 9.48. The lower-frequency station sounds exceeded the ambient by at least 10–20 dB. In addition, the sound in the 31.5-Hz octave band exceeded 75 dB, a level at which complaints are sometimes made about vibration in a house. A suggested noise-control goal for daytime operation is also shown in Fig. 9.48. Reductions of 10–13 dB in the 31.5- and 63-Hz octave bands are suggested to alleviate the community complaint problem.

Similar data obtained inside the nearest residence, 300 m from the station, are shown in Fig. 9.49. These data are plotted with NC curves, which can be used to rate or judge an acoustic environment for various activities.

Narrowband analysis of the data tape-recorded at the station and at the nearest house indicated that the sound energy leading to the complaint was contained primarily in the range of about 18–75 Hz. To reduce this low-frequency noise, a tuned dissipative muffler was designed and added to each of the existing muffled stacks.

Control Description. The dominant radiation path for the low-frequency noise was from the open top of the six exhaust stacks. An ini-

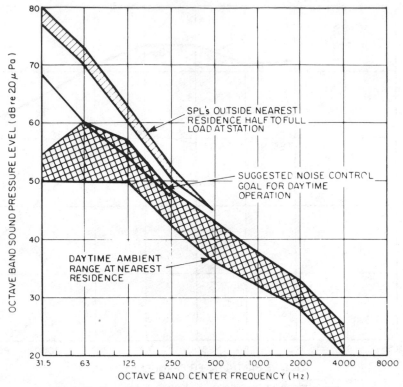

Fig. 9.48. Sound-pressure levels outside nearest residence at 300 m.

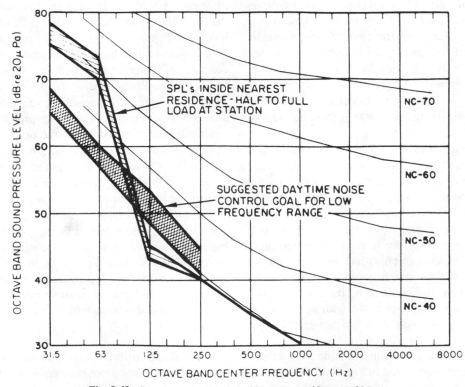

Fig. 9.49. Sound-pressure levels inside nearest residence at 300 m.

tial concept design was prepared of a tuned dissipative muffler section to be inserted in the lower end of the stacks. Acoustic model tests were performed of numerous configurations to optimize the muffler's insertion loss in the frequency range of interest. Aerodynamic model tests were also conducted to ensure that the additional pressure losses through the new muffler section would not be excessive (high pressure losses would reduce the generating capacity of the gas turbine unit.) Other considerations included fabrication cost and time, installation cost, aesthetics, self-noise, structural integrity, and weight.

As a result of these investigations, a prototype exhaust muffler was designed, fabricated, and installed. The muffler, 5 m in diameter and 8 m long, was installed at the lower section of the existing stack. The original stack was reinstalled above the new muffler. Field measurements were conducted to evaluate the muffler's low-frequency insertion loss, and five additional mufflers were subsequently fabricated and installed.

Result. Sound-pressure-level measurements near an exhaust stack with and without the new muffler section indicated an insertion loss of 11 and 12 dB in the 31.5- and 63-Hz octave bands. Outside and inside the nearest residence, the measured insertion loss was 8–9 dB in the 31.5-Hz octave band and 7–11 dB in the 63-Hz octave band. These favorable results indicate the success of this noise control project.

Case History 18. Pneumatic Grinder (OSHA Noise Problem)

Austin E. Morgan
Berkmont Industries, Inc.
Unicast Division, Union Mfg. Division
Box 527
Boyertown, Pennsylvania 19512
(215) 367-2116

Douglas H. Martin
Douglas Martin & Associates, Inc.
119 Heather Drive   R.D. No. 7
Allentown, Pennsylvania 02138
(215) 435-2400

Problem Description. This case history concerns operation of hand-held pneumatic grinders, devices often used throughout industry to clean, smooth, or otherwise improve surface features of metal parts. In this operation, the air tool noise was cited, in part, for contributing to OSHA noise overexposures in a gray iron foundry.

Problem Analysis. Analysis of the problem indicated that the tool was a major continuing noise source. Sound levels measured at the operator's ear ranged between 100 dB(A) and 109 dB(A) when the various tools were held in the free-spinning mode. Close-in measurements indicated most noise originated at the tool exhaust, and hence an exhaust muffler was considered to alleviate the problem. Metal prototypes of the muffler were designed and evaluated. Eventually, rubber mufflers were developed.

Control Description. The muffler is essentially a "rubber band" that fits over the tool exhaust parts. Porous muffler stuffing slows the air stream and dissipates the energy of the moving air before it is exhausted.

Results. Sound levels at the operator's ear are reduced to the 84-dB(A) to 88-dB(A) range for the free-spinning tool, depending on the tool tested. The tool treatment, coupled with other noise controls currently being implemented in the plant, will reduce noise exposures to levels in compliance with OSHA standards.

Comments. In many cases of pneumatic tool usage, tool noise dominates the noise exposures. In other cases, especially when light structures are worked on, workpiece-induced vibrations become more important than tool noise. In the latter situation, mufflers such as described above should be considered only partial treatment and should be coupled with enclosure (using glove-box-type controls), covering (using a heavy "blanket"), or other forms of noise control.

Note that tool manufacturers' claims for quieted air tools should be examined carefully. Although their quieted tools are indeed less noisy than original models, ANSI measurement standards specify a 1-m distance from the tool for making the measurement. In practice, an operator's ear may be closer than 1 m to the tool and, hence, his noise exposure higher than would be expected on the basis of tool manufacturers' promotional literature.

Case History 19. Wood Planer* (OSHA Noise Problem)

Problem Description. Wood planers use a high-speed rotating cutter head to produce lumber with a finished surface. Sound levels near the operator are high.

There are apparently many noise sources for investigation:

1. The board, excited by cutter knife impacts;
2. The heavy structure under the cutter head, excited by vibration transmitted through the board;
3. Modulation of air flow by cutter knife chopping at the chip collector air stream;
4. Motor windage, hum;
5. Dust collector blower, vibration noise;
6. Machine surfaces excited by impacts.

Problem Analysis. Analysis resulted in the following possibilities for control of planer noise:

(1) Restrain the board from vibrating. Feed belts on both sides can be used with considerable backup mass and pressure. This would require a radical machine design change.

(2) Contact the board by means that add damping, to reduce resonant vibration. If this is done as an add-on, it must occur beyond the feed and delivery ports of the planer. This it would be helpful only for long lengths of board.

(3) Use a helical-knife cutter head, which will also reduce idling noise. A helix angle larger than is commonly available would be desirable.

(4) Enclose the planer and board. This is a brute-force method that depends for its success on controlling the amount of sound that escapes from the feed and delivery areas; most of the acoustic energy contributing to the sound level is between 500 and 5000 Hz.

Results. The result achieved by the helical-knife cutter head is shown in Fig. 9.50: reduction from 106 dB(A) to 93 dB(A). Fig. 9.51 shows the operator sound level related to length of board planed, comparing the helical-knife cutter with the straight-knife cutter. The helical knife is by far the quieter.

Comments. To meet OSHA operator sound levels for full-day operation, the plant would need a further sound-level reduction, perhaps by the design of a total enclosure with an acoustic lined tunnel for the infeed and outfeed. This should not be tried until it has indeed been determined that the openings are the chief sources. In many mills, however, the planar is not operated on a full-time basis, thus allowing a higher sound level for the shorter time period that an operator is present. At 93 dB(A), 5.3 hr are permitted.

Case History 20. Textile Braiding Machines** (OSHA Noise Problem)

Problem Description. In braiding operations, a bobbin of thread is rotated on a carrier base in a special slotted cam. This cam revolves as it is rotated around the machine, with several other carriers and cams. The carriers are thrown from one cam to another. With steel carriers, the major source of the intense noise present is the resulting metal-to-metal impact. The manufacturer was willing to consider machine modifications to reduce noise in the case history reported here.

Problem Analysis and Control Description. In a laboratory study, the metal-to-metal contact was easily identified as the chief noise source. It was recognized that a carrier with inherent damping properties should reduce the noise. Replacement of the carrier by a nonme-

---

*Steward, J. S. and F. D. Hart, 1972. "Analysis and Control of Wood Planer Noise." *Sound and Vibration*, **6**(3), 24.

**Cudworth, A. L. and J. E. Stahl, 1972. Noise control in the textile industry. Proc. Inter-Noise. 72:177.

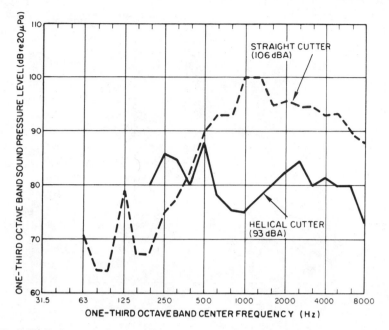

Fig. 9.50. Before-and-after one-third-octave-band sound-pressure levels for wood planer.

tallic one was thus considered. Of the several materials tried, the material that provided the best combination of strength, light weight, and damping was an injection moldable polyurethane.

Result. The carriers were installed in a 13-carrier braider operating at a handle speed of 340 rpm. With the microphone 10 in. above the

top plate of the braider and 18 in. out, the sound-pressure levels were as shown in Fig. 9.52. A reduction of 11 dB was obtained.

The above results were obtained in the laboratory. For an inplant test, a row of 84 braiders was converted to plastic carriers. The adjacent row was left with steel carriers; other rows of braiders were operating. The microphone was 3 ft from the centerline between the test rows, and 3 ft above the floor. The sound levels for various combinations of machines are shown below (an x indicates on).

Sound level, dB(A)	97	97	90	85
Steel test row	x	x		
Plastic test row	x		x	
All other	x	x	x	x

Residual noise from the motor cooling system remained and limited the noise reduction to the 7 dB achieved in this production test.

Comments. Since this study, it has been found that the plastic carriers are not strong enough for some operations requiring heavy yarn (or wire). This finding suggests consideration of a composite carrier with a steel core for strength and a cladding of heavy polyur-

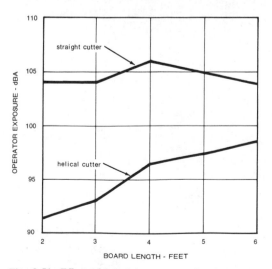

Fig. 9.51. Effect of board length on noise from wood planer.

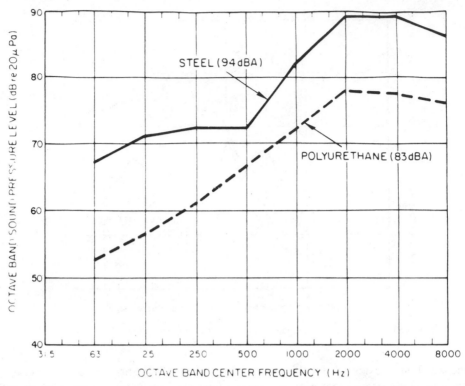

Fig. 9.52. Textile braiding machine: comparison of sound pressure levels from steel carriers and from polyurethane carriers.

ethane for damping. To our knowledge, this concept has not yet been tried. This case emphasizes the need for considering nonacoustical parameters along with the acoustical.

Case History 21. Steam-Line Regulators* (OSHA Noise Problem)

Problem Description. Steam lines with regulators are used in many industries and can be a problem noise source if they are in an area occupied by employees.

Control Description. The method used here, which can also be used to regulate other gas flows, was to modify the design of the main valve plug. The redesigned valve plug has throttling vanes, as shown in Fig. 9.53, to reduce the noise source—the turbulence of the steam flowing through the space between the regulator's main valve and its valve seat.

Results. For a $2\frac{1}{2}$-in. steam line handling 50,000 lb/hr through a reduction from 555 to 100 psig, the redesigned valve reduced pipe line noise from 97 dB(A) to 85 dB(A).

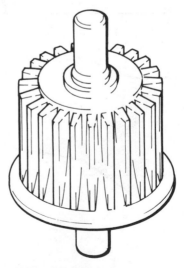

Fig. 9.53. Main valve plug with throttling vanes to reduce noise in steam line regulator.

*From *Electrical World*, January 1973.

Case History 22. Barley Mill (OSHA Noise Problem)

Problem Description. Excessive sound levels existed around the Moorspeed and Ross barley mills (rolls 8 in. diameter, 15 in. long), a hay shredder, and a control operator's chair in cattle-feed grinding mill. The objective was to reduce the sound level at the operator's position for OSHA compliance.

Problem Analysis. A- and C-weighted sound levels and octave-band sound-pressure level measurements were made between the Moorspeed and the Ross mills and at the hay shredder with both mills in normal continuous operation. With $L_C - L_A = 9$ dB, excessive low-frequency sound levels were predicted. These were confirmed by octave-band sound-pressure level measurements. Octave-band sound-pressure level measurements at the control operator's chair, the mills, and at the hay shredder are shown in Fig. 9.54. Fig. 9.55 is a sketch of the room, showing the relative location of the equipment.

Roller crushing actions produced high sound levels, and correction by machine redesign was believed to be too costly a method for solving this problem. When the source is too difficult or uneconomical to attempt to correct, working on the noise path will often result in a more economical solution. Therefore, a partial enclosure, open at the top, was chosen.

Control Description and Design. Although walls can be of solid construction with a minimum of access doors, in this case access was needed for adjustment, maintenance, repair, and roll replacement. For roll replacement, a forklift truck entry was required. For ease of quick access, a fixed barrier wall was discarded in favor of a lead-vinyl curtain wall extending, if required, up to the 17-ft height of the roof support beams. All three noise sources could be enclosed by two curtain walls at the corner of the building, as shown in Fig. 9.55. The curtains run on rails for easy sliding back and are held together by Velcro closures.

Figure 9.54 shows that, if the sound-pres-

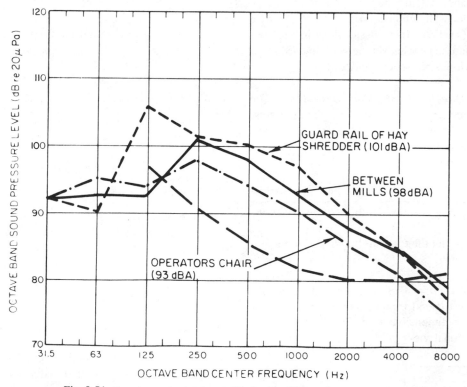

Fig. 9.54. Sound-pressure levels at mills, hay shredder, and operator's chair.

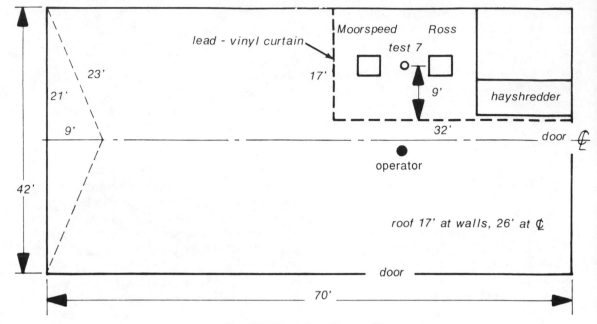

Fig. 9.55. Floor plan of barley mill.

sure levels from 250 Hz up are reduced by at least 14 dB, the resulting A-weighted sound-level readings would be less than 90 dB(A) for compliance outside the curtain walls.

Barrier-wall attenuation is limited in three ways: (1) direct transmission loss in each octave band, (2) noise over the wall, and (3) room absorption, noise-source side.

(1) Direct Transmission Loss (TL). The manufacturer of lead-vinyl fiberglass curtains, 0.75 lb/ft^2, was chosen. Manufacturer's literature gave the transmission loss in each octave band as follows:

	125 Hz	250 Hz	500 Hz	1000 Hz	2000 Hz
TL	11 dB	16 dB	20 dB	26 dB	31 dB

It is seen that the transmission loss is not a limiting factor.

(2) Noise over Wall. Barrier-wall attenuation can be estimated from data in Beranek (1971)* using the dimensions from Fig. 9.55 and from the sectional view in Fig. 9.56.

---

*Beranek, L. L. 1971. *Noise and Vibration Control*, New York: McGraw-Hill, p. 178.

$$N = \frac{2}{\lambda} (A + B - D)$$

$$= \frac{2}{\lambda} (16.6 + 16.6 - 18)$$

$$N = \frac{30.4}{\lambda} \text{ (Fresnel number)}$$

	125 Hz	250 Hz	500 Hz	1000 Hz	2000 Hz
λ	9.6 ft	4.8	2.4	1.2	0.6
N	3.2	6.3	12.6	25.2	50.4
Attenuation, dB	14	16	18	20	20

(Beranek, 1971, graph on page 178). In practical situations, the attenuation is limited to about 20 dB.

By a rough first-approximation procedure, we can obtain an estimate of the reduction afforded by the curtain walls. In the listing below, we start with the worst-case octave-band sound-pressure levels of Fig. 9.54 and then list the transmission loss and barrier effects just calculated. Subtracting the minimum of these two reduction mechanisms yields a tentative spectrum of the resulting sound in the room. After A-weighting and combining of sound pressure levels, the predicted reduced room sound level is 85 dB(A).

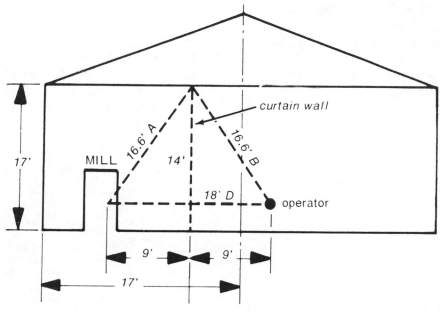

Fig. 9.56. Sectional view of barley mill.

Octave bands	125	250	500	1000	2000
Noise source	106	101	98	97	90
Direct TL	11	16	20	26	31
Over wall	14	16	18	20	20
Reduced sound-pressure levels	95	85	80	77	70
A-weighting	−16	−9	−3	0	1
A-weighted	79	76	77	77	71
A-weighted sound level	84 dBA				

For visual access, the enclosure can have 10 × 20-in. plastic windows placed to order; use only the minimal number. To reduce leaks, the curtains should be long enough to drag a bit on the floor. Some rerouting of power, steam, and air lines may be required.

The approximate 1973 costs were: $4.00/ft² for curtains made to order with grommets; Velcro fasteners, $3.00/ft; track, $1.50/ft; rollers (one per grommet), about $2.50 each; windows, $25.00 each; total cost about $4,000.

The preceding simplified treatment neglects an important fact: We have not gotten rid of the noise, but have merely redistributed it. Thus, the total sound power from the machines escapes from the topless enclosure and spreads throughout the room. Close to the curtains, there should be some reduction, but very little farther away. Absorption is required for actual reduction of the sound power. This was considered next.

(3) Absorption, Noise-Source Side of Wall. When noise sources are confined to a space with less absorption than before, they may build up higher sound levels because of reverberation. The sound barrier curtain material can be obtained with sections of sound absorbent on the inside, to counteract this effect. In the barley mill, however, this choice was not recommended as the porous open material could easily become dust-clogged. Shortly after this noise control job was completed, absorbents covered with a plastic film became available. At the time, the recommendation was for an easily installed and maintained material, Owens-Corning Fiberglas Noise Stop Baffles.* These are 23 × 48 × 1.5-in. baffles, which comprise an absorbent board wrapped in a washable, noncombustible plastic film; each baffle is supplied with two wires through the 23-in. dimension. These wires terminate in hooks; to install, stretch wires, 3 ft on center, parallel to the line joining the two mills and about flush with the top of the enclosure rails.

*These are no longer sold by OCF, but can be readily fabricated from acoustical insulation board.

The enclosure developed by the curtain walls is, in effect, a separate small room, and the noise reduction can be estimated from the relationship of total absorption before and after adding the sound absorption panels. This relationship is

$$\text{dB Attenuation} = 10 \log A_2/A_1,$$

where: A_2 is new total absorption,
A_1 is original absorption.*

Original absorption, A_1:

	Area	Coefficient = Absorption (Sabins)
Long wall	32 ft × 17 ft × 2 × 0.02	22
End wall	17 ft × 17 ft × 2 × 0.02	11
Roof	17 ft × 20 ft × 1 × 0.02	7
		40 ft²-Sabin

Absorption by adding 100 panels 2 × 4 ft:

100 × 2 ft × 4 ft × 2 sides × 0.8 (average A-weighted absorption coefficient of panel) =	1280
Original absorption	40
New total absorption	1320 ft²-Sabin

$$\text{dB attenuation} - 10 \log \frac{A_2}{A_1}$$

$$= 10 \log \frac{1320}{40}$$

$$= 15.2 \text{ dB}.$$

Resultant level = measured level − reduction

$$= 86 \text{ dB(A)}.$$

Result. The measured final sound level was 87 dB(A), a reduction of 7 dB. This level was 3 dB lower than the maximum desired sound level, and was the result of paying careful attention to elimination of leaks. The room formed by the curtain did not realize such a reduction, but since these machines required no attention while running, the noise exposure of personnel was significantly reduced below unity. The major remaining path is reflection from the ceiling.

Comments. Barrier walls of various heights can often be used between a noise source and a machine operator. A major pitfall is that, in a room with a high level of reverberant noise, the partial barrier will be short-circuited by the reflected noises from walls, ceilings, and other surfaces. In such cases, attenuation based on the partial wall theory will not be obtained, and the result may often be no attenuation at all in highly reverberant rooms. Curtain walls must be kept closed to get attenuation. Sound-absorbing units must be kept clean to be efficient.

Even in a semireverberant room, a reduced barrier height can be used. In this case, a 7-ft barrier should ideally reduce the level to 89 dB(A) at the receiving location. However, since the semireverberant conditions will introduce more reflected sound with the lower barrier, the high wall used in this case history is recommended because the added absorption within the barrier area has, in effect, made a separate small room and created the condition on which the barrier wall theory was based.

Case History 23. Punch Press** (OSHA Noise Problem)

Problem Description. Punch presses constitute a most troublesome source of industrial noise, both because of their number and because of their high noise output.

**American Industrial Hygiene Association. 1966. *Industrial Noise Manual*. Examples 11.C, 11.EE. Detroit, MI: AIHA.

Allen, C. H., and R. C. Ison, 1974. "A Practical Approach to Punch Press Quieting." *Noise Control Eng.*, **3**(1), 18.

Bruce, R. D. 1971. "Noise Control of Metal Stamping Operations." *Sound and Vibration*, **5**(11), 41.

Shinaishin, O. A. 1972. "On Punch Press Diagnostics and Noise Control." *Proc. Inter-Noise 72*, 243.

Shinaishin, O. A. 1974. "Sources and Control of Noise in Punch Presses." *Proc. Purdue University Conference on Reduction of Machine Noise*, 240.

Stewart, N. D., J. A. Daggerbart, and J. R. Bailey. "Identification and Reduction of Punch Press Noise." *Proc. Inter-Noise 74*, 225.

*Harris, C. M., ed. 1951. *Handbook of Noise Control*, pp. 18–19. New York: McGraw-Hill.

Problem Analysis. From various papers on the subject of punch presses, the following list of noise sources has been gathered. These may not all be present on any one press but are listed as a guide to specified press noise source analysis.

1. Shock excitation of the workpiece, machine guards, floor and building
2. Gears, drive, bearings, and components, such as clutch and brake mechanism and drive shaft;
3. Plunger resonance;
4. Air ejection, air jet cleaning, and air cylinder exhausts;
5. Die design;
6. Stripper plate design;
7. Ejection of parts leaving press on chute or bin;
8. Vibration of sheet metal being fed to the press;
9. Start and stop of automatic feed to the press;
10. Building acoustics.

Control Description. *Shock Excitation of Surrounding Structures.* This effect can be minimized by properly designed vibration mounts for the entire press to reduce excitation of floors, walls, and other equipment.

Drives, etc. Good maintenance can contribute to noise reduction. The noise of drive gears can be reduced by damping the gear body, improving gear surface quality and tolerances, precision installation and bearings, better lubrication, and/or changing gear material for a better damped material. On existing equipment, many of the above aids cannot be added at reasonable expense, but gear drives are often enclosed in a boxlike structure whose surfaces radiate noise. These surfaces can be damped with off-the-shelf materials, or the drive unit, if space is available, can be enclosed fully or partially. Heat dissipation should be considered. Solid metal or plastic guards can be changed to expanded metal or wire mesh for less noise, or the guard surface can be vibration damped. The entire guard, if solid, should be vibration-isolated from the vibrating machine.

Plunger Resonance. If a hollow plunger or ram is a Helmholtz-resonant type of noise source, its noise radiation can often be reduced by covering the hole in the plunger. See Case History 5.

Air Ejection of Punched Parts. If possible, substitute mechanical ejection to eliminate a large noise source. One comparison, shown in Fig. 9.57, (AIHA 1966), resulted in an 8-dB reduction in sound level. Multiple jet nozzles are also available for reduced noise. Reduce the air velocity used for ejection to a minimum (since sound level is related to velocity) by reducing the air pressure available. Achieve better air jet efficiency by accurate setting and aiming where needed.

Shield the area of punch-air ejection from the operator. An example of the result of this method, in Fig. 9.58, shows the sound levels of a press with and without a 24 × 48-in. shield to protect operator from air ejection noise.

Die Design. Changes in die design can reduce noise by spreading the punching action, slanting the blanking punch or die, or other means of promoting consecutive shear action instead of instant action. Shinaishin reported the results of a slanted die, as shown in Fig. 9.59. Changes in die materials can reduce noise. As presses produce sound energy from vibration of metal plates upon impact, the velocity of impact can be reduced by using hard rubber mounts (snubbers). Another possibility is a laminated and more massive plate, reducing the size of the plate and radiating area.

A change of work stock material from steel to a lead-steel composition has also reduced impact noise; Shinaishin reported a 14-dB reduction with this test method. Noise radiation can be lessened by reducing plate area by cutting out surface areas that perform no function.

These comments emphasize that the tool engineer must now consider designing for noise reduction as well as for mechanical performance. Within such a general framework as outlined, any improvements in sound level will come by experiment and testing results.

Stripper Plates. Stripper plates in some dies contribute to sound levels because of metal-to-

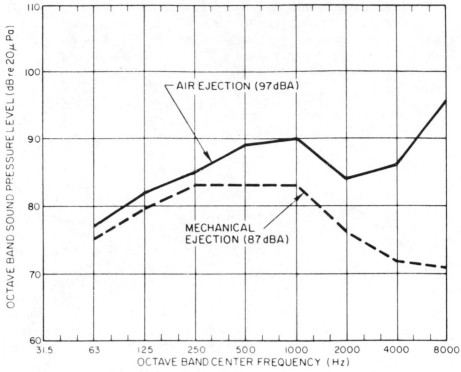

Fig. 9.57. Comparison of punch press sound-pressure levels with air ejection and with mechanical ejection.

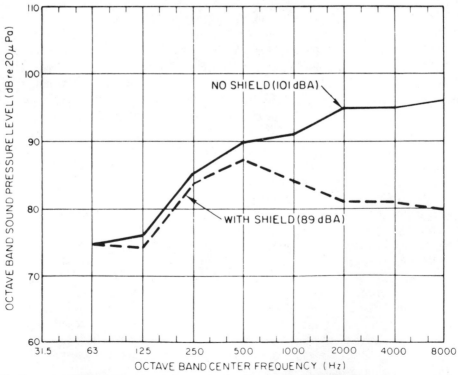

Fig. 9.58. Comparison of punch press sound-pressure levels with and without a shield between operator and air ejection noise.

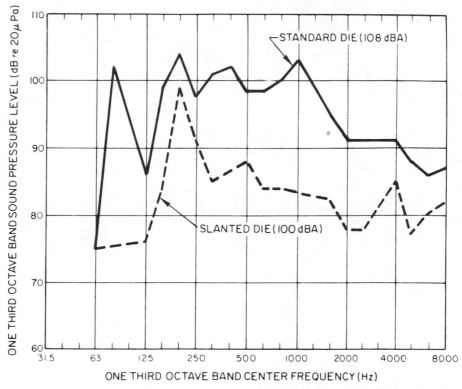

Fig. 9.59. Comparison of punch press sound-pressure levels: standard die versus slanted die.

metal contact, which could be changed to plastic or elastomeric contact with better damping and reduced noise.

Ejection of Parts to Chute or Bin. Sound levels can be reduced by damping metal chutes, using damping materials on the market or making a constrained layer design.

Vibration of Sheet Metal Being Fed to Press. Sound levels can be reduced by preventing vibration, such as by adding a hold-down conveyor. The noise can also be constrained by using an acoustic tunnel infeed, or the operator can be shielded by properly designed barriers.

Start-and-Stop Feed Mechanisms. Noise can be reduced by redesign. Substitute with plastic contact areas where possible; enclose the noise source partially; or add barriers between noise source and operator.

Building Acoustics. In a room with many noise sources, the operator may be in the reverberant field. Such noise can be reduced by adding absorption. From Bruce, an example of use of absorption to reduce noise in a press room is shown in Fig. 9.60, 30 ft from presses. Closer to presses, noise reduction would be less—with probably no more than 2–3 dB at the operator position. The press area can also be enclosed or walled off from the rest of the plant.

Results. Allen and Ison reported a partial enclosure of ram, die, infeed, and ejection on a 50-ton test press. A sound level reduction of 13 dB was obtained for an enclosure; see Fig. 9.61. The model enclosure was made of cardboard, $\frac{1}{2}$ lb/ft^2, lined with 1 in. of polyurethane foam. Later a steel enclosure was installed, for durability.

Total enclosures with opening via an acoustic tunnel may be required.

Comments. The remaining radiation came chiefly from the flywheel cover, which was neither damped nor vibration-isolated. Diagnostic measurements should indicate the rela-

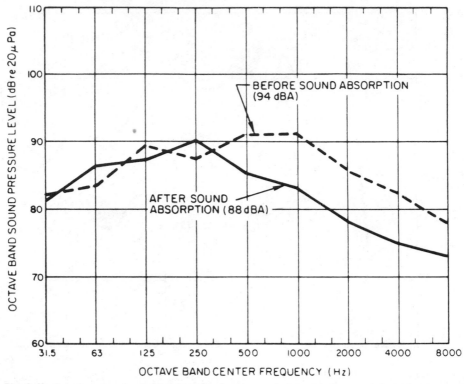

Fig. 9.60. Sound-pressure levels 30 ft from bench press area before and after sound absorption treatment.

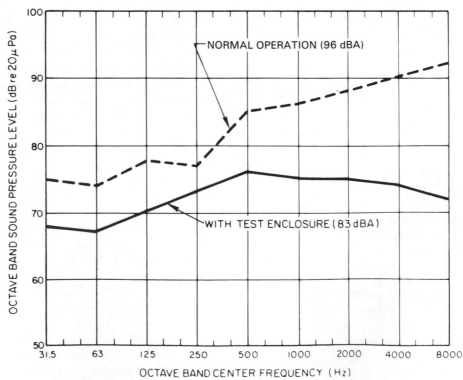

Fig. 9.61. Data 30 in. from punch press before and after test cardboard enclosure.

tive contributions from each source, so that the residual noise will be known.

Case History 24. Chemical Process Plants* (OSHA Noise Problem)

Problem Description. Existing chemical process plant noise reduction requires source analysis to determine the method of noise reduction.

Problem Analysis and Control Description. As a result of this study, a list of noise sources is shown in Table 9.16, with recommended methods of noise reduction. Some specific examples and results obtained by each noise-control method are cited in Figs. 9.62 through 9.65. The attenuation attained is shown in each figure.

Case History 25. Vibration Table (Hearing Conservation Noise Problem)

Lars Holmberg
IFM Akustikbyrån AB
Warfvinges väg 26
S-112 51 Stockholm, Sweden
(08) 131220

Problem Description. Product compaction is a necessity in the manufacture of prefabricated concrete building elements. In certain cases, the compaction can be achieved only by external application of vibrations to the molds. This case history concerns vibration tables used in the production of a product called well rings. Sound levels as high as 104 dB(A), containing a strong low-frequency tone, were measured at operator stations, approximately 1 m from the approximately 2-m-diameter mold, during vibration. Vibration table noise takes place intermittently about 4 hr a day, and operators can also be exposed to noise from several other machines 10–40 m away. The operators control the filling of the molds.

Problem Analysis. This problem was analyzed by measuring and plotting operator po-

---

*From Judd, S. H. 1971. "Noise Abatement in Process Plants." *Chemical Engineering*, January 11.

sition sound-pressure levels during mold vibration on octave-band graph paper that included five curves, each representing maximum recommended daily exposure time in accordance with International Standards Organization guidelines for industrial noise exposure. Results, shown in Fig. 9.66, indicate the 4 hr of daily exposure are greater than indicated by the penetrated curve on the plot. (Note that our OSHA regulation would allow between 1 and 2 hr/day of exposure to 104-dB(A) sounds.) A noise reduction of approximately 10 dB is called for in this case.

Although detailed analysis of noise-producing mechanisms would be desirable to identify quantitatively the relative contributions of the table vibrator, table vibrations, and mold vibrations, such data were not obtained. However, some qualitative determinations were made, based on observations.

Low-frequency emissions from the vibrator and broader-band emissions from resonances induced in the mold structure and the table were identified as the major noise sources. The rattle of the loose parts of the molds also contributed to the overall noise environment.

Several possibilities exist for reducing noise exposures in this type of process:

- Reducing the vibrated surface area (i.e., by vibration of the bottom of the mold instead of the sides, or damped mold sides);
- Using alternative methods of compaction;
- Optimizing vibration components (frequency, amplitude, time) according to properties of the concrete used (e.g., initiating vibration after the mold is partly filled, adjusting vibration amplitude and/or frequency to obtain maximum compaction for minimal noise emission);
- Eliminating unnecessary impacts between the vibration table and the mold;
- Containing noise emissions by use of shields or enclosure.

Prior studies had revealed that some of these possibilities have yielded good results:

- Elimination of rattles provided between 3 and 10 dB of noise reduction.
- Vibration isolation of the mold from the table had provided up to 20 dB of noise

Table 9.16. Sources of Noise and Methods of Noise Reduction.

Equipment	Source of noise	Methods of noise reduction
Heaters	Combustion of burners	Acoustic plenum* (10 Bwg. plate)
		Seals around control rods and over sight holes
	Inspiration of premix air at burners	Inspirating intake silencer
	Draft fans	Intake silencer or acoustic plenum
	Ducts	Lagging
Motors	TEFC cooling air fan	Intake silencer
		Undirectional fan
	WP II cooling system	Absorbent duct liners
	Mechanical and electrical	Enclosure
Airfin coolers	Fan	Decrease rpm (increasing pitch)
		Tip and hub seals
		Increase number of blades**
		Decrease static pressure drop**
		Add more fin tubes**
	Speed changer	Belts in place of gears
	Motors	Quite motor
		Slower motor
	Fan shroud	Streamline airflow
		Stiffening and damping (reducing vibration)
Compressors	Discharge piping and expansion joint	Inline silencer and/or lagging
	Antisurge bypass	Use quiet valves and enlarge and streamline piping**
		Inline silencers
	Intake piping and suction drum	Lagging
	Air intake	Silencer
	Discharge to air	Silencer
	Timing gears (axial)	Enclosure (or constrained damping on case)
		Silencers on discharge and lagging
	Speed changers	Enclosure (or constrained damping on case)
Engines	Exhaust	Silencer (muffler)
	Air intake	Silencer
	Cooling fan	Enclose intake or discharge or both
		Use quieter fan
Miscellaneous	Turbine steam discharge	Silencer
	Air and steam vents	Silencer
		Use quiet valve
	Eductors	Lagging
	Piping	Limit velocities
		Avoid abrupt changes in size and direction
		Lagging
	Valves	Limit pressure drops and velocities
		Limit mass flow
		Use constant velocity or other quiet valve
		Divide pressure drop
		Size adequately for total flow
		Size for control range
	Pumps	Enclosure

*If oil fired, provide for drainage of oil leaks and inspection.
**Usually limited to replacement or new facilities.

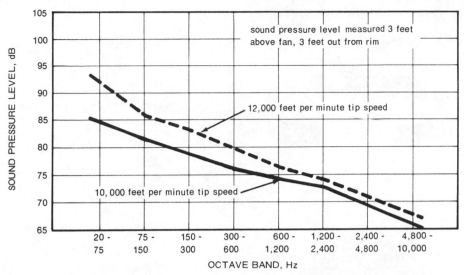

Fig. 9.62. Noise reduction achieved by reducing fan speed, using increased blade pitch to offset decrease in speed (measured 3 ft above fan, 3 ft out from rim).

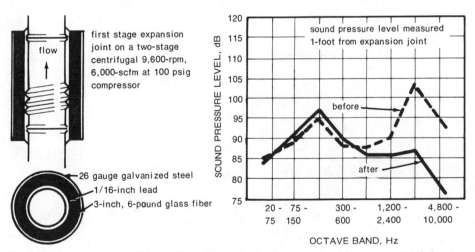

Fig. 9.63. Compressor discharge noise reduction achieved by lagging expansion joint (measured 1 ft from expansion joint).

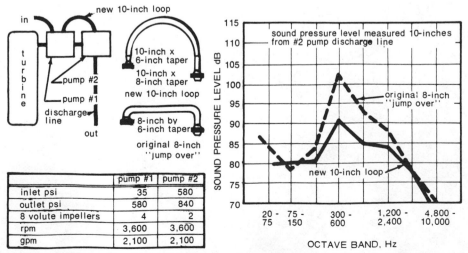

	pump #1	pump #2
inlet psi	35	580
outlet psi	580	840
8 volute impellers	4	2
rpm	3,600	3,600
gpm	2,100	2,100

Fig. 9.64. Noise reduction achieved by redesigning pump bypass loop (measured 10 in. from No. 2 pump discharge line).

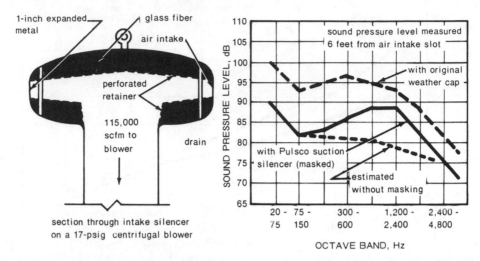

Fig. 9.65. Noise reduction achieved by adding silencer to air blower intake (measured 6 ft from air intake slot).

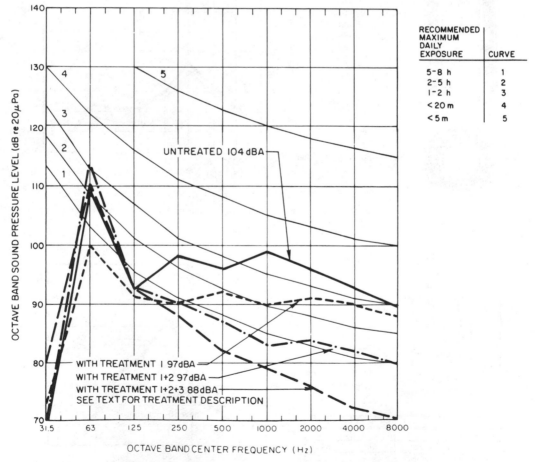

Fig. 9.66. Results of measurement of operator position sound-pressure levels.

reduction, at the expense of requiring additional vibration time.

- Other methods of compaction are considerably quieter. In particular, internal vibration (using devices that can be held in place inside the mold) produces sound levels in the 85-dB(A) to 95-dB(A) range at a distance of 1 m.

Because alternative methods of compaction would be too costly to install and because several of the remaining noise-control possibilities require considerable experimentation and study, it was decided, first, to implement vibration isolation of the mold and then, if necessary, containment of the generated sounds.

Control Description.   A vibration table was quieted with the three-phase program of noise control depicted in Fig. 9.67.

1. A rubber ring was mounted on the table below the guide ring.
2. A rubber ring was mounted between the guide ring and the mold.
3. A screen was constructed around the mold.

Rings were made of 4-mm rubber. The screen that encloses the 6-ft-diameter mold was constructed of 3-mm steel (outside) and perforated steel plate (inside), sandwiching 100-mm mineral wool. Rubber sheeting completed a seal at floor level.

Results.   Noise at the vibration table was reduced to 97 dB(A) after installation of the first two phases of noise control and to 88 dB(A) when all three phases were completed. Fig. 9.67 summarizes the reductions obtained.

9.6. GENERAL TEXTS, HANDBOOKS, AND MANUALS*

Architectural Acoustics. M. David Egan. New York: McGraw-Hill, 1988.

This book is written for architects, interior designers, engineers and all others concerned with the design and construction of buildings—who need to know the basics of architectural acoustics, but who do not have the time to peruse in-depth presentations. The book contains numerous checklists of design aids, data tables of sound absorption and sound isolation properties for a wide variety of modern building materials, case study examples, and step-by-step practical problem solutions.

Control of Noise, 3rd ed. Des Plaines, IL: American Foundrymen's Society, 1972.

The engineering section of this manual was prepared by an experienced consultant. It is written without equations, but with many charts, graphs, and tables. Although the many

*From U.S. NIOSH, *Industrial Noise Control Manual,* with additions by the present author. See Section 9.5.

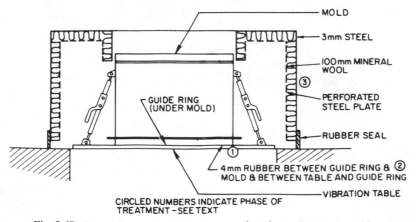

Fig. 9.67. Three-phase program of noise control used to quiet vibration table.

examples are taken from foundry technology, the control techniques are quite general in application. There are many compact case histories, together with data on the noise reductions obtained. The point of view is very practical.

Criteria for a Recommended Standard: Occupational Exposure to Noise, Department of Health Education and Welfare, NIOSH No. 73-11001, August 14, 1972.

Fundamentals of Industrial Noise Control. L. H. Bell. Trumbull, CT: Harmony Publications, 1973.

This practical book is written from the point of view of the practicing noise-control engineer/consultant. A minimum amount of mathematics is used; many examples and exercises are given. The chapters on enclosures, fans, gears, silencers, and vibration control are quite useful. A feature of the book is the compact case histories, for which photographs and drawings amply describe the techniques used.

Guidelines to Noise. American Petroleum Institute, Medical Research Report EA 7301, 1973.

This commissioned report summarizes measurement instruments and procedures, explicit noise reduction techniques, new plant design for low noise, and source characteristics. The appendices have detailed information on noise control materials, levels from machinery, and addresses of suppliers of noise control materials.

Handbook of Acoustic Noise Control. W. A. Rosenblith and K. N. Stevens. WAD Tech. Publ. No. 52-204, 1953.

Handbook of Noise Control. C. M. Harris, ed. New York: McGraw-Hill, 1957.

Although old, this is still the fundamental reference handbook for the noise-control engineer. Of particular interest are these sections: 13, vibration isolation; 14, vibration damping; 21, acoustical filters and mufflers; 23, gear noise, 24, bearing noise; 25, fan noise; 26, noise in water and steam systems; 27, heating and ventilating system noise; and 30, electric motor and generator noise. Of course, recent developments in acoustical materials and measuring equipment are missing, but the fundamentals are unchanged.

Handbook of Noise Measurement, 7th ed., A. P. G. Peterson and E. E. Gross. Concord, MA: GenRad, Inc., 1972.

This book is an excellent source of data on measurement of sound pressure and calculation of sound power levels. Valuable details are given on sound analysis techniques, characteristics of many types of acoustical instruments, and a summary of noise reduction procedures. An especially useful section covers precautions to be observed to ensure that valid data are required.

Industrial Noise and Vibration Control. Irwin and Graf. Englewood Cliffs, NJ: Prentice-Hall, 1979.

Good practical book, with much emphasis on dynamic systems and measurements.

Industrial Noise Control Handbook. P. N. and P. P. Cheremisinoff, eds. Ann Arbor, MI: Ann Arbor Science Publishers, 1977.

This book is a practical guide to industrial noise and vibration control. The text is well illustrated and discusses the important topics with a minimum of mathematical treatment. The text suffers a bit from imbalance—some topics are discussed only briefly, whereas others are discussed in depth. Information contained in the detailed sections, particularly those on the use of glass and lead materials, contains a good deal of valuable data. The reader will benefit from the discussions on noise legislation and personal safety devices. This book also contains a number of illustrative case histories pertaining to, for example, electric utility and refinery noise, paper rewinders, jet engine test cells, and several other common noise problems.

Industrial Noise Manual, 2nd ed. Detroit: American Industrial Hygienists Association, 1966.

Although the instrument section is outdated, the described measurement techniques are still applicable. Considerable data are given on ear plugs and muffs. The chapter on engineering control is very practical; it is copiously illustrated and describes many useful techniques. A most valuable section on examples presents compact, illustrated case histories in which the noise reduction obtained is given, usually with octave-band spectra.

Machinery Acoustics. G. M. Diehl. New York: John Wiley & Sons, 1973.

The chief contribution of this book is a detailed description of practical techniques, backed by analysis, for the in-situ measurements required for calculating sound power. Every professional noise-control engineer should be aware of these techniques. The sections on noise sources and reduction procedures have a great deal of directly useful information, especially for enclosure design.

Noise and Its Control. Pollution Engineering.

This reprint of very readable 1973 articles summarizes characteristics of machine noise sources and noise-control techniques. It will provide a general background to the problems.

Noise and Vibration Control. L. L. Beranek, ed. New York: NY: McGraw-Hill, 1971.

This is the major modern reference source for the noise-control engineer. The treatment is often mathematical, but there are plenty of illustrative worked-out problems. Especially useful are the treatments of transmission loss of simple and complex panels; mufflers and silencers; sound in rooms; vibration isolation; and sound power measurement.

Noise and Vibration Control for Industrialists. S. A. Petrusewicz and D. K. Longmore, eds. New York: American Elsevier, 1974.

This book contains a good deal of technical information on acoustics, noise control, and especially vibration and vibration control. However, there is also much clearly written practical advice in the text on principles of noise and vibration control and measurement techniques. Readers may find the sections on criteria and hearing conservation particularly enlightening and useful. A case history for new plant installation is included as the final section of the text.

Noise and Vibration Control in Buildings. Robert S. Jones. New York: McGraw-Hill, 1984.

This book provides a unique guide for selecting seismic protection for all mechanical and electrical equipment commonly used in buildings. It also covers installation of various types of vibration isolators for HVAC, plumbing, and noise generating equipment such as escalators, elevators, and conveyors.

Noise Control. R. Taylor, ed. Rupert Taylor and Partners Ltd., 114 Westbourne Grove, London, W2 4UP, England.

Noise Control Approaches. M. V. Crocker, *Proc. Inter-Noise 72*, Tutorial, 1972.

Excellent summary of procedures.

Noise Control for Engineers. Lord, Gately and Evensen. New York: McGraw-Hill, 1980.

This book is comprehensive, with an emphasis upon use as teaching textbook.

Secrets of Noise Control. A. Thumann and R. K. Miller. Atlanta: Fairmont Press, 1974.

This book presents much practical noise-control information in graphs and tables, with a minimum of mathematics. Especially useful are data on cost estimating, a listing of suppliers of noise control products, means of source location, silencers, and checklists for management of noise control. There are many useful worked-out problems. A comprehensive list is supplied for all the standard methods of measurement that a professional noise control engineer should use.

Sound, Noise, and Vibration Control. L. F. Yerges. New York: Van Nostrand Reinhold, 1969.

This practical book has almost no mathematics and relies almost completely on tables, charts, and graphs for its data. The author, an experienced acoustical consultant, provides a great deal of directly useful information on materials selection, noise characteristics of machinery, design of noise control means, and translation of subjective reactions to noise into causes and solutions.

What to Do About Noise. T. J. Schultz. Washington, D. C.: U.S. Department of Housing and Urban Development, 1973. Also, Report No. 2549, Bolt Beranek and Newman Inc., 50 Moulton Street, Cambridge, MA 02138.

CREDITS FOR FIGURES AND TABLES, WITH PERMISSIONS (Where Applicable)

Chapter 1

Fig. 1.4. Sinusoidal signal showing various measures of signal amplitude. (From *Acoustic Noise Measurements*, by Bruel and Kjaer, Naerum, Denmark, 1979—with permission.)

Fig. 1.5. The dispersion of sound from a point source. (From *Acoustic Noise Measurements* by Bruel and Kjaer, Naerum, Denmark, 1979—with permission.)

Fig. 1.6. The transformation of vibrations into waves (a) by a vibrating pointer on a moving band (b) by a vibrating piston in a fluid medium. (From *Acoustic Noise Measurements*, by Bruel and Kjaer, Naerum, Denmark, 1979—with permission.)

Chapter 2

Fig. 2.1. The main parts of the ear. (From *Acoustic Noise Measurements*, by Bruel and Kjaer, Naerum, Denmark, 1979—with permission.)

Fig. 2.2. Eardrum and three auditory ossicles. (From *Anatomy for Speech and Hearing*, by J. W. Palmer. New York: Harper & Row, 1972—with permission.)

Fig. 2.3. Cochlea. (From ''The Ear,'' by G. Von Bekesy. *Scientific American*, **197,** 69, 1957.)

Fig. 2.4. Longitudinal section of the cochlea showing the positions of response maxima. (From *Acoustic Noise Measurements*, by Bruel and Kjaer, Naerum, Denmark, 1979—with permission.)

Fig. 2.8. Typical sound-pressure levels of common noise sources. (From *Acoustic Noise Measurements*, by Bruel and Kjaer, Naerum, Denmark, 1979—with permission.)

Fig. 2.9. · · · subjective perception of short impulses. (From *Acoustic Noise Measurements*, by Bruel and Kjaer, Naerum, Denmark, 1979—with permission.)

Fig. 2.10. Average shifts with age of threshold of hearing for pure tones. (From Spoor, *International Audiology*, **6**(1), July 1967.)

Fig. 2.11. The development of noise induced hearing loss. (From *Acoustic Noise Measurements*, by Bruel and Kjaer, Naerum, Denmark, 1979—with permission.)

Chapter 3

Figs. 3.1, 3.4, 3.5, 3.8, 3.9, 3.11, 3.12, 3.13a, 3.13b, 3.14, 3.15, 3.16, 3.17, 3.18, 3.19, 3.20, 3.25, 3.27, 3.28, 3.29, 3.30, 3.31, 3.32, 3.33, 3.35, 3.36, 3.37, 3.41 through 3.59, 3.61, 3.62. (From publications—Refs. 7, 11, 15—application notes, literature, by Bruel and Kjaer, Naerum, Denmark—with permission.)

Figs. 3.2, 3.3, 3.6, 3.11, 3.34. (From General Radio Co. (GenRad) general literature—with permission.)

Fig. 3.10. (From Quest Electronics general literature—with permission.)

Chapter 4

Fig. 4.8. Directivity pattern, directivity index and directivity factor. (From *Noise Reduction*, by L. L. Beranek. New York: McGraw-Hill, 1960—with permission.)

Fig. 4.12. Chart for determining sound-pressure level in a large irregular enclosure at a distance r from the centre of a source of directivity factor Q_θ. (From *Noise and Vibration Control*, by L. L. Beranek. New York: McGraw-Hill, 1971—with permission.)

Figs. 4.9, 4.10, 4.13–4.24. (From *Acoustic Noise Measurements*, by Bruel and Kjaer, Naerum, Denmark, 1979—with permission.)

Figs. 4.25–4.29. (From ''Environmental Noise Assessment in Land Use Planning,'' Ontario Ministry of the Environment, Toronto, February 1987—with permission.)

Chapter 5

Fig. 5.2. Absorption coefficients of porous fiberglass and polyurethane foam. (From *Noise and Vibration Control*, by L. L. Beranek. New York: McGraw-Hill, 1971—with permission.)

Fig. 5.3. Random sound absorption of polyurethane foam with Mylar facing. (From *Noise and Vibration Control*, by L. L. Beranek. New York: McGraw-Hill, 1971—with permission.)

Fig. 5.4(a). Space sound absorbers. (From *Environmental Acoustics*, by L. L. Douelle. New York: McGraw-Hill, 1972—with permission.)

Fig. 5.4(b). Use of sound-absorbing baffles. (From *Noise Control* by Bruel and Kjaer, Naerum, Denmark, 1982—with permission.)

Fig. 5.6. Example of absorption of different panel configurations and mounting. (From *Noise Control*, by Bruel and Kjaer, Naerum, Denmark, 1982—with permission.)

Fig. 5.7. Resonance frequency of sound-absorbing flexible panels. (From *Noise and Noise Control*, Vol. I, by Crocker and Price. Boca Raton, FL: CRC Press Inc., 1975—with permission.)

Fig. 5.8. The effect on absorption coefficient of thick sound-absorptive blanket behind flexible plywood panel. (From *Noise and Noise Control*, Vol. I, by Crocker and Price. Boca Raton, FL: CRC Press Inc., 1975—with permission.)

Fig. 5.9. Application of panel resonance. (From *Noise Control*, by Bruel and Kjaer, Ltd., Naerum, Denmark—with permission.)

Fig. 5.10. Schematic of some types of dissipative silencers. (From *Noise and Vibration Control*, by L. L. Beranek. New York: McGraw-Hill, 1971—with permission.)

Table 5.2. Absorption coefficient of duct liner and attenuation of lined ducts. (From *Noise Control for Engineers*, by Lord et al. New York: McGraw-Hill, 1980—with permission.)

Table 5.3. Attenuation of round and rectangular 90° lined bends. (From *Noise Control for Engineers*, by Lord et al. New York: McGraw-Hill, 1980—with permission.)

Fig. 5.12. Design curves for lined 180° bends. (From *Handbook of Noise Control*, C. M. Harris, ed. New York: McGraw-Hill, 1957—with permission.)

Fig. 5.13. Plenum for reduction of airborne duct noise. (From "Compendium of Materials," National Institute for Occupational Safety and Health, June 1975.)

Fig. 5.14. Isometric drawing of a jet engine test cell muffler. (From *Noise and Vibration Control*, by L. L. Beranek. New York: McGraw-Hill, 1971—with permission.)

Fig. 5.15. Performance curves for three lengths of absorbing units. (From *Noise and Vibration Control*, by L. L. Beranek. New York: McGraw-Hill, 1971—with permission.)

Fig. 5.16. Types of absorbent material used as duct liners and baffles. (From *Noise Control*, by Bruel and Kjaer, Naerum, Denmark, 1982—with permission.)

Fig. 5.17. Use of baffles in silencing gas turbine noise. (From *Noise Control*, by Bruel and Kjaer, Naerum, Denmark, 1982—with permission.)

Fig. 5.18(a). Artist's sketch of a low-priced, low-performance straight-through type of muffler. (From *Noise and Vibration Control*, by L. L. Beranek. New York: McGraw-Hill, 1971—with permission.)

Fig. 5.18(b). Artist's sketch of two high-performance mufflers. (From Nelson Muffler Corporation, U.S.A.)

Fig. 5.20. Transmission loss of an expansion chamber. (From *Noise and Vibration Control*, by L. L. Beranek. New York: McGraw-Hill, 1971—with permission.)

Figs. 5.21, 5.22, 5.24. Comparisons of theoretical and experimental values of reactive mufflers. (From *Handbook of Noise Control*, C. M. Harris, ed. New York: McGraw-Hill, 1957—with permission.)

Fig. 5.23. Diagram of a volume (side branch) resonator. (From *Noise and Vibration Control*, by L. L. Beranek. New York: McGraw-Hill, 1971—with permission.)

Fig. 5.27. Typical Soundblox units used as individual cavity resonators. (From *Environmental Acoustics*, by L. L. Douelle. New York: McGraw-Hill, 1972—with permission.)

Fig. 5.28. Typical installation of a perforated panel resonator using various types of perforated facings and with an isolation blanket in the air space. (From *Environmental Acoustics*, by L. L. Douelle. New York: McGraw-Hill, 1972—with permission.)

Fig. 5.29. Sound absorption of perforated panel resonators with isolation blanket in air space. (From *Environment Acoustics*, by L. L. Douelle. New York: McGraw-Hill, 1972—with permission.)

Fig. 5.30(a). View of acoustical treatment with open brick and wood slats in auditorium. (From *Environmental Acoustics*, by L. L. Douelle. New York: McGraw-Hill, 1972—with permission.)

Fig. 5.30(b). Details of typical applications of separated brick, steel channel and wooden slat facings with absorbent backing. (From *Environmental Acoustics*, by L. L. Douelle. New York: McGraw-Hill, 1972—with permission.)

Fig. 5.31. Slotted concrete blocks faced with fiberglass and covered with a perforated metal. (From *Noise and Noise Control*, Vol. I, by Crocker and Price. Boca Raton, FL: CRC Press Inc., 1975—with permission.)

Fig. 5.40. Standing wave apparatus used to determine normal incidence absorption coefficient and complex impedance. (From *Noise and Noise Control*, Vol. I, by Crocker and Price. Boca Raton, FL: CRC Press Inc., 1975—with permission.)

Fig. 5.42. Transmission loss of panels. (From *Noise Reduction*, by L. L. Beranek. New York: McGraw-Hill, 1960—with permission.)

Fig. 5.43. Coincidence effect in panels. (From *Noise Reduction*, by L. L. Beranek. New York: McGraw-Hill, 1960—with permission.)

Fig. 5.45. Examples to illustrate the effects of some noise control measures. (From *Handbook of Noise Measurement*, Peterson and Gross. Concord, MA: General Radio Co. (GenRad).)

Fig. 5.46. Representative transmission loss data and corresponding STC contour. (From Owens-Corning Fiberglass Corporation literature, Toledo, OH.)

Fig. 5.47. Relative transmission loss for typical construction. (From *An Introduction to Architectural Acoustics*, by Kingsbury and Albright. Pennsylvania State University, 1965.)

Fig. 5.48 Effects of openings on transmission loss. (From *Industrial Noiose Control Manual*. National Institute for Occupational Safety and Health, 1975.)

Chapter 6

Fig. 6.13. Recommended isolation efficiencies and static deflections for various applications. (From Vibration Mountings and Controls, Inc., U.S.A.)

Figs. 6.22(b), 6.23, 6.24, 6.25. (From *Acoustic Noise Measurements*, by Bruel and Kjaer, Naerum, Denmark, 1979—with permission.)

Fig. 6.27. Currently available vibration criteria. (From *HVAC Systems and Applications*. Atlanta, GA: American Society of Heating, Refrigerating and Air Conditioning Engineers, Inc., 1987—with permission.)

Fig. 6.28. Homogeneous and constrained-layer damping. (From *Noise Control for Engineers*, by Lord et al. New York: McGraw-Hill, 1980—with permission.)

Fig. 6.29. Methods for measuring the loss factor of a damped structure. (From *Noise and Vibration Control*, by L. L. Beranek. New York: McGraw-Hill, 1980—with permission.)

Fig. 6.30. Effects of temperature, thickness ratio, and frequency on effectiveness of homogeneous damping. (From "Composite Materials for Noise and Vibration

Control," by J. Emme. *Sound and Vibration*, July 1970—with permission.)

Chapter 7

Fig. 7.1. Indoor noise criteria (NC) curves. (From *Noise Reduction*, by L. L. Beranek. New York: McGraw-Hill, 1960—with permission.)

Fig. 7.2. Preferred noise criteria (PNC) curves. (From *Noise and Vibration Control*, by L. L. Beranek. New York: McGraw-Hill, 1971—with permission.)

Fig. 7.3. RC (room criterion) curves for specifying design level in terms of a balanced spectrum shape. (From *HVAC Systems and Applications*. Atlanta, GA: American Society of Heating, Refrigerating and Air Conditioning Engineers, Inc., 1987—with permission.)

Fig. 7.4. Rating chart for determining speech communication capability from speech-interference levels. (From *Proceedings of the Conference on Noise as a Public Hazard*, by J. C. Webster, 1979.)

Table 7.2. Indoor sound level limits. (From "Model Municipal Noise Control By-Law," Ontario Ministry of the Environment, 1978.)

Table 7.3. Maximum acceptable levels of road and rail traffic noise in dwellings and outdoor areas. (From "Road and Rail Noise: Effects on Housing," Canada Central Mortgage and Housing Corporation, 1978.)

Fig. 7.5. Curves for the computation of the articulation index, AI. (From *Handbook of Noise Control*, D. N. May, ed. New York: Van Nostrand Reinhold, 1978—with permission.)

Fig. 7.6. Speech communication with various screen configurations. (From *Handbook of Noise Control*, D. N. May, ed. New York: Van Nostrand Reinhold, 1978—with permission.)

Fig. 7.7. Segment of typical open-plan office. (From *Handbook of Noise Control*, D. N. May, ed. New York: Van Nostrand Reinhold, 1978—with permission.)

Table 7.4. Assessment of speech privacy in an open-plan office. (From *Handbook of Noise Control*, D. N. May, ed. New York: Van Nostrand Reinhold, 1978—with permission.)

Fig. 7.8. Risk of noise induced hearing damage vs years of exposure. (From "Hazardous Exposure to Intermittent and Steady-State Noise," National Academy of Science and National Research Council Committee on Hearing, Bioacoustics and Biomechanics (CHABA), January 1965.)

Fig. 7.10. Recommended time intervals of noise "on" and "off" for recovery from preceding noise and preparation for next exposure. (From "Hazardous Exposure to Intermittent and Steady-State Noise," National Academy of Science and National Research Council Committee on Hearing, Bioacoustics and Biomechanics (CHABA), January 1965.)

Fig. 7.11.   Two samples of outdoor noise in normal suburban neighborhood. (From *Noise Control for Engineers*, by Lord et al. New York: McGraw-Hill, 1980—with permission.)

Fig. 7.12.   Statistical portrayal of community noise within a 24-hour period. (From *Noise Control for Engineers*, by Lord et al. New York: McGraw-Hill, 1980—with permission.)

Table 7.8.   Sound-level limits for outdoor recreational areas (0700 to 2300 hours). (From "Model Municipal Noise Control By-Law," Ontario Ministry of the Environment, 1978.)

Table 7.9.   Sound level limits for outdoor areas (2300 to 0700 hours). (From "Model Municipal Noise Control By-Law," Ontario Ministry of the Environment, 1978.)

Fig. 7.13.   Community reaction to intrusive noise measured in L_{dn} and normalized by factors set out in Table 7.10. (From EPA Report 550/9-74-004.)

Table 7.10.   Corrections to be added to the measured day-night sound level (L_{dn}) of intruding noise to obtain normalized L_{dn}. (From EPA Report 550/9-74-004.)

Fig. 7.14.   Family of curves used to determine the noise-level rank (in assessing CNR_c). (From "Handbook of Noise Rating," by K. S. Pearsons and R. L. Bennett. NASA Report CR-2376, National Aeronautics and Space Administration, Washington, D.C., 1974.)

Fig. 7.15.   Family of curves used to determine the correction number for background noise (in assessing CNR_c). (From "Handbook of Noise Rating" by K. S. Pearsons and R. L. Bennett. NASA Report CR-2376, National Aeronautics and Space Administration, Washington, D.C., 1974.)

Fig. 7.16.   Proposed correction numbers for repetitiveness of the noise when the source operates on a reasonably regular daily schedule. (From "Handbook of Noise Rating," by K. S. Pearsons and R. L. Bennett. NASA Report CR-2376, National Aeronautics and Space Administration, Washington, D.C., 1974.)

Table 7.11.   Corrections for background noise (in assessing CNR_c). (From "Handbook of Noise Rating," by K. S. Pearsons and R. L. Bennett, NASA Report CR-2376, National Aeronautics and Space Administration, Washington, D.C., 1974.)

Table 7.12.   List of correction numbers to be applied to noise-level rank to give CNR_c. (From "Handbook of Noise Rating" by K. S. Pearsons and R. L. Bennett. NASA Report CR-2376, National Aeronautics and Space Administration, Washington, D.C., 1974.)

Fig. 7.17.   Estimated community response vs composite noise rating (CNR_c), modified. (From "Handbook of Noise Rating," by K. S. Pearsons and R. L. Bennett, NASA Report CR-2376, National Aeronautics and Space Administration, Washington, D.C., 1974.)

Chapter 8

All of the figures, diagrams and tables of Chapter 8 have been taken from *Noise Control—Principles and Practice*, 1st ed., 1982, by Bruel and Kjaer Ltd., Naerum, Denmark—with permission.

Chapter 9

Figs. 9.1–9.4, Table 9.1.   (From *Industrial Noise Control Manual*, U.S. National Institute for Occupational Safety and Health, 1974; reprinted by Canadian Acoustical Association, 1984.)

Figs. 9.5–9.9, Tables 9.2–9.6.   (From *Noise Control for Engineers*, by Lord, Gately, and Evensen. New York: McGraw-Hill, 1980—with permission.)

All figures, diagrams and tables in Section 9.5 have been taken from *Industrial Noise Control Manual*, U.S. National Institute for Occupational Safety and Health, 1974; reprinted by Canadian Acousticial Association, 1984.

Appendix V

"Guideline for Regulatory Control of Occupational Noise Exposure and Hearing Conservation," *Proceedings of 12th International Congress on Acoustics*, Vol. 1, C7-1, Toronto, Canada, July, 1986.

Appendix VI

"Buyer's Guide to Products for Noise and Vibration Control." (From *Sound and Vibration Magazine*, July, 1988, Acoustical Publications, Bay Village, OH—with permission.)

Appendix VII

"Dynamic Measurement Instrumentation Buyer's Guide." (From *Sound and Vibration Magazine*, March, 1988, Acoustical Publications, Bay Village, OH—with permission.)

Appendix I

GLOSSARY OF ACOUSTICAL TERMS

This is a glossary of the acoustical terms most frequently used in this book. Explanations rather than definitions are used in order to avoid highly technical terminology which could confuse those without acoustics experience. Readers may obtain further explanations for these terms by using the index to find the various pages on which a given term appears. The text of the book may discuss these terms in contexts making for easier understanding.

In using this glossary to refer to terms beginning with either "noise" or "sound," bear in mind that these words are often interchangeable, so that a term not found under one of these words may be found under the other.

Absorption. A process, occurring generally at a solid surface but also in a fluid, resulting in sound energy being dissipated. When absorption occurs at a solid surface, an incident sound wave is not fully reflected.

AI. See *Articulation Index.*

Ambient Sound Level. The sound level in the absence of the sound under study.

Annoyance. In acoustics, the overall unwantedness of sound heard in a real-life (as opposed to laboratory) situation.

Articulation Index. An index which describes the suitability of a noise environment for distinguishing speech signals.

Attenuation. Reduction in sound level.

Audiogram. A graph showing, as a function of frequency, the amount in decibels by which a person's threshold of hearing differs from a standard.

Audiometry. The measurement of hearing.

Audiometer. An instrument for measuring the sensitivity of hearing.

A-Weighted Sound Level. A level of sound pressure in which the sound pressure levels of the various frequency bands have been weighted to accord roughly with human aural system frequency sensitivity. The A weighting is defined in standards such as IEC 179 (1973) and ANSI S1.4-1971. (Similarly for B, C, and D weightings.)

Background Noise. Noise other than that from the source being studied.

Bandwidth. The size of a frequency band.

CNR$_c$. See *Composite Noise Rating.*

Composite Noise Rating. The composite noise rating for assessment of community noise; it has been used mainly to assess the influence of various intruding noises such as traffic noise, industrial noise, and aircraft noise on the community.

Continuous Noise. Noise which continues without interruption over a period of time.

Criterion. A standard by which a noise is judged. For example, criteria for an acceptable noise are that it causes no annoyance, interferes with no form of behavior, and damages no being or object.

Day–Night Sound Level. A statistical descriptor of the sound over a 24-hour period taking account of the fact that sounds are more annoying at night than during the day. Calculated by determining the equivalent sound level over a 24-hour period after adding 10 dB(A) to the sound levels occurring in the period 10 pm to 7 am.

dB. See *Decibel.*

dB(A). The unit of A-weighted sound level. (Similarly, dB(B), dB(C), dB(D).)

dB(NP). The unit of Noise Pollution Level.

Deafness. Complete impairment of hearing.

Decibel. In acoustics, the unit of level for sound pressure squared, sound power, sound intensity, etc. Most commonly used for sound pressure level, when it is ten times the logarithm to the base ten of the ratio of rms sound pressure squared and a reference pressure squared.

Direct Wave. The sound wave arriving at a point without prior impingement on a solid object.

Directivity. The directional nature of the sound from a given source.

Drive-By Test. A test in which the source is driven in a prescribed manner past a fixed measuring position.

Effective Perceived Noise Level. A level describing the noisiness of a single aircraft flyover assessed at a number of points near an airport.

EPNdB. Unit of Effective Perceived Noise Level.

EPNL. See *Effective Perceived Noise Level.*

Equivalent Continuous Perceived Noise Level. An index which describes the noise in a period of aircraft flyovers.

Equivalent Sound Level. The level of a constant sound having the same sound energy as an actual time-varying sound over a given period. An energy-averaged sound level, usually but not always of the A-weighted energy.

Excess Attenuation. Attenuation of sound with distance other than that caused by the divergence of the sound wave.

Frequency. The repetition rate of the sound pressure. The characteristic of a sound which influences our perception of it as high or low in pitch.

Frequency Band. A continuum of frequencies from a lower frequency to a higher.

Hertz (Hz). Unit of frequency, in cycles/sec.

Impact Noise. See *Impulse Noise.*

Impulse (Impulsive) Noise. Noise of short duration, typically less than a second. Also called transient or impact noise.

Infrasound. Sound of frequencies below the normal human audible range, often taken to begin at 16 Hz.

Intermittent Noise. Noise which falls below measurable levels one or more times over a given period.

L_{dn}. See *Day–Night Sound Level.*

L_{eq}. See *Equivalent Sound Level.*

Limits. Prohibitions, legal or otherwise, on the amount of sound. Amount may be expressed in terms of sound level, sound exposure, sound duration or any other characteristic of the sound output.

L_n. The sound level exceeded $n\%$ of the time over a given period. Thus L_1, L_{10}, L_{50}, L_{90}.

Loudness. The subjective magnitude of sound.

Masking. The process by which one sound has a reduced loudness when heard in the presence of another.

Misfeasance. The doing of a lawful act in a wrongful manner. In acoustics, the infliction of a sound which could be avoided, attenuated, confined to certain hours, etc., and therefore a product of the mindlessness of the noisemaker.

NC. See *Ranking Curves.*

NEF. See *Noise Exposure Forecast.*

Noise. Unwanted sound. However, "sound" and "noise" are often used virtually synonymously.

Noise and Number Index. An index which describes the noise in a period of aircraft flyovers.

Noise Dose. See *Sound Exposure.*

Noise Exposure Forecast. An index which describes the noise in a period of aircraft flyovers.

Noisiness. The unwantedness of a sound heard in isolation from real-life situations. Also known as perceived noisiness.

Noise Reduction Coefficient. An index of the sound absorptivity of a material. Calculated by averaging the sound absorption coefficients at the frequencies 250, 500, 1000, and 2000 Hz.

Noise Pollution Level. An index describing the noise at a given point over a period of time, which takes account of the equivalent sound level and the standard deviation of the sound level over the period in question.

Notational Background Level. A background sound level established without measurement.

Noy. A unit of noisiness.

NPL. See *Noise Pollution Level.*

NR. See *Ranking Curves.*

Octave Band. A band of frequencies the highest frequency of which has twice the value of the lowest.

One-Third Octave Band. A band of frequencies the highest frequency of which is $(2)^{1/3}$ or $10^{0.1}$ greater than the lowest. There are three such bands in an octave band.

Phon. A unit of loudness.

PNC. See *Ranking Curves.*

Preferred Speech Interference Level. An index which describes the suitability of a noise environment for distinguishing speech signals. Calculated by averaging the sound pressure levels in the octave bands centered at 500, 1000, and 2000 Hz.

PSIL. See *Preferred Speech Interference Level.*

Psychoacoustics. The science of investigating acoustical matters from the standpoint of psychology.

Pure Tone. A sound of a single (i.e., discrete) frequency. Perceived as a "pure" note, e.g., whine, buzz, ring, squeal.

Ranking Curves. Curves which evaluate the suitability of a sound spectrum experienced inside a room or a vehicle. Include the NC, PNC, and NR curves.

Reverberation. The continuation of a sound in an enclosed space after the source has stopped; a result of reflections from the surfaces of the enclosure.

Reverberation Time. The time taken for the sound pressure level in a room to decay by 60 dB after the source has stopped.

SIL. See *Preferred Speech Interference Level.*

Sone. A unit of loudness.

Sonic Boom. The impulse sound from a supersonic flyover of an aircraft.

Sound. A fluctuation of particle displacement or pressure, particularly one resulting in a hearing sensation.

Sound Exposure. The cumulative acoustic stimulation at the ear of a person or persons over a period of time.

Also known as Noise Dose when the exposure of one individual is described.

Sound Level. A frequency-weighted sound pressure level. Assumed to be the A-weighted sound level unless otherwise stated.

Sound Pressure Level. A level of sound pressure expressed in decibels.

Sound Wave. A disturbance consisting of vibrating particles moving through a medium.

Speech Interference Level. See *Preferred Speech Interference Level.*

SPL. See *Sound Pressure Level.*

Threshold of Hearing. The minimum sound pressure level that a person can hear in certain prescribed conditions.

Tone Correction. A number to be added to a scale of noisiness to account for the noisiness of pure tones.

Traffic Noise Index. An index which describes the noise of traffic over a period of time.

Transient Noise. See *Impulse Noise.*

Ultrasound. Sound of frequencies of above the normal human audible range, which is often taken to end at 20 kHz.

Waveform. The graph of pressure versus time at a point through which a wave passes.

CONVERSION FACTORS

To convert	Into	Multiply by	Conversely, multiply by
atm	lb/in^2	14.70	6.805×10^{-2}
	N/m^2	1.0132×10^5	9.872×10^{-6}
°C	°F	$(°C \times 9/5) + 32$	$(°F - 32) \times 5/9$
cm	in	0.3937	2.540
	ft	3.281×10^{-2}	30.48
cm^2	in^2	0.1550	6.452
	ft^2	1.0764×10^{-3}	929
cm^3	in^3	0.06102	16.387
	ft^3	3.531×10^{-5}	2.832×10^4
dyn	lb (force)	2.248×10^{-6}	4.448×10^5
	N	10^{-5}	10^5
dyn/cm^2	lb/ft^2 (force)	2.090×10^{-3}	478.5
	N/m^2	10^{-1}	10
erg	$ft \cdot lb$ (force)	7.376×10^{-8}	1.356×10^7
	J	10^{-7}	10^7
erg/s	W	10^{-7}	10^7
	$ft \cdot lb/s$	7.376×10^{-8}	1.356×10^7
ft	in	12	0.08333
	cm	30.48	3.281×10^{-2}
	m	0.3048	3.281
hp ($550 \, ft \cdot lb/s$)	$ft \cdot lb/min$	3.3×10^4	3.030×10^{-5}
	W	745.7	1.341×10^{-3}
	kW	0.7457	1.341
in	ft	0.0833	12
	cm	2.540	0.3937
	m	0.0254	39.37
$\log_e n$, or $\ln n$	$\log_{10} n$	0.4343	2.303
m	in	39.371	0.02540
	ft	3.2808	0.30481
	yd	1.0936	0.9144
	cm	10^2	10^{-2}
μbars (dyn/cm^2)	lb/in^2	1.4513×10^{-5}	6.890×10^4
	lb/ft^2	2.090×10^{-3}	478.5
	N/m^2	10^{-1}	10
mi (statute)	ft	5280	1.894×10^{-4}
	km	1.6093	0.6214
mi/h	ft/min	88	1.136×10^{-2}
	km/min	2.682×10^{-2}	37.28
	km/h	1.6093	0.6214
N	lb (force)	0.2248	4.448
	dyn	10^5	10^{-5}
N/m^2	lb/in^2 (force)	1.4513×10^{-4}	6.890×10^3
	lb/ft^2 (force)	2.090×10^{-2}	47.85
	dyn/cm^2	10	10^{-1}
lb (force)	N	4.448	0.2248
lb/in^2 (force)	lb/ft^2 (force)	144	6.945×10^{-3}
	N/m^2	6894	1.4506×10^{-4}
W	erg/s	10^7	10^{-7}
	hp ($550 \, ft \cdot lb/s$)	1.341×10^{-3}	745.7

Appendix III

ACOUSTICAL STANDARDS ORGANIZATIONS AND STANDARD DOCUMENTS

The two international standards organizations of most importance in acoustics are the International Organization for Standardization (ISO) and the International Electrotechnical Commission (IEC). Other major organizations which provide guidelines and standards in acoustics, sound analysis, noise control and noise abatement are the American National Standards Institute (ANSI), the Canadian Standards Association (CSA), the American Society for Testing and Materials (ASTM), the Society of Automotive Engineers (SAE), and the American Society of Heating, Refrigerating and Air-Conditioning Engineers (ASHRAE).

ISO and IEC standards may be ordered from: each nation's standards organization; (in Canada, they may be ordered from the Standards Council of Canada, 350 Sparks Street, Ottawa, Canada K1R 7S8). Standards from ANSI, ASTM, SAE and ASHRAE may be ordered from each organization (see listings and addresses below); in Canada, they may be ordered from the Standards Council of Canada. CSA standards may be ordered from the Canadian Standards Association (see listing and address below).

International standards are often identical to certain national standards (i.e., the national standard is an endorsement of the international standard—in particular, ANSI and CSA standards).

In addition to international and national standards, various industrial and professional institutes and associations have their own acoustical standards or test methods. For instance, the Acoustical Society of America has established certain standards re methods for the measurement of hearing protectors and noise emitted by engine-powered equipment. The addresses and standards of these associations are given below for reference.

INTERNATIONAL

International Electrotechnical Commission (IEC)

(Available from American National Standards Institute, 1430 Broadway, New York, NY 10018)

IEC 58-08 (1960)	International Electrotechnical Vocabulary 08: Electro-Acoustics
IEC 118 (1959)	Measurements of the Electro-Acoustical Characteristics of Hearing Aids
IEC 123 (1961)	Recommendations for Sound Level Meters
IEC 124 (1960)	Rated Impedances and Dimensions for Loudspeakers
IEC 126 (1961)	IEC Reference Coupler for the Measurement of Hearing Aids Using Earphones Coupled to the Ear by Means of Ear Inserts
IEC 177 (1965)	Pure Tone Audiometers for General Diagnostic Purposes
IEC 178 (1965)	Pure Tone Screening Audiometers
IEC 179 (1965)	Precision Sound Level Meters (revised 1973)
IEC 184 (1965)	Specifying the Characteristics of Electromechanical Transducers for Shock and Vibration Measurements
IEC 200 (1966)	Measurement of Loudspeakers
IEC 222 (1966)	Specifying the Characteristics of Auxiliary Equipment for Shock and Vibration Measurement
IEC 225 (1966)	Octave, Half-Octave and Third-Octave Band Filters Intended for the Analysis of Sound and Vibration
IEC 263 (1968)	Scales and Sizes for Plotting Frequency Characteristics
IEC 268 (1968)	Sound System Equipment
IEC 303 (1970)	Provisional Reference Coupler for the Calibration of Earphones Used in Audiometry
IEC 318 (1970)	Artificial Ear for the Calibration of Earphones Used in Audiometry
IEC 327 (1971)	Precision Method for Pressure Calibration of One-Inch Standard Condenser Microphones by the Reciprocity Technique
IEC 534 PT8 (1986)	Laboratory Measurement of Noise Generated by Aerodynamic Flow through Industrial Process Control Valves, First Edition
IEC 537-76	Frequency Weighting for the Measurement of Aircraft Noise (D-Weighting), First Edition

IEC 561-76	Electro-Acoustical Measuring Equipment for Aircraft Noise Certification, First Edition
IEC 651-79	Sound Level Meters, First Edition
IEC 704 P+1 (1982)	Test Code for the Determination of Airborne Acoustical Noise Emitted by Household and Similar Electrical Appliances, Part 1: General Requirements, First Edition
IEC 805-85	Integrating-Averaging Sound Level Meters, First Edition
IEC 578-77	Multichannel Amplitude Analyzers Types: Main Characteristics and Technical Requirements, First Edition

International Organization for Standardization (ISO)

(Available from American National Standards Institute, 1430 Broadway, New York, NY 10018)

Note: ISO documents prefixed "R" are recommendations; some may already have become standards due to a continual upgrading of documents. Some drafts may now have become standards.

ISO 31/VII (1965)	Quantities and Units of Acoustics, First Edition, Erratum 1981, Amended (1) 1985
ISO 16-1975	Standard Tuning Frequency (Standard Musical Pitch)
ISO/R 131 (1959)	Expression of the Physical Magnitudes of Sound and Noise
ISO/R 140 (1960)	Field and Laboratory Measurement of Airborne and Impact Sound Transmission
ISO/R 226 (1961)	Normal Equal Loudness Contours for Pure Tones and Normal Threshold of Hearing under Free Field Listening Conditions
ISO 266-1975	Preferred Frequencies for Acoustical Measurements
ISO/R 354 (1963)	Measurements of Absorption Coefficients in a Reverberation Room
ISO/R 357 (1963)	Expression of the Power and Intensity Levels of Sound or Noise
ISO/R 362 (1964)	Measurement of Noise Emitted by Vehicles
ISO 389-1985	Standard Reference Zero for the Calibration of Pure-Tone Air Conduction Audiometers (adden. 1986)
ISO 454-1975	Relation Between Sound Pressure Levels of Narrow Bands of Noise in a Diffuse Field and in a Frontally-Incident Free Field for Equal Loudness
ISO/R 495 (1966)	General Requirements for the Preparation of Test Codes for Measuring the Noise Emitted by Machines

ISO/R 507 (1970)	Procedure for Describing Aircraft Noise Around an Airport
ISO/R 512 (1966)	Sound Signalling Devices on Motor Vehicles
ISO 532-1975	Method for Calculating Loudness Level
ISO/R 717 (1968)	Rating of Sound Insulation for Dwellings
ISO 1680 (1970)	Test Code for the Measurement of Airborne Noise Emitted by Rotating Electrical Machinery
ISO 1683-1983	Preferred Reference Quantities for Acoustic Levels, First Edition
ISO/R 1761 (1970)	Monitoring Aircraft Noise Around an Airport
ISO/R 1996 (1971)	Assessment of Noise with Respect to Community Response
ISO 1996 Part 1 (1982)	Description and Measurement of Environmental Noise
ISO 1996 Part 2 (1983)	Acquisition of Data Pertinent to Land Use
ISO 1996 Part 3 (1983)	Application to Noise Limits
ISO 1999-1975	Assessment of Occupational Noise Exposure for Hearing Conservation Purposes
ISO 2017-1972	Specifying Characteristics for Mechanical Isolation
ISO 2151-1972	Measurement of Airborne Noise Emitted by Compressors (prime mover units intended for outdoor use)
ISO 2204-1973	Guide to the Measurement of Airborne Acoustical Noise and Evaluation of its Effect on Man, 2nd Edition 1979
ISO 2249-1973	Description and Measurement of Physical Properties of Sonic Booms
ISO 2922-1975	Measurement of Noise Emitted by Vessels on Inland Waterways and Harbours
ISO 2923-1975	Measurement of Noise on Board Vessels
ISO 3095-1975	Measurement of Noise Emitted by Railbound Vehicles
ISO 3352-1975	Assessment of Noise with Respect to its Effect on the Intelligibility of Speech
ISO 3382-1975	Measurement of Reverberation Time in Auditoria
ISO 3740-1975	Determination of Sound Power Levels of Noise Sources— Guidelines for the Use of Basic Standards and for the Preparation of Noise Test Codes
ISO 3741-1975	Determination of Sound Power Levels of Noise Sources— Precision Method for Broad Band Sound Sources Operating in Reverberation Rooms

ISO 3742-1975   Determination of Sound Power Levels of Noise Sources—Precision Methods for Discrete Frequency and Narrow Band Sound Sources Operating in Reverberation Rooms

ISO 3743-1976   Determination of Sound Power Levels of Noise Sources—Engineering Methods for Special Reverberant Test Rooms

ISO 3744-1976   Determination of Sound Power Levels of Noise Sources—Engineering Methods for Free-Field Conditions over a Reflecting Plane

ISO 3745-1977   Determination of Sound Power Levels of Noise Sources—Precision Methods for Sources Operating in Anechoic Rooms

ISO 3746-1977   Determination of Sound Power Levels of Noise Sources—Survey Method

ISO 2880   Determination of Sound Power Emitted by Small Noise Sources in Reverberation Rooms. Part 1: Broad Band Sources

ISO 2946   Determination of Sound Power Emitted by Small Noise Sources in Reverberation Rooms. Part 2: Discrete Frequencies and Narrow Band Sources

ISO 4869-1981   Measurement of Sound Attenuation of Hearing Protectors—Subjective Method, First Edition

ISO 4871-1984   Noise Labelling of Machinery and Equipment, First Edition

ISO 5135-1984   Determination of Sound Power Levels of Noise from Air Terminal Devices, High/Low Velocity/Pressure Assemblies, Dampers and Valves by Measurement in a Reverberation Room, First Edition

ISO 7029-1984   Threshold of Hearing by Air Conduction as a Function of Age and Sex for Otologically Normal Persons, First Edition

ISO 7574-1985 Part 1   Statistical Methods for Determining and Verifying Stated Noise Emission Values of Machinery and Equipment—Part 1: General Considerations and Definitions, First Edition

ISO 7574-1985 Part 2   Statistical Methods for Determining and Verifying Stated Noise Emission Values of Machinery and Equipment—Part 2: Methods for Stated Values for Individual Machines, First Edition

ISO 7574-1985 Part 3   Statistical Methods for Determining and Verifying Stated Noise Emission Values of Machinery and Equipment—Part 3: Simple (Transition) Method for Stated Values for Batches of Machines, First Edition

ISO 7574-1985 Part 4   Statistical Methods for Determining and Verifying Stated Noise Emission Values of Machinery and Equipment—Part 4: Methods for Stated Values for Batches of Machines, First Edition

ISO 140-1985   Measurements of Sound Insulation in Buildings and of Building Elements—Part 9: Laboratory Measurement of Room-to-Room Airborne Sound Insulation of a Suspended Ceiling with a Plenum Above, First Edition

ISO 717-1982   Rating of Sound Insulation in Buildings and of Building Elements
Part 1: Airborne Sound Insulation in Buildings and of Interior Building Elements, First Edition
Part 2: Impact Sound Insulation, First Edition
Part 3: Airborne Sound Insulation of Facade Elements and Facades, First Edition

ISO 3381-1978   Measurement of Noise Inside Rail-Bound Vehicles, First Edition

ISO 4872-1978   Measurement of Airborne Noise Emitted by Construction Equipment Intended for Outdoor Use—Method for Determining Compliance with Noise Limits, First Edition (Erratum 1979)

ISO 5128-1980   Acoustic Measurement of Noise Inside Motor Vehicles, First Edition

ISO 354-1985   Measurement of Sound Absorption in a Reverberation Room, First Edition

ISO 6394-1985   Measurement of Airborne Noise Emitted by Earth Moving Machinery—Operator's Position—Stationary Test Condition, First Edition

ISO 7188-1985   Measurement of Noise Emitted by Passenger Cars under Conditions Representative of Urban Driving, First Edition

ISO 226-1987   Normal Equal Loudness Level Contours, First Edition

ISO 2603-83   General Characteristics and Equipment for Booths for Simultaneous Interpretation

ISO 6189-83   Pure-Tone Air Conduction Threshold Audiometry for Hearing Conservation Purposes

ISO 7029-84	Threshold of Hearing by Air Conduction as a Function of Age and Sex for Otologically Normal Persons
ISO 7566-87	Standard Reference Zero for the Calibration of Pure-Tone Bone Conduction Audiometers

UNITED STATES OF AMERICA

Acoustical and Board Products Association (ABPA)

205 West Tuohy Avenue
Park Ridge, IL 60068

Acoustical Society of America (ASA)

American Institute of Physics
333 East 45th Street
New York, NY 10017

ASA STD1-1975	Method for the Measurement of Real-Ear Protection of Hearing Protectors and Physical Attenuation of Earmuffs
ASA STD3-1975	Test-Site Measurement of Noise Emitted by Engine-Powered Equipment
ASA STD4-1975	Method for Rating the Sound-Power Spectra of Small Stationary Noise Sources (ANSI S3.17-1975)
ASA STD5-1976	Method for the Designation of Sound Power Emitted by Machinery and Equipment
ASA 66-86	Methods for the Evaluation of the Potential Effect on Human Hearing of Sounds with Peak A-Weighted Sound Pressure Levels above 120 Decibels and Peak C-Weighted Sound Pressure Levels below 140 Decibels (Draft ANSI S3-28-1986)
ASA S3.20-73	Psychoacoustical Terminology (rev. 1978)
ASA S3-W-39 1960	Effects of Shock and Vibration on Man
ASA 10-79	Guidelines for Use of Sound Power Standards and for the Preparation of Noise Test Codes (rev. 1985)
ASA 14-80	Engineering Methods for the Determination of Sound Power Levels of Noise Sources for Essentially Free-Field Conditions over a Reflecting Plane (ANSI S1.34)
ASA 16-79	Survey Methods for the Determination of Sound Power Levels of Noise Sources (rev. 1985) (ANSI S1.36-79)

ASA 21-77	Rating Noise with Respect to Speech Interference (ANSI S3.14-77)
ASA 22-80	Sound Level Descriptors for Determination of Compatible Land Use (ANSI S3.23-80)
ASA 25-78	Specs. for Personal Noise Dosimeters (ANSI S1.25-78)
ASA 38-79	Guide for Evaluation of Human Exposure to Whole-Body Vibration (ANSI S3.18-79)
ASA 47-83	Spec. for Sound Level Meters (ANSI S1.4A-1985)
ASA 49-83	Guidelines for Preparation of Standard Procedures to Determine the Noise Emission from Sources (ANSI S12.1-83)
ASA 57-85	Statistical Methods for Determining and Verifying Stated Noise Emission Values of Machinery and Equipment (ANSI S12.3-85)
ASA 64-86	Design Response of Weighting Networks for Acoustical Measurements (ANSI S1.42-86)
ASA S1.8-69	Preferred Reference Quantities for Acoustical Levels (rev. 1974)
ASA S1.13-71	Methods for Measurement of Sound Pressure Levels (rev. 1976)
ASA 9-77	Criteria for Permissible Ambient Noise during Audiometric Testing (ANSI S3.1-77)
ASA 19-78	Methods for Manual Pure-Tone Threshold Audiometry (ANSI S3.21-78)
ASA 41-81	Reference Equivalent Threshold Force Levels for Audiometric Bone Vibrators (ANSI S3.26-81)
ASA S3.6-69	Specs. for Audiometers (rev. 1973)
ASA 49-83	Guidelines for Preparation of Standard Procedures to Determine the Noise Emission from Sources (ANSI S12.1-83)

Air Conditioning and Refrigeration Institute (ARI)

1815 North Fort Meyer Drive
Arlington, VA 22209

ARI 575-87	Method of Measuring Machinery Sound within an Equipment Space
ARI 270-67	Sound Rating of Outdoor Unitary Equipment
ARI 275-69	Application of Sound Rated Outdoor Unitary Equipment
ARI 443-66	Rooms Fan-Cool Air Conditioner
ARI 446-68	Sound Rating of Room Air-Induction Units
ARI 579-73	Standard for Measuring Machinery Sound within Equipment Rooms

Air Diffusion Council (ADC)

435 North Michigan Avenue
Chicago, IL 60611

AD-63   Measurement of Room-to-Room Sound
Transmission through Plenum Air
Systems

1062-R2   Equipment Test Code

Air Moving and Conditioning Association (AMCA)

30 West University Drive
Arlington Heights, IL 60004

AMCA 300-67   Test Code for Sound Rating of Air
Moving Devices

AMCA Bul.   Application of Sound Power Rating
303   for Ducted Air Moving Devices
(1965)

AMCA Pub.   Certified Sound Ratings Program for
311-67   Air Moving Devices

American Boat and Yacht Council

15 East 26th Street
New York, NY 10010

ABYC H17   Practices and Standards Covering
Insulating, Soundproofing and
Sheathing Materials and Fire Retardant
Coatings

American Gear Manufacturers Association (AGMA)

1330 Massachusetts Avenue, N.W.
Washington, DC 20005

295.02   AGMA Standard Specification for
Measurement of Sound in High-Speed
Helical and Herringbone Gear Units
(1965)

299.01.1980   Sound Manual Section II: Sources,
Specifications and Levels of Gear
Sound

American National Standards Institute (ANSI)

1430 Broadway
New York, NY 10018

S1.1-1960   Acoustical Terminology (including
(R 1971)   mechanical shock and vibration)

S1.2-1962   Physical Measurement of Sound
(R 1971)   (revised by S1.13-1971 and S1.21-
1972)

S1.4-1971   Specifications for Sound Level Meters
(replaced by ASA 47-83)

S1.5-1963   Loudspeaker Measurements
(R 1971)

S1.6-1967   Preferred Frequencies and Band
(R 1971)   Numbers for Acoustical
Measurements (replaced by ASA 53-
84)

S1.7-1970   Method of Test for Sound Absorption
of Acoustical Materials in
Reverberation Rooms

S1.8-1969   Preferred Reference Quantities for
Acoustical Levels (rev. 1974)

S1.10-1966   Calibration of Microphones (Method)
(R 1971)

S1.11-1966   Specifications for Octave, Half-Octave
(R 1971)   and Third-Octave Band Filter Sets

S1.12-1967   Laboratory Standard Microphones
(R 1972)

S1.13-1971   Methods for the Measurement of
Sound-Pressure Levels (revision of
S1.2-1962)

S1.20-1972   Calibration of Underwater
Electroacoustic Transducers

S1.21-1972   Methods for the Determination of
Sound-Power Levels of Small
Sources in Reverberation Rooms
(revision of S1.2-1962)

S1.23-1972   Method for Designating the Sound
Power Emitted by Machinery and
Equipment

S2.2-1959   Calibration of Shock and Vibration
(R 1971)   Pickups

S2.3-1964   High-Impact Shock Machine for
(R 1970)   Electronic Devices

S2.4-1960   Specifying the Characteristics of
(R 1971)   Auxiliary Equipment for Shock and
Vibration Measurements

S2.5-1962   Specifying the Performance of
(R 1971)   Vibrating Machines

S2.6-1963   Specifying the Mechanical Impedance
(R 1971)   of Structures

S2.7-1964   Terminology for Balancing Rotating
(R 1971)   Machines

S2.8-1972   Describing the Characteristics of
Resilient Mountings

S2.10-1971   Methods for Analysis and Presentation
of Shock and Vibration Data

S2.11-1969   Selection of Calibrations and Tests for
Electrical Transducers Used for
Measuring Shock and Vibration

S2.15-1972   Design Construction and Operation of
Class HI (High-Impact) Shock-
Testing Machines for Lightweight
Equipment

S3.1-1960   Criteria for Background Noise in
(R 1971)   Audiometer Rooms

S3.2-1960   Method for Measurement of
(R 1971)   Monosyllabic Word Intelligibility

S3.2-1960   Method for Measurement of
(R 1971)   Electroacoustical Characteristics of
Hearing Aids

S3.4-1968   Procedure for the Computation of the
(R 1980)   Loudness of Noise (replaced by ASA
37-80)

S3.5-1969 (R 1973) — Methods for the Calculation of the Articulation Index

S3.6-1969 — Specifications for Audiometers

S3.7-1973 — Method for Coupler Calibration of Earphones

S3.8-1967 (R 1971) — Method of Expressing Hearing Aid Performance

S3.13-1977 — Artificial Head-Bone for the Calibration of Audiometer Bone Vibrators

S3.15 — Method for the Measurement of Community Noise

S3.17-1975 — Method for Rating the Sound Power Spectra of Small Stationary Noise Sources

S3.19-1974 — Method for the Measurement of Real-Ear Protection of Hearing Protectors and Physical Attenuation of Earmuffs (revision and redesignation of Z24.22-1957 ASA Std. 1-1975)

S3.20-1973 — Psychoacoustical Terminology

S3-W-39-60 — Effects of Shock and Vibration on Man

S5.1-1971 — Test Code for the Measurement of Sound from Pneumatic Equipment

S6.2-1973 — Exterior Sound Levels for Snowmobiles

S6.3-1973 — Sound Levels for Passenger Cars and Light Trucks

S6.4-1973 — Definitions and Procedures for Computing the Effective Perceived Noise Level for Flyover Aircraft Noise

S9.1-1975 — Guide for the Selection of Mechanical Devices Used in Monitoring Acceleration Induced by Shock

Y10.11-1953 — Letter Symbols for Acoustics

Y32.18-1968 — Symbols for Mechanical and Acoustical Elements Used in Schematic Diagrams

Z24.9-1949 (R 1971) — Coupler Calibration of Earphones

Z24.21-1957 (R 1971) — Specifying the Characteristics of Pickups for Shock and Vibration Measurement

Z24.22-1957 (R 1971) — Measurement of the Real-Ear Attenuation for Ear Protectors at Threshold

Z24-X-2 — The Relations of Hearing Loss to Noise Exposure

American Society of Heating, Refrigerating and Air-Conditioning Engineers (ASHRAE)

345 East 47th Street
New York, NY 10017

36-62 — Measurement of Sound Power Radiated from Heating, Refrigerating and Air-Conditioning Equipment

36A-63 — Method of Determining Sound Power Levels of Room Air-Conditioners and Other Ductless, Through-the-Wall Equipment

36B-63 — Method of Testing for Rating the Acoustic Performance of Air Control and Terminal Devices and Similar Equipment

36-72 — Method of Testing for Sound Rating Heating, Refrigerating and Air Conditioning Equipment

— — *ASHRAE Guide and Data Book.* Chapter on Assessment of Noise from Ventilating and Air-Conditioning Systems (not a standard, but a useful design guide—a regular publication of ASHRAE)

CH 32-84 — Sound and Vibration Control (*ASHRAE Handbook 1984:* Systems)

CH 7-85 — Sound and Vibration Fundamentals (*ASHRAE Handbook 1985:* Fundamentals)

American Society of Mechanical Engineers

345 East 47th Street
New York, NY 10017

ASME Y10.11 1984 — Letter Symbols and Abbreviations for Quantities Used in Acoustics

ASME Y32.18 1972 — Symbols for Mechanical and Acoustical Elements Used in Diagrams

ASME PTC36 1985 — Measurement of Industrial Sound (Performance Test Code)

American Society for Testing and Materials (ASTM)

1916 Race Street
Philadelphia, PA 19103

C367-57 — Strength Properties of Prefabricated Architectural Acoustical Materials

C384-58 — Standard Method of Test for Impedance and Absorption of Acoustical Materials by the Tube Method (Reapproved 1972)

C423-81A — Standard Method of Test for Sound Absorption and Absorption Coefficient of Acoustical Materials in Reverberation Rooms (ANSI S1.7-1970—Reapproved 1984)

C522-69 — Airflow Resistance of Acoustical Materials

C634-69 — Definitions of Terms Relating to Acoustical Tests of Building Construction and Materials (rev. 1973)

E90-75 — Standard Recommended Practice for Laboratory Measurement of Airborne Sound Transmission Loss of Building Partitions

E336-71 — Standard Recommended Practice for Measurement of Airborne Sound Insulation in Buildings

E413-73 — Standard Classification for Determination of Sound Transmission Class

C636-69   Installation of Metal Ceiling Suspension Systems for Acoustical Tile and Lay-In Panels

C634-86   Definition of Terms Relating to Environmental Acoustics

RM14-3   Steady-State Determination of Changes in Sound Absorption of a Room

RM14-4   Laboratory Measurement of Impact Sound Transmission through Floor–Ceiling Assemblies Using the Tapping Machine

E569-85   Practice for Acoustic Emission Monitoring of Structures during Controlled Stimulation

E1110-86   Classification for Determination of Articulation Class

E1130-86   Standard Test Method for Objective Measurement of Speech Privacy in Open Offices Using Articulation Index

E596-86   Standard Method for Laboratory Measurement of Noise Reduction of Sound-Isolating Enclosures

E750-80   Measuring the Operating Characteristics of Acoustic Emission Instrumentation, Practice for

E1014-84   Guide for Measurement of Outdoor A-Weighted Sound Levels

E1124-86   Standard Test Method for Field Measurement of Sound Power Level by the Two-Surface Method

E1139-87   Standard Practice for Continuous Monitoring of Acoustic Emission from Metal Pressure Boundaries

B615-76   Measuring Electrical Contact Noise in Sliding Electrical Contacts, Practice for

E717-80   Guide for the Preparation of the Accreditation Section of Acoustical Test Standards

American Textile Machinery Association (ATMA)

1730 M Street, Northwest
Washington, DC 20036

Anti-Friction Bearing Manufacturers Association (AFBMA)

60 East 42nd Street
New York, NY 10017

AFBMA Std.   Roller Bearing Vibration and Noise
Sect. 13
(1968)

Association of Home Appliance Manufacturers

20 North Wacker Drive
Chicago, IL 60606

RAC-2-SR   Room Air-Conditioning Sound Rating (1971)

California Redwood Association

617 Montgomery Street
San Francisco, CA 94111

Compressed Air and Gas Institute (CAGI)

2130 Keith Building
Cleveland, OH 44115

CAGI Test Code   Measurement of Sound from
S5.1-1971   Pneumatic Equipment

Diesel Engine Manufacturers Association (DEMA)

2130 Keith Building
Cleveland, OH 44115

Electronic Industries Association (EIA)

2001 I Street Northwest
Washington, DC 20006

Factory Mutual Systems

184 High Street
Boston, MA 02110

Loss Prevention   Insulating and Acoustical Materials
Data 1-11
(1952)

Federal Specifications

Specification Sales
Building 197, Washington Navy Yard
General Services Administration
Washington, DC 20407

HH-I-545   Insulation, Thermal and Acoustical
(1970)   (Mineral Fibre, Duct Lining Material)
SS-S-111a   Sound Controlling Materials (Trowel and
Amend.   Spray Applications)
1 (1968)
SS-S-118a   Sound Controlling Blocks and Boards
Amend.
1 (1967)

General Services Administration

Washington, DC

PBS-C.1   Direct Measurement of Speech-Privacy Potential

PBS-C.2   Sufficient Verification of Speech-Privacy Potential

Hearing Aid Industry Conference, Inc.

75 East Wacker Drive
Chicago, IL 60001

61-1   Standard Method of Expressing Hearing-Aid Performance

65-1   Interim Bone Conduction Thresholds for
        Audiometry

Home Ventilating Institute (HVI)

230 North Michigan Avenue
Chicago, IL 60601

HVI Test   Air Flow Test Procedure
Pra.
(1968)

Industrial Silencer Manufacturers Association (ISMA)

c/o Burgess Industries
P.O. Box 47146
Dallas, TX 75247

ISMA Test   Standard Laboratory Test Procedure for
Procedures   Insertion Loss Measurement of Intake
                and Exhaust Silencers for
                Reciprocating Engines (1974)
                Insertion Loss of Pressure Reduction
                and Regulator Valve Silencers

Institute of Electrical and Electronic Engineers (IEEE)

445 Hoes Lane
Piscataway, NJ 08854

IEEE 85   Airborne Noise Measurements on
               Rotating Electric Machinery (1973)
IEEE 219   Loudspeaker Measurements (ANSI
               S1.5-1963)
IEEE 258   Methods of Measurement for Close-
               Talking Pressure Type
               Microphones (1965)
IEEE 269   Method for Measuring Transmission
               Performance of Telephone Sets
               (1966)
IEEE 685-1985 Measurement of Audible Noise from
               High Voltage Transmission Lines
IEEE C37.082   Measurement of Sound Pressure
1982   Levels of AC Power Circuit
               Breakers, Methods of (ANSI/
               IEEE)
IEEE 748-1979 Spectrum Analyzers

Instrument Society of America (ISA)

400 Stanwix Street
Pittsburgh, PA 15222

RP37.2   Guide for Specifications and Tests for
(1964)   Piezoelectric Accelerometer Transducers
            for Aerospace Testing
S37.10   Specifications and Tests for Piezoelectric
(1969)   Pressure and Sound Pressure Transducers

International Conference of Building Officials (ICBO)

5360 South Warkman Mill Road
Whittier, CA 90601

STC   Laboratory Determination of Airborne
  UBC35-1   Sound Transmission Class (STC)

Military Specifications (U.S.)

Commanding Officer
Naval Publications and Forms Center
5801 Tabor Avenue
Philadelphia, PA 19120

A-8806A and   Acoustical Noise Level in Aircraft
  Amend. 1
  (1966)
N-83155A and   Noise Suppressor System, Aircraft
  Amend. 1   Turbine Engine Ground Run-Up
  (1970)
N-83158A   Noise Suppressor Systems, Engine
  (1970)   Test Stand A/F32T-2 and A/F32T-
                  3; for Turbojet and Turbofan
                  Engines
S-3151a   Sound Level Measuring and
  (1967)   Analyzing Equipment
S-008806B   Sound Pressure Levels in Aircraft
  (1970)

Motor Vehicles Manufacturers Association (MVA)

320 New Center Building
Detroit, MI 48202

National Electrical Manufacturers Association (NEMA)

815 15th Street, Suite 438
Washington, DC 20005

SM33-1964   Gas Turbine Sound and Its Reduction
MG 1   Motors and Generators (Noise)
Sect. 4.3.2 Method of Measuring Machine Noise

National Fluid Power Association (NFPA)

3333 North Mayfair Road
Milwaukee, WI 53222

National Machine Tool Builders Association (NMTBA)

7901 West Park Drive
McLean, VA 22101

NMTBA Std.   Noise Measurement Techniques
  1970

National School Supply and Equipment Association (NSSEA)

Folding Partition Subsection
1500 Wilson Boulevard
Arlington, VA 22209

NSSEA Test Proc. (1966)	Testing Procedures for Measuring Sound Transmission Loss through Movable and Folding Walls

Power Saw Manufacturers Association (PSMA)

Box 7256
Belle View Station
Alexandria, VA 22307

N.1-66	Noise Levels
N.1-67	Noise Octave Band Measurement

Radio Manufacturers Association

1317 F Street North West
Washington, DC 20004

SE 105 (1949)	Microphones for Sound Equipment

Society of Automotive Engineers (SAE)

400 Commonwealth Drive
Warrendale, PA 15096

SAEJ34	Exterior Sound Level Measurements for Pleasure Motorboats
SAEJ47	Maximum Sound Level Potential for Motorcycles
SAEJ57	Sound Level of Highway Truck Tires
SAEJ87	Exterior Sound Level for Powered Mobile Construction Equipment
SAEJ88a	Exterior Sound Level Measurement for Powered Mobile Construction Equipment
SAEARP 1307-79	Measurement of Exterior Noise Produced by Aircraft Auxiliary Power Units and Associated Equipment during Ground Operation
SAEAIR 1935-85	Method of Controlling Distortion of Inlet Air Flow during Static Acoustical Tests of Turbofan Engines and Fan Rigs
SAEJ88-86	Sound Measurement—Earth Moving Machinery—Exterior
SAEJ184-85	Recommended Practice for Qualifying a Sound Data Acquisition System
SAEJ247-80	Recommended Practice for Instrumentation for Measuring Acoustic Impulses within Vehicles
SAEJ919-86	Sound Measurement—Earth Moving Machinery—Operator—Singular Type
SAEJ1008-87	Recommended Practice, Sound Measurement—Self Propelled Agricultural Equipment—Exterior
SAEJ1074-84	Recommended Practice, Engine Sound Level Procedure
SAEJ1075-87	Sound Measurement—Construction Site
SAEJ1166-87	Sound Measurement—Earth Moving Machinery—Operator Work Cycle
SAEJ1174-85	Operator Ear Sound Level Measurement Procedure for Small Engine-Powered Equipment
SAEJ1175-85	Bystander Sound Level Measurement Procedure for Small Engine-Powered Equipment
SAEJ1262-85	Sound Measurement—Trenching Machines
SAEJ1477-86	Measurement of Interior Sound Levels of Light Vehicles
SAEHS-184-79	Surface Vehicle Sound Measurement Procedures
SAEJ192a	Exterior Sound Levels for Snowmobiles
SAEJ331a	Sound Levels for Motorcycles (1975)
SAEJ336a	Sound Levels for Truck Cab Interiors
SAEJ336b	Exterior Sound Level Evaluation for Heavy Trucks and Buses
SAEJ337	Performance of Vehicle Traffic Horns
SAEJ919a	Sound Level Measurements at Operator Station
SAEJ952b	Sound Levels for Engine Powered Equipment
SAEJ986a	Sound Level for Passenger Cars and Light Trucks (ANSI S6.3-1973)
SAEJ994a	Criteria for Back-Up Alarm Devices
SAESP373	Truck Tire Noise
SAEJ1046	Exterior Sound Level Measurement Procedures for Small Engine-Powered Equipment
SAEJ1060	Subjective Rating Scale for Evaluation of Noise and Ride Comfort Characteristics Related to Motor Vehicle Tires
SAEJ1077	Exterior Sound Level for Trucks with Auxiliary Equipment
SAEJ1105	Performance, Test and Application Criteria of Electrically Operated Forward Warning Horn for Mobile Construction Machinery
SAEAIR817	A Technique for Narrow Band Analysis of a Transient
SAEAIR852	Methods of Comparing Aircraft Takeoff and Approach Noises
SAEAIR902	Determination of Minimum Distance from Ground Observer to Aircraft for Acoustic Tests

SAEAIR923   Method for Calculating the Attenuation of Aircraft Ground-to-Ground Noise Propagation During Takeoff and Landing

SAEAIR1151   Evaluation of Headphones for Demonstration of Aircraft Noise

SAEAIR1216   Comparison of Ground Runup and Flyover Noise

SAEARP796   Measurements of Aircraft Exterior Noise in the Field

SAEARP865a   Definitions and Procedures for Computing the Perceived Noise Level of Aircraft Noise

SAEARP866   Standard Values of Atmospheric Absorption as a Function of Temperature and Humidity for Use in Evaluating Aircraft Flyover Noise

SAEARP876   Jet Noise Prediction

SAEARP1071   Definitions and Procedures for Computing the Effective Perceived Noise Level for Flyover Aircraft Noise (1973)

Steel Door Institute (SDI)

2130 Keith Building
Cleveland, OH 44115

Woodworking Machinery Manufacturers Association (WMMA)

1900 Arch Street
Philadelphia, PA 19103

AUSTRALIA

Standards Association of Australia

Standards House
80–86 Arthur Street
North Sydney, NSW 2060

CANADA

Canadian Standards Association

178 Rexdale Boulevard
Rexdale, Ontario M9W 1R3

CSAZ107.0
1984   Definitions of Common Acoustical Terms Used in CSA Standards; (Gen. Instr. 1)

CSAZ107.51
M1980   Procedure for In Situ Measurement of Noise from Industrial Equipment (rev. 1985); (Gen. Inst. 1)

CSAZ107.53
M1982   Procedure for Performing a Survey of Sound Due to Industrial, Institutional or Commercial Activity (Gen. Inst. 1)

CSAZ107.55
M1986   Recommended Practice for the Prediction of Sound Levels Received at a Distance from an Industrial Plant; (Gen. Instr. 1)

CSAZ107.56
M1986   Procedures for the Measurement of Occupational Noise Exposure

CSAZ107.4
M1986   Pure Tone Air Conduction Audiometers for Hearing Conservation and for Screening; (Gen. Instr. 1)

CSA Task Force on Community Noise 84   Statement of General Principles and Recommendations

CSAZ107.21
M1977   Procedure for Measurement of the Maximum Exterior Sound Level of Pleasure Motorboats

CSAZ107.22
M1977   Procedure for the Measurement of the Maximum Exterior Sound Level of Stationary Trucks with Governed Diesel Engines

CSAZ107.23
M1977   Procedure for the Measurement of the Maximum Interior Sound Level in Trucks with Governed Diesel Engines

CSAZ107.25
M1983   Procedure for the Measurement of the Exhaust Sound Level of Stationary Motorcycles (rp: 11/83)

CSAZ107.31
M1986   Test Procedures for the Measurement of Sound Levels from Agricultural Machines

CSAZ107.32
M1986   Test Procedures for the Measurement of Sound Emitted from Construction, Forestry and Mining Machines to the Operator Station and Exterior of the Machine (rp: 02/87)

CSAZ107.52
M1983   Recommended Practice for the Prediction of Sound Pressure Levels in Large Rooms Containing Sound Sources

CSAZ107.54
M1985   Procedure for the Measurement of Sound and Vibration Due to Blasting Operations

CSAZ107.71
M1981   Measurement and Rating of the Noise Output of Consumer Appliances

FRANCE

Association Française de Normalization

Tour Europe
Cedex 7
92080 Paris—La Defense

FEDERAL REPUBLIC OF GERMANY

Deutscher Normenausschuss

4–7 Burggrafenstrasse
1 Berlin 30

INDIA

Indian Standards Institution

Manak Bhavan
9 Bahadur Shah Zafar Marg
New Delhi 110001

ITALY

Ente Nazionale Italiano di Unificazione

Piazza Armando Diaz 2
1 20123 Milano

JAPAN

Japanese Industrial Standards Committee

Ministry of International Trade and Industry
3-1, Kasumigasekil, Chiyodaku
Tokyo

THE NETHERLANDS

Nederlands Normalisatie-Institut

Polakweg 5
Rijswijk (ZH)-2108

SWEDEN

Sveriges Standardiseriugskomission

Box 3295
S-103 66 Stockholm 3

SWITZERLAND

Association Suisse de Normalisation

Kirchenweg 4
Postfach 8032 Zurich

UNITED KINGDOM

British Standards Institution

2 Park Street
London W1A 2BS

USSR

Gosudarstvennyj Komitet Standartov

Soveta Ministrov SSSR
Leninsky Prospekt 9b
Moskva 1170 49

USEFUL ACOUSTICS PERIODICALS

Acoustics in Canada
Canadian Acoustical Association
P.O. Box 3651, Station C
Ottawa, Ontario
K1Y 4J1

Journal of Acoustical Society of America
America Institute of Physics
335 East 45th Street
New York, NY 10017

Journal of Sound and Vibration
Academic Press Inc.
111 Fifth Avenue
New York, NY 10003
Also:
24–28 Oval Road
London NW1 7DX
England

Noise Control Engineering
9 Saddle Road
Cedar Knolls, NJ 07927

Noise Control Report
Business Publishers, Inc.
P.O. Box 1067
Blair Station
Silver Springs, MD 20910

Noise/News
Institute of Noise Control Engineering
P.O. Box 3206
Arlington Branch
Poughkecpsie, NY 12603

Noise Regulation Reporter
Bureau of National Affairs, Inc.
1231 25th Street North West
Washington, DC 20037

Proceedings of Inter-Noise (Annual Conference)
Institute of Noise Control Engineering
P.O. Box 3206
Arlington Branch
Poughkeepsie, NY 12603

Proceedings of Noisexpo (Annual Conference)
Acoustical Publications Inc.
27101 East Oviatt Road
Bay Village, OH 44140

Sound and Vibration
Acoustical Publications Inc.
27101 East Oviatt Road
Bay Village, OH 44140

GUIDELINE FOR REGULATORY CONTROL OF OCCUPATIONAL NOISE EXPOSURE AND HEARING CONSERVATION*

INTRODUCTION

The Federal/Provincial Advisory Committee on Environmental and Occupational Health is composed of senior officials from the health, labour, and environment departments of the federal, ten provincial, and two territorial governments in Canada. It is charged with advising Ministers and Deputy Ministers of Health on all matters of environmental and occupational health. Under the auspices of this Advisory Committee, a working group was established in September 1982 with specific terms of reference, namely ''to prepare for publication a guideline for regulatory control of occupational noise exposure and hearing conservation with due regard for respective jurisdictions and responsibilities''. The membership consisted of two groups. The first was a core group of seven professionals representing Nova Scotia, Ontario, Manitoba, Saskatchewan, Alberta, British Columbia and Health and Welfare Canada, who attended all meetings and who were responsible for specific assignments and for writing the various sections of the final report. In this way, the working group was kept to an economical size for the purpose of attending meetings and preparing documents. The individuals comprising the core group are all active in the field of noise exposure and hearing conservation, and possess expertise in acoustics, occupational health, engineering, industrial hygiene, audiology, physics, and government regulations, with direct access to medical, legal and psychoacoustics expertise as well. The second group comprising the working group were corresponding members from the remaining provincial and territorial governments who were briefed regularly on progress and who could provide input or review documents as appropriate. In this way, the working group maintained a full federal/provincial character.

The working group convened its first meeting in December 1982 to consider the terms of reference, establish the work programme and allocate assignments and responsibilities. Five additional meetings were subsequently held,

the last in November 1985, to discuss the documents and assignments prepared by members of the core group. Therefore the working group's report, represents the consensus of the core group. In addition, an interim report by the working group, dated April 1984, was circulated to the corresponding members and other interested professionals and organizations. Therefore, the final report which incorporated the comments received has input from the entire working group as well as outside experts.

At the time of writing this paper (January 1986), the final report has not yet been presented to or approved by the Advisory Committee, and hence it has a draft status at this stage. The final report consists of two parts. Part 1 is the Model Regulation, whose main components are:
- A Framework for a Regulation Respecting Noise Exposure and Hearing Conservation;
- Codes of Practice for Audiometry, Hearing Protectors, and Noise Measurement; and
- A Rationale for the Framework.

Part 2 is a COMPLIANCE GUIDE whose components are:
- Guides on Audiometry, Hearing Protectors, and Noise Measurement; and
- A Bibliography.

The final report is described in general terms below.

MODEL REGULATION: THE FRAMEWORK AND ITS RATIONALE

In the Model Regulation (Part 1 of the final report), the Framework is structured like a regulation for noise exposure and hearing conservation. Therefore, it provides a guideline which could serve to promote uniformity in such regulations across Canada. However, much as it would be desirable to achieve this objective, it must be recognized that, in Canada, occupational health matters are within provincial jurisdiction, and different agencies have different regulatory approaches and responsibilities. In addition, the resources required to implement the various aspects of a regulation may not always be available in a given agency. Given these considerations, a degree of flexibility must be built into the Framework to allow an agency to modify it appropriately, as required. This flexibility is provided by the Rationale which contains explanatory information on

*S. Gewurtz
Program Development Unit, Ontario Ministry of Labour, 400 University Ave., 9th Floor, Toronto, Canada M7A 1T7

all aspects and every clause of the Framework, as well as alternatives and the factors to be considered in implementation.

Application Of The Regulation

Since the Framework is potentially a regulatory document whose provisions must be clearly understood by all affected parties, both technical and administrative terms used in it must be defined. This is done in the opening section with the proviso that certain administrative terms of a policy nature must be left for an agency to define. One of the key definitions is the one for noise which is an A-weighted sound pressure level equal to or greater than 80 dBA. This definition, when used in section 2 of the Framework, establishes that the regulation applies to every employer and worker at a work place where any worker is exposed to 80 dBA or greater for a significant period during a work day, taking into account the prescribed exposure limits. The application of the regulation is specified in terms of an employer and a worker rather than in terms of noise at a work place, because both have duties and responsibilities and because this is a health-related regulation directed towards worker exposure. In this regard, two conditions must co-exist in order that the regulation apply, namely:

- noise (i.e., sound levels equal to or greater than 80 dBA) must be present in the work place; and
- workers are exposed to this noise.

The phrase "for a significant period during a work day" emphasizes these conditions further by specifying that the regulation only applies when the noise to which a worker is exposed exceeds 80 dBA for a time period rather than instantaneously. The phrase "taking into account the prescribed exposure limits" is intended to give some guidance as to what constitutes a "significant period". It should also be noted that noise measurements are not required for determining whether the regulation applies since qualitative criteria exist to indicate whether the sound levels are likely to be above 80 dBA for a significant period of time during a work day.

Since the application of the regulation, as described above, is broad, an agency may wish to specify exemptions. For example, it could be argued that specific industries such as construction and trucking do not lend themselves to provisions which apply to the general industrial establishment. The Framework provides for exemptions to be specified in section 2 at the discretion of the agency.

Closely associated with section 2 regarding the application of the regulation is section 3 which addresses the need for consultation between the employer and workers in the implementation of the regulation. It is by such consultation that workers, through their health and safety representatives or committee, receive information on matters pertinent to the regulation and have input into the assessment process and the establishment of a hearing conservation program. Such joint participation of employer and worker, sometimes referred to as an internal responsibility system, must exist if the hazard presented by occupational noise exposure is to be addressed effectively.

Assessment Of Noise Exposure

The initial requirement of the regulation is an assessment of noise exposure at the work place with its inherent positive benefits, and since the application of the regulation is broad, many employers will have to comply with this requirement. As a first step, the assessment can also serve to determine which employers must comply with the more onerous provisions of the regulation. In this regard, there are two conflicting aspects to the assessment. The assessment should be qualitative to avoid the need for mandatory noise measurement in all work places. On the other hand, the results of the assessment should be precise to allow credible decisions to be made.

The novel approach in the Framework is to ascribe these conflicting aspects to two separate components of the assessment. The first is a Screening Assessment of the actual or potential exposure of a worker to noise for a significant period during a work day taking into account the prescribed exposure limits. This is to be accomplished by an examination of the work place, not based on noise measurement but taking into account the sources of noise and the nature and extent to which workers are exposed. Clearly, a screening assessment should also be required whenever a change in the work place is made which may alter the noise environment. The conclusions of the screening assessment will indicate whether a more extensive Noise Exposure Assessment, based on noise measurement, is required to determine the noise exposures of workers. This represents the second stage of the assessment process, and is designed to determine if other provisions of the regulation should be implemented. Both the screening assessment and the noise exposure assessment are to be conducted formally, in writing, under employer/worker consultation, and subject to record keeping. However, the screening assessment remains a light burden on employers while providing the benefits of knowledge, awareness and education with respect to noise exposure at the work place.

Hearing Conservation Program

If the noise exposures of workers, measured in the assessment, exceed a specified value and hence constitute an occupational hazard, the employer should take certain measures to protect workers from the noise hazard. The required protective measures constitute a hearing conservation program which can be identified and referred to by all parties.

After careful consideration, the working group recommends as the action level for the hearing conservation program, a time-weighted average exposure (TWA) value of 85 dBA for an 8-hour work day, to be determined with a 3 dB exchange rate taking all types of noise into account.

Clearly the hearing conservation program has to be unique to each specific workplace. It may include any, some, or all of the monitoring surveys, controls, hearing protectors, audiometric testing, record keeping, and worker education, as required by other provisions of the regulation.

Noise Exposure Limits And Control

The regulation places a duty on the employer to ensure that the noise exposure of any worker does not exceed specified exposure limits, of which there are two types. The first is a time-weighted average (TWA) exposure limit of 90 dBA for an 8-hour work day. In determining the TWA exposure of a worker, a 3 dB exchange rate must be used and all types of noise, including impulse noise must be measured. Since the TWA exposure limit is logarithmic, it serves as a maximum exposure limit for non-impulse noise as well; for example, 90 dBA for 8 hours is equivalent to 114 dBA for 1.5 minutes. However, for impulse noise, the regulation sets a second type of exposure limit, namely a peak sound pressure of 140 dB.

The regulation requires that the primary methods of meeting these limits are engineering controls and work practices. However, where it can be demonstrated that the required controls are not practicable, the employer is permitted to control noise exposure by the use of hearing protectors. In addition, the employer is required to provide a hearing protector upon request by the worker if the worker's TWA exposure exceeds 85 dBA for an 8 hour work day. The regulation places a duty on the worker to use the hearing protector so provided. Both the hearing protectors and their usage must comply with the requirements set out in the code of practice.

Audiometric Tests And Reviews

The employer must provide audiometric tests to all workers covered under a hearing conservation program, and such tests must comply with a code of practice for audiometry. Only the test results determined by the audiometric technician to be abnormal are to be referred to a physician or an audiologist for professional review. If it determined that the abnormal test result is due to noise exposure, both the employer and worker must be advised that remedial action is necessary. Audiometric test results are to be treated as confidential medical information, not accessible by the employer. In addition, the regulation prohibits the employer from taking discriminatory action against a worker whose hearing appears to be affected by noise exposure.

MODEL REGULATION: CODES OF PRACTICE

There are three Codes of Practice included in the model regulation, namely for Audiometry, Hearing Protectors, and Noise Measurements. These codes are short documents and include only mandatory requirements. However an equivalency clause at the beginning of each code allows the employer to vary from the code procedures if the protection afforded to the workers or factors of accuracy are not compromised. Since the codes are referenced in the regulation, they are legally enforceable.

COMPLIANCE GUIDE

A greater degree of flexibility can be realized by dealing with advisory information outside the model regulation. Thus Part 2 of the working group's report contains three guides on Audiometry, Hearing Protectors, and Noise Measurement. These guides are detailed and include recommended practices and descriptive information. Thus they add guidance to the affected parties in the areas addressed by the codes.

CONCLUSION

It is hoped that the model noise regulation, included in the working group report will set the stage for a uniform approach by agencies in Canada for the identification and control of occupational noise exposure. Like any new regulatory initiative, the regulation will be subjected to interpretation. Its provisions will not address all the issues or questions. The resulting dialogue should give rise to a greater appreciation by all concerned parties of the hazards to a worker from noise exposure, and the benefits to be gained from a comprehensive noise regulation which addresses such exposure.

BUYER'S GUIDE TO PRODUCTS FOR NOISE AND VIBRATION CONTROL

This Buyer's Guide lists manufacturers and suppliers of products for the control of noise, vibration and mechanical shock grouped into the following nine categories:

☐ Sound Absorptive Materials
☐ Sound Absorptive Systems
☐ Sound Barrier Materials
☐ Sound Barrier Systems
☐ Composite Materials
☐ Composite Systems
☐ Vibration Damping Materials
☐ Vibration Isolation Systems
☐ Silencers

Each category contains a series of numbered classifications for specific product identification.

To locate the noise and vibration control product of interest first refer to the main categories listed above then go to the detailed listings to locate the manufacturers that offer that particular product. The numbers after each company listing indicate the products that they offer. Boldface listings are advertisers in this issue.

Listings followed with an asterisk (*) are Members of The Noise Control Products and Materials Association (NCPMA).

Sound Absorptive Materials
1. Felts
2. Foams
3. Glass Fiber
4. Mineral Fiber
5. Perforated Sheet Metal
6. Spray-On Coatings
7. Wall Treatments

Acoustic Sciences Corporation, P.O. Box 1189 Eugene, OR 97440, (800) 272-8823 - (7)
Acoustic Standards, Inc., 1350-G W. Collins Ave., Orange, CA 92667, (714) 997-8283 - (7)
Air-Loc Products, 5 Fisher St., P.O. Box 269, Franklin, MA 02038, (617) 528-0022 - (1)
Allied Witan Company, 13805 Progress Parkway, Cleveland, OH 44133, (216) 237-9630 - (5)
Amber/Booth Co. Inc., 1403 N. Post Oak, Houston, TX 77055, (713) 688-1228 - (2,3,5)
American Acoustical Products, 9 Cochituate St., Natick, MA 01770, (617) 655-0870 - (2,4)
Antiphon Inc., 2655 Woodward Ave., Suite 350, Bloomfield Hills, MI 48013, (313) 332-0267 - (1-3)*
Barley-Earhart Co., 233 Divine Hwy., Portland, MI 48875, (517) 647-4117 - (2,3,5,7)
Barrier Corporation, 9908 S.W. Tigard St., Tigard, OR 97223, (503) 639-4192 - (2-5,7)
Louis P. Batson Company, P.O. Box 3978, Greenville, SC 29608, (803) 242-5262 - (1)
H. L. Blachford, Inc., 1855 Stephenson Hwy., P.O. Box 397, Troy, MI 48007, (313) 689-7800 - (2,3,7)*
The Branford Companies P.O. Box 713, Shelton, CT 06484, (203) 735-6415 - (2-4,6)*
BRD Noise and Vibration Control, 112 Fairview Ave., Wind Gap, PA 18091, (215) 863-6300 - (2-5,7)
Brunswick Corp. Technetics Div., 2000 Brunswick Lane, Deland, FL 32724, (904) 736-1700 - (1)
Chestnut Ridge Foam, Inc., P.O. Box 781, Latrobe, PA 15650, (412) 537-9000 - (2)
Comprador, Inc., 3 Dewalt Road, Newark, DE 19711, (302) 368-5089 - (2,3)
Controlled Acoustics Corporation, 12 Wilson St., Hartsdale, NY 10530, (914) 428-7740 - (1-7)
Craxton Acoustical Products Inc., 2838 Vicksburg Lane, Plymouth, MN 55447, (612) 553-9890 - (3,7)
The E. J. Davis Company, 10 Dodge Ave., North Haven, CT 06473, (203) 239-5391 - (4)
Designer Acoustics, Inc., 1671 Penfield Road, Rochester, NY 14625, (716) 385-3320 - (3,7)
E-A-R Division, Cabot Corporation, 7911 Zionsville Rd., Indianapolis, IN 46268, (317) 872-1111 - (2,3,7)

Eckel Industries, Inc., 155 Fawcett St., Cambridge, MA 02138, (617) 491-3221 - (2-5,7)*
Engineering Aural Research, 1777-G17 Woodlawn St. Upland, CA 91786, (714) 946-0108 - (1,2,6)
Engineering Sales & Services, P.O. Box 558, Dayton, OH 45405, (513) 277-2047 - (2,5-7)
Enviro-Acoustics Company, Inc., 6150 Olson Memorial Highway, Minneapolis, MN 55422, (612) 542-8665 - (7)
Essi Acoustical Products Co., 10580 Berea Rd., Cleveland, OH 44102, (216) 961-3111 - (3,4,7)
Ferguson Perforating & Wire Co. Inc., 133 Ernest St., Providence, RI 02905, (401) 941-8876 - (5)
Frommelt Industries, 4343 Chavenelle, Dubuque, IA 52001 (319) 556-2020 - (2)
General Acoustics Corp., 12248 Santa Monica Blvd., Los Angeles, CA 90025, (213) 820-1531 - (5,7)
Carroll George, Inc., 15th Street South, P.O. Box 144, Northwood, IA 50459, (515) 324-2231 - (2)
The Harrington & King Perforating Co., Inc., 5655 W. Fillmore St., Chicago, IL 60644, (312) 626-1800 - (5)
Higgott-Kane Industrial Noise Controls Ltd., 1085 Bellamy Road North, Suite 214, Scarborough, Ont, Canada M1H 3C7 (416) 431-0641 - (5)
Hitco, P.O. Box 1097, Gardena, CA 90249, (213) 321-8080 - (3,4)
Illbruck, 3800 Washington Ave. N., Minneapolis, MN 55412, (612) 521-3555 - (2,7)*
Imi-Tech Corp., 701 Fargo Ave., Elk Grove Village, IL 60007, (312) 981-7610 - (2)
Industrial Acoustics Company, 1160 Commerce Ave., Bronx, NY 10462, (212) 931-8000 - (5,7)
Industrial Noise Control, Inc., 1411 Jeffrey Dr., Addison, IL 60101, (312) 620-1998 - (2-7)
Insul-Art Acoustics Corporation, 107 Allen Blvd., Farmingdale, NY 11735, (800) 526-0908 - (3,7)
Interior Acoustics, Inc., 176 Route 206 South, Somerville, NJ 08876, (800) 221-0580 - (7)
International Cellulose Corp., P.O. Box 450006, Houston, TX 77245, (713) 433-6701 - (6,7)
King Noise Control, Inc., 1506 Magnolia Dr., Cincinnati, OH 45215, (513) 733-1222 - (2,5-7)
George Koch Sons, Inc., 10 South Eleventh Ave., Evansville, IN 47744, (812) 426-9880 - (3-5)
Korfund Dynamics Co., 35 South Service Road, Plainview, NY 11803, (516) 752-2453 - (2,6)
Linear Products Corp., P.O. Box 902, Cranford, NJ 07016, (201) 272-2211 - (2-7)
Mason Industries Inc., 350 Rabro Dr., Hauppauge, NY 11788, (516) 348-0282 - (6)
MBI Products Co., 5309 Hamilton Ave., Cleveland, OH 44114, (216) 431-6400 - (3,4,7)
Midwest Acoust-A-Fiber Inc., 7790 Marysville Rd., Ostrander, OH 43061, (614) 666-2241 - (2-4,7)
MPC, Inc., 835 Canterbury Rd., Westlake, OH 44145, (216) 835-1405 - (3-5,7)
National Perforating Corp., Parker St., Clinton, MA 01510, (800) 225-9050 - (5)
Noise Control Associates, Inc., P.O. Box 261, Montclair, NJ 07042, (201) 746-5181 - (2,3,5)
Noise Control Company, Div. of Noise Control Products, 443 Oak Glenn Dr., Bartlett, IL 60103, (312) 830-0090 - (2-7)
Noise Control Products, Inc., 700 Chettic Ave., Copiague, NY 11726, (516) 226-6100 - (2,7)
Noise Reduction Corporation, Route 2, Box 152, Redwood Falls, MN 56283, (507) 644-3067 - (2,3)
Noise Suppression Systems, Inc., 2886 Metropolitan Place, Pomona, CA 91767, (714) 596-6710 - (5)
O'Brien Partition Co., Inc., 5301 E. 59th St., Kansas City, MO 64130, (800) 821-3595 - (7)
Patterson Associates, 935 Summer St., Lynnfield, MA 01940, (617) 246-0888 - (2-4,7)
Peabody Noise Control, 6300 Irelan Place, Dublin, OH 43017, (614) 889-0480 - (2,3,5,7)
Presto Manufacturing Co., Inc., 2 Franklin Ave., Brooklyn, NY 11211, (718) 852-0187 - (2)
Pyrok, Inc., 136 Prospect Park West, Brooklyn, NY 11215, (718) 788-1225 - (6,7)
Singer Safety Company, 2300 W. Logan Blvd., Chicago, IL 60647, (312) 235-2100 - (2,3,7)
The Soundcoat Company, Inc., One Burt Drive, Deer Park, NY 11729, (516) 242-2200 - (2,3,7)*
Soundown Corporation, 45 Congress St., Salem, MA 01970, (617) 598-4248 - (2,3)
Specialty Composites Corp., Delaware Industrial Park, Newark, DE 19713, (800) 544-5180 - (2)*
Stark Ceramics, Inc., P.O. Box 8880, Canton, OH 44711, (800) 321-0662 - (7)
Tech Products Corporation, 5030 Linden Ave., Dayton, OH 45432, (513) 252-3661 - (1-3,6,7)
Technicon, 4412 Republic Dr., Concord, NC 28025, (704) 788-1131 - (2,3,6)
Tectum Inc., 105 S. 6th St., Newark, OH 43055, (614) 345-9691 - (7)
United McGill Corporation, 1501 Kalamazoo Dr., Griffin GA 30223, (404) 228-9864 - (2-5,7)
United Process Inc., 279 Silver St., P.O. Box 545, Agawam, MA 01001, (413) 789-1770 - (2,3,7)

Vibration Mountings & Controls, 113 Main St., P.O. Box 37, Bloomingdale, NJ 07403, (201) 838-1780 - (2,6)
Vibron Limited, 1720 Meyerside Dr., Mississauga, Ont., Canada L5T 1A3, (416) 677-4922 - (5)
Voltek Division of Sekisui America Corp., 100 Shepard Street, Lawrence, MA 01843, (617) 685-2557 - (2)

Sound Absorptive Systems
1. Ceiling Systems
2. Masking Noise Generators
3. Panels
4. Unit Absorbers
5. Wall Treatments

Acon, Inc., 4600 Webster St. Dayton, OH 45414, (513) 276-2111 - (3)
Acoustic Sciences Corporation, P.O. Box 1189 Eugene, OR 97440, (800) 272-8823 - (1,3-5)
Acoustic Standards, Inc., 1350-G W. Collins Ave., Orange, CA 92667, (714) 997-8283 - (3-5)
The Aeroacoustic Corp., 4876 Victor St., Jacksonville, FL 32207, (904) 731-3577 - (3)
Amber/Booth Co. Inc., 1403 N. Post Oak, Houston, TX 77055, (713) 688-1228 - (3)
American Acoustical Products, 9 Cochituate St., Natick, MA 01770, (617) 655-0870 - (5)
Antiphon Inc., 2655 Woodward Ave., Suite 350, Bloomfield Hills, MI 48013, (313) 332-0267 - (5)*
Armstrong World Industries, Inc., P.O. Box 3001, Lancaster, PA 17604, (717) 397-0611 - (1,3,5)
Barley-Earhart Co., 233 Divine Hwy., Portland, MI 48875, (517) 647-4117 - (1,3)
BRD Noise and Vibration Control, 112 Fairview Ave., Wind Gap, PA 18091, (215) 863-6300 - (1-5)
Controlled Acoustics Corporation, 12 Wilson St., Hartsdale, NY 10530, (914) 428-7740 - (1-5)
Craxton Acoustical Products Inc., 2838 Vicksburg Lane, Plymouth, MN 55447, (612) 553-9890 - (1,3,5)
The E. J. Davis Company, 10 Dodge Ave., North Haven, CT 06473, (203) 239-5391 - (4)
Designer Acoustics, Inc., 1671 Penfield Road, Rochester, NY 14625, (716) 385-3320 - (1,3,5)
E-A-R Division, Cabot Corporation, 7911 Zionsville Rd., Indianapolis, IN 46268, (317) 872-1111 - (3-5)
Eckel Industries, Inc., 155 Fawcett St., Cambridge, MA 02138, (617) 491-3221 - (1,3,5)*
Engineering Aural Research, 1777-G17 Woodlawn St. Upland, CA 91786, (714) 946-0108 - (3,4)
Engineering Sales & Services, P.O. Box 558, Dayton, OH 45405, (513) 277-2047 - (2-5)
Enviro-Acoustics Company, Inc., 6150 Olson Memorial Highway, Minneapolis, MN 55422, (612) 542-8665 - (1,3-5)
Essi Acoustical Products Co., 10580 Berea Rd., Cleveland, OH 44102, (216) 961-3111 - (1-5)
General Acoustics Corp., 12248 Santa Monica Blvd., Los Angeles, CA 90025, (213) 820-1531 - (1,3,5)
Higgott-Kane Industrial Noise Controls Ltd., 1085 Bellamy Road North, Suite 214, Scarborough, Ont, Canada M1H 3C7 (416) 431-0641 - (1,3,5)
Hitco, P.O. Box 1097, Gardena, CA 90249, (213) 321-8080 - (2-5)
I.D.E. Processes Corporation, Noise Control Div., 106 81st Ave., Kew Gardens, NY 11415, (718) 544-1177 - (3)
Illbruck, 3800 Washington Ave. N., Minneapolis, MN 55412, (612) 521-3555 - (1,3,5)
Imi-Tech Corp., 701 Fargo Ave., Elk Grove Village, IL 60007, (312) 981-7610 - (1,4,5)
Industrial Acoustics Company, 1160 Commerce Ave., Bronx, NY 10462, (212) 931-8000 - (1,3,5)
Industrial Noise Control, Inc., 1411 Jeffrey Dr., Addison, IL 60101, (312) 620-1998 - (1,3,5)
Insul-Art Acoustics Corporation, 107 Allen Blvd., Farmingdale, NY 11735, (800) 526-0908 - (1-5)
Interior Acoustics, Inc., 176 Route 206 South, Somerville, NJ 08876, (800) 221-0580 - (1-5)
International Cellulose Corp., P.O. Box 450006, Houston, TX 77245, (713) 433-6701 - (1,5)
King Noise Control, Inc., 1506 Magnolia Dr., Cincinnati, OH 45215, (513) 733-1222 - (1,3,5)
George Koch Sons, Inc., 10 South Eleventh Ave., Evansville, IN 47744, (812) 426-9880 - (3)
Korfund Dynamics Co., 35 South Service Road, Plainview, NY 11803, (516) 752-2453 - (3,5)
Linear Products Corp., P.O. Box 902, Cranford, NJ 07016, (201) 272-2211 - (2-5)
Mason Industries Inc., 350 Rabro Dr., Hauppauge, NY 11788, (516) 348-0282 - (1)
MBI Products Co., 5309 Hamilton Ave., Cleveland, OH 44114, (216) 431-6400 - (1,3,5)
Midwest Acoust-A-Fiber Inc., 7790 Marysville Rd., Ostrander, OH 43061, (614) 666-2241 - (3,5)
MPC, Inc., 835 Canterbury Rd., Westlake, OH 44145, (216) 835-1405 - (1,3,5)

Neiss Corp., P.O. Box 478, Rockville, CT 06066, (203) 872-8528 - (3)
Noise Control Associates, Inc., P.O. Box 261, Montclair, NJ 07042, (201) 746-5181 - (2,5)
Noise Control Company, Div. of Noise Control Products, 443 Oak Glenn Dr., Bartlett, IL 60103, (312) 830-0090 - (1,3-5)
Noise Control Products, Inc., 700 Chettic Ave., Copiague, NY 11726, (516) 226-6100 - (3,5)
O'Brien Partition Co., Inc., 5301 E. 59th St., Kansas City, MO 64130, (800) 821-3595 - (3,5)
Patterson Associates, 935 Summer St., Lynnfield, MA 01940, (617) 246-0888 - (3-5)
Peabody Noise Control, 6300 Irelan Place, Dublin, OH 43017, (614) 889-0480 - (1,3-5)
Gordon J. Pollock & Associates, Inc., 19120 Detroit Rd., Rocky River, OH 44116, (216) 333-8710 - (1,3-5)
The Proudfoot Company, Inc., P.O. Box 1537, Greenwich, CT 06836, (203) 869-9031 - (4)
Rathbone Products, P.O. Box 10356, Columbus, OH 43201, (614) 297-7782 - (3,5)
Singer Safety Company, 2300 W. Logan Blvd., Chicago, IL 60647, (312) 235-2100 - (1,3,5)

Sound Fighter Systems Inc., P.O. Box 6075, Shreveport, LA 71136, (318) 861-6640 - (3)
The Soundcoat Company, Inc., One Burt Drive, Deer Park, NY 11729, (516) 242-2200 - (1,4,5)*
Soundown Corporation, 45 Congress St., Salem, MA 01970, (617) 598-4248 - (3)
Stark Ceramics, Inc., P.O. Box 8880, Canton, OH 44711, (800) 321-0662 - (5)
Starrco Company, Inc., 1515 Fairview Ave., St. Louis, MO 63132, (314) 429-5650 - (3)
Target Enterprises Inc., Roxbury, VT 05669, (802) 728-3081 - (4)
Tech Products Corporation, 5030 Linden Ave., Dayton, OH 45432, (513) 252-3661 - (3-5)
Technicon, 4412 Republic Dr., Concord, NC 28025, (704) 788-1131 - (4,5)
Tectum Inc., 105 S. 6th St., Newark, OH 43055, (614) 345-9691 - (1,5)
Transco Products Inc., 55 E. Jackson Blvd., Chicago, IL 60604, (312) 427-2818 - (3)
United McGill Corporation, 1501 Kalamazoo Dr., Griffin, GA 30223, (404) 228-9864 - (1-5)
United Process Inc., 279 Silver St., P.O. Box 545, Agawam, MA 01001, (413) 789-1770 - (1,3-5)

Vibration Mountings & Controls, 113 Main St., P.O. Box 37, Bloomingdale, NJ 07403, (201) 838-1780 - (3,5)
Vibron Limited, 1720 Meyerside Dr., Mississauga, Ont., Canada L5T 1A3, (416) 677-4922 - (3)
Wenger Corporation, 555 Park Drive, Owatonna, MN 55060, (507) 455-4100 - (1,3,5)

Sound Barrier Materials
1. Pipe Lagging
2. Plain and Mass-Loaded Plastics
3. Sealants and Sealing Tapes
4. Sheet Glass, Metal and Plastic

Amber/Booth Co. Inc., 1403 N. Post Oak, Houston, TX 77055, (713) 688-1228 - (1,2)
American Acoustical Products, 9 Cochituate St., Natick, MA 01770, (617) 655-0870 - (1,2)
Antiphon Inc., 2655 Woodward Ave., Suite 350, Bloomfield Hills, MI 48013, (313) 332-0267 - (2)*
Arvinyl Division of Arvin Industries, Inc., 1531 13th St., Columbus, IN 47201, (812) 379-3000 - (4)
Barley-Earhart Co., 233 Divine Hwy., Portland, MI 48875, (517) 647-4117 - (2,3)
Barrier Corporation, 9908 S.W. Tigard St., Tigard, OR 97223, (503) 639-4192 - (1,2)
H. L. Blachford, Inc., 1855 Stephenson Hwy, P.O. Box 397, Troy, MI 48007, (313) 689-7800 - (1-4)*
The Branford Companies P.O. Box 713, Shelton, CT 06484, (203) 735-6415 - (1,2)*
BRD Noise and Vibration Control, 112 Fairview Ave., Wind Gap, PA 18091, (215) 863-6300 - (1,2)
Comprador, Inc., 3 Dewalt Road, Newark, DE 19711, (302) 368-5089 - (2)
Controlled Acoustics Corporation, 12 Wilson St., Hartsdale, NY 10530, (914) 428-7740 - (1,2,4)
The E. J. Davis Company, 10 Dodge Ave., North Haven, CT 06473, (203) 239-5391 - (1)
E-A-R Division, Cabot Corporation, 7911 Zionsville Rd., Indianapolis, IN 46268, (317) 872-1111 - (1-3)
Eckel Industries, Inc., 155 Fawcett St., Cambridge, MA 02138, (617) 491-3221 - (4)*
Engineering Aural Research, 1777-G17 Woodlawn St. Upland, CA 91786, (714) 946-0108 - (3)
Engineering Sales & Services, P.O. Box 558, Dayton, OH 45405, (513) 277-2047 - (1,2)

Enviro-Acoustics Company, Inc., 6150 Olson Memorial Highway, Minneapolis, MN 55422, (612) 542-8665 - (1,2)
Higgott-Kane Industrial Noise Controls Ltd., 1085 Bellamy Road North, Suite 214, Scarborough, Ont, Canada M1H 3C7(416) 431-0641 - (1,2)
Hitco, P.O. Box 1097, Gardena, CA 90249, (213) 321-8080 - (1)
Industrial Noise Control, Inc., 1411 Jeffrey Dr., Addison, IL 60101, (312) 620-1998 - (2)
The Kennedy Company, P.O. Box 1216, Scottsboro, AL 35768, (205) 259-4436 - (1,2)
King Noise Control, Inc., 1506 Magnolia Dr., Cincinnati, OH 45215, (513) 733-1222 - (1)
George Koch Sons, Inc., 10 South Eleventh Ave., Evansville, IN 47744, (812) 426-9880 - (4)
Korfund Dynamics Co., 35 South Service Road, Plainview, NY 11803, (516) 752-2453 - (1,2)
Linear Products Corp., P.O. Box 902, Cranford, NJ 07016, (201) 272-2211 - (1-2)
Mason Industries Inc., 350 Rabro Dr., Hauppauge, NY 11788, (516) 348-0282 - (3)
Midwest Acoust-A-Fiber Inc., 7790 Marysville Rd., Ostrander, OH 43061, (614) 666-2241 - (4)
Noise Control Associates, Inc., P.O. Box 261, Montclair, NJ 07042, (201) 746-5181 - (1,2)
Noise Control Company, Div. of Noise Control Products, 443 Oak Glenn Dr., Bartlett, IL 60103, (312) 830-0090 - (1,2)
Noise Reduction Corporation, Route 2, Box 152, Redwood Falls, MN 56283, (507) 644-3067 - (1,3)
Pan Lam Industries, 27741 Calle Valdes, Mission Viejo, CA 92692, (714) 380-7547 - (4)
Patterson Associates, 935 Summer St., Lynnfield, MA 01940, (617) 246-0888 - (2,3)
Peabody Noise Control, 6300 Irelan Place, Dublin, OH 43017, (614) 889-0480 - (1,2)
Gordon J. Pollock & Associates, Inc., 19120 Detroit Rd., Rocky River, OH 44116, (216) 333-8710 - (1)
Singer Safety Company, 2300 W. Logan Blvd., Chicago, IL 60647, (312) 235-2100 - (1,2)
Sound Fighter Systems Inc., P.O. Box 6075, Shreveport, LA 71136, (318) 861-6640 - (1)
The Soundcoat Company, Inc., One Burt Drive, Deer Park, NY 11729, (516) 242-2200 - (1,2,4)*
Soundown Corporation, 45 Congress St., Salem, MA 01970, (617) 598-4248 - (1,2)
Specialty Composites Corp., Delaware Industrial Park, Newark, DE 19713, (800) 544-5180 - (2,3)*
Tech Products Corporation, 5030 Linden Ave., Dayton, OH 45432, (513) 252-3661 - (1-4)
Technicon, 4412 Republic Dr., Concord, NC 28025, (704) 788-1131 - (1,2)
Transco Products Inc., 55 E. Jackson Blvd., Chicago, IL 60604, (312) 427-2818 - (1)

United McGill Corporation, 1501 Kalamazoo Dr., Griffin, GA 30223, (404) 228-9864 - (1-3)
United Process Inc., 279 Silver St., P.O. Box 545, Agawam, MA 01001, (413) 789-1770 - (1,2)
Vibration Mountings & Controls, 113 Main St., P.O. Box 37, Bloomingdale, NJ 07403, (201) 838-1780 - (1,2)
Vibron Limited, 1720 Meyerside Dr., Mississauga, Ont., Canada L5T 1A3, (416) 677-4922 - (4)

Sound Barrier Systems
1. Curtains
2. Doors
3. Operable Partitions
4. Panels
5. Seals
6. Transportation Noise Barriers
7. Walls
8. Windows

Acon, Inc., 4600 Webster St. Dayton, OH 45414, (513) 276-2111 - (2-4,7)
Acoustic Sciences Corporation, P.O. Box 1189 Eugene, OR 97440, (800) 272-8823 - (4,7,8)
Acoustic Standards, Inc., 1350-G W. Collins Ave., Orange, CA 92667, (714) 997-8283 - (1,4,7)
The Aeroacoustic Company, 4876 Victor St., Jacksonville, FL 32207, (904) 731-3577 - (2,4,7,8)
Amber/Booth Co. Inc., 1403 N. Post Oak, Houston, TX 77055, (713) 688-1228 - (4,5)
Amweld Building Products, Inc., 1500 Amweld Drive, Garrettsville, OH 44231, (800) 248-6116 - (2)
Barley-Earhart Co., 233 Divine Hwy., Portland, MI 48875, (517) 647-4117 - (1,4,7)
Barrier Corporation, 9908 S.W. Tigard St., Tigard, OR 97223, (503) 639-4192 - (1)
Louis P. Batson Company, P.O. Box 3978, Greenville, SC 29608, (803) 242-5262 - (2)
H. L. Blachford, Inc., 1855 Stephenson Hwy., P.O. Box 397, Troy, MI 48007, (313) 689-7800 - (6)*

The Branford Companies P.O. Box 713, Shelton, CT 06484, (203) 735-6415 - (1,3,4)*
BRD Noise and Vibration Control, 112 Fairview Ave., Wind Gap, PA 18091, (215) 863-6300 - (1-5,7)
Controlled Acoustics Corporation, 12 Wilson St., Hartsdale, NY 10530, (914) 428-7740 - (1-8)
The E. J. Davis Company, 10 Dodge Ave., North Haven, CT 06473, (203) 239-5391 - (1)
Designer Acoustics, Inc., 1671 Penfield Road, Rochester, NY 14625, (716) 385-3320 - (4)
E-A-R Division, Cabot Corporation, 7911 Zionsville Rd., Indianapolis, IN 46268, (317) 872-1111 - (1,4,5)
Eckel Industries, Inc., 155 Fawcett St., Cambridge, MA 02138, (617) 491-3221 - (1-4,6)*
Engineering Aural Research, 1777-G17 Woodlawn St. Upland, CA 91786, (714) 946-0108 - (1,5)
Engineering Sales & Services, P.O. Box 558, Dayton, OH 45405, (513) 277 2047 - (1-4,6-8)
Enviro-Acoustics Company, Inc., 6150 Olson Memorial Highway, Minneapolis, MN 55422, (612) 542-8665 - (2,4,7,8)
The Fanwall Corporation, 1700 N. Moore St., Suite 2200, Arlington, VA 22209, (703) 524-5301 - (6)
Frommelt Industries, 4343 Chavenelle, Dubuque, IA 52001 (319) 556-2020 - (1,4)
General Acoustics Corp., 12248 Santa Monica Blvd., Los Angeles, CA 90025, (213) 820-1531 - (1,2,4,6)
General American Window Corp., P.O. Box 55250, Stockton, CA 95205, (209) 462-3732 - (8)
Carroll George, Inc., 15th Street South, P.O. Box 144, Northwood, IA 50459, (515) 324-2231 - (6)
Guardian Architectural Visions, Sitelines, Inc., 2301 Raymer Ave., Fullerton, CA 92633, (800) 237-9055 - (8)
Higgott-Kane Industrial Noise Controls Ltd., 1085 Bellamy Road North, Suite 214, Scarborough, Ont, Canada M1H 3C7(416) 431-0641 - (2-4,6,7)
Hitco, P.O. Box 1097, Gardena, CA 90249, (213) 321-8080 - (4,6,7)
Imi-Tech Corp., 701 Fargo Ave., Elk Grove Village, IL 60007, (312) 981-7610 - (4,6)
Industrial Acoustics Company, 1160 Commerce Ave., Bronx, NY 10462, (212) 931-8000 - (2-4,6-8)
Industrial Noise Control, Inc., 1411 Jeffrey Dr., Addison, IL 60101, (312) 620-1998 - (1-4)
International Cellulose Corp., P.O. Box 450006, Houston, TX 77245, (713) 433-6701 - (7)
Jamison Door Company, P.O. Box 70, Hagerstown, MD 21741, (800) 532-3667 - (2,8)
The Kennedy Company, P.O. Box 1216, Scottsboro, AL 35768, (205) 259-4436 - (1,6)
King Noise Control, Inc., 1506 Magnolia Dr., Cincinnati, OH 45215, (513) 733-1222 - (1,2,4,7)
George Koch Sons, Inc., 10 South Eleventh Ave., Evansville, IN 47744, (812) 426-9880 - (2,4,7,8)
Korfund Dynamics Co., 35 South Service Road, Plainview, NY 11803, (516) 752-2453 - (1,4)
Krieger Steel Products Co., 4896 Gregg Rd., P.O. Box 308, Pico Rivera, CA 90660, (213) 695-0645 - (2,4,5,8)

Linear Products Corp., P.O. Box 902, Cranford, NJ 07016, (201) 272-2211 - (1-4,6-8)
Mason Industries Inc., 350 Rabro Dr., Hauppauge, NY 11788, (516) 348-0282 - (7)
MBI Products Co., 5309 Hamilton Ave., Cleveland, OH 44114, (216) 431-6400 - (1,4)
Midwest Acoust-A-Fiber Inc., 7790 Marysville Rd., Ostrander, OH 43061, (614) 666-2241 - (1)
MPC, Inc., 835 Canterbury Rd., Westlake, OH 44145, (216) 835-1405 - (4,7)
Neiss Corp., P.O. Box 478, Rockville, CT 06066, (203) 872-8528 - (2-4,7,8)
Noise Control Associates, Inc., P.O. Box 261, Montclair, NJ 07042, (201) 746-5181 - (1,4)
Noise Control Company, Div. of Noise Control Products, 443 Oak Glenn Dr., Bartlett, IL 60103, (312) 830-0090 - (1-5,7,8)
Noise Control Products, Inc., 700 Chettic Ave., Copiague, NY 11726, (516) 226-6100 - (2,4,7)
Noise Reduction Corporation, Route 2, Box 152, Redwood Falls, MN 56283, (507) 644-3067 - (1)
O'Brien Partition Co., Inc., 5301 E. 59th St., Kansas City, MO 64130, (800) 821-3595 - (4,7)

Overly Manufacturing Company, 574 W. Otterman St., P.O. Box 70, Greensburg, PA 15601, (412) 834-7300 - (2,4,5,7)
Pan Lam Industries, 27741 Calle Valdes, Mission Viejo, CA 92692, (714) 380-7547 - (8)
Patterson Associates, 935 Summer St., Lynnfield, MA 01940, (617) 246-0888 - (1-4,7)
Peabody Noise Control, 6300 Irelan Place, Dublin, OH 43017, (614) 889-0480 - (1,2,4,7)
Pioneer Industries, 401 Washington Ave., Carlstadt, NJ 07072, (201) 933-1900 - (2)
Gordon J. Pollock & Associates, Inc., 19120 Detroit Rd., Rocky River, OH 44116, (216) 333-8710 - (1-4,6,7)

The Proudfoot Company, Inc., P.O. Box 1537, Greenwich, CT 06836, (203) 869-9031 - (1,4,6)
Pyrok, Inc., 136 Prospect Park West, Brooklyn, NY 11215, (718) 788-1225 - (4,6)
Richards-Wilcox Mfg. Co., 174 Third St., Aurora, IL 60507, (312) 897-6951 - (2-4)
Singer Safety Company, 2300 W. Logan Blvd., Chicago, IL 60647, (312) 235-2100 - (1,3,4)
Sound Fighter Systems Inc., P.O. Box 6075, Shreveport, LA 71136, (318) 861-6640 - (2,4,6)

The Soundcoat Company, Inc., One Burt Drive, Deer Park, NY 11729, (516) 242-2200 - (1,5)*
Stark Ceramics, Inc., P.O. Box 8880, Canton, OH 44711, (800) 321-0662 - (6)
Tech Products Corporation, 5030 Linden Ave., Dayton, OH 45432, (513) 252-3661 - (1-5,7,8)
Technicon, 4412 Republic Dr., Concord, NC 28025, (704) 788-1131 - (1,4,5)
Transco Products Inc., 55 E. Jackson Blvd., Chicago, IL 60604, (312) 427-2818 - (2,4,6-8)
United McGill Corporation, 1501 Kalamazoo Dr., Griffin, GA 30223, (404) 228-9864 - (1,2,4-8)
United Process Inc., 279 Silver St., P.O. Box 545, Agawam, MA 01001, (413) 789-1770 - (1,4,6-8)
Vibration Mountings & Controls, 113 Main St., P.O. Box 37, Bloomingdale, NJ 07403, (201) 838-1780 - (1,4)
Vibron Limited, 1720 Meyerside Dr., Mississauga, Ont., Canada L5T 1A3, (416) 677-4922 - (1-4,7)
Wenger Corporation, 555 Park Drive, Owatonna, MN 55060, (507) 455-4100 - (2,4,7,8)

Composite Materials
1. Barrier/Fiber Composites
2. Barrier/Foam Composites
3. Masonry Units

Acoustic Sciences Corporation, P.O. Box 1189 Eugene, OR 97440, (800) 272-8823 - (1)
Amber/Booth Co. Inc., 1403 N. Post Oak, Houston, TX 77055, (713) 688-1228 - (1,2)
American Acoustical Products, 9 Cochituate St., Natick, MA 01770, (617) 655-0870 - (1,2)
Antiphon Inc., 2655 Woodward Ave., Suite 350, Bloomfield Hills, MI 48013, (313) 332-0267 - (1,2)*
Barley-Earhart Co., 233 Divine Hwy., Portland, MI 48875, (517) 647-4117 - (1,2)
Barrier Corporation, 9908 S.W. Tigard St., Tigard, OR 97223, (503) 639-4192 - (1,2)
H. L. Blachford, Inc., 1855 Stephenson Hwy., P.O. Box 397, Troy, MI 48007, (313) 689-7800 - (1,2)*
The Branford Companies P.O. Box 713, Shelton, CT 06484, (203) 735-6415 - (1,2)*
BRD Noise and Vibration Control, 112 Fairview Ave., Wind Gap, PA 18091, (215) 863-6300 - (1,2)
Chestnut Ridge Foam, Inc., P.O. Box 781, Latrobe, PA 15650, (412) 537-9000 - (2)
Comprador, Inc., 3 Dewalt Road, Newark, DE 19711, (302) 368-5089 - (1,2)
Controlled Acoustics Corporation, 12 Wilson St., Hartsdale, NY 10530, (914) 428-7740 - (1,2)
The E. J. Davis Company, 10 Dodge Ave., North Haven, CT 06473, (203) 239-5391 - (1,2)
Designer Acoustics, Inc., 1671 Penfield Road, Rochester, NY 14625, (716) 385-3320 - (1)
Duracote Corporation, 350 N. Diamond St., P.O. Box 512, Ravenna, OH 44266, (216) 296-9600 - (2)
E-A-R Division, Cabot Corporation, 7911 Zionsville Rd., Indianapolis, IN 46268, (317) 872-1111 - (1,2)
Eckel Industries, Inc., 155 Fawcett St., Cambridge, MA 02138, (617) 491-3221 - (2)*
Engineering Aural Research, 1777-G17 Woodlawn St. Upland, CA 91786, (714) 946-0108 - (2)
Engineering Sales & Services, P.O. Box 558, Dayton, OH 45405, (513) 277-2047 - (1,2)
Enviro-Acoustics Company, Inc., 6150 Olson Memorial Highway, Minneapolis, MN 55422, (612) 542-8665 - (1,2)
Frommelt Industries, 4343 Chavenelle, Dubuque, IA 52001 (319) 556-2020 - (1)
Carroll George, Inc., 15th Street South, P.O. Box 144, Northwood, IA 50459, (515) 324-2231 - (2)
Higgott-Kane Industrial Noise Controls Ltd., 1085 Bellamy Road North, Suite 214, Scarborough, Ont, Canada M1H 3C7(416) 431-0641 - (3)
Hitco, P.O. Box 1097, Gardena, CA 90249, (213) 321-8080 - (1)
Imi-Tech Corp., 701 Fargo Ave., Elk Grove Village, IL 60007, (312) 981-7610 - (2)
Industrial Noise Control, Inc., 1411 Jeffrey Dr., Addison, IL 60101, (312) 620-1998 - (1,2)
Interior Acoustics, Inc., 176 Route 206 South, Somerville, NJ 08876, (800) 221-0580 - (1)
The Kennedy Company, P.O. Box 1216, Scottsboro, AL 35768, (205) 259-4436 - (2)
King Noise Control, Inc., 1506 Magnolia Dr., Cincinnati, OH 45215, (513) 733-1222 - (1,2)
George Koch Sons, Inc., 10 South Eleventh Ave., Evansville, IN 47744, (812) 426-9880 - (1,3,6)
Korfund Dynamics Co., 35 South Service Road, Plainview, NY 11803, (516) 752-2453 - (1-3)
Linear Products Corp., P.O. Box 902, Cranford, NJ 07016, (201) 272-2211 - (1-3)
Mason Industries Inc., 350 Rabro Dr., Hauppauge, NY 11788, (516) 348-0282 - (3)
MBI Products Co., 5309 Hamilton Ave., Cleveland, OH 44114, (216) 431-6400 - (1)
Midwest Acoust-A-Fiber Inc., 7790 Marysville Rd., Ostrander, OH 43061, (614) 666-2241 - (1,2)
Noise Control Associates, Inc., P.O. Box 261, Montclair, NJ 07042, (201) 746-5181 - (1,2)

Noise Control Company, Div. of Noise Control Products, 443 Oak Glenn Dr., Bartlett, IL 60103, (312) 830-0090 - (1,2)
Noise Control Products, Inc., 700 Chettic Ave., Copiague, NY 11726, (516) 226-6100 - (2)
Noise Reduction Corporation, Route 2, Box 152, Redwood Falls, MN 56283, (507) 644-3067 - (1,2)
Patterson Associates, 935 Summer St., Lynnfield, MA 01940, (617) 246-0888 - (1,2)
Peabody Noise Control, 6300 Irelan Place, Dublin, OH 43017, (614) 889-0480 - (1,2)
The Proudfoot Company, Inc., P.O. Box 1537, Greenwich, CT 06836, (203) 869-9031 - (3)
Rathbone Products, P.O. Box 10356, Columbus, OH 43201, (614) 297-7782 - (2,4)
Singer Safety Company, 2300 W. Logan Blvd., Chicago, IL 60647, (312) 235-2100 - (1,2)
The Soundcoat Company, Inc., One Burt Drive, Deer Park, NY 11729, (516) 242-2200 - (1,2)*
Soundown Corporation, 45 Congress St., Salem, MA 01970, (617) 598-4248 - (1,2)
Specialty Composites Corp., Delaware Industrial Park, Newark, DE 19713, (800) 544-5180 - (2)*
Stark Ceramics, Inc., P.O. Box 8880, Canton, OH 44711, (800) 321-0662 - (3)
Tech Products Corporation, 5030 Linden Ave., Dayton, OH 45432, (513) 252-3661 - (1,2)
Technicon, 4412 Republic Dr., Concord, NC 28025, (704) 788-1131 - (1,2)
Transco Products Inc., 55 E. Jackson Blvd., Chicago, IL 60604, (312) 427-2818 - (1)
United McGill Corporation, 1501 Kalamazoo Dr., Griffin, GA 30223, (404) 228-9864 - (1,2)
United Process Inc., 279 Silver St., P.O. Box 545, Agawam, MA 01001, (413) 789-1770 - (1,2)
Vibration Mountings & Controls, 113 Main St., P.O. Box 37, Bloomingdale, NJ 07403, (201) 838-1780 - (1,2)

Composite Systems
1. Curtains
2. Enclosures/Quiet Rooms
3. Open-Plan Partitions
4. Panels
5. Quilted Composites
6. Roof Decks

Acon, Inc., 4600 Webster St. Dayton, OH 45414, (513) 276-2111 - (2,4)
Acoustic Sciences Corporation, P.O. Box 1189 Eugene, OR 97440, (800) 272-8823 - (2-4)
Acoustic Standards, Inc., 1350-G W. Collins Ave., Orange, CA 92667, (714) 997-8283 - (1,2,4)
Acoustics Development Corporation, 3800 South 48th Terrace, St. Joseph, MO 64503, (816) 233-8061 - (2)
The Aeroacoustic Corp., 4876 Victor St., Jacksonville, FL 32207, (904) 731-3577 - (2,4)
Amber/Booth Co. Inc., 1403 N. Post Oak, Houston, TX 77055, (713) 688-1228 - (2,4)
American Acoustical Products, 9 Cochituate St., Natick, MA 01770, (617) 655-0870 - (5)
Antiphon Inc., 2655 Woodward Ave., Suite 350, Bloomfield Hills, MI 48013, (313) 332-0267 - (5)*
Automation Devices Inc., P.O. Box AD, Fairview, PA 16415, (814) 474-5561 - (2)
Barley-Earhart Co., 233 Divine Hwy., Portland, MI 48875, (517) 647-4117 - (1,3,5)
Barrier Corporation, 9908 S.W. Tigard St., Tigard, OR 97223, (503) 639-4192 - (1,2,5)
Bowman Metal Deck Div., Cyclops Corporation, P.O. Box 260, Pittsburgh, PA 15230, (412) 429-7508 - (6)
The Branford Companies P.O. Box 713, Shelton, CT 06484, (203) 735-6415 - (1,5)*
BRD Noise and Vibration Control, 112 Fairview Ave., Wind Gap, PA 18091, (215) 863-6300 - (1-5)
Cadillac Plastic & Chemical Co., P.O. Box 7035, Troy, MI 48007, (313) 583-1200 - (2)
Controlled Acoustics Corporation, 12 Wilson St., Hartsdale, NY 10530, (914) 428-7740 - (1-5)
Craxton Acoustical Products Inc., 2838 Vicksburg Lane, Plymouth, MN 55447, (612) 553-9890 - (4)
Designer Acoustics, Inc., 1671 Penfield Road, Rochester, NY 14625, (716) 385-3320 - (4)
Digisonix, P.O. Box 200, Highway 51 West, Stoughton, WI 53589, (608) 873-1500 - (2)
E-A-R Division, Cabot Corporation, 7911 Zionsville Rd., Indianapolis, IN 46268, (317) 872-1111 - (1,4,5)
Eckel Industries, Inc., 155 Fawcett St., Cambridge, MA 02138, (617) 491-3221 - (1,2,4)*
Engineering Aural Research, 1777-G17 Woodlawn St. Upland, CA 91786, (714) 946-0108 - (2)
Engineering Sales & Services, P.O. Box 558, Dayton, OH 45405, (513) 277-2047 - (1,2,4,5)
Enviro-Acoustics Company, Inc., 6150 Olson Memorial Highway, Minneapolis, MN 55422, (612) 542-8665 - (1,2,4-6)
Frommelt Industries, 4343 Chavenelle, Dubuque, IA 52001 (319) 556-2020 - (1)
General Acoustics Corp., 12248 Santa Monica Blvd., Los Angeles, CA 90025, (213) 820-1531 - (1,2,4,5)
Higgott-Kane Industrial Noise Controls Ltd., 1085 Bellamy Road North, Suite 214, Scarborough, Ont, Canada M1H 3C7(416) 431-0641 - (2,4)
Hitco, P.O. Box 1097, Gardena, CA 90249, (213) 321-8080 - (2,4,5)
I.D.E. Processes Corporation, Noise Control Div., 106 81st Ave., Kew Gardens, NY 11415, (718) 544-1177 - (2,4)
Imi-Tech Corp., 701 Fargo Ave., Elk Grove Village, IL 60007, (312) 981-7610 - (2,4)
Industrial Acoustics Company, 1160 Commerce Ave., Bronx, NY 10462, (212) 931-8000 - (2,4)
Industrial Noise Control, Inc., 1411 Jeffrey Dr., Addison, IL 60101, (312) 620-1998 - (1,2,4,5)
Interior Acoustics, Inc., 176 Route 206 South, Somerville, NJ 08876, (800) 221-0580 - (2,4,5)
International Cellulose Corp., P.O. Box 450006, Houston, TX 77245, (713) 433-6701 - (2,6)
King Noise Control, Inc., 1506 Magnolia Dr., Cincinnati, OH 45215, (513) 733-1222 - (1,2,4,5)
George Koch Sons, Inc., 10 South Eleventh Ave., Evansville, IN 47744, (812) 426-9880 - (2-4,6)
Korfund Dynamics Co., 35 South Service Road, Plainview, NY 11803, (516) 752-2453 - (1,2,4,5)
Linear Products Corp., P.O. Box 902, Cranford, NJ 07016, (201) 272-2211 - (1-5)
MBI Products Co., 5309 Hamilton Ave., Cleveland, OH 44114, (216) 431-6400 - (4,6)
Midwest Acoust-A-Fiber Inc., 7790 Marysville Rd., Ostrander, OH 43061, (614) 666-2241 - (3-5)
Neiss Corp., P.O. Box 478, Rockville, CT 06066, (203) 872-8528 - (2-4)
Noise Control Associates, Inc., P.O. Box 261, Montclair, NJ 07042, (201) 746-5181 - (1,5)
Noise Control Company, Div. of Noise Control Products, 443 Oak Glenn Dr., Bartlett, IL 60103, (312) 830-0090 - (1-6)
Noise Control Products, Inc., 700 Chettic Ave., Copiague, NY 11726, (516) 226-6100 - (2,4)
Noise Reduction Corporation, Route 2, Box 152, Redwood Falls, MN 56283, (507) 644-3067 - (1,2)
O'Brien Partition Co., Inc., 5301 E. 59th St., Kansas City, MO 64130, (800) 821-3595 - (2-4)
Patterson Associates, 935 Summer St., Lynnfield, MA 01940, (617) 246-0888 - (1-5)
Peabody Noise Control, 6300 Irelan Place, Dublin, OH 43017, (614) 889-0480 - (1,2,4,5)
Gordon J. Pollock & Associates, Inc., 19120 Detroit Rd., Rocky River, OH 44116, (216) 333-8710 - (1-5)
Pyrok, Inc., 136 Prospect Park West, Brooklyn, NY 11215, (718) 788-1225 - (4)
Singer Safety Company, 2300 W. Logan Blvd., Chicago, IL 60647, (312) 235-2100 - (1,2,4,5)
Sound Fighter Systems Inc., P.O. Box 6075, Shreveport, LA 71136, (318) 861-6640 - (2)
The Soundcoat Company, Inc., One Burt Drive, Deer Park, NY 11729, (516) 242-2200 - (1,5)*
Soundown Corporation, 45 Congress St., Salem, MA 01970, (617) 598-4248 - (5)
Starrco Company, Inc., 1515 Fairview Ave., St. Louis, MO 63132, (314) 429-5650 - (2-4)
Tech Products Corporation, 5030 Linden Ave., Dayton, OH 45432, (513) 252-3661 - (1,3-5)
Technicon, 4412 Republic Dr., Concord, NC 28025, (704) 788-1131 - (1,5)
Tectum Inc., 105 S. 6th St., Newark, OH 43055, (614) 345-9691 - (6)
Transco Products Inc., 55 E. Jackson Blvd., Chicago, IL 60604, (312) 427-2818 - (2,4,5)
United McGill Corporation, 1501 Kalamazoo Dr., Griffin, GA 30223, (404) 228-9864 - (1,2,4,5)
United Process Inc., 279 Silver St., P.O. Box 545, Agawam, MA 01001, (413) 789-1770 - (1-6)
Vibration Mountings & Controls, 113 Main St., P.O. Box 37, Bloomingdale, NJ 07403, (201) 838-1780 - (1,2,4,5)
Vibron Limited, 1720 Meyerside Dr., Mississauga, Ont., Canada L5T 1A3, (416) 677-4922 - (1,2,4,6)
Wenger Corporation, 555 Park Drive, Owatonna, MN 55060, (507) 455-4100 - (2,4)

Vibration Damping Materials
1. Adhesives
2. Constrained-Layer Composites
3. Mastics
4. Sheets
5. Tapes

Air-Loc Products, 5 Fisher St., P.O. Box 269, Franklin, MA 02038, (617) 528-0022 - (4)
Amber/Booth Co. Inc., 1403 N. Post Oak, Houston, TX 77055, (713) 688-1228 - (4)
American Acoustical Products, 9 Cochituate St., Natick, MA 01770, (617) 655-0870 - (1-4)
Antiphon Inc., 2655 Woodward Ave., Suite 350, Bloomfield Hills, MI 48013, (313) 332-0267 - (2-4)*
Arvinyl Division of Arvin Industries, Inc., 1531 13th St., Columbus, IN 47201, (812) 379-3000 - (2)
Automation Devices Inc., P.O. Box AD, Fairview, PA 16415, (814) 474-5561 - (1)
Barley-Earhart Co., 233 Divine Hwy., Portland, MI 48875, (517) 647-4117 - (2)
Barry Controls/a Unit of Barry Wright, 700 Pleasant St., Watertown, MA 02172, (617) 923-1150 - (2,4)
H. L. Blachford, Inc., 1855 Stephenson Hwy., P.O. Box 397, Troy, MI 48007, (313) 689-7800 - (1-5)*

Bostik Division of Emhart, Boston St., Middleton, MA 01949, (617) 777-0100 - (2)
The Branford Companies P.O. Box 713, Shelton, CT 06484, (203) 735-6415 - (1,3)*
BRD Noise and Vibration Control, 112 Fairview Ave., Wind Gap, PA 18091, (215) 863-6300 - (3-5)
Comprador, Inc., 3 Dewalt Road, Newark, DE 19711, (302) 368-5089 - (1,2)
Controlled Acoustics Corporation, 12 Wilson St., Hartsdale, NY 10530, (914) 428-7740 - (1-5)
E-A-R Division, Cabot Corporation, 7911 Zionsville Rd., Indianapolis, IN 46268, (317) 872-1111 - (2,4,5)
Eckel Industries, Inc., 155 Fawcett St., Cambridge, MA 02138, (617) 491-3221 - (4)*
Engineering Aural Research, 1777-G17 Woodlawn St. Upland, CA 91786, (714) 946-0108 - (5)
Engineering Sales & Services, P.O. Box 558, Dayton, OH 45405, (513) 277-2047 - (3)
Fabreeka International, Inc., 1190 Adams St., Boston, MA 02124, (617) 296-6700 - (2,4)
Frommelt Industries, 4343 Chavenelle, Dubuque, IA 52001 (319) 556-2020 - (2)
Imi-Tech Corp., 701 Fargo Ave., Elk Grove Village, IL 60007, (312) 981-7610 - (2)
Industrial Noise Control, Inc., 1411 Jeffrey Dr., Addison, IL 60101, (312) 620-1998 - (1,2,4)
King Noise Control, Inc., 1506 Magnolia Dr., Cincinnati, OH 45215, (513) 733-1222 - (1,2,4)
Korfund Dynamics Co., 35 South Service Road, Plainview, NY 11803, (516) 752-2453 - (4)
Linear Products Corp., P.O. Box 902, Cranford, NJ 07016, (201) 272-2211 - (1,2,4)
Mason Industries Inc., 350 Rabro Dr., Hauppauge, NY 11788, (516) 348-0282 - (3)
MB Dynamics, Inc., 25865 Richmond Rd., Cleveland, OH 44146, (216) 292-5850 - (4)
Midwest Acoust-A-Fiber Inc., 7790 Marysville Rd., Ostrander, OH 43061, (614) 666-2241 - (1,3,4)
National Starch and Chemical Corp., Finderne Ave., Bridgewater, NJ 08807, (201) 685-5000 - (1)
Noise Control Associates, Inc., P.O. Box 261, Montclair, NJ 07042, (201) 746-5181 - (2,4)
Noise Control Company, Div of Noise Control Products, 443 Oak Glenn Dr., Bartlett, IL 60103, (312) 830-0090 - (3,4)
Noise Reduction Corporation, Route 2, Box 152, Redwood Falls, MN 56283, (507) 644-3067 - (2,5)
Patterson Associates, 935 Summer St., Lynnfield, MA 01940, (617) 246-0888 - (2-4)
Peabody Noise Control, 6300 Irelan Place, Dublin, OH 43017, (614) 889-0480 - (2-5)
Pre Finish Metals Inc., 2300 E. Pratt Blvd., Elk Grove Village, IL 60007, (312) 439-2210 - (2)*
Presto Manufacturing Co., Inc., 2 Franklin Ave., Brooklyn, NY 11211, (718) 852-0187 - (1,4,5)
Rogers Corporation, One Technology Drive, Rogers, CT 06263, (203) 774-9605 - (4)
Singer Safety Company, 2300 W. Logan Blvd., Chicago, IL 60647, (312) 235-2100 - (2,3)
Sorbothane, Inc., 2144 State Route 59, Kent, OH 44240, (216) 678-9444 - (2,4)
The Soundcoat Company, Inc., One Burt Drive, Deer Park, NY 11729, (516) 242-2200 - (1,2,4)*
Specialty Composites Corp., Delaware Industrial Park, Newark, DE 19713, (800) 544-5180 - (2,3)*
Tech Products Corporation, 5030 Linden Ave., Dayton, OH 45432, (513) 252-3661 - (1-5)
Technicon, 4412 Republic Dr., Concord, NC 28025, (704) 788-1131 - (3,4)

United McGill Corporation, 1501 Kalamazoo Dr., Griffin, GA 30223, (404) 228-9864 - (1,2,4)
United Process Inc., 279 Silver St., P.O. Box 545, Agawam, MA 01001, (413) 789-1770 - (2,4)
Vibra-Check Inc., Bldg. 32, Endicott St., Norwood, MA 02062, (617) 782-2800 - (4)
Vibra-Damp Corp., 1055 N.W. 54th Street, Fort Lauderdale, FL 33309, (800) 237-6974 - (4)
Vibration Mountings & Controls, 113 Main St., P.O. Box 37, Bloomingdale, NJ 07403, (201) 838-1780 - (3,4)
Voltek Division of Sekisui America Corp., 100 Shepard Street, Lawrence, MA 01843, (617) 685-2557 - (4)

Vibration Isolation Systems
1. Active Isolators
2. Bases
3. Elastomeric
4. Floating Floors
5. Machinery Mounts
6. Pipe Connectors
7. Pneumatic
8. Seismic
9. Steel Spring
10. Vibration Dampers

Active Noise and Vibration Technologies, 1755 W. University, Suite 6, Tempe, AZ 85281, (602) 829-1868 - (1)
Aeroflex International, 35 S. Service Rd., Plainview, NY 11803, (516) 694-6700 - (2,9,10)
Air-Loc Products, 5 Fisher St., P.O. Box 269, Franklin, MA 02038, (617) 528-0022 - (3,10)
Akzo Industrial Systems Company, One North Pack Sq., P.O. Box 7249, Asheville, NC 28802, (704) 258-5050 - (4)
Amber/Booth Co. Inc., 1403 N. Post Oak, Houston, TX 77055, (713) 688-1228 - (1-7)
Anamet Inc./Anaconda Metal Hose, 698 S. Main St., P.O. Box 2618, Waterbury, CT 06725, (203) 574-8500 - (6,10)
Barry Controls/a Unit of Barry Wright, 700 Pleasant St., Watertown, MA 02172, (617) 923-1150 - (1-10)
Louis P. Batson Company, P.O. Box 3978, Greenville, SC 29608, (803) 242-5262 - (5,10)
BRD Noise and Vibration Control, 112 Fairview Ave., Wind Gap, PA 18091, (215) 863-6300 - (1-6,8-10)
Bushings, Inc., 4358 Coolidge Hwy., P.O. Box 189, Royal Oak, MI 48068, (313) 564-6357 - (3,5,10)
Controlled Acoustics Corporation, 12 Wilson St., Hartsdale, NY 10530, (914) 428-7740 - (1-10)
E-A-R Division, Cabot Corporation, 7911 Zionsville Rd., Indianapolis, IN 46268, (317) 872-1111 - (3,5,10)
Engineering Aural Research, 1777-G17 Woodlawn St. Upland, CA 91786, (714) 946-0108 - (1,4,5)
Engineering Sales & Services, P.O. Box 558, Dayton, OH 45405, (513) 277-2047 - (3-5,10)
Fabreeka International, Inc., 1190 Adams St., Boston, MA 02124, (617) 296-6700 - (1-10)
Firestone Industrial Products Co., 1700 Firestone Blvd., Noblesville, IN 46060, (317) 773-0650 - (7)
Higgott-Kane Industrial Noise Controls Ltd., 1085 Bellamy Road North, Suite 214, Scarborough, Ont, Canada M1H 3C7 (416) 431-0641 - (2,5,9)
Industrial Acoustics Company, 1160 Commerce Ave., Bronx, NY 10462, (212) 931-8000 - (4)
Industrial Noise Control, Inc., 1411 Jeffrey Dr., Addison, IL 60101, (312) 620-1998 - (10)
King Noise Control, Inc., 1506 Magnolia Dr., Cincinnati, OH 45215, (513) 733-1222 - (2,5,9,10)
Korfund Dynamics Co., 35 South Service Road, Plainview, NY 11803, (516) 752-2453 - (2-10)
L-A-B Div. of Mechanical Technology, Inc., Onondaga St., Skaneateles, NY 13152, (315) 685-5781 - (7)
Linear Products Corp., P.O. Box 902, Cranford, NJ 07016, (201) 272-2211 - (1-10)
Lord Corporation, Industrial Products Div., 1952 West Grandview Blvd., Erie, PA 16514 (814) 864-5424 - (3,5,10)
M/RAD Corporation, 71 Pine St., Woburn, MA 01801, (617) 935-5940 - (1-4,7-10)
Mason Industries Inc., 350 Rabro Dr., Hauppauge, NY 11788, (516) 348-0282 - (1-10)
MB Dynamics, Inc., 25865 Richmond Rd., Cleveland, OH 44146, (216) 292-5850 - (1,2)
Mercer Rubber Company, 136 Mercer St., Hamilton Square, NJ 08690, (800) 443-5764 - (3,6,7)
Noise Cancellation Technology, 98 Cutter Mill Drive, Suite 275N, Great Neck, NY 11021, (516) 466-1060 - (1,4,5,10)
Noise Control Company, Div. of Noise Control Products, 443 Oak Glenn Dr., Bartlett, IL 60103, (312) 830-0090 - (1-10)
Peabody Noise Control, 6300 Irelan Place, Dublin, OH 43017, (614) 889-0480 - (1-5,7-10)
Rogers Corporation, One Technology Drive, Rogers, CT 06263, (203) 774-9605 - (3)

Royal Produts, 210 Oser Ave., Hauppauge, NY 11788, (800) 645-4174 - (5)
SigmaFlex, Inc., 961 Decker Rd., Walled Lake, MI 48088, (313) 624-2588 - (4,5)
Sorbothane, Inc., 2144 State Route 59, Kent, OH 44240, (216) 678-9444 - (3)
Stock Drive Products, 2101 Jericho Turnpike, New Hyde Park, NY 11040, (516) 328-3300 - (1-3,6,7,10)
Sunnex Inc., 87 Crescent Rd., Needham, MA 02194, (617) 444-4730 - (5)
Target Enterprises Inc., Roxbury, VT 05669, (802) 728-3081 - (3)

Tech Products Corporation, 5030 Linden Ave., Dayton, OH 45432, (513) 252-3661 - (1-10)
Unisorb Machinery Installation Systems, P.O. Box 1000, Jackson, MI 49204, (517) 764-6060 - (5)
United McGill Corporation, 1501 Kalamazoo Dr., Griffin, GA 30223, (404) 228-9864 - (5)
United Process Inc., 279 Silver St., P.O. Box 545, Agawam, MA 01001, (413) 789-1770 - (6)
Vibra-Check Inc., Bldg. 32, Endicott St., Norwood, MA 02062, (617) 782-2800 - (2,5,10)
Vibra-Damp Corp., 1055 N.W. 54th Street, Fort Lauderdale, FL 33309, (800) 237-6974 - (3,5,10)
Vibration Eliminator Co. Inc., 10-28 47th Ave., Long Island City, NY 11101, (718) 729-2500 - (1-10)
Vibration Mountings & Controls, 113 Main St., P.O. Box 37, Bloomingdale, NJ 07403, (201) 838-1780 - (2-10)
The Vibration Isolation Co., Inc., 225 Grand St., Paterson, NJ 07501, (201) 345-8282 - (1-6,8-10)
Vibro/Dynamics Corporation, 500 E. Plainfield Rd., Countryside, IL 60525, (312) 482-8220 - (5)
Vibron Limited, 1720 Meyerside Dr., Mississauga, Ont., Canada L5T 1A3, (416) 677-4922 - (1-6,8-10)
Vlier Engineering, a unit of Barry Wright Corp., 2333 Valley St., Burbank, CA 91505, (818) 843-1922 - (5)
Voltek Division of Sekisui America Corp., 100 Shepard Street, Lawrence, MA 01843, (617) 685-2557 - (4)
Wenger Corporation, 555 Park Drive, Owatonna, MN 55060, (507) 455-4100 - (4)

Silencers
1. Active Attenuators
2. Ducts
3. Duct Silencers
4. Electric Motor Silencers
5. Fan Silencers
6. Filter Silencers
7. General Industrial Silencers
8. High-Pressure Discharge Silencers
9. Intake and Exhaust Silencers
10. Pulsation Dampers
11. Splitter/Louvre Silencers

Acoustic Sciences Corporation, P.O. Box 1189 Eugene, OR 97440, (800) 272-8823 - (3,9,11)
Active Noise and Vibration Technologies, 1755 W. University, Suite 6, Tempe, AZ 85281, (602) 829-1868 - (1,3-7,9)
The Aeroacoustic Corp., 4876 Victor St., Jacksonville, FL 32207, (904) 731-3577 - (3-9,11)
Airsan Corporation, 4554 W. Woolworth Ave., Milwaukee, WI 53218, (414) 353-5800 - (3,11)
Allied Witan Company, 13805 Progress Parkway, Cleveland, OH 44133, (216) 237-9630 - (1,5-10)
Amber/Booth Co. Inc., 1403 N. Post Oak, Houston, TX 77055, (713) 688-1228 - (3-8,10)
Arrow Pneumatics, Inc., 500 Oakwood Rd., Lake Zurich, IL 60047, (312) 438-9100 - (9)
Atlas Minerals & Chemicals Inc., Farmington Rd., Mertztown, PA 19539, (215) 682-7171 - (9)
BRD Noise and Vibration Control, 112 Fairview Ave., Wind Gap, PA 18091, (215) 863-6300 - (1-11)
Brunswick Corp. Technetics Div., 2000 Brunswick Lane, DeLand, FL 32724, (904) 736-1700 - (3,5,8,9,11)
Burgess-Manning, Inc., French & Old Union Rds., Buffalo, NY 14224, (716) 668-8111 - (3,6-11)
Controlled Acoustics Corporation, 12 Wilson St., Hartsdale, NY 10530, (914) 428-7740 - (2,3,5,7,9,11)
Digisonix, P.O. Box 200, Highway 51 West, Stoughton, WI 53589, (608) 873-1500 - (1,3,5,9)
Duracote Corporation, 350 N. Diamond St., P.O. Box 512, Ravenna, OH 44266, (216) 296-9600 - (3-11)
Eckel Industries, Inc., 155 Fawcett St., Cambridge, MA 02138, (617) 491-3221 - (5,11)*

Engineering Aural Research, 1777-G17 Woodlawn St. Upland, CA 91786, (714) 946-0108 - (3)
Engineering Sales & Services, P.O. Box 558, Dayton, OH 45405, (513) 277-2047 - (4-9)
Enviro-Acoustics Company, Inc., 6150 Olson Memorial Highway, Minneapolis, MN 55422, (612) 542-8665 - (1-9,11)

Exair Corporation, 1250 Century Circle North, Cincinnati, OH 45246, (513) 671-3322 – (7,8)

Flowmaster, Inc., 2975 Dutton Ave., 3, Santa Rosa, CA 95407, (707) 544-4761 – (7,9)

Fluid Kinetics Corp., P.O. Box CE, Ventura, CA 93002, (805) 644-5587 – (3,5,7-10)

General Acoustics Corp., 12248 Santa Monica Blvd., Los Angeles, CA 90025, (213) 820-1531 – (1-9)

Higgott-Kane Industrial Noise Controls Ltd., 1085 Bellamy Road North, Suite 214, Scarborough, Ont. Canada M1H 3C7(416) 431-0641 – (2-9,11)

Hitco, P.O. Box 1097, Gardena, CA 90249, (213) 321-8080 – (2-4,7,9)

I.D.E. Processes Corporation, Noise Control Div., 106 81st Ave., Kew Gardens, NY 11415, (718) 544-1177 – (3,5,7-9)

Industrial Acoustics Company, 1160 Commerce Ave., Bronx, NY 10462, (212) 931-8000 – (3-9,11)

Industrial Noise Control, Inc., 1411 Jeffrey Dr., Addison, IL 60101, (312) 620-1998 – (3)

King Noise Control, Inc., 1506 Magnolia Dr., Cincinnati, OH 45215, (513) 733-1222 – (3-5,7-9)

Korfund Dynamics Co., 35 South Service Road, Plainview, NY 11803, (516) 752-2453 – (3,5,7)

MBI Products Co., 5309 Hamilton Ave., Cleveland, OH 44114, (216) 431-6400 – (3,5)

Noise Cancellation Technology, 98 Cutter Mill Drive, Suite 275N, Great Neck, NY 11021, (516) 466-1060 – (1,4,5,7,9,10)

Noise Control Company, Div. of Noise Control Products, 443 Oak Glenn Dr., Bartlett, IL 60103, (312) 830-0090 – (2,3,5-11)

Noise Control Products, Inc., 700 Chettic Ave., Copiague, NY 11726, (516) 226-6100 – (3,5)

Noise Suppression Systems, Inc., 2886 Metropolitan Place, Pomona, CA 91767, (714) 596-6710 – (1-11)

Patterson Associates, 935 Summer St., Lynnfield, MA 01940, (617) 246-0888 – (2,3,7,9)

Quietflo Noise Control Div., Flaregas Corp., 100-B Airport Executive Park, Spring Valley, NY 10977, (914) 352-8877 – (2-11)

Riley-Beaird, Inc., P.O. Box 31115, Shreveport, LA 71130, (318) 865-6351 – (5-9)

Schrader Bellows, 200 W. Exchange St., P.O. Box 631, Akron, OH 44309, (216) 375-5202 – (7,9)

Schreter Associates, 930 Oakhaven Dr., Roswell, GA 30075, (404) 641-1843 – (3,5,7,9)

Sound Fighter Systems Inc., P.O. Box 6075, Shreveport, LA 71136, (318) 861-6640 – (3,5,7,9,11)

The Soundcoat Company, Inc., One Burt Drive, Deer Park, NY 11729, (516) 242-2200 – (5,11)*

Spencer Turbine Co., 600 Day Hill Rd., Windsor, CT 06095, (203) 688-8361 – (4-6,9)

Stoddard Silencers, Inc., P.O. Box 397, Grayslake, IL 60030, (312) 223-8636 – (3,5-9,11)

Sunnex Inc., 87 Crescent Rd., Needham, MA 02194, (617) 444-4730 – (9)

Tech Products Corporation, 5030 Linden Ave., Dayton, OH 45432, (513) 252-3661 – (2,3,7-9)

TLT-Babcock, Inc., 3480 W. Market St., Akron, OH 44313, (216) 867-8540 – (3,5,8,9,11)

Transco Products Inc., 55 E. Jackson Blvd., Chicago, IL 60604, (312) 427-2818 – (2,3,5,7-9)

United McGill Corporation, 1501 Kalamazoo Dr., Griffin, GA 30223, (404) 228-9864 – (2,3,5,7,9)

United Process Inc., 279 Silver St., P.O. Box 545, Agawam, MA 01001, (413) 789-1770 – (3)

Vibration & Noise Engineering Corp., 2655 Villa Creek Dr., 185, Dallas, TX 75234, (214) 243-1951 – (3,5-11)

Vibration Mountings & Controls, 113 Main St., P.O. Box 37, Bloomingdale, NJ 07403, (201) 838-1730 – (3,5,7)

Vibron Limited, 1720 Meyerside Dr., Mississauga, Ont., Canada L5T 1A3, (416) 677-4922 – (3,5-9,11)

Vlier Engineering, a unit of Barry Wright Corp., 2333 Valley St., Burbank, CA 91505, (818) 843-1922 – (9)

DYNAMIC MEASUREMENT INSTRUMENTATION BUYER'S GUIDE

This Instrumentation Buyer's Guide lists manufacturers and suppliers of dynamic measurement instrumentation grouped into the following ten categories:

☐ Analyzers
☐ Calibrators
☐ Computer-Based Systems
☐ Generators
☐ Machinery Monitoring Systems
☐ Meters
☐ Recorders and Displays
☐ Signal Conditioners
☐ Special Purpose Systems
☐ Transducers

Each category contains a series of numbered subclassifications for specific product identification.

All instrumentation systems and components listed are for the measurement of dynamic phenomena or the control of dynamic parameters. These phenomena include vibration, noise, shock, strain, force, and pressure. The frequency range of the instrumentation is generally within the limits of dc to 100 kHz. Note that measurement systems that consist of hardware and/or software adaptations of general purpose computers are listed under Computer-Based Systems. Dedicated hardware systems are listed in other categories. Instrumentation systems including analyzers, meters, and transducers primarily used for machinery monitoring, diagnostics, and preventive maintenance are listed under Machinery Monitoring Systems.

To locate the instrumentation of interest, first refer to the main categories listed above then go to the detailed listings to locate the manufacturers and suppliers who offer that particular product. The numbers after each company listing indicate the products that they offer. Boldface company listings are advertisers in this issue.

Analyzers
1. Acoustic Emission
2. FFT, Multi-Channel
3. FFT, Single-Channel
4. Frequency Spectrum
5. Noise Level
6. Octave Band
7. 1/3-Octave Band
8. Shock Spectrum
9. Sound Intensity
10. Sound Power

A&D Engineering, Inc., 1555 McCandless Dr., Milpitas, CA 95035, (408) 263-5333 - (2-7,9,10)
Acoustic Emission Technology Corp., 1824J Tribute Rd., Sacramento, CA 95815, (916) 927-3861 - (1-4)
Balmac Inc., 4010 Main St., Hilliard, OH 43026, (614) 876-1295 - (4)
Bruel & Kjaer Instruments, Inc., 185 Forest St., Marlborough, MA 01752, (617) 481-7000 - (2-7,9,10)
Cetec Ivie, 1366 W. Center St., Orem, UT 84057, (801) 224-1800 - (5-7)
Cirrus Research Ltd., 1350-G West Collins, Orange, CA 92667, (714) 997-7430 - (4-7)
Data Physics Corporation, 1210 S. Bascom Ave., #224, San Jose, CA 95128, (408) 977-0800 - (2-4)

Data Precision, Div. of Analogic, 16 Electronics Ave., Danvers, MA 01923, (617) 246-1600 - (2)
Difa Measuring Systems, 37620 Hills Tech Dr., Farmington Hills, MI 48331, (313) 489-8588 - (2,4)
Endevco, 30700 Rancho Viejo Rd., San Juan Capistrano, CA 92675, (714) 493-8181 - (2,8)
GenRad, Inc., 510 Cottonwood Drive, Milpitas, CA 95035, (408) 432-1000 - (2,4-10)
GHI Systems, Inc., 916 N. Western Ave., #201, San Pedro, CA 90732, (213) 548-6544 - (2,4,8)
Hewlett-Packard Co., 5161 Lankershim Blvd., N. Hollywood, CA 91601, (818) 505-5600 - (2,3)
Kay Elemetrics Corp., 12 Maple Ave., Pine Brook, NJ 07058, (201) 227-2000 - (2,4)
Klark-Teknik Electronics, Inc., 30B Banfi Plaza North, Farmingdale, NY 11735, (516) 249-3660 - (7)
Larson-Davis Laboratories, 280 South Main, Pleasant Grove, UT 84062, (801) 785-8352 - (2,3,5-7,9,10)
LeCroy Corporation, 700 Chestnut Ridge Rd., Chestnut Ridge, NY 10977, (800) 553-2769 - (1-4,8)
Lucas CEL Instruments, 5500 New King St., Troy, MI 48098, (313) 879-1920 - (4-7)
Metrosonics Inc., P.O. Box 23075, Rochester, NY 14692, (716) 334-7300 - (4-6)
Mott Associates, Inc., P.O. Box 42546, Houston, TX 77042, (713) 777-5773 - (3,5-7)
Multigon Industries, Inc., 559 Gramatan Ave., Mt. Vernon, NY 10552, (914) 664-7300 - (2-4)
NF Electronic Instruments, Inc., 3303 Harbor Blvd., Suite D-6, Costa Mesa, CA 92626, (714) 557-1684 - (1)
Northern Sound, 5155 East River Rd. NE, #413, Minneapolis, MN 55421, (612) 572-1515 - (5-7)
One Sokki Technology, Inc., 1120 Tower Lane, Bensenville, IL 60106, (312) 595-9292 - (2-7,9,10)
Panasonic Industrial Co., Two Panasonic Way, Secaucus, NJ 07094, (201) 392-4050 - (2,3)
Quan-Tech, Div. Stansbury Industries, Inc., P.O. Box 28, Route 206, Flanders, NJ 07836, (201) 927-1766 - (4)
Quest Electronics, 510 S. Worthington St., Oconomowoc, WI 53066, (414) 567-9157 - (1,5-7,9)
Rockland Scientific Corporation, 10 Volvo Dr., Rockleigh, NJ 07647, (201) 767-7900 - (2,3,7)
Scantek Inc./Norwegian Electronics, 51 Monroe Street, #1606, Rockville, MD 20850, (301) 279-9308 - (1,4-7,9,10)
Schlumberger Technologies, Instruments Div., 20 North Ave., Burlington, MA 01720, (800) 225-5765 - (2)
Scientific-Atlanta, Spectral Dynamics Div., 4255 Ruffin Road, San Diego, CA 92123, (619) 268-7100 - (2-4,6-10)
Scott Instrument Laboratories, 13 Sagamore Park Rd., Hudson, NH 03051, (800) 343-0683 - (5,6)
Systolic Systems, Inc., 2240 N. First St., San Jose, CA 95131, (408) 435-1760 - (2)
Technology for Energy Corp., P.O. Box 22996 Knoxville, TN 37933, (615) 966-5856 - (3)
Trig-Tek Inc., 423 S. Brookhurst St., Anaheim, CA 92804, (714) 956-3593 - (4)
Unholtz-Dickie Corporation, 6 Brookside Dr., Wallingford, CT 06492, (203) 265-3929 - (2,8)
White Instruments, Div. C VAN R, Inc., 5408 U.S. Highway 290 West, Austin, TX 78735, (512) 892-0752 - (4,5,7,9,10)
Zonic Corporation, 2000 Ford Circle, Milford, OH 45150, (513) 248-1911 - (2,4,6,7,9,10)

Calibrators
1. Accelerometer
2. Acoustic Emission
3. Hydrophone
4. Microphone
5. Proximity Sensor
6. Sound Power

Acoustic Emission Technology Corp., 1824J Tribute Rd., Sacramento, CA 95815, (916) 927-3861 - (2)
Alpha-M Corporation, 913 Highland Dr., Euless, TX 76040, (817) 267-7022 - (1)
Bently Nevada Corp., P.O. Box 157, Minden, NV 89423, (702) 782-3611 - (5)
Bruel & Kjaer Instruments, Inc., 185 Forest St., Marlborough, MA 01752, (617) 481-7000 - (1,3,4)
Chadwick Helmuth Co., Inc., 4601 N. Arden Dr., El Monte, CA 91731, (818) 575-6161 - (1)
Channel Industries, Inc., 839 Ward Dr., Santa Barbara, CA 93111, (805) 967-0171 - (3)
Cirrus Research Ltd., 1350-G West Collins, Orange, CA 92667, (714) 997-7430 - (1,4)
Dytran Instruments, Inc., 21592 Marilla St., Chatsworth, CA 91311, (818) 700-7818 - (1)
Endevco, 30700 Rancho Viejo Rd., San Juan Capistrano, CA 92675, (714) 493-8181 - (1)

Entran Devices, Inc., 10 Washington Ave., Fairfield, NJ 07006, (800) 635-0650 - (1)
Ilg Industries, Inc., 2850 N. Pulaski, Chicago, IL 60641, (312) 725-8016 - (6)
IQS Inc., 12812-J Garden Grove Blvd., Garden Grove, CA 92643, (714) 539-7842 - (1,4)
IRD Mechanalysis, Inc., 6150 Huntley Rd., Columbus, OH 43229, (614) 885-5376 - (1,5)
Kistler Instrument Corp., 75 John Glenn Dr., Amherst, NY 14120, (716) 691-5100 - (1)
Larson-Davis Laboratories, 280 South Main, Pleasant Grove, UT 84062, (801) 785-8352 - (4)
Lucas CEL Instruments, 5500 New King St., Troy, MI 48098, (313) 879-1920 - (1,4)
MB Dynamics, 25865 Richmond Rd., Cleveland, OH 44146, (216) 292-5850 - (1,5)
Metrix Instrument Co., 1711 Townhurst Dr., Houston, TX 77043, (713) 461-2131 - (5)
Metrosonics Inc., P.O. Box 23075, Rochester, NY 14692, (716) 334-7300 - (4)
Mott Associates, Inc., P.O. Box 42546, Houston, TX 77042, (713) 777-5773 - (1,4)
Northern Sound, 5155 East River Rd. NE, #413, Minneapolis, MN 55421, (612) 572-1515 - (4)
PCB Piezotronics, Inc., 3425 Walden Ave., Depew, NY 14043, (716) 684-0001 - (1)
Quest Electronics, 510 S. Worthington St., Oconomowoc, WI 53066, (414) 567-9157 - (4)
Scientific-Atlanta, Spectral Dynamics Div., 4255 Ruffin Road, San Diego, CA 92123, (619) 268-7100 - (1,5)
Specialty Engineering, 481 Sinclair Frontage Rd., Milpitas, CA 95035, (408) 946-9779 - (3)
Trig-Tek Inc., 423 S. Brookhurst St., Anaheim, CA 92804, (714) 956-3593 - (1)
Unholtz-Dickie Corporation, 6 Brookside Dr., Wallingford, CT 06492, (203) 265-3929 - (1)
H. M. Wilson Co., 11501 Chimney Rock, Houston, TX 77035, (713) 729-1183 - (1)

Computer-Based Systems
1. Acoustics Analysis
2. Control
3. Data Acquisition
4. Data Transmission
5. Machinery Analysis
6. Predictive Maintenance
7. Signal Analysis
8. Sound Intensity Analysis
9. Waveform Synthesis

Anatrol Corporation, 10895 Indeco Dr., Cincinnati, OH 45241, (513) 793-8844 - (3,7,8)
Bently Nevada Corp., P.O. Box 157, Minden, NV 89423, (702) 782-3611 - (3,5,6).·

Cirrus Research Ltd., 1350-G West Collins, Orange, CA 92667, (714) 997-7430 - (1-4,6,7)
Computational Systems Inc., 1900 N. Winston Rd., Knoxville, TN 37919, (615) 693-0551 - (5-7)
COSMIC; The University of Georgia, 382 East Broad St., Athens, GA 30602, (404) 542-3265 - (1,2,5-9)
Creare Inc., Etna Rd. P.O. Box 71, Hanover, NH 03755, (603) 643-3800 - (3,7)
Dallas Instruments, Inc., P.O. Box 38139, Dallas, TX 75238, (214) 349-1180 - (3)
Data Capture Technology, 10837 E. Marshall, Tulsa, OK 74116, (918) 438-5368 - (3,9)
Data Physics Corporation, 1210 S. Bascom Ave., #224, San Jose, CA 95128, (408) 977-0800 - (3,7,9)
Datel Inc., 11 Cabot Blvd., Mansfield, MA 02048, (617) 339-3000 - (3)
Daytronic Corporation, 2589 Corporate Place, Miamisburg, OH 45342, (513) 866-3300 - (2,3)
Difa Measuring Systems, 37620 Hills Tech Dr., Farmington Hills, MI 48331, (313) 489-8588 - (3,7,9)
DLI Engineering Corporation, 253 Winslow Way West, Bainbridge Island, WA 98110, (206) 842-7656 - (5,6)
DSP Technology Inc., 48500 Kato Road, Fremont, CA 94538, (415) 657-7555 - (2,3,7 9\

Dynamics Research Corp., 112 Killewald Ave., Tonawanda, NY 14150, (716) 692-1777 - (5)
Encore Electronics, RD #2, Route 50, Saratoga Springs, NY 12866, (518) 584-5354 - (3)
Endevco, 30700 Rancho Viejo Rd., San Juan Capistrano, CA 92675, (714) 493-8181 - (3)
Entek Scientific Corporation, 4480 Lake Forest Dr., Suite 316, Cincinnati, OH 45242, (513) 563-7500 - (6)
Exact Electronics, Div. of Dynatech Nevada Inc. 2000 Arrowhead Drive, Carson City, NV 89701, (800) 648-7952 - (9)
Floating Point Systems, Inc., 3601 SW Murray Blvd., Beaverton, OR 97005, (503) 641-3151 - (7)

GenRad, Inc., 510 Cottonwood Drive, Milpitas, CA 95035, (408) 432-1000 - (1-3,5-9)

GHI Systems, Inc., 916 N. Western Ave., #201, San Pedro, CA 90732, (213) 548-6544 - (3)

Grozier Technical Systems, Inc., 157 Salisbury Rd., Brookline, MA 02146, (617) 277-1133 - (1,3)

HEM Data Corporation, 17025 Crescent Drive, Southfield, MI 48076, (313) 559-5607 - (1-3,5-9)

Hessler Associates, Inc./TPM Software, 6400 Wishbone Terrace, Cabin John, MD 20818, (301) 229-4901 - (1)

Hewlett-Packard Co., 5161 Lankershim Blvd., N. Hollywood, CA 91601, (818) 505-5600 - (3,5-7)

Honeywell Test Instruments Div., P.O. Box 5227, Denver, CO 80217, (303) 773-4700 - (2,3)

IMO Delaval Inc., CEC Instruments Div., 325 Halstead St., Pasadena, CA 91109, (818) 351-4241 - (3)

Industrial Computer Source, 5466 Complex St., #208, San Diego, CA 92123, (619) 279-0084 - (2,3,5,7)

Infotek Systems, 1045 S. East St., Anaheim, CA 92805, (714) 956-9300 - (2,3,7,9)

IQS Inc., 12812-J Garden Grove Blvd., Garden Grove, CA 92643, (714) 539-7842 - (1-3,5,7,9)

Kay Elemetrics Corp., 12 Maple Ave., Pine Brook, NJ 07058, (201) 227-2000 - (1,7)

Kelmet, Inc., 569 W. Covina Blvd., San Dimas, CA 91773, (714) 599-0954 - (3,5,6)

Kinemetrics/Systems, 222 Vista Ave., Pasadena, CA 91107, (818) 795-2220 - (3)

Larson-Davis Laboratories, 280 South Main, Pleasant Grove, UT 84062, (801) 785-6352 - (1,6)

LeCroy Corporation, 700 Chestnut Ridge Rd., Chestnut Ridge, NY 10977, (800) 553-2769 - (1,3,7-9)

Lucas CEL Instruments, 5500 New King St., Troy, MI 48098, (313) 879-1920 - (3)

Masscomp, One Technology Way, Westford, MA 01886, (617) 692-6200 - (1-5,7-9)

MB Dynamics, 25865 Richmond Rd., Cleveland, OH 44146, (216) 292-5850 - (2,3)

Mechanical Technology, 968 Albany-Shaker Rd., Latham, NY 12110, (518) 785-2211 - (5,6)

MetraByte Corp., 440 Myles Standish Blvd., Taunton, MA 02780, (617) 880-3000 - (2,3,9)

Metrosonics Inc., P.O. Box 23075, Rochester, NY 14692, (716) 334-7300 - (1-3)

Mott Associates, Inc., P.O. Box 42546, Houston, TX 77042, (713) 777-5773 - (5,6)

Multigon Industries, Inc., 559 Gramatan Ave., Mt. Vernon, NY 10552, (914) 664-7300 - (7)

Northern Sound, 5155 East River Rd. NE, #413, Minneapolis, MN 55421, (612) 572-1515 - (1)

Optim Electronics Corp., 12401 Middlebrook Dr., Germantown, MD 20874, (301) 428-7200 - (3)

Pacific Instruments, Inc., 215 Mason Circle, Concord, CA 94520, (415) 827-9010 - (3)

Quantitative Technology Corp., 8700 S.W. Creekside Place, Suite D, Beaverton, OR 97005, (503) 626-3081 - (1,7)

Quest Electronics, 510 S. Worthington St., Oconomowoc, WI 53066, (414) 567-9157 - (1,3,8)

R.C. Electronics, 5386-D Hollister Ave., Santa Barbara, CA 93111, (805) 964-6708 - (1,3,7,9)

Rapid Systems, Inc., 433 N. 34th St., Seattle, WA 98103, (206) 547-8311 - (1,3,7)

Scantek Inc./Norwegian Electronics, 51 Monroe Street, #1606, Rockville, MD 20850, (301) 279-9308 - (1,2,8)

Schenck-Trebel Corp., 535 Acorn St., Deer Park, NY 11729, (516) 242-4010 - (5,6)

Schlumberger Technologies, Instruments Div., 20 North Ave., Burlington, MA 01720, (800) 225-5765 - (3)

Scientific Recording Associates, 59 Princeton Terrace, Watertown, CT 06795, (203) 274-7761 - (3,7)

Signal Technology, Inc., 5951 Encina Rd., Goleta, CA 93117, (805) 683-3771 - (1,3,7-9)

Soltec Corporation, Sol Vista Park, San Fernando, CA 91340, (818) 365-0800 - (3)

Spanta, Inc., 795 Franklin Ave., Franklin Lakes, NJ 07417, (201) 848-0033 - (3-6)

Specialty Engineering, 481 Sinclair Frontage Rd., Milpitas, CA 95035, (408) 946-9779 - (3,7)

SPM Instrument Inc., 359 N. Main St., Marlborough, CT 06447, (203) 295-8241 - (5,6)

Structural Measurement Systems, Inc., 651 River Oaks Pkwy., San Jose, CA 95134, (408) 263-2200 - (6)

Synergistic Technology Inc., 20065 Stevens Creek Blvd., Cupertino, CA 95014, (408) 253-5800 - (2,3,7)

Systolic Systems, Inc., 2240 N. First St., San Jose, CA 95131, (408) 435-1760 - (2,3)

TEAC Corporation of America, 7733 Telegraph Rd., Montebello, CA 90640, (213) 726-0303 - (3)

Technology for Energy Corp., P.O. Box 22996, Knoxville, TN 37933, (615) 966-5856 - (5,6)

Techron, 1718 W. Mishawaka Rd., Elkhart, IN 46517, (219) 294-3000 - (1,7)

Tektronix, Signal Analysis Unit, 1350 Dell Ave., Unit #104, Campbell, CA 95008, (408) 374-6464 - (3,7,9)

Timewave Technology, Inc., 2401 Pilot Knob Rd., Suite 134, St. Paul, MN 55120, (612) 452-5939 - (1,3,8)

Unholtz-Dickie Corporation, 6 Brookside Dr., Wallingford, CT 06492, (203) 265-3929 - (2)

University Software Systems Inc., 250 N. Nash St., El Segundo, CA 90245, (213) 640-7616 - (1,7)

Vibration Engineering Consultants, Inc., 10 State St., Woburn, MA 01801, (617) 932-9585 - (7)

Vibration Instruments Co., 1614 Orangethorpe Way, Anaheim, CA 92801, (714) 879-6021 - (3)

Vibro-Meter Corporation, 1 Progress Rd., Billerica, MA 01822, (617) 663-7322 - (3)

Vitec, Inc., 23600 Mercantile Rd., Cleveland, OH, 44122, (216) 464-4670 - (5-7)

Wavetek San Diego Inc., 9045 Balboa Ave., San Diego, CA 92123, (619) 279-2200 - (3)

White Instruments, Div. C VAN R, Inc., 5408 U.S. Highway 290 West, Austin, TX 78735, (512) 892-0752 - (1,7,8)

Wolfe Instrumentation Co., 10760 Burbank Blvd., North Hollywood, CA 91601, (800) 554-1224 - (3)

Zonic Corporation, 2000 Ford Circle, Milford, OH 45150, (513) 248-1911 - (1,3,5-9)

Generators
1. Acoustic Impulse
2. Arbitrary Function
3. Audio
4. Random Noise
5. Sine Wave

A&D Engineering, Inc., 1555 McCandless Dr., Milpitas, CA 95035, (408) 263-5333 - (3)

Acoustilog, Inc., 19 Mercer St., New York, NY 10013, (212) 925-1365 - (1)

Alpha-M Corporation, 913 Highland Dr., Euless, TX 76040, (817) 267-7022 - (5)

Cetec Ivie, 1366 W. Center St., Orem, UT 84057, (801) 224-1800 - (1)

Cirrus Research Ltd., 1350-G West Collins, Orange, CA 92667, (714) 997-7430 - (4)

Data Physics Corporation, 1210 S. Bascom Ave., #224, San Jose, CA 95128, (408) 977-0800 - (2,4,5)

Data Precision, Div. of Analogic, 16 Electronics Ave., Danvers, MA 01923, (617) 246-1600 - (2,4,5)

Difa Measuring Systems, 37620 Hills Tech Dr., Farmington Hills, MI 48331, (313) 489-8588 - (2,4,5)

DLI Engineering Corporation, 253 Winslow Way West, Bainbridge Island, WA 98110, (206) 842-7656 - (2,5)

Exact Electronics, Div. of Dynatech Nevada Inc. 2000 Arrowhead Drive, Carson City, NV 89701, (800) 648-7952 - (3,5)

Grozier Technical Systems, Inc., 157 Salisbury Rd., Brookline, MA 02146, (617) 277-1133 - (1)

Krohn-Hite Corporation, 255 Bodwell St., Avon, MA 02322, (617) 580-1660 - (2,3,5)

LeCroy Corporation, 700 Chestnut Ridge Rd., Chestnut Ridge, NY 10977, (800) 553-2769 - (2)

Lucas CEL Instruments, 5500 New King St., Troy, MI 48098, (313) 879-1920 - (4)

MB Dynamics, 25865 Richmond Rd., Cleveland, OH 44146. (216) 292-5850 - (2,4,5)

Mott Associates, Inc., P.O. Box 42546, Houston, TX 77042, (713) 777-5773 - (4,5)

NF Electronic Instruments, Inc., 3303 Harbor Blvd., Suite D-6, Costa Mesa, CA 92626, (714) 557-1684 - (2,4,5)

Ono Sokki Technology, Inc., 1120 Tower Lane, Benseaville, IL 60106, (312) 595-9292 - (2,4,5)

Scantek Inc./Norwegian Electronics, 51 Monroe Street, #1606, Rockville, MD 20850, (301) 279-9308 - (1,3,4)

Techron, 1718 W. Mishawaka Rd., Elkhart, IN 46517, (219) 294-3000 - (3-5)

Tektronix, Signal Analysis Unit, 1350 Dell Ave., Unit #104, Campbell, CA 95008, (408) 374-6464 - (2,4,5)

Timewave Technology, Inc., 2401 Pilot Knob Rd., Suite 134, St. Paul, MN 55120, (612) 452-5939 - (4,5)

Trig-Tek Inc., 423 S. Brookhurst St., Anaheim, CA 92804, (714) 956-3593 - (4,5)

Unholtz-Dickie Corporation, 6 Brookside Dr., Wallingford, CT 06492, (203) 265-3929 - (4,5)

Wavetek San Diego Inc., 9045 Balboa Ave., San Diego, CA 92123, (619) 279-2200 - (2-5)

White Instruments, Div. C VAN R, Inc., 5408 U.S. Highway 290 West, Austin, TX 78735, (512) 892-0752 - (4,5)

Wilson Co., H. M. 11501 Chimney Rock, Houston, TX 77035, (713) 729-1183 - (3)

Zonic Corporation, 2000 Ford Circle, Milford, OH 45150, (513) 248-1911 - (2-5)

Machinery Monitoring Systems
1. Alignment Systems
2. Balancing Systems
3. Bearing Testers
4. Limit Switches
5. Phase Monitors
6. Portable Data Collectors
7. Position Monitors
8. Remote Monitoring Systems
9. Signal Conditioning Systems
10. Stroboscopes
11. Tachometers
12. Transducers
13. Vibration Analyzers
14. Vibration Meters

Alpha-M Corporation, 913 Highland Dr., Euless, TX 76040, (817) 267-7022 - (10,14)

Balmac Inc., 4010 Main St., Hilliard, OH 43026, (614) 876-1295 - (2-4,8,10,12-14)

Bently Nevada Corp., P.O. Box 157, Minden, NV 89423, (702) 782-3611 - (1-3,5-9,11-14)

Binsfeld Engineering Inc., 8944 County Road 675, Maple City, MI 49664, (616) 334-4383 - (9)

Bruel & Kjaer Instruments, Inc., 185 Forest St., Marlborough, MA 01752, (617) 481-7000 - (2,10,12-14)

Chadwick Helmuth Co., Inc., 4601 N. Arden Dr., El Monte, CA 91731, (818) 575-6161 - (2,9-11,13)

Cirrus Research Ltd., 1350-G West Collins, Orange, CA 92667, (714) 997-7430 - (13,14)

Computational Systems Inc., 1900 N. Winston Rd., Knoxville, TN 37919, (615) 693-0551 - (2,5,6,8,13,14)

Dallas Instruments, Inc., P.O. Box 38139, Dallas, TX 75238, (214) 349-1180 - (6)

Data Physics Corporation, 1210 S. Bascom Ave., #224, San Jose, CA 95128, (408) 977-0800 - (13)

Davis Instruments Mfg. Co. Inc., 513 E. 36th St., Baltimore, MD 21218, (301) 243-4301 - (5,10,11,13,14)

Daytronic Corporation, 2589 Corporate Place, Miamisburg, OH 45342, (513) 866-3300 - (6,8,9,12)

DLI Engineering Corporation, 253 Winslow Way West, Bainbridge Island, WA 98110, (206) 842-7656 - (6,13,14)

DSP Technology Inc., 48500 Kato Road, Fremont, CA 94538, (415) 657-7555 - (6,9)

Dynamics Research Corp., 112 Killewald Ave., Tonawanda, NY 14150, (716) 692-1777 - (1,12-14)

Encore Electronics, RD #2, Route 50, Saratoga Springs, NY 12866, (518) 584-5354 - (9,11)

Endevco, 30700 Rancho Viejo Rd., San Juan Capistrano, CA 92675, (714) 493-8181 - (5,8,9,12)

Entran Devices, Inc., 10 Washington Ave., Fairfield, NJ 07006, (800) 635-0650 - (9,12-14)

GenRad, Inc., 510 Cottonwood Drive, Milpitas, CA 95035, (408) 432-1000 - (10,13)

Hewlett-Packard Co., 5161 Lankershim Blvd., N. Hollywood, CA 91601, (818) 505-5600 - (6)

IMO Delaval Inc., CEC Instruments Div., 325 Halstead St., Pasadena, CA 91109, (818) 351-4241 - (9)

Industrial Computer Source, 5466 Complex St., #208, San Diego, CA 92123, (619) 279-0084 - (8,9)

IRD Mechanalysis, Inc., 6150 Huntley Rd., Columbus, OH 43229, (614) 885-5376 - (2-7,10-14)

Kaman Instrumentation Corp., 1500 Garden of the Gods Rd., Colorado Springs, CO 80907, (303) 599-1892 - (1,4,7,12)

Kernco Instruments Co., Inc., 420 Kenazo Ave., El Paso, TX 79927, (915) 852-3375 - (2,3,10,11,13,14)

Krohn-Hite Corporation, 255 Bodwell St., Avon, MA 02322, (617) 580-1660 - (5,9)

Larson-Davis Laboratories, 280 South Main, Pleasant Grove, UT 84062, (801) 785-6352 - (13)

MB Dynamics, 25865 Richmond Rd., Cleveland, OH 44146, (216) 292-5850 - (14)

Mechanical Technology, Inc., 968 Albany-Shaker Rd., Latham, NY 12110, (518) 785-2211 - (2,8,13)

Metrix Instrument Co., 1711 Townhurst Dr., Houston, TX 77043, (713) 461-2131 - (2-10,12-14)

Metrosonics Inc., P.O. Box 23075, Rochester, NY 14692, (716) 334-7300 - (6)

Monarch Instrument, Columbia Dr., Amherst, NH 03031, (603) 883-3390 - (11,12)

Motion Analysis Corporation, 3650 North Laughlin Rd., Santa Rosa, CA 95403, (800) 841-4555 - (13)

Mott Associates, Inc., P.O. Box 42546, Houston, TX 77042, (713) 777-5773 - (3,4,6,8,9,12-14)

Ono Sokki Technology, Inc., 1120 Tower Lane, Benseaville, IL 60106, (312) 595-9292 - (11-13)

PCB Piezotronics, Inc., 3425 Walden Ave., Depew, NY 14043, (716) 684-0001 - (12,14)

Philtec, Inc., P.O. Box 359, Arnold, MD 21012, (301) 757-4404 - (3,12)

PMC/Beta, 4 Tech Circle, Natick, MA 01760, (617) 237-6920 - (2,8,13,14)

PrimeLine, Div. of Soltec Corporation, P.O. Box 670, San Fernando, CA 91341, (818) 365-0101 - (6)

Quest Electronics, 510 S. Worthington St., Oconomowoc, WI 53066, (414) 567-9157 - (13,14)

Rochester Instrument Systems, 255 N. Union St., Rochester, NY 14605, (716) 263-7700 - (8,9,13)

Schenck-Trebel Corp., 535 Acorn St., Deer Park, NY 11729, (516) 242-4010 - (2,3,6,12-14)

Schlumberger Technologies, Instruments Div., 20 North Ave., Burlington, MA 01720, (800) 225-5765 - (13)

Scientific-Atlanta, Spectral Dynamics Div., 4255 Ruffin Road, San Diego, CA 92123, (619) 268-7100 - (1,2,6-13)

Sensotec, Inc., 1200 Chesapeake Ave., Columbus, OH 43212, (614) 486-7723 - (9,12,14)

Spanta, Inc., 795 Franklin Ave., Franklin Lakes, NJ 07417, (201) 848-0033 - (6)

SPM Instrument Inc., 359 N. Main St., Marlborough, CT 06447, (203) 295-8241 - (3,6,8,11,12,14)

TEAC Corporation of America, 7733 Telegraph Rd., Montebello, CA 90640, (213) 726-0303 - (3,6,8,9,13)

Technology for Energy Corp., P.O. Box 22996 Knoxville, TN 37933, (615) 966-5856 - (1,2,5,6,13,14)

Texas Analytical Controls, Inc., 4434 Bluebonnet Dr., Stafford, TX 77477, (713) 240-4160 - (3,11,14)

Trig-Tek Inc., 423 S. Brookhurst St., Anaheim, CA 92804, (714) 956-3593 - (13,14)

Turvac Alignment Engineering Services, 11991 Mill Rd., Cincinnati, OH 45240, (513) 851-1439 - (1)

Unholtz-Dickie Corporation, 6 Brookside Dr., Wallingford, CT 06492, (203) 265-3929 - (13,14)

Vibration Instruments Co., 1614 Orangethorpe Way, Anaheim, CA 92801, (714) 879-6021 - (3)

Vibration Specialty Corp., 100 Geiger Rd., Philadelphia, PA 19115, (215) 698-0800 - (5,6,9,13)

Vibro-Meter Corporation, 1 Progress Rd., Billerica, MA 01822, (617) 663-7322 - (9,12,13)

Vitec, Inc., 23600 Mercantile Rd., Cleveland, OH,44122, (216) 464-4670 - (2,3,6,8,9,12-14)

Wavetek San Diego Inc., 9045 Balboa Ave., San Diego, CA 92123, (619) 279-2200 - (6)

Wolfe Instrumentation Co., 10760 Burbank Blvd., North Hollywood, CA 91601, (800) 554-1224 - (6)

Ya-Man Ltd., 2005 Hamilton Ave., Suite 100, San Jose, CA 95125, (408) 559-9100 - (1,12)

Yokogawa Corporation of America, P.O. Box 2719, Peachtree City, GA 30269, (404) 487-1471 - (11,12)

Zonic Corporation, 2000 Ford Circle, Milford, OH 45150, (513) 248-1911 - (2,5,8,13)

Meters
1. Human Vibration Response
2. Integrating Sound Level
3. Integrating Vibration
4. Personal Noise Dosimeters
5. Phase
6. Reverberation Time
7. Seismographs
8. Sound Level
9. Vibration

ACO Pacific, Inc. 2604 Read Ave., Belmont, CA 94002, (415) 595-8588 - (8)

Acoustilog, Inc., 19 Mercer St., New York, NY 10013, (212) 925-1365 - (6)

Alpha-M Corporation, 913 Highland Dr., Euless, TX 76040, (817) 267-7022 - (9)

Bently Nevada Corp., P.O. Box 157, Minden, NV 89423, (702) 782-3611 - (3,5,9)

Bruel & Kjaer Instruments, Inc., 185 Forest St., Marlborough, MA 01752, (617) 481-7000 - (1-5,8,9)

Cetec Ivie, 1366 W. Center St., Orem, UT 84057, (801) 224-1800 - (2,6,8)

Cirrus Research Ltd., 1350-G West Collins, Orange, CA 92667, (714) 997-7430 - (1-4,8,9)

Dallas Instruments, Inc., P.O. Box 38139, Dallas, TX 75238, (214) 349-1180 - (7)

Davis Instruments Mfg. Co. Inc., 513 E. 36th St., Baltimore, MD 21218, (301) 243-4301 - (12)

DuPont Instrument Systems, P.O. Box 10, Kennett Square, PA 19348, (800) 344-4900 - (4)

Dytran Instruments, Inc., 21592 Marilla St., Chatsworth, CA 91311, (818) 700-7818 - (9)

GenRad, Inc., 510 Cottonwood Drive, Milpitas, CA 95035, (408) 432-1000 - (2,9)

IMO Delaval Inc., CEC Instruments Div., 325 Halstead St., Pasadena, CA 91109, (818) 351-4241 - (9)

Kinemetrics/Systems, 222 Vista Ave., Pasadena, CA 91107, (818) 795-2220 - (7)

Klark-Teknik Electronics, Inc., 30B Banfi Plaza North, Farmingdale, NY 11735, (516) 249-3660 - (8)

Krohn-Hite Corporation, 255 Bodwell St., Avon, MA 02322, (617) 580-1660 - (5)

Larson-Davis Laboratories, 280 South Main, Pleasant Grove, UT 84062, (801) 785-6352 - (1-4,6,8,9)

Lucas CEL Instruments, 5500 New King St., Troy, MI 48098, (313) 879-1920 - (2-4,6,8,9)

MB Dynamics, 25865 Richmond Rd., Cleveland, OH 44146, (216) 292-5850 - (3,9)

Metrosonics Inc., P.O. Box 23075, Rochester, NY 14692, (716) 334-7300 - (2,4,8)

Mott Associates, Inc., P.O. Box 42546, Houston, TX 77042, (713) 777-5773 - (2,8,9)

Northern Sound, 5155 East River Rd. NE, #413, Minneapolis, MN 55421, (612) 572-1515 - (2,4,8)

PCB Piezotronics, Inc., 3425 Walden Ave., Depew, NY 14043, (716) 684-0001 - (9)

Quest Electronics, 510 S. Worthington St., Oconomowoc, WI 53066, (414) 567-9157 - (2,4,8,9)

Scantek Inc./Norwegian Electronics, 51 Monroe Street, #1606, Rockville, MD 20850, (301) 279-9308 - (1,2,6,8,9)

Scott Instrument Laboratories, 13 Sagamore Park Rd., Hudson, NH 03051, (800) 343-0683 - (8,9)

Trig-Tek Inc., 423 S. Brookhurst St., Anaheim, CA 92804, (714) 956-3593 - (9)

TSI Incorporated, 500 Cardigan Rd., St. Paul, MN 55126, (612) 483-0900 - (9)

Unholtz-Dickie Corporation, 6 Brookside Dr., Wallingford, CT 06492, (203) 265-3929 - (9)

Vibra-Metrics, Inc., 1014 Sherman Avenue, Hamden, CT 06514, (203) 288-6158 - (9)

Vibration Instruments Co., 1614 Orangethorpe Way, Anaheim, CA 92801, (714) 879-6021 - (5)

Wiltron Company, 490 Jarvis Drive, Morgan Hill, CA 95037, (415) 778-2000 - (5)

Recorders and Displays
1. CRT Displays
2. Data Acquisition Systems
3. Data Recorders
4. Digital Cassette Recorders
5. Digital Oscilloscopes
6. Environmental Noise Monitors
7. Graphic Level Recorders
8. Graphics Recorders
9. Noise Exposure Recorders
10. Transient Recorders
11. Waveform Recorders
12. X-Y Recorders

Acoustic Emission Technology Corp., 1824J Tribute Rd., Sacramento, CA 95815, (916) 927-3861 - (2,5,10,11)

Allen Datagraph, Inc., 2 Industrial Way, Salem, NH 03079, (603) 893-1983 - (12)

Balmac Inc., 4010 Main St., Hilliard, OH 43026, (614) 876-1295 - (12)

BBC-Metrawatt/Goerz, 2150 West 6th Ave., Broomfield, CO 80020, (303) 469-5231 - (2,3,5,8,10-12)

Bently Nevada Corp., P.O. Box 157, Minden, NV 89423, (702) 782-3611 - (2,3,10,11)

Bruel & Kjaer Instruments, Inc., 185 Forest St., Marlborough, MA 01752, (617) 481-7000 - (4,6-10,12)

Cirrus Research Ltd., 1350-G West Collins, Orange, CA 92667, (714) 997-7430 - (2,6,9)

Dallas Instruments, Inc., P.O. Box 38139, Dallas, TX 75238, (214) 349-1180 - (2,3)

Data Capture Technology, 10837 E. Marshall, Tulsa, OK 74116, (918) 438-5366 - (3,8,11)

Data Precision, Div. of Analogic, 16 Electronics Ave., Danvers, MA 01923, (617) 246-1600 - (5,11)

Datel Inc., 11 Cabot Blvd., Mansfield, MA 02048, (617) 339-3000 - (2,3)

Davis Instruments Mfg. Co. Inc., 513 E. 36th St., Baltimore, MD 21218, (301) 243-4301 - (12)

Daytronic Corporation, 2589 Corporate Place, Miamisburg, OH 45342, (513) 866-3300 - (1-3)

DSP Technology Inc., 48500 Kato Road, Fremont, CA 94538, (415) 657-7555 - (2,3,10,11)

Endevco, 30700 Rancho Viejo Rd., San Juan Capistrano, CA 92675, (714) 493-8181 - (10)

Fairchild Weston Systems, Inc., P.O. Box 3041, Sarasota, FL 34230, (813) 377-5521 - (2-4)

GHI Systems, Inc., 916 N. Western Ave., #201, San Pedro, CA 90732, (213) 548-6544 - (2,3,10,11)

Gould Inc., Recording Systems Div., 3631 Perkins Ave., Cleveland, OH 44114, (216) 361-3315 - (2,3,5,7,8,10-12)

Hewlett-Packard Co., 5161 Lankershim Blvd., N. Hollywood, CA 91601, (818) 505-5600 - (2)

Industrial Computer Source, 5466 Complex St., #208, San Diego, CA 92123, (619) 279-0084 - (1)

Kay Elemetrics Corp., 12 Maple Ave., Pine Brook, NJ 07058, (201) 227-2000 - (2,11,12)

Kinemetrics/Systems, 222 Vista Ave., Pasadena, CA 91107, (818) 795-2220 - (2,3)

Kyowa Dengyo Corp., 10 Reuten Dr., Closter, NJ 07624, (201) 784-0500 - (3,4)

Larson-Davis Laboratories, 280 South Main, Pleasant Grove, UT 84062, (801) 785-6352 - (6)

LeCroy Corporation, 700 Chestnut Ridge Rd., Chestnut Ridge, NY 10977, (800) 553-2769 - (2,3,5,10,11)

Lucas CEL Instruments, 5500 New King St., Troy, MI 48098, (313) 879-1920 - (6,7,9)

Masscomp, One Technology Way, Westford, MA 01886, (617) 692-6200 - (2,10,11)

Memodyne Corporation, 200 Reservoir St., Needham Heights, MA 02194, (617) 444-7000 - (4)

Metrix Instrument Co., 1711 Townhurst Dr., Houston, TX 77043, (713) 461-2131 - (3)

Metrosonics Inc., P.O. Box 23075, Rochester, NY 14692, (716) 334-7300 - (2,3,6,9)

Mott Associates, Inc., P.O. Box 42546, Houston, TX 77042, (713) 777-5773 - (3,7-9,11)

NF Electronic Instruments, Inc., 3303 Harbor Blvd., Suite D-6, Costa Mesa, CA 92626, (714) 557-1684 - (10)

Nicolet Test Instruments Div., P.O. Box 4288, 5225 Verona Rd., Madison, WI 53711, (608) 273-5000 - (5)

Northern Sound, 5155 East River Rd. NE, #413, Minneapolis, MN 55421, (612) 572-1515 - (6,7,9)

Optim Electronics Corp., 12401 Middlebrook Dr., Germantown, MD 20874, (301) 428-7200 - (2,3)

Pacific Instruments Inc., 215 Mason Circle, Concord, CA 94520, (415) 827-9010 - (10)

Panasonic Industrial Co., Two Panasonic Way, Secaucus, NJ 07094, (201) 392-4050 - (5)

PrimeLine, Div. of Soltec Corporation, P.O. Box 670, San Fernando, CA 91341, (818) 365-0101 - (3,8,12)

Quest Electronics, 510 S. Worthington St., Oconomowoc, WI 53066, (414) 567-9157 - (6,7,9)

Racal Recorders, Inc., 4 Goodyear St., Irvine, CA 92718, (714) 380-0900 - (3)

Rapid Systems, Inc., 433 N. 34th St., Seattle, WA 98103, (206) 547-8311 - (2,3,5,10-12)

Rochester Instrument Systems, 255 N. Union St., Rochester, NY 14605, (716) 263-7700 - (10)

Scantek Inc./Norwegian Electronics, 51 Monroe Street, #1606, Rockville, MD 20850, (301) 279-9308 - (2,6,7,10)

Schlumberger Technologies, Instruments Div., 20 North Ave., Burlington, MA 01720, (800) 225-5765 - (2)

Scientific Recording Associates, 59 Princeton Terrace, Watertown, CT 06795, (203) 274-7761 - (5,10,11)

Scientific-Atlanta, Spectral Dynamics Div., 4255 Ruffin Road, San Diego, CA 92123, (619) 268-7100 - (1,2)

Soltec Corporation, Sol Vista Park, San Fernando, CA 91340, (818) 365-0800 - (2,3,8,10-12)

Spanta, Inc., 795 Franklin Ave., Franklin Lakes, NJ 07417, (201) 848-0033 - (2,3)

SPM Instrument Inc., 359 N. Main St., Marlborough, CT 06447, (203) 295-8241 - (2,3)

TEAC Corporation of America, 7733 Telegraph Rd., Montebello, CA 90640, (213) 726-0303 - (1-4,10,11)

Vibration Specialty Corp., 100 Geiger Rd., Philadelphia, PA 19115, (215) 698-0800 - (2)

Vibro-Meter Corporation, 1 Progress Rd., Billerica, MA 01822, (617) 663-7322 - (2)

Vitec, Inc., 23600 Mercantile Rd., Cleveland, OH,44122, (216) 464-4670 - (12)

Wavetek San Diego Inc., 9045 Balboa Ave., San Diego, CA 92123, (619) 279-2200 - (2,3)

Western Graphtec, Inc., 11 Vanderbilt, Irvine, CA 92718, (800) 854-8385 - (2,3,8,12)

Wolfe Instrumentation Co., 10760 Burbank Blvd., North Hollywood, CA 91601, (800) 554-1224 - (2,3)

Yokogawa Corporation of America, P.O. Box 2719, Peachtree City, GA 30269, (404) 487-1471 - (2,3,10-12)

Zonic Corporation, 2000 Ford Circle, Milford, OH 45150, (513) 248-1911 - (1,2,10,11)

Signal Conditioners
1. Analog/Digital Converters
2. Gating Systems
3. Measuring Amplifiers
4. Power Amplifiers
5. Preamplifiers
6. Programmable Filters
7. Signal Delay Units
8. Signal Translators
9. Spectrum Shapers
10. Telemetry Systems
11. Tracking Filters
12. Tunable Filters

A&D Engineering, Inc., 1555 McCandless Dr., Milpitas, CA 95035, (408) 263-5333 - (11)

ACO Pacific, Inc. 2604 Read Ave., Belmont, CA 94002, (415) 595-8588 - (5)

Acoustic Emission Technology Corp., 1824J Tribute Rd., Sacramento, CA 95815, (916) 927-3861 - (5)

Alpha-M Corporation, 913 Highland Dr., Euless, TX 76040, (817) 267-7022 - (4,11)

Balmac Inc., 4010 Main St., Hilliard, OH 43026, (614) 876-1295 - (11,12)

Bently Nevada Corp., P.O. Box 157, Minden, NV 89423, (702) 782-3611 - (3,6,11,12)

Binsfeld Engineering Inc., 8944 County Road 675, Maple City, MI 49664, (616) 334-4383 - (10)

Bruel & Kjaer Instruments, Inc., 185 Forest St., Marlborough, MA 01752, (617) 481-7000 - (3-5,11,12)
Cirrus Research Ltd., 1350-G West Collins, Orange, CA 92667, (714) 997-7430 - (3,5)
Data Capture Technology, 10837 E. Marshall, Tulsa, OK 74116, (918) 438-5366 - (1)
Data Precision, Div. of Analogic, 16 Electronics Ave., Danvers, MA 01923, (617) 246-1600 - (1,5)
Datel Inc., 11 Cabot Blvd., Mansfield, MA 02048, (617) 339-3000 - (1,6,12)
Daytronic Corporation, 2589 Corporate Place, Miamisburg, OH 45342, (513) 866-3300 - (1)
Difa Measuring Systems, 37620 Hills Tech Dr., Farmington Hills, MI 48331, (313) 489-8588 - (1,5,6,11)
DSP Technology Inc., 48500 Kato Road, Fremont, CA 94538, (415) 657-7555 - (1,3-6)
Ectron Corporation, 8159 Engineer Road, San Diego, CA 92111, (619) 278-0600 - (3)
Encore Electronics, RD #2, Route 50, Saratoga Springs, NY 12866, (518) 584-5354 - (3-5,12)
Endevco, 30700 Rancho Viejo Rd., San Juan Capistrano, CA 92675, (714) 493-8181 - (3,6,11,12)
Frequency Devices, Inc., 25 Locust St., Haverhill, MA 01830, (617) 374-0761 - (6,12)
Gould Inc., Recording Systems Div., 3631 Perkins Ave., Cleveland, OH 44114, (216) 361-3315 - (1,3,5,10)
IMO Delaval Inc., CEC Instruments Div., 325 Halstead St., Pasadena, CA 91109, (818) 351-4241 - (6)
Industrial Computer Source, 5466 Complex St., #208, San Diego, CA 92123, (619) 279-0084 - (1)
Infotek Systems, 1045 S. East St., Anaheim, CA 92805, (714) 956-9300 - (1)

Ithaco, Inc., P.O. Box 6437, Ithaca, NY 14851, (607) 272-7640 - (3,5,12)
Kay Elemetrics Corp., 12 Maple Ave., Pine Brook, NJ 07058, (201) 227-2000 - (1)
Kinemetrics/Systems, 222 Vista Ave., Pasadena, CA 91107, (818) 795-2220 - (3)
Klark-Teknik Electronics, Inc., 30B Banfi Plaza North, Farmingdale, NY 11735, (516) 249-3660 - (7,9,12)
Krohn-Hite Corporation, 255 Bodwell St., Avon, MA 02322, (617) 580-1660 - (6,11,12)
Kyowa Dengyo Corp., 10 Routen Dr., Closter, NJ 07624, (201) 784-0500 - (1,6)
Larson-Davis Laboratories, 280 South Main, Pleasant Grove, UT 84062, (801) 785-6352 - (5)
LeCroy Corporation, 700 Chestnut Ridge Rd., Chestnut Ridge, NY 10977, (800) 553-2769 - (1-3,5)
MB Dynamics, 25865 Richmond Rd., Cleveland, OH 44146, (216) 292-5850 - (1,3-6,9)
Metrosonics Inc., P.O. Box 23075, Rochester, NY 14692, (716) 334-7300 - (1)
Mott Associates, Inc., P.O. Box 42546, Houston, TX 77042, (713) 777-5773 - (3,5,6,12)

NF Electronic Instruments, Inc., 3303 Harbor Blvd., Suite D-6, Costa Mesa, CA 92626, (714) 557-1684 - (3-6,11,12)
Optim Electronics Corp., 12401 Middlebrook Dr., Germantown, MD 20874, (301) 428-7200 - (3-5)
Precision Filters, Inc., 240 Cherry St., Ithaca, NY 14850, (607) 277-3550 - (5,6)
Rapid Systems, Inc., 433 N. 34th St., Seattle, WA 98103, (206) 547-8311 - (1)
Rochester Instrument Systems, 255 N. Union St., Rochester, NY 14605, (716) 263-7700 - (1,8)
Rockland Scientific Corporation, 10 Volvo Dr., Rockleigh, NJ 07647, (201) 767-7900 - (11)
Scantek Inc./Norwegian Electronics, 51 Monroe Street, #1606, Rockville, MD 20850, (301) 279-9308 - (6,9)
Scientific-Atlanta, Spectral Dynamics Div., 4255 Ruffin Road, San Diego, CA 92123, (619) 268-7100 - (6,11)
Sensotec, Inc., 1200 Chesapeake Ave., Columbus, OH 43212, (614) 486-7723 - (1,3)

Soltec Corporation, Sol Vista Park, San Fernando, CA 91340, (818) 365-0800 - (3)
Specialty Engineering, 481 Sinclair Frontage Rd., Milpitas, CA 95035, (408) 946-9779 - (5)
Stanford Research Systems, 1290D Reamwood Ave., Sunnyvale, CA 94089, (408) 744-9040 - (6,12)
Systolic Systems, Inc., 2240 N. First St., San Jose, CA 95131, (408) 435-1760 - (11)
TEAC Corporation of America, 7733 Telegraph Rd., Montebello, CA 90640, (213) 726-0303 - (1)
Timewave Technology, Inc., 2401 Pilot Knob Rd., Suite 134, St. Paul, MN 55120, (612) 452-5939 - (5,6,8,10)
Trig-Tek Inc., 423 S. Brookhurst St., Anaheim, CA 92804, (714) 956-3593 - (11,12)
Unholtz-Dickie Corporation, 6 Brookside Dr., Wallingford, CT 06492, (203) 265-3929 - (1,4,5,11)
Vibration Instruments Co., 1614 Orangethorpe Way, Anaheim, CA 92801, (714) 879-6021 - (6,10,11)
Vitec, Inc., 23600 Mercantile Rd., Cleveland, OH,44122, (216) 464-4670 - (6,12)
Wavetek San Diego Inc., 9045 Balboa Ave., San Diego, CA 92123, (619) 279-2200 - (6,12)
White Instruments, Div. C VAN R Inc., 5408 U.S. Highway 290 West, Austin, TX 78735, (512) 892-0752 - (9)

Wilcoxon Research, Inc., 2096 Gaither Rd., Rockville, MD 20850, (301) 330-8811 - (4,5)
Zonic Corporation, 2000 Ford Circle, Milford, OH 45150, (513) 248-1911 - (1,3,6,9)

Special Purpose Systems
1. Absorption Factor
2. Acoustical Scale Modeling
3. Audiological Testing
4. Audiometer Calibration
5. Building Acoustics
6. Damping Factor
7. Electroacoustic Measurement
8. Hearing Aid Testing
9. Holographic Testing
10. Human Vibration Response
11. Motion Analysis
12. Optical Displacement
13. Speech Transmission Index
14. Strain
15. Time Delay Spectrometry

ACO Pacific, Inc. 2604 Read Ave., Belmont, CA 94002, (415) 595-8588 - (7)
Acoustilog, Inc., 19 Mercer St., New York, NY 10013, (212) 925-1365 - (7)
Anatrol Corporation, 10895 Indeco Dr., Cincinnati, OH 45241, (513) 793-8844 - (6,10)
Apollo Lasers, 20977 Knapp St., Chatsworth, CA 91311, (818) 407-3000 - (9)
Bently Nevada Corp., P.O. Box 157, Minden, NV 89423, (702) 782-3611 - (11)
Bruel & Kjaer Instruments, Inc., 185 Forest St., Marlborough, MA 01752, (617) 481-7000 - (5,7,10,13,15)
Cirrus Research Ltd., 1350-G West Collins, Orange, CA 92667, (714) 997-7430 - (4,5,8)
Difa Measuring Systems, 37620 Hills Tech Dr., Farmington Hills, MI 48331, (313) 489-8588 - (14)
Grason-Stadler, Inc., 537 Great Rd., P.O. Box 1400, Littleton, MA 01460, (617) 486-3514 - (3)
Grozier Technical Systems, Inc., 157 Salisbury Rd., Brookline, MA 02146, (617) 277-1133 - (2,15)
Larson-Davis Laboratories, 280 South Main, Pleasant Grove, UT 84062, (801) 785-6352 - (4,5,10)
Laser Technology, Inc., 1055 W. Germantown Pike, Norristown, PA 19403, (215) 631-5043 - (9)
Lucas CEL Instruments, 5500 New King St., Troy, MI 48098, (313) 879-1920 - (5)
Motion Analysis Corporation, 3650 North Laughlin Rd., Santa Rosa, CA 95403, (800) 841-4555 - (11)
Mott Associates, Inc., P.O. Box 42546, Houston, TX 77042, (713) 777-5773 - (4)
Newport Corporation, 18235 Mt. Baldy Circle, Fountain Valley, CA 92708, (714) 963-9811 - (9,11)
NTE TV-Holography for the Workplace, 3411 Candelaria Rd. NE, Albuquerque, NM 87107, (505) 884-6222 - (9)
Ometron Incorporated, 380 Herndon Parkway, Suite 300, Herndon, VA 22070, (703) 435-9799 - (14)
Optron Corp., 30 Hazel Terrace, Woodbridge, CT 06525, (203) 389-5384 - (11,12)
Philtec, Inc., P.O. Box 359, Arnold, MD 21012, (301) 757-4404 - (12)
Quest Electronics, 510 S. Worthington St., Oconomowoc, WI 53066, (414) 567-9157 - (4,8)
Scantek Inc./Norwegian Electronics, 51 Monroe Street, #1606, Rockville, MD 20850, (301) 279-9308 - (1,5,6,10,13)
Soltec Corporation, Sol Vista Park, San Fernando, CA 91340, (818) 365-0800 - (14)
Techron, 1718 W. Mishawaka Rd., Elkhart, IN 46517, (219) 294-3000 - (2,5,7,13,15)
Timewave Technology, Inc., 2401 Pilot Knob Rd., Suite 134, St. Paul, MN 55120, (612) 452-5939 - (5,10)

Tracor Instruments, 6500 Tracor Lane, Bldg. 27, Austin, TX 78725 (512) 929-2027 - (3,4)
Vibro-Meter Corporation, 1 Progress Rd., Billerica, MA 01822, (617) 663-7322 - (10)
Ya-Man Ltd., 2005 Hamilton Ave., Suite 100, San Jose, CA 95125, (408) 559-9100 - (12)
Zonic Corporation, 2000 Ford Circle, Milford, OH 45150, (513) 248-1911 - (6)

Transducers
1. Accelerometers
2. Acoustic Emission
3. Displacement
4. Electro-Optical
5. Building Acoustics
6. Force
7. Hydrophones
8. Impedance
9. Microphones
10. Pressure
11. Proximity
12. Torque
13. Velocity Pickups

A&D Engineering, Inc., 1555 McCandless Dr., Milpitas, CA 95035, (408) 263-5333 - (1,6)
ACO Pacific, Inc. 2604 Read Ave., Belmont, CA 94002, (415) 595-8588 - (9)
Acoustic Emission Technology Corp., 1824J Tribute Rd., Sacramento, CA 95815, (916) 927-3861 - (2,12)
Alpha-M Corporation, 913 Highland Dr., Euless, TX 76040, (817) 267-7022 - (3)
Balmac Inc., 4010 Main St., Hilliard, OH 43026, (614) 876-1295 - (1,13)
Bently Nevada Corp., P.O. Box 157, Minden, NV 89423, (702) 782-3611 - (1,3,10,11,13)
Binsfeld Engineering Inc., 8944 County Road 675, Maple City, MI 49664, (616) 334-4383 - (11)
Bruel & Kjaer Instruments, Inc., 185 Forest St., Marlborough, MA 01752, (617) 481-7000 - (1,2,5,6,8)
Cetec Ivie, 1366 W. Center St., Orem, UT 84057, (801) 224-1800 - (8)
Chadwick Helmuth Co., Inc., 4601 N. Arden Dr., El Monte, CA 91731, (818) 575-6161 - (1)
Channel Industries, Inc., 839 Ward Dr., Santa Barbara, CA 93111, (805) 967-0171 - (1-3,5,6,10,12)
Cirrus Research Ltd., 1350-G West Collins, Orange, CA 92667, (714) 997-7430 - (1,8)
Daytronic Corporation, 2589 Corporate Place, Miamisburg, OH 45342, (513) 866-3300 - (1,3,5,9,11,13)
Dytran Instruments, Inc., 21592 Marilla St., Chatsworth, CA 91311, (818) 700-7818 - (1,5,7,9)
Encore Electronics, RD #2, Route 50, Saratoga Springs, NY 12866, (518) 584-5354 - (10)
Endevco, 30700 Rancho Viejo Rd., San Juan Capistrano, CA 92675, (714) 493-8181 - (1,8,9)
Entran Devices, Inc., 10 Washington Ave., Fairfield, NJ 07006, (800) 635-0650 - (1,5,9)
Gulton Industries, 1644 Whittier Ave., Costa Mesa, CA 92627, (714) 642-2400 - (1,3,8-10)
IMO Delaval Inc., CEC Instruments Div., 325 Halstead St., Pasadena, CA 91109, (818) 351-4241 - (1,3,9,13)
IQS Inc., 12812-J Garden Grove Blvd., Garden Grove, CA 92643, (714) 539-7842 - (8)
Katt & Associates, P.O. Box 623, Zoar, OH 44697, (216) 874-4586 - (1)
Kelmet, Inc., 569 W. Covina Blvd., San Dimas, CA 91773, (714) 599-0954 - (1)
Kinemetrics/Systems, 222 Vista Ave., Pasadena, CA 91107, (818) 795-2220 - (1)
Kistler Instrument Corp., 75 John Glenn Dr., Amherst, NY 14120, (716) 691-5100 - (1,5,9,11)
Kulite Semiconductor Products, Inc., 1039 Hoyt Ave., Ridgefield, NJ 07657, (201) 945-3000 - (1,5,6,9)
Larson-Davis Laboratories, 280 South Main, Pleasant Grove, UT 84062, (801) 785-6352 - (8)
Lucas CEL Instruments, 5500 New King St., Troy, MI 48098, (313) 879-1920 - (1,8)
MB Dynamics, 25865 Richmond Rd., Cleveland, OH 44146, (216) 292-5850 - (1,9)
Mechanical Technology Inc., 968 Albany-Shaker Rd., Latham, NY 12110, (518) 785-2211 - (10)
Metrosonics Inc., P.O. Box 23075, Rochester, NY 14692, (716) 334-7300 - (8)
Mott Associates, Inc., P.O. Box 42546, Houston, TX 77042, (713) 777-5773 - (1,8)
NF Electronic Instruments, Inc., 3303 Harbor Blvd., Suite D-6, Costa Mesa, CA 92626, (714) 557-1684 - (2)
Northern Sound Instruments, 5155 East River Rd. NE, #413 Minneapolis, MN 55421, (612) 572-1515 - (8)
Ono Sokki Technology, Inc., 1120 Tower Lane, Bensenville, IL 60106, (312) 595-9292 - (3,4,8)
Optron Corp., 30 Hazel Terrace, Woodbridge, CT 06525, (203) 389-5384 - (4)
PCB Piezotronics, Inc., 3425 Walden Ave., Depew, NY 14043, (716) 684-0001 - (1,5,7-9)
Philtec, Inc., P.O. Box 359, Arnold, MD 21012, (301) 757-4404 - (3,4)

Polytec Optronics, Inc., 3001 Redhill Ave., Bldg. 5-104, Costa Mesa, CA 92626, (714) 850-1835 - (3,4)
Rochester Instrument Systems, 255 N. Union St., Rochester, NY 14605, (716) 263-7700 - (1)
Scantek Inc./Norwegian Electronics, 51 Monroe Street, #1606, Rockville, MD 20850, (301) 279-9308 - (8)

Scientific-Atlanta, Spectral Dynamics Div., 4255 Ruffin Road, San Diego, CA 92123, (619) 268-7100 — (1,3,9,10,13)

Sensotec, Inc., 1200 Chesapeake Ave., Columbus, OH 43212, (614) 486-7723 — (1,3,5,9,11)

Specialty Engineering, 481 Sinclair Frontage Rd., Milpitas, CA 95035, (408) 946-9779 — (6,12)

SPM Instrument Inc., 359 N. Main St., Marlborough, CT 06447, (203) 295-8241 — (1,12,13)

Sundstrand Data Control, Inc., 15001 N.E. 36th Street, Redmond, WA 98073, (206) 885-3711 — (1)

Texas Analytical Controls, Inc., 4434 Bluebonnet Dr., Stafford, TX 77477, (713) 240-4160 — (10)

Unholtz-Dickie Corporation, 6 Brookside Dr., Wallingford, CT 06492, (203) 265-3929 — (1)

Vibra-Metrics, Inc., 1014 Sherman Avenue, Hamden, CT 06514, (203) 288-6158 — (1,13)

Vibro-Meter Corporation, 1 Progress Rd., Billerica, MA 01822, (617) 663-7322 — (1,9-11)

Vitec, Inc., 23600 Mercantile Rd., Cleveland, OH, 44122, (216) 464-4670 — (1,3,10)

Wilcoxon Research, Inc., 2096 Gaither Rd., Rockville, MD 20850, (301) 330-8811 — (1,5-7,13)

Ya-Man Ltd., 2005 Hamilton Ave., Suite 100, San Jose, CA 95125, (408) 559-9100 — (3,4)

Zonic Corporation, 2000 Ford Circle, Milford, OH 45150, (513) 248-1911 — (5,11)

INDEX

INDEX

WIDENER UNIVERSITY
WOLFGRAM
LIBRARY
CHESTER, PA.

DATE DUE

OCT 0 9 1991		
SEP 1 6 1991		
DEC 1 3 1991		
DEC 1 3 1991		
OCT 3 1 1994		
DEC 2 2 1995		

Demco, Inc. 38-293